AF181992

PRODUCED WATER 2

Environmental Issues and Mitigation Technologies

ENVIRONMENTAL SCIENCE RESEARCH

Recent Volumes in this Series

PRODUCED WATER 2

Environmental Issues and Mitigation Technologies

Edited by

Mark Reed

IKU Petroleum Research
Trondheim, Norway

and

Ståle Johnsen

STATOIL Research and Development Centre
Trondheim, Norway

Plenum Press • New York and London

Library of Congress Cataloging-in-Publication Data

Produced water 2 : environmental issues and mitigation technologies / edited by Mark Reed and Ståle Johnsen.
p. cm. -- (Environmental science research ; v. 52)
"Proceedings of the 1995 International Produced Water Seminar, held September 25-28, 1995, in Trondheim, Norway"--T.p. verso.
Includes bibliographical references and index.
ISBN-13:978-1-4613-8025-2 e-ISBN-13:978-1-4613-0379-4
DOI: 10.1007/978-1-4613-0379-4

1. Oil field brines--Congresses. I. Reed, Mark, 1947- II. Johnsen, Ståle. III. International Produced Water Seminar (1995 : Trondheim, Norway) IV. Series.
TN871.P7175 1996
622'.33819--dc20 96-26471
CIP

Proceedings of the 1995 International Produced Water Seminar, held September 25–28, 1995, in Trondheim, Norway

ISBN-13:978-1-4613-8025-2

Softcover reprint of the hardcover 1st edition 1996

A Division of Plenum Publishing Corporation
233 Spring Street, New York, N.Y. 10013

10 9 8 7 6 5 4 3 2 1

To Unni and Gretha

for their support during the lengthy process
of which this volume is the final result

PREFACE

This volume constitutes the proceedings of the Produced Water Seminar held in Trondheim, Norway, in September 1995. Hosted by Statoil Research and Development and IKU Petroleum Research, the seminar was an update of the 1992 seminar of the same title held in San Diego, California (Ray and Engelhardt, 1992).

Produced water remains the largest volume waste stream from oil and gas production offshore. In the North and Norwegian Seas, produced water volumes are projected to increase significantly over the coming decades, as oil reservoirs near depletion. These releases are therefore the focus of continuing environmental concern.

The purpose of this seminar was to provide a forum for scientists, legislators, and industrial and environmental representatives to share recent information and research results, and to encourage cooperative pursuit of solutions in the future.

The success of the seminar, and the quality of this volume, are due in large part to the many authors from around the world who presented almost 50 posters and papers focused on environmental issues and mitigation technologies. In addition, we wish to acknowledge the contributions of the local and international organizing committees.

Local Committee

Asbjørg Øverli and Heidi Torp, Statoil
Egil Wanvik and Laila S. Olden, IKU Petroleum Research

International Committee

James P. Ray, Shell Chemical and Petroleum Products Companies
Alexis E. Steen, American Petroleum Institute
Theodor C. Sauer, Battelle Ocean Sciences
Steven A. Flynn, British Petroleum
Martin C. Th. Scholten, TNO
Kjell Lohne, Statoil
Ingvild Martinsen, Norwegian Pollution Control Authority.

Mark Reed
Ståle Johnsen
Trondheim, Norway

REFERENCE

Ray, J. P., and R. Engelhardt, 1992. *Produced Water*. Proceedings of the 1992 Seminar. Plenum Press, NY.

CONTENTS

Section I. Legislative and Management Issues

Section II. Chemistry and Toxicity Issues

Section III. Environmental Effects

Section IV. Dispersion Models

Section V. Environmental Modelling

Section VI. U.S. DOE: Gulf of Mexico Studies

Section VII. Technology for Mitigation

Section I

LEGISLATIVE AND MANAGEMENT ISSUES

1

SESSION SUMMARY

James P. Ray

Although the 1995 International Seminar on Produced Water was primarily a scientific conference to present and discuss the latest advances in our knowledge of the treatment, discharge, and environmental fate and effects of produced water, the lead-off session still reflected the basis for most of the discussion: Legislative and Management Issues. This topic is the driving impetus (and funding) for most of the scientific findings presented over the several day conference. As noted in a wide range of papers presented in this proceedings volume, produced water is a generally high volume waste water stream which is separated from oil and gas that is produced from subsurface formations. It is separated from the oil and gas, and in most cases, discharged to the environment. Because of its long contact with the subsurface oil and gas, as well as the geological formations, produced water contains a variety of organics, inorganics. Some of these compounds and elements are of environmental concern.

As is appropriate when any material is discharged into environments where they (i.e., produced water) do not normally occur, the potential for negative environmental effects is present. Because of this inherent factor, environmental regulatory agencies from oil producing countries have sought scientific information on the composition, discharge characteristics, and fate and effects of produced waters. This information is fundamental in formulating scientifically based environmental regulations to provide adequate protection to the marine environment.

One of the interesting dilemmas is the diverse approach to research and regulations in different parts of the world. As such, it was most appropriate to discuss the varied international approaches so that the conference attendees could better understand the driving factors behind the differing environmental studies conducted in different geographical regions.

One common thread that appeared in regional discussions was that produced water oil and grease (O&G) content was a commonly measured and regulated parameter. The differences appear in the analytical methodology, e.g., Freon extraction gravimetric versus infrared spectrometry, and in the regulatory limits. The Norwegian sector of the North Sea regulates to 40 mg/l, while the United States standard is currently 29 mg/l (recently changed from 42 mg/l). The Arabian Gulf, Mediterranean, and far east also seem to regulate in the approximate monthly average range of 40 mg/l. The lead off speaker, Erik Syversten (Norway SFT) reviewed the Norwegian approach to regulation.

The basic principles for regulation are: 1) pollution is forbidden. Discharges are considered exceptions; 2) precaution; 3) polluter pays principle; 4) priority is to prevent

Produced Water2: Environmental Issues and Mitigation Technologies
Edited by M. Reed and S. Johnsen, Plenum Press, New York, 1996

rather than to cure; and 5) cost effectiveness across sectors. Norway is currently assessing techniques to substitute for the currently used Freon extraction technique.

Dr. Ray, in the last presentation of the session, also pointed out the current shift in the United States away from the Freon extraction technique. The U.S. Environmental Protection Agency (EPA) has recently published Method 1664 which outlines a new technique using hexane as the extractant. The American Petroleum Institute is currently planning followup studies to compare freon versus hexane extraction with different effluents, and to assess commercial laboratory performance with the new Method 1664 (hexane).

The presentations and ensuing discussions on the analytical methods focused on questions related to comparability between methods, what the tests were actually measuring, and what the new tests results would mean to the companies in relation to their ability to be in compliance. Testing is currently in progress in the U.S. to compare the new methods vs. the old methods to determine how the regulatory numbers might vary depending on the type of effluent tested (produced waters and other discharges). In one of the discussions, a point was made that if proper calibrations are conducted, comparisons can be made between the different methodologies.

A contentious subject has always been the identification of sources and estimates of quantities of oil input into the marine environment. The differences of opinion are alive and well as demonstrated during the questions and answers session. Dr. Syvertsen stated that produced water related oil and grease input to the Norwegian sector was approximately 1.6% of the overall oil entering the North Sea. An attendee from Denmark quoted a figure as high as 23%. A representative from E&P Forum countered with yet another set of figures which showed an estimate for the entire North Sea of approximately 3%. Estimates for the U.S. are also low, in the single percent range. The primary sources still appear to be nonpoint sources such as rivers and land runoff. The real number will probably always be elusive, as there is no reasonable means to quantitate input from a broad variety of sources.

The U.S. has also evolved their regulations to include a toxicity limit linked to a discharge dispersion model. Two species of marine organisms are tested using five different test measures (chronic). These are used to determine a No Observable Effects Concentration (NOEC) which must be met at the 100 meter boundary from the discharge point. There is also a trend in the U.S. towards zero discharge in the wetlands areas, and possibly shallow coastal waters. This is still not completely resolved.

Toril Inga Røe of Norsk Hydro discussed experimental work that is currently underway in the North Sea to compare uptake of hydrocarbons into artificial Semi-Permeable Membrane Devices (SPMD's) in relation to similar uptake up in blue mussels. Their experimental work is still underway and the final data is not yet available. The future role of these sort of passive, nonliving monitoring systems in a regulatory role are uncertain. Similar experimental work is ongoing in the U.S. The largest challenge is still the extrapolation of SPMD data to real world exposures and varying species.

In wrapping up the session, Dr. Ray summarized current U.S. regulations (mentioned above). Under the current regulatory requirements of toxicity and discharge rate limitations (tied to NOEC levels and dispersion rates as determined by model), the offshore regulatory picture seems to have leveled off for the near future. The necessary protection against negative environmental impacts seems to be in place, and the current technologies for produced water treatment appear to be adequate for most operators to meet the regulatory limits. The one remaining unresolved issue is that of produced waters with elevated levels of soluble organics which are picked up in the oil and grease tests, but cannot be removed by current treatment technologies. These represent a small percentage of discharges. Previous laboratory toxicity testing has shown these solubles to have very little contribution to toxicity.

The issue of naturally occurring radioactive materials (NORM) is still fairly new. The U.S. regulations currently require all operators to analyze for NORM and report the data to EPA. No pending regulatory limits are currently on the horizon. Studies conducted by EPA, the Department of Energy (DOE), and the petroleum industry on the fate and effect of NORM from discharges would suggest that there is not an accumulation or bioaccumulation in the receiving environment. Recent environmental and human health risk assessments also show a very low probability of environmental harm.

Some discussion related to the importance of compositional information on produced water. This includes both original source produced water, and the treatment chemicals that are added in the production process. Progress on obtaining chemical composition information for additives has progressed in Europe, with much of the information being incorporated in the model CHARM. Similar progress in collecting compositional information cannot be reported for the U.S. Vendors of the various products used are reluctant to provide detailed information due to concerns over proprietary formulations and competition between manufacturers. Without this information, it will continue to be difficult to make educated modifications to treatment systems to reduce toxicity of the final, discharged produced water stream.

Despite the significant funding of environmental studies, the apparent lack of significant measurable environmental damage due to produced water discharges in the open ocean would suggest that current regulatory limits are appropriate.

2

REGULATION OF PRODUCED WATER ON THE NORWEGIAN CONTINENTAL SHELF

Erik E. Syvertsen

Norwegian Pollution Control Authority
Section for Offshore Activities and Monitoring
Strømsveien 96
P.O.Box 8100Dep, N-0032 Oslo
Oslo, Norway

ABOUT THE SFT

The *Norwegian Pollution Control Authority* (SFT) is under the auspices of the Ministry of the Environment, and plays an important role in the effort to combat pollution, waste problems and noise.

SFT is responsible for measures against industrial pollution, oversees the national oil pollution emergency services and administers the Pollution Control Act and the Product Control Act. SFT acts as a technical advisory agency for the Ministry of the Environment, and is generally responsible for monitoring atmospheric and aquatic pollution. SFT is organized in 8 departments, one of which is concerned with the industry.

The *Department of Industry* handles applications for discharge permits in connection with industrial and offshore activities. The department is also responsible for following up permits by reviewing internal control reports and participating in inspections and audits run by the Pollution Control Department. The department administers monitoring of areas which are polluted by industrial and offshore activities, and follows up programmes to develop cleaner technology. The department participates in international efforts to reduce pollution from industry and offshore activities. The Department of Industry consists of 5 sections, one of which is reponsible for offshore activities.

The *Section for Offshore Activities and Monitoring* handles applications for offshore discharge permits, monitoring of offshore pollution and marine pollution in general and acts as an advisor to the Ministry of the Environment in connection with opening of new areas for offshore activities, evaluation of offshore environmental impact assessments etc. The division has professional competence on the impact side (marine biologists) and the technical side (engeneers). In addition it may draw on the competence of the other divisions of the organization whenever needed (law, economics etc).

Produced Water2: Environmental Issues and Mitigation Technologies
Edited by M. Reed and S. Johnsen, Plenum Press, New York, 1996

The responsibilities of the SFT border on those of the Norwegian Petroleum Directorate (Dept. of Industry and Energy) and the relationship between the two directorates is regulated by a coordination directive.

DISCHARGES OF PRODUCED WATER

The Norwegian offshore industry is responsible for a limited but increasing part of the oil entering the North Sea (≈ 1.6% in 1994). However, as the petroleum activities move northwards on the shelf, the offshore discharges will contribute more to the oil contamination of the sea. At the same time the industry is moving into ever more environmentally sensitive areas.

The total Norwegian offshore oil discharges to the North Sea reached a maximum of close to 4.000 tonnes as early as in 1984, after 13 years of activity. The bulk (≈ 3.500 tonnes) came from discharges of oily cuttings. The main environmental impact of these discharges was contamination of the sea bed affecting primarily the bottom fauna. At some of the fields, as much as 100 km^2 of seabottom is contaminated (e.g. Valhall), and recent assessments show that the bottom fauna is affected in close to the whole contaminated area. Over the years stricter regulations have been implemented on the discharges of oily cuttings, until they were in practical terms banned from 1 January 1993. Oily cuttings are now either reinjected or brought to land for treatment.

As steadily more fields have been developed and the fields first put into production have grown older, the discharges of produced water have escalated. In 1993 produced water for the first time became the largest source of offshore oil dicharges to the sea, reaching 585 tonnes out of a total 783 tonnes. While oil from all other offshore sources are estimated to decline towards the turn of the century, oil discharges from produced water are expected to reach 2.200 tonnes by the year 2000. This has made the environmental concern of the authorities shift to become more focused on produced water and the water column, rather than on discharged cuttings and the sea floor, even if the latter will still be closely monitored.

REGULATION OF PRODUCED WATER BY SFT

Discharges of produced water on the Norwegian shelf are regulated by the Pollution Control Act of 13 March 1981, a set of specific regulations and international conventions.

There is a strong political agreement that Norwegian environmental protection policy shall have high ambitions. As an advisory agency for the Ministry of the Environment, SFT aims at contributing to fulfil these ambitions. Some main principles underlying Norwegian regulations are:

- Pollution is in general not allowed. Discharge permits are only to be regarded as exceptions and the petroleum industry has accordingly been regulated by individual decisions. SFT is regulating the exploration drilling activities by conditions in the Norwegion petroleum Directorates handling of applications for consent in areas not considered particularly environmentally sensitive. Exploration drilling in sensitive areas is regulated on an individual basis by the issue of discharge permits. Production activities on oil and gas fields are regulated by discharge permits.
- The precautionary principle. If there is a risk of serious or irremediable harm to humans or the environment, the lack of an indisputable scientific proof shall not be used as an argument against prevention or abatement of pollution.

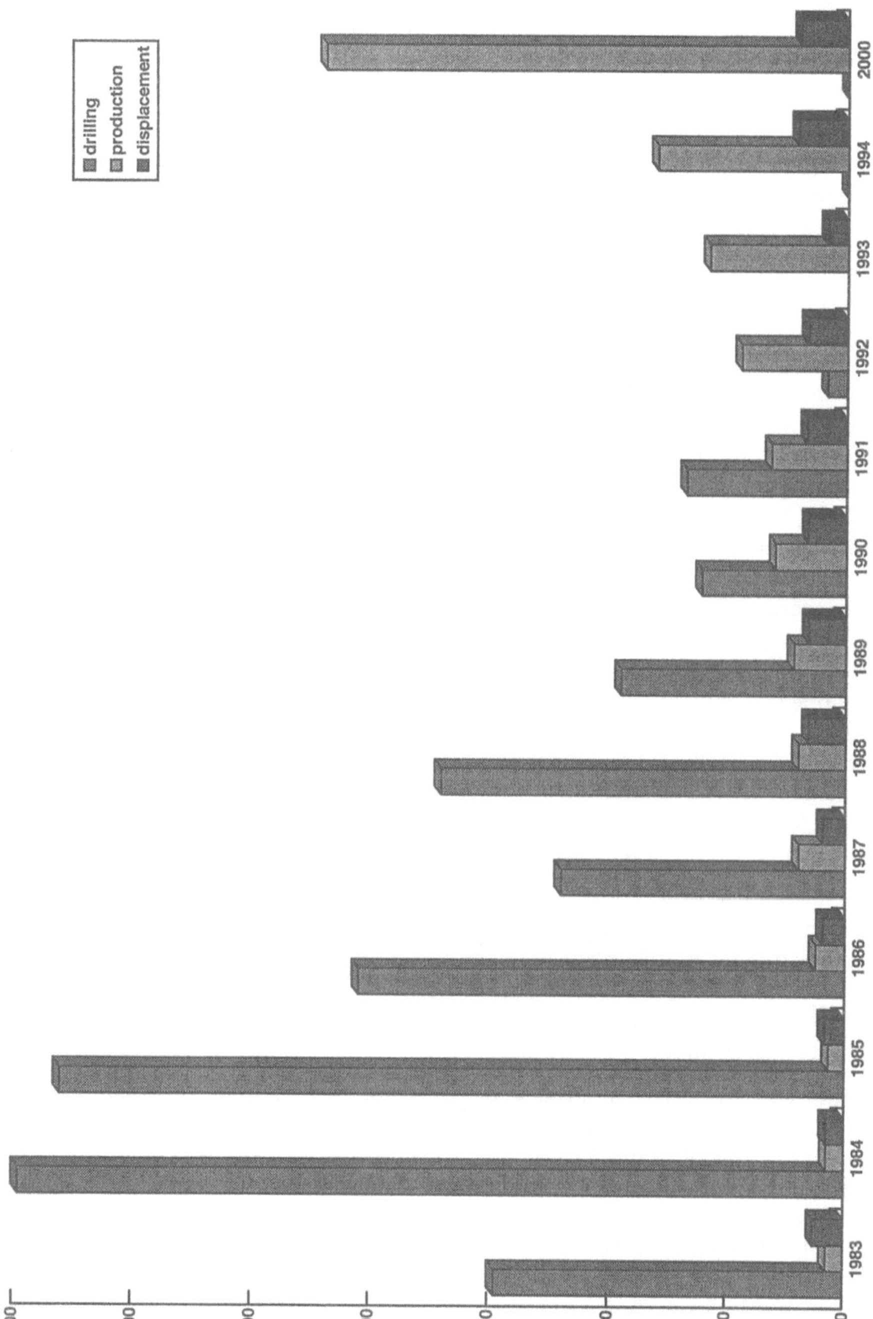

Figure 1. Continual discharges of oil on the Norwegian continental shelf, 1983-1994 (2000). Drilling for oil and gas, oil production and displacement during storage. Tonnes.

- Polluter pays. The costs of preventing and limiting pollution and treating waste shall be borne by the polluter.
- Best available technology / cost-effectiveness. The prevention and abatement of pollution and waste problems shall be based on the technology which, on the basis of a total appraisal of present and future use of the environment and of economic considerations, gives the best result.
- Strong emphasis on international agreements.

Norwegian authorities give priority to the work to reach international agreement on environmental issues. Relevant to produced water discharges are the Oslo (OSCOM) and Paris Conventions (PARCOM), now joined in the OSPAR Convention.

The "Convention for the Prevention of Marine Pollution from Land-based Sources" - the Paris Convention (PARCOM) was opened for signature in Paris from 4 June 1974, and came into force on 6 May 1978. It followed on from the Oslo Convention (Convention for the Prevention of Marine Pollution by Dumping from Ship and Aircraft) concluded in Oslo in February 1972. Contracting parties to PARCOM were the North Sea states, and the European Community and Luxembourg were signatories. The contracting parties undertake to reduce, if necessary to eliminate by steps, pollution of the maritime area from landbased and offshore sources.

In 1989, PARCOM adopted the two main principles upon which its programmes and measures are based: *The Precautionary Principle* and the use of *Best Available Technology*-BAT. The Contracting Parties have thus agreed that it may be necessary to act upon the scientifically-based presumption of a causal link between emissions and effects, even if no definitive evidence of such a link has been established. BAT is taken to mean the the latest stage of development (state of the art) of processes, facilities or methods of operation which indicate the practical suitability of a particular measure for limiting emissions.

In 1992, the ministers of the North Sea countries responsible for the protection of the environment of the North Sea and the members of the CEC responsible for environmental protection, signed the "Convention for the Protection of the Marine Environment of the North East Atlantic" in Paris, merging the Oslo and Paris Conventions into the OSPAR Convention. This convention represents important progress, particularly in:

- i) the obligation accepted by all North Sea States and the European community to adopt the *precautionary principle* and the *polluter pays principle* in their work of protecting the marine environment;
- ii) the possibility of adressing through the Convention all matters relating to protection of the marine environment, especially in a comprehensive assessment of its quality status.

At present discharges of produced water are regulated in accordance with the guidelines of the Paris Convention, and implemented in the discharge permit. The guidelines imply that the oil content of discharged water shall not exceed 40 mg l^{-1}. From originally being a goal for the mean value of the oil content in discharged water, it has developed into becoming a limit for the maximum acceptable concentration of oil in discharged water (Cardiff 1987). If the limit is exceeded it has to be reported nationally as well as internationally.

Recent studies have shown that the *acute* effects of produced water discharges are restricted to the near surroundings of the outlet. The studies also show that water-soluble aromatic hydrocarbon components, phenols and PAHs seem to contribute most to the acute toxicity of produced water.

However, some field and laboratory studies indicate that sublethal effects may occur at very low produced water concentrations (1 ppm), even after very short exposure times (minutes). Early life stages seem to be particularly susceptible and there seems to be considerable differences in sensitivity between species. Some authors seem to believe that these kinds of effects may have the potential to alter the biodiversity over large areas.

According to some investigators sublethal effects may be attributed to certain aromatic hydrocarbons, alkylated phenols and heavy metals, components that are persistent and have a potential for bioaccumulation. These components are *not* covered by the present regulations (neither national nor international). Instead, focus is still placed on the fraction of dispersed aliphatic hydrocarbons commonly accepted as having low environmental impacts at concentrations below 40 mg l^{-1}.

The objective of the SFT is that future analyses and regulations shall include the components considered to be most environmentally harmful. A change in the regulation strategy should not imply a change in the present discharge limit for aliphatic hydrocarbons. However, new discharge limits should be assessed on the background of expected effects, available technology, cost-effectiveness etc.

Even if the relative discharge limit for aliphatic hydrocarbons does not change, the analysis method will have to be phased out within a short time due to the present use of freon CFC 113 as a solvent. According to the Montreal protocol production and use of CFC will be forbidden after 1 January 1996. Even if oil-in-water analyses are nominated as an area of essential use and thus entitled to exemption from the rules of the protocol, it would be a paradox and an awkward signal if the environmental authorities imposed on the petroleum industry the use of CFC products that are strictly regulated internationally. SFT is actively working to replace the existing method with a new standard, and has invited the industry to paricipate in this work.

Future solution of the produced water problem?

During the last couple of years there has been a trend on the Norwegian shelf to include reinjection of produced water in the plans for the development of new fields. Even fields in production are reconsidered for produced water reinjection. Produced water is in these cases used as part of a pressure stabilizing medium in the formation.

British Petroleum was the first company to show that reinjection of produced water may be a cost-effective solution on the Norwegian shelf (Ula and Gyda fields). Now, others are following. Most potential problems with reinjection of just a few years back appear to have been overcome, and reinjection seems to be a cost-effective option in many cases.

SFT greatly appreciates this farsighted attitude of the operating companies and believes that reinjection may be an important means of slowing down the expected growth in produced water discharges. For field development in environmentally sensitive areas of the Norwegian Sea and further north on the shelf, SFT are considering severe restrictions on all discharges to the sea.

3

DISCHARGES OF PRODUCED WATER TO THE NORTH SEA

Effects in the Water Column

T. I. Røe,[1] S. Johnsen,[2] and The Norwegian Oil Industry Association (OLF)[3]

[1] Norsk Hydro, Research Centre
Section of Analytical Chemistry
Pb. 2560, N-3901 Porsgrunn, Norway
[2] Statoil
Research Centre
Postuttak, N-7004 Trondheim, Norway
[3] Stavanger, Norway

INTRODUCTION

Knowledge related to toxic effects of produced water in areas adjacent to the discharge points has been improved through several studies during the recent years (Neff, 1987, Schiff et al., 1992; Sauer et al., 1992; Brendehaug et al., 1992). However more knowledge is required, especially with regard to possibilities of long term effects. It is difficult to do toxicity tests in the laboratory due to the complex and variable composition of produced water (Middleditch, 1984; Higashi et al., 1992), and field assessments are complicated by other types of production activities that confound effects of produced water (Carney, 1987; Spies, 1987). It is also necessary to improve and verify dispersion models and to improve correlations between produced water dispersion, presence and distribution of biological resources and potential environmental effects. The Norwegian Oil Industry Association (OLF), in association with Statoil and Norsk Hydro, is presently undertaking a project to clarify the relationships between produced water discharges and potential environmental effects. The project is titled "Discharges of produced water to the North Sea - effects in the water column". The ultimate vision of the project is to develop a model that may assist the industry in predicting such effects. The programme was started in 1993 and is expected to continue throughout 1997. Phase I, which has now been completed, has primarily been a fact-finding and status-collection effort which will form basis for the work to be performed in the subsequent phases. Phase II, which started in January 1995, is primarily concerned with biological degradation and availability of produced water components, and development of techniques for monitoring the produced water components in the

Produced Water2: Environmental Issues and Mitigation Technologies
Edited by M. Reed and S. Johnsen, Plenum Press, New York, 1996

recipient. This paper summarises the results of phase I, and presents the work to be done, and the results obtained so far in phase II.

MAPPING OF PRODUCED WATER DISCHARGE SOURCES AND EVALUATION OF DISPERSION MODELS

Produced Water Discharges in the North Sea

The locations of produced water discharges have been mapped. All discharge points in the North Sea, including the Norwegian, UK, Danish, and Dutch sector, have been identified with regards to location and quantity. The locations of the discharge points are shown in figure1. The total discharges to the North Sea are steadily increasing as the producing oil fields are becoming more mature. The discharges were 187 million m^3 in 1993

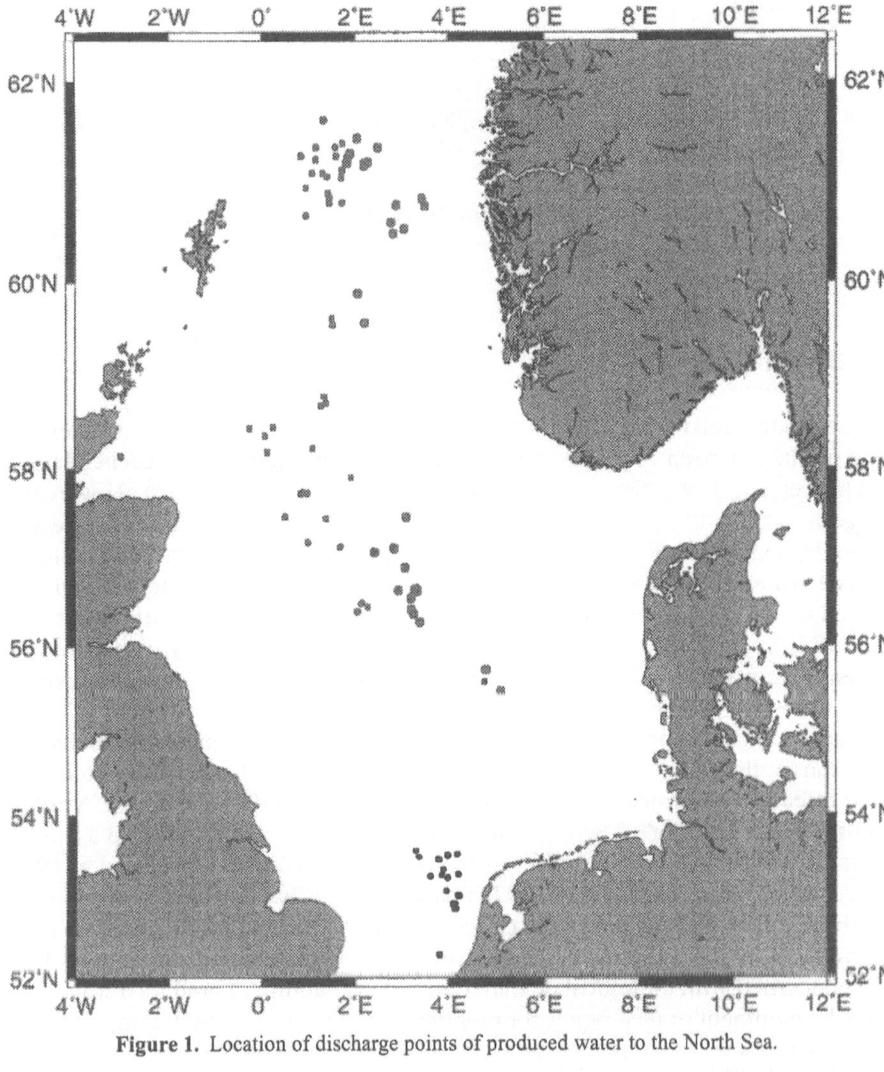

Figure 1. Location of discharge points of produced water to the North Sea.

and are expected to culminate at a level of 340 million m^3 in 1997. The discharges from the Norwegian sector were 26 million m^3 in 1993 and are expected to grow to approximately 90 million m^3 in the year 2000.

Dispersion Models

In recent years several field studies to investigate dispersion and deposition of produced water in the water column have been performed (Smith, 1993; Statoil, 1993). Experiments have shown that the diffusive processes are very efficient in the vicinity of the outlet, where the flow is dominated by the initial momentum (within typical 10 metres). During this phase the concentration is normally reduced approximately by two orders of magnitude (Furnes, 1994). Factors which are important for the dilution are the shape and direction of the outlet and the local oceanographic conditions (current, waves, variations in density of the water, and depth). The depth of deposition is dependent of the density of the discharge in proportion to the density of sea water.

When evaluating toxicological effects on marine organisms, the exposure time and the concentration gradient are very important. This can be derived from the dispersion model, when integrating the concentration over time for an imagined particle that is dropped in the discharge point, and follows the produced water in the recipient giving the actual exposure dose for a given specimen (Furnes, 1994). Most evaluation of potential toxic effects have so far been based on the concentration of pollutants while a real life situation implies that biota in the influence area will experience a variable concentration over a limited time window.

Model simulations (plume type) of produced water have so far mostly been concerned with the near field dispersion and discharge from individual sources, and to a less extent with the contribution fields resulting from the total number of discharges. The problem of computing dispersion of passive substances within fluids has traditionally been approached by two methods:

- Solving the advection-diffusion equation

$$\frac{\partial C}{\delta t} + \vec{V} \circ \nabla C = \nabla \circ (D \nabla C) + Q$$

  by a finite difference or a finite element technique. In this equation C represents the concentration in space (x,y,z) and time (t), D is the diffusivity which also varies in space and time and ∇ is the gradient operator. The term Q expresses source and sink processes.
- An alternative to the solution of the above equation on a spatial grid is the particle-tracking method in which the substance is modelled by an assemblage of discrete particles, which may be subject to advection, diffusion and various decay processes. Using this method the advection is simulated by a translation of the individual particles in the local fluid velocity field. The turbulent diffusion is simulated by a random walk technique which displaces each particle a certain distance derived from the variance of a given distribution function.

The second method offers some advantages over the first one, since the advection can be computed by a simpler and more accurate formulation. With a detailed description of the velocity field the particles may be advected exactly through that field by a simple integration. An interesting aspect of the particle-tracking method is the possibility to tag each particle with specific attributes such as age, density and size. This property opens a number of possibilities as e.g. to compute exposure of individual particles drifting in a concentration

Table 1. Chemical composition of produced water from Norwegian North Sea platforms

Fields	Unit	Statfjord	Gullfaks	Ekofisk 2/4B-K	Ekofisk 2/4T	Tor	Ula
Compounds							
TOC	mg/l	850	61	180		85.5	71
THC	mg/l	15	35				50
Sum Aromatics	mg/l	6.00	9.56	5.67	66.95		15
BTX	mg/l	4	5	5.41	66.90	1.1	12
Naphthalenes	mg/l	0.942	2.16	0.247	0.052	0.597	
Naphthalene	mg/l	0.261	0.398	0.157	0.038	0.073	
C1-naph	mg/l	0.35	0.629	0.062	0.012	0.17	
C2-naph	mg/l	0.199	0.584	0.018	0.002	0.204	
C3-naph	mg/l	0.132	0.55	0.010	0.0005	0.155	
Phenanthrenes	µg/l	45	90	6.26	0.28	135	
Phenanthrene	µg/l			2.09	0.08		
C1-phenanthrene	µg/l			2.43	0.12		
C2-phenanthrene	µg/l			1.74	0.08		
C3-phenantrene	µg/l			n.d.	n.d.		
Dibenzothiophenes	µg/l	8.6	22.7	1.39	0.15	10	
Dibenzothiophene	µg/l			n.d.	n.d.		
C1-dibenzothiophene	µg/l			1.39	0.03		
C2-dibenzothiophene	µg/l			n.d.	0.12		
C3-dibenzothophene	µg/l			n.d.	n.d.		
Sum NPD	mg/l	1.00	2.27	0.254	0.055	0.74	
Acenaphtylene	µg/l			0.89	0.02		
Acenaphthene	µg/l	0.001	0.001	n.d.	0.04	0	
Fluorene	µg/l	12	11.3	n.d.	0.33	8.1	
Fluoranthene	µg/l	0.0854	0.195	n.d.	n.d.	0.24	
Pyrene	µg/l	0.0897	0.194	n.d.	0.08	0.42	
Chrysene	µg/l	0.226	0.398			0	
Benz(a)anthracene	µg/l	0.0193	0.311	n.d.	n.d.	0.23	
Benzo(a)pyrene	µg/l	0.001	0.001	n.d.	n.d.	0	
Benzo(ghi)perylene	µg/l	0.001	0.001	n.d.	n.d.	1.35	
Benzo(k)fluoranthene	µg/l	0.0197	0.0528	n.d.	n.d.	0.016	
Sum PAH 3-6 ring	µg/l	66.04	125.15	0.89	0.47	155.36	
Sum phenol	mg/l	8.3	2.7	1.03	2.65	3.62	0.09
Phenol	mg/l	5.1	0.8	0.61	0.97	2.19	0.033
C1-phenol	mg/l	2.5	0.86	0.19	0.83	1.1	0.028
C2-phenol	mg/l	0.4	0.6	0.14	0.57	0.254	0.02
C3-phenol	mg/l	0.13	0.18	0.06	0.26	0.0316	0.0006
C4-phenol	mg/l	0.026	0.1	0.03	0.02		
C5-phenol	mg/l	0.016	0.065	n.d.	n.d		
C6-phenol	mg/l	0.013	0.11	n.d.	n.d.		
C7-phenol	mg/l	0.005	0.012	n.d.	n.d.		
Sum organic acids	mg/l	895	55	323	577	234	
Formic acid	mg/l			148	275		
Acetic acid	mg/l	732	15.6	132	267	104	9.5
Propionic acid	mg/l	106	8.9	35.2	27.4	10	1.2
Butylic acid	mg/l	39	14.1	6.35	5.18		1.5
Valeric acid	mg/l	18	8.2	1.61	2.17		0.6
Caprioic acid	mg/l	9	8.2	n.d.	0.09		
Organic acids > C6	mg/l			n.d.	n.d.		
Methanol	mg/l			6.3	33.9		
Salinity Cl-	mg/l			30400		90500	40440
Amonium	mg/l	25.4	26.9				0.1
Lead	µg/l	50	50	n.d		80	270
Copper	µg/l	2	2	20		600	20

Fields	Unit	Statfjord	Gullfaks	Ekofisk 2/4B-K	Ekofisk 2/4T	Tor	Ula
Iron	mg/l			4		8.9	23
Barium	mg/l			28.2		42.1	12
Cr-VI	µg/l	10	10	6		0.08	40
Mercury	µg/l	1.9	1.9	n.d.			9
Zinc	µg/l	6.8	13	13		200	0.26
Cadmium	mg/l	10	10	n.d.			0.02
H2S	mg/l	0.12	0.17				
Total radioactivity	Bql						
40K	Bql						
226Ra	Bql						

field consisting of a known substance. The exposure is then obtained through simple integration. Taking the particles to represent a biological resource (fish, egg, larvae, etc.), this type of simulation is particularly useful. It is important, however, to be aware of the differences between passive drifting organisms and organisms who can swim in and out of the plume. The particle-tracking method does not include numerical dispersion in the classical sense, which makes it better suited to resolve sharp gradients. The main disadvantage of the method is the random fluctuation of computed concentrations. The relative size of the statistical fluctuations can be reduced by increasing the number of particles used, but will in turn increase the computational effort.

CHEMICAL COMPOSITION OF PRODUCED WATER

The chemical composition of produced water has been described by several authors, and Middleditch (1984) and Stephenson (1992) have presented comprehensive reviews of the studies performed during the last two decades. The composition is strongly field dependent, and counts a variety of inorganic and organic compounds. In addition to dispersed oil, the organic content includes dissolved hydrocarbons, organic acids, phenols and traces of chemicals added in the production/separation line. Results from analyses of produced water from different Norwegian platforms are given in Table 1. Carboxylic acids represent the highest concentration of organic components in the produced water. Low molecular carboxylic acids, such as acetic acid and propionic acid, are most abundant. The carboxylic acids represent the majority of the total organic carbon content (TOC) of produced water, and they are not considered to be environmentally harmful. The second largest group of dissolved organic compounds in produced water is generally the phenols. Other organic components are denoted "total aromatics", which in this case are mainly volatile aromatic hydrocarbons such as benzene, toluene and xylene (BTX). The Polynuclear Aromatic Hydrocarbon (PAH) compounds consist mainly of naphtalenes and phenylbenzene. Analysis of the aromatic hydrocarbons shows large differences from field to field. Water from gas production generally shows a higher content of these components than that from oil production. On the other hand, the water volumes are much lower. The total discharges of these compounds into the North Sea from produced water have not been calculated. However, based on measured concentrations of these compounds the following orders of magnitude may be indicated:

Total aromatics:	100 - 1000 tonnes/year
PAH:	10 tonnes/year

DISTRIBUTION AND SUSCEPTIBILITY OF BIOLOGICAL RESOURCES

The North Sea

The surface area of the North Sea, including the Skagerrak, is 575.300 km^2. The volume is 42.294 km^3 and the mean depth is 74 m. It is one of the most productive fishing areas in the world, and covers vast sources of oil and natural gas. It is also an important area for shipping to and from the surrounding countries. The water quality is not only affected by the discharges, but also by the ability of the North Sea to mix and renew its water masses. The inflows may be observed in the central North Sea, where water from five different sources can be demonstrated (Nordic Council, 1989):

- Water of Atlantic origin (the Gulf Stream), enters the North Sea via the Orkney-Shetland area. Clear water of high salinity and low nutrient content.
- Water of Atlantic origin (the Gulf Stream) enters through the English Channel. Clear water of high salinity and low nutrient content.
- Brackish water from the Baltic flows into the northernmost section of the North Sea via the Skagerrak. Fairly turbid water of low salinity and high nutrient content.
- Brackish water - from rivers and land run-off - from the coastal areas of the UK. Highly turbid water of low salinity and high nutrient content.
- Brackish water - from rivers and land run-off - from the continental coasts. Highly turbid water of low salinity and high nutrient content.

Water circulation varies greatly in various parts of the North Sea. In general, it takes about 36 months for the water in the German Bight to leave the North Sea. In the northern North Sea, the water remains for only about six months. In general, long-term circulation in the North Sea may be said to be anti-clockwise. The main dispersion of water out of the North Sea takes place along the Norwegian and Swedish coasts in the Norwegian Current, which is a continuation of the Jutland Current from the Kattegat and the Danish Straits. Studies of the water circulation and currents in the North Sea have shown, in particular, that water discharged from British east-coast estuaries spreads throughout the North Sea. Water discharged from continental estuaries tends to be more confined to the eastern parts of the North Sea and flows along the coast up to the Skagerrak, where it joins the Norwegian Current and exits northwards along the Norwegian coast. The volume and flow at the latitude of Utsira has been estimated to 1 million m^3/sec. Two potential areas in the Norwegian sector of the North Sea were focused in the present study; the Ekofisk area (central North Sea) with a produced water discharge of 4 mill. m^3 in 1993, and the Statfjord/Gullfaks/Oseberg area (northern North Sea) having a produced water discharge of 22 mill. m^3 in 1993. These two influence areas dominate with regards to produced water discharges in the Norwegian sector of the North Sea. Both regions are covered in the Quality Status Reports from the North Sea Task Force (NSTF, 1993) which was the basis source of information to the present study. The organisms studied comprise phytoplankton, zooplankton and pelagic fish, with a special reference to the early life stages, reproduction and potential conflicts with commercial fisheries in the two regions.

The Ekofisk Region

This area has northern and southern borders following the 57°N and the 56°N parallels respectively, and western and eastern borders between 2°E and 4°E (figure 1). No exceptional phytoplankton blooms have been observed in the area and the primary production is estimated to be moderate. Large amounts of zooplankton are found in the area during late spring, consisting mostly of *Calanus sp.* (mainly *Calanus finmarchicus*). The stocks are transported by surface currents, and therefore some reach the shallow North Sea, where their biomass can increase vastly in summer. Large stocks of haddock, Norway pout and herring are found in the central North Sea area. Other species which are represented are cod, saithe, whiting and mackerel. National fishing fleets operate in different parts of the region using fishing gear such as trawls, gill nets and long lines. A shift in the commercially exploited fish stocks in this part of the North Sea has taken place during recent years. Cod and Haddock stocks have been depleted while whiting and herring stocks have increased (NSTF, 1993). Intensive fishing is the heaviest stress to which fish in the North Sea are exposed. Any subtle effect from pollution on the overall fish stocks is completely masked by such factors as fishing intensity, annual variation in breeding success and growth. Most fish stocks are overexploited, but as a whole they continue to maintain their yield (Nordic Council, 1989).

The Statfjord/Gullfaks/Oseberg Region

This area has northern and southern borders following the 62°N and the 60°N parallels respectively, and western and eastern borders between 1°E and 4°E. This region in the northern North Sea has water depths between 100 and 300 metres. A main part of the area is referred to as the Northern North Sea plateau. The Atlantic water flowing through the region is the main source of nutrients to the North Sea. This fact, combined with a wind climate causing an upward flux of nutrients, is the main mechanism determining the North Sea productivity. A range of phytoplankton species are present in the area, but relative dominance varies through the year. During winter, when phytoplankton biomass is low, the most important group is naked flagellates. During the spring bloom the dominant species are planktonic diatoms immediately succeeded by a rapid increase in small flagellates. During summer and early autumn the most dominant algae are the dinoflagellates. *Calanus finmarchicus* is the dominant copepod species in the area both with regard to numbers and in terms of biomass. Another important species contributing to the zooplankton biomass is the *Thysanoessa inermis* (krill). Mackerel and horse mackerel feed in the area in the summer and autumn seasons but are less abundant during winter and spring. Variable influence of Atlantic water masses between seasons and years cause environmental changes that influence the fish distribution. In summer and autumn both mackerel and horse mackerel are feeding in the area, while they are less abundant during winter and spring. In the upper layer (>50 m) of the water column there are typically species like mackerel, herring, haddock and whiting.

Susceptibility of Biological Resources

Tests of chronic toxicity (with growth as endpoint) of produced water from North Sea platforms to algae show a reduction in growth (EC_{50}) at a concentration of produced water of 45-535 ml/l (2 days) for oil platforms (Brendehaug et al., 1992) and 0.9-36 ml/l (4 days) for gas platforms (Jacobs, 1987; Jacobs et al., 1991). Sommerville et al. (1987) reported an LC_{50} of 100 ml/l produced water to the copepod *Calanus finmarchicus* in a 1 day duration test. Krause et al. (1992) found that brief (10 min) exposure of sea urchin (*Strongylocentrotus purpuratus*) gametes and longer (48 hr) exposure of zygotes to concentrations of Carpinteria produced water as low as 0.0001% significantly slowed embryonic development. Exposures of red abalone larvae (*Haliotis rufescens*) to a plume of produced water in the field reduced the probability that a larva would successfully complete the transition to the adult body form (Raimondi et al., 1992). For fish, there is a great variation in the response in toxicity tests due to among others the exposure media (E&P Forum, 1994). The lowest value registered of LC50 is for guppy, *Poecilia reticulata*, at a value of 7.5-423 ml/l (Jacobs, 1987; Jacobs et al., 1991). As mentioned before, there are indications that produced water from gas platforms may be more toxic than water from oil platforms.

ENVIRONMENTAL IMPORTANCE OF DISPERSED OIL IN PRODUCED WATER

A study was initiated by one of the OLF member companies to test the toxic effects of variable concentrations of dispersed oil in water on different marine organisms. The initial tests demonstrated that this approach was not feasible due to unstable oil-in-water emulsions throughout the test period. The task was therefore approached from a more theoretical angle, combining available detailed produced water composition data, knowledge on dilution factors, and literature data on toxic effects on marine organisms. A risk evaluation based on produced water composition from the Statfjord and Gullfaks fields was performed as a parameter study with an oil-in-water concentration of 15 and 30 µg/l, and with and without production chemicals. The modelling showed that the risk reduction obtained by reducing the dispersed oil content in produced water from 30 to 15 µg/l is marginal compared to the effect of dilution, but that production chemicals increase the risk of effects. However, the study was based on fixed concentration of produced water components in the exposure area. A comparative study where dose-related exposure was applied showed that the true risk of effects is probably considerably lower (Furuholt, 1995).

FURTHER WORK: PHASE II OF THE PROJECT

The objective of phase II of the project "Discharges of produced water to the North Sea — effects in the water column" will be to clarify any relationship between persistent components in produced water discharges and effects on the biological resources in the recipient. The results from the work done in phase I indicate that these relationships are complex. The following tasks are focused in phase II:

Identification of Produced Water Components in the Recipient

One of the major issues in assessing the environmental impacts of produced water discharges is the potential bioaccumulation and possible biomagnification of components by indigenous organisms. Caged mussel monitoring (the mussel watch approach) has been widely used as a time integrated method to identify and quantify the presence of bioavailable petroleum hydrocarbons and other pollutants in the water column. Semipermeable polymeric membrane devices (SPMD), which consist of low density polyethene tubings filled with fibres of lipid, have been shown to hold considerably promise as non-living, time integrated concentrators of non polar organic compounds in aquatic environments. It is therefore a possible alternative to using living organisms such as mussels for estimating bioavailability and bioconcentration factors for organic chemical contaminants. Their application can be especially useful in environments where natural incidence of mortality and physically induced stress to mussels are high. A field study at Statfjord, where the SPMD and mussel method were applied in a side-by-side test for monitoring of hydrocarbon bioaccumulation potentials, has been conducted in April-July 1995 (see figure 2 and 3). The objective was to evaluate the performance and suitability of the two methods. The chemical analyses are not yet completed, and the results will be published later.

Figure 2. Semipermeable polymeric membrane devices (SPMDs) and blue mussels were applied in a side-by-side test for monitoring hydrocarbon bioaccumulation potentials.

Figure 3. SPMDs and blue mussels were placed on sampling sites 100 and 300 metres downstream the discharge point at Statfjord C.

At the same sampling sites at Statfjord, water samples of 50 litres were taken and extracted *at line* through a polystyrene divinylbenzene extraction disk. Results from the chemical analyses of these samples will be published together with the mussel / SPMD-results.

Chemical Composition of Produced Water

Analysis of produced water samples from the platforms Oseberg C, Oseberg Feltsenter, Brage and Troll will be carried out during October-November 1995. The results will be reported in the database of produced water composition from the Norwegian sector.

Establishment of Concentration Fields of Produced Water Components from Multiple Discharge Sources

The dispersion model will be improved with regard to current fields in the North Sea by the Norwegian Meteorological Institute. Concentration fields of produced water components from discharge sources in both Norwegian, British and Dutch sector will then be established.

Establishing Rate of Chemical and Biological Degradation of Persistent Components

Bacteria have a large biomass, short generation time, fast adaptation of the metabolism to degradation of organic compounds, and are assumed to be largely responsible for degradation of produced water components in the North Sea. A study of three different ^{14}C-labelled produced water components in a dynamic flow system is in progress. This system is simulating as much as possible the *in situ* conditions regarding temperature, light, salinity and density of organisms in the recipient. In addition, this system makes it possible to simulate time-dependent dilution of the produced water components and a realistic time of exposure. Aerobic, heterotrophic bacteria are used as test organisms, and the ratio component : bacteria is changed during the experiment to simulate *in situ* conditions. Comparative investigations of intake, accumulation and degradation of produced water components in planktonic bacteria, algae and zooplankton will be conducted. The degree and rate of degradation of the selected produced water components in the different groups of organisms will be investigated, and degradation products will be identified.

Assessing Compartmentalisation of Persistent Components

In an attempt to map the ecotoxicological effects of discharges of produced water, it is important to examine which parts of the ecosystem that are exposed to produced water components. This is done by experiments in a multi-chamber dynamic flow system, where different groups of organisms are present together. Three ^{14}C labelled selected produced water components are used in these experiments. The results will hopefully be presented during spring of 1996.

CONCLUSIONS

Total discharges of produced water from all installations in the North Sea were 187 millions m^3 in 1993 and are expected to culminate at a level of 340 million m^3 in 1997. The discharges from the Norwegian sector were 26 million m^3 in 1993 and are expected to grow to approximately 90 million m^3 in the year 2000. The composition is strongly field dependent, and includes a variety of inorganic and organic compounds. In addition to dispersed oil, the organic content includes dissolved hydrocarbons, organic acids, phenols and traces of chemicals added in the production/separation line. A database of chemical analyses of produced water from Norwegian platforms was established. A status description of the recipient, the North Sea, was compiled, comprising oceanographical data as well as a summary of the distribution and susceptibility of biological resources in influenced areas. The two areas with the largest produced water discharges in the Norwegian part of the North Sea, the Ekofisk area, and the Statfjord/Gullfaks/ Oseberg area were specifically covered. A risk evaluation based on produced water composition from the Statfjord and Gullfaks fields, showed that the toxicity of produced water discharges seem to be dominated by the dilution and not by the oil content in the discharged water. More investigations are necessary to clarify biological degradation of the produced water components, and to identify which species that are exposed to the discharges, the dose of exposure, and the long term effects in the food webs. These are the main objects of phase II of the project, which is currently in progress.

REFERENCES

BP, v/ Sande, A., 1994, Personal communication.

Brandsma, M.G., Smith, J.P., O'Reilly, J.E., Ayers, R.C.Jr., and Holmquist, A.L., 1992, Modeling offshore discharges of produced water, in: Ray, J.P. and Engelhart, F.R.: Produced water. Technological/ Environmental Issues and Solutions. Plenum Press, New York.

Brendehaug, J., Johnsen, S., Bryne, K.H., Gjøse, A.L., Eide, T.H., and Aamot, E., 1992, Toxicity testing and chemical characterization of produced water - a prelimnary study. In: Ray, J.P. and Engelhart, F.R.: Produced water. Technological/Environmental Issues and Solutions. Plenum Press, New York 245-256.

Carney, R.S., 1987, A review of study designs for the detection of long-term environmental effects of offshore petroleum activities, in: Long term environmental effects of offshore oil and gas development, Boesch, D.F.,and Rabalais, N.N., Elsevier applied Science, New York.

Cofino, W.P., Slager, L.K., and Van Hattum, B., 1993, Environmental aspects of produced water discharges from oil and gas production on the Dutch Continental Shelf, Part 1: Overview of surveys on the composition of produced waters conducted on the Dutch Continental Shelf, NOGEPA, The Hague, ISBN 90-5383-218-1.

Davies, J.M., Hay, S.J., Gamble, J.C., and Dow, K., 1987, The Ecological effects of produced water discharges from offshore oil platforms in the Northern North Sea, Marine Environmental Research.

E&P Forum, 1988, Water soluble organics in produced water, meeting in London, 16 March 1988.

E&P Forum, 1994, North Sea produced water: fate and effects in the marine environment, Report no. 2.62/204, May 1994.

Furnes, G.,1994, Discharges of produced water from production platforms in the North Sea, Norsk Hydro, Research Centre, Bergen, Report No. R-064641

Furuholt, E., 1995, Environmental effects of discharge and reinjection of produced water, 1995 International Seminar on Produced Water, Trondheim

Higashi, R.M., Cher, G.N., Bergens, C.A., and Fan, T.W-M., 1992, An approach to toxicant isolation from a produced water source in tne Santa Barbara Channel, In: Ray, J.P. and Engelhart, F.R.: Produced water. Technological/Environmental Issues and Solutions. Plenum Press, New York.

Jacobs, R.P.W.M., 1987, Environmental impact assessment of effluents discharged from gas/condensate platforms, NOGEPA, Report 6, The Hague, The Netherlands.

Jacobs, R.P.W.M., and Marquenie, J.M., 1991, Produced water discharges from gas/ condensate platforms, Environmental considerations, Proc. First Int.Conf. HSE in oil and gas expl. and prod., SPE, Richardson, Texas, vol.2, 89-96.

Krause, P.R., Osenberg, C.W., Schmitt, R.J., 1992, Effects of produced water on early life stages of a sea urchin: stage-specific responses and delayed expression, In: Ray, J.P. and Engelhart, F.R.: Produced water. Technological/ Environmental Issues and Solutions. Plenum Press, New York, Marine Chemistry, 1971, National Academy of Sciences, Washington, D.C.

Middleditch, B.S., 1984, Ecological effects of produced water discharges from offshore oil and gas production platforms. Final report on API Project no. 248, American Petroleum Institute, Washington D.C.

Neff, J.M., Rabalais, N.N., and Boesch, D.F., 1987, Offshore oil and gas development activities potentially causing long-termenvironmental effects., In: Boesch, D.F. and Rabalais, N.N. (eds), Long-term environmental effects of offshore oil and gas development, Elsevier, London.

Nordic Council (1989), Northern Europe's seas, northern Europe's environment, Report to the Nordic Council's International Conference on the pollution of the Seas, ISBN 91-38-12246-4.

NSTF, North Sea Task Force, 1993, North Sea Assessment Report, subregion 1 and 7a, State Pollution Control Authority (SFT), Norway.

Phillips Petroleum, v/ Wolden, S., 1994, Personal communication.

Raimondi, P.T. and Schmitt, R.J., 1992, Effects of produced water on settlement of larvae: field tests using red abalone, In: Ray, J.P. and Engelhart, F.R.: Produced water. Technological/ Environmental Issues and Solutions. Plenum Press, New York.

Riksheim, H. and Johnsen, S., 1994, Determination of produced water consistuents in the vicinity of offshore production fields in the North Sea, SPE paper No. 27138, Djakarta.

Sauer, T.C., Ward, T.J., Brown, J.S., O'Nil, M., and Wade, M.J., 1992, Identification of toxicity in low-TDS produced water, In: Ray, J.P. and Engelhart, F.R.: Produced water. Technological/Environmental Issues and Solutions. Plenum Press, New York, 223.

Schiff, K.C., Reish, D.J., Anderson, J.W., and Bay, S.M., 1992, A comparative evaluation of produced water toxicity, Paper first presented at first Int. Produced water symposium, 4-7 February, 1992, San Diego, California.

Slager, L.K., Van Hattum, B., Tromp Meesters, M., Sholten, M.C.Th., Kaag, N.H.B.M., Cofino, W.P., and Feenstra, J.F., 1992. Environmental aspects of produced water discharges from oil and gas production on the Dutch Continental Shelf, Part 3: Environmental effects, NOGEPA, The Hague, ISBN 90-5192-021-0.

Sommerville, H.J., Bennet, D., Davenport, J.N., Holt, M.S., Lynes, A., Malieu, A., McCourt, B., Parker, J.B., Stephenson, R.R., Watkinson, R.J., and Wilkinson, T.G., 1987, Environmental effect of produced water from North Sea Oil Operations., Mar. Poll. Bull., 16, 10, 549.

Spies, R.B., 1987, The biological effects of petroleum hydrocarbons in the sea: assessments from the field and microcosmos, in: Long term environmental effects of offshore oil and gas development, Boesch, D.F.,and Rabalais, N.N., Elsevier applied Science, New York.

Statoil F&U, v/ Johnsen, S., 1994, Pers. Comm.

Stephenson, M.T., 1992, Components of produced water: A compilation of industry studies, JPT, 548-603.

Sørstrøm, S.E., 1992, Produced water. Chemical and toxicity study. IKU report no. 222059.00/0102 prepared for OLF, Stavanger, Norway.

Van Hattum, B., Cofino, W.P., and Feenstra, J.F., 1992, Environmental aspects of produced water discharges from oil and gas production on the Dutch Continental Shelf, Part 2: A literature review of characteristics of produced water from offshore platforms, NOGEPA, The Hague, ISBN 90-5383-200-9.

4

REGULATION OF PRODUCED WATER BY THE U.S. ENVIRONMENTAL PROTECTION AGENCY

Allison Wiedeman

Project Manager
Engineering and Analysis Division
U.S. EPA Office of Water

INTRODUCTION

Among its many responsibilities, the Environmental Protection Agency (EPA) is charged with implementing the Clean Water Act established by Congress in 1972. The major objective of the Clean Water Act is to eliminate or control the discharge of pollutants, particularly toxic pollutants, into the Nation's waterways. One of the ways EPA is meeting this objective is to develop national technology-based effluent limitations guidelines for the oil and gas exploration and production industry. The development of these effluent guidelines required first an analysis of the performance capability of wastestream pollution control technology, followed by a determination of costs and economic achievability of the implementation of such technology. The final effluent limitations developed are a result of EPA's determination of the best available technology that is economically achievable for the industry. This paper will describe the effluent limitations regulations specifically for produced water generated by the oil and gas industry in the United States, and will include a description of the technologies that serve as a basis for the regulation, the actual effluent limitations requirements, and the costs of compliance.

OVERVIEW OF THE U.S. NATIONAL EFFLUENT GUIDELINES PROGRAM

The U.S. Clean Water Act (CWA) mandates that national technology-based effluent limitations guidelines and standards be developed for all industrial facilities which discharge into waterways of the United States, or which discharge or may discharge into publicly owned treatment works. The national guidelines and standards specify the achievable effluent pollutant reduction attainable based upon performance of treatment technologies actually employed within an industrial category. The program was established with a phased approach whereby industry would first be required to meet interim limitations based on *Best*

Produced Water2: Environmental Issues and Mitigation Technologies
Edited by M. Reed and S. Johnsen, Plenum Press, New York, 1996

Practicable Control Technology Currently Available (BPT) for existing sources. The second level of effluent limitations to be attained by industry was *Best Available Technology Economically Achievable* (BAT), which was established primarily for the control of toxic pollutants.

The 1977 CWA Amendments established an additional category of effluent limitations for maximum control of conventional pollutants such as Biochemical Oxygen Demand (BOD) and Total Suspended Solids (TSS) based on cost comparisons between existing and higher levels of treatment by the industry and municipalities. This level of treatment was called *Best Conventional Pollutant Control Technology* (BCT). In addition to the guidelines for existing sources, the act also establishes *Pretreatment Standards for Existing Sources* (PSES) and *Pretreatment Standards for New Sources* (PSNS) for indirect dischargers which send their wastewater to municipal treatment plans, referred to as POTWs (Publicly Owned Treatment Works).

Implementation of the national effluent guidelines program is delegated to the Engineering and Analysis Division within the EPA's Office of Water. This Division serves as the focal point of technical expertise regarding industrial processes and wastewater treatment controls. It is responsible for the development of the national industrial guidelines and standards which are incorporated, via the National Pollutant Discharge Elimination System (NPDES) program, into permit requirements for industrial facilities.

The Clean Water Act requires consideration of different factors in setting the various effluent limitations guidelines and standards. Among these factors are evaluation of costs and the economic effects of the regulations upon affected parties, usually industries. The act requires that costs and economic impacts be evaluated differently according to the level of regulation (BPT, BAT, etc.).

The regulatory process evolving since 1972 focuses on the subcategorization of the industries, usually by products, processes or waste discharge loadings; identification of pollutant constituents and their treatment technologies; the development of analytical methods for detecting and quantifying toxic pollutants; and the establishment of national effluent regulations. To date, effluent regulations have been promulgated for over fifty industrial categories, the oil and gas industry being among them, including 27 major industry categories for which toxic pollutant controls have been evaluated.

EFFLUENT GUIDELINES DEVELOPMENT

The task of developing effluent regulations involves an in-depth investigation of an industry's wastestreams and treatment technologies. The regulations are based on complex engineering and economic studies that determine a subcategorization scheme for each industrial category and the wastewater characteristics and treatment capabilities of each industrial category and/or subcategory. This requires examination of the industry processes, products, raw materials, and byproducts including such factors as location, age of equipment and plant size; the identification of the types and amounts of pollutants discharged; the evaluation of control technology performance and cost; the financial status of the industry; and the impacts of the regulations on other media such as air pollution and solid waste (sludge disposal).

Industrial Categorization

Developing effluent regulations is accomplished first by identifying the various industrial point source categories. Categorical regulations and standards are required by the CWA and are based on the principle that what is best practicable control technology for one industry is not necessarily the best for another. This categorization process, therefore, considers specific factors unique to a particular type of industry; e.g., the types of manufacturing processes and the types and quantities of pollutants generated. The factors assist in distinguishing between the classes, types, and sizes of various industries. Combined with issuing regulations on a national level, they remove any regional or economic advantage to establishing site-specific pollution control requirements.

Level of Technology-Based Limitations and Standards

The effluent limitations and standards contain limits on substances that can be discharged directly into public waterways (direct discharge) or indirectly into public sewer systems (indirect discharge). The limits are based on the treatment or removal of the substances and are determined from the performance of technologies available and, in some cases, from process equipment modifications. These limits are applied uniformly to every facility within the industrial category, regardless of the condition of the receiving water to which the effluent is discharged. More stringent site-specific limits can be applied to facilities based upon receiving water quality considerations.

Development of the specific effluent limits requires detailed evaluations of wastewater and the sources and types of pollutants generated at a facility. The evaluations enable the Agency to characterize various controls and treatment options, determine their costs, both capital and annualized, and identify pollutant removals. Effluent limitations are specific numerical limits on the allowable discharge concentration of a specific pollutant. They may be either concentration based (mg/l) or mass based (lbs/day). The regulations do not require any particular technology, but they do require plants to achieve the effluent limitations which reflect the proper operation of model technologies developed by EPA as a basis for calculating the limits.

Once the collected data and information about an industry have been evaluated, the Agency determines the appropriate levels of technology to be used as a basis for establishing the effluent limits. The major levels of technology for direct dischargers are Best Practicable Technology (BPT), Best Available Technology (BAT), Best Conventional Pollution Control Technology (BCT), and New Source Performance Standards (NSPS). For indirect dischargers (facilities discharging into a publicly owned treatment works (POTW), Performance Standards for existing Standards (PSES) and Performance Standards for New Sources (PSNS) are developed. (There are few, if any, indirect dischargers for the oil and gas industry).

BPT represents the average of the best existing performance of plants. BAT represents best economically achievable performance of plants and is the major method of controlling the direct discharge of Priority Pollutants and non-conventional pollutants to navigable waters (see below for explanation of pollutants). BCT is essentially BAT for the control of conventional pollutants. NSPS is the best available demonstrated technology aimed to reduce pollution to the maximum extent feasible.

Pollutants Regulated

Another major component of developing technology-based regulations is the selection of individual pollutant parameters for regulation. The selection process involves a thorough analysis of literature and collected industry data, and conducting detailed sampling, screening, and verification programs, e.g., analyzing raw and treated wastewater streams from several discrete operations generating wastewater. These analyses determine which pollutant is present, the performance of control technologies available that will remove the pollutant, and where the most effective removal can take place (either in-plant or end-of-pipe). Additionally, various pollution control systems are used for the removal of different classes of pollutants; e.g., metals vs. organics. Moreover, depending upon the different matrices of wastewater conditions, limitations established for a particular pollutant may vary among the industrial categories and subcategories.

There are three groups of industrial pollutants for which effluent limitations, standards, and guidelines are established: conventional, toxic, and nonconventional. Each group is described below.

Group I. Conventional Pollutants

"Conventional" pollutants represent the most familiar group of pollutants generally accepted by professionals in the field of wastewater treatment. This group consists of the following five parameters which are generic in character in that they are not specific compounds but measure classes of substances which have similar physical, chemical, or biological responses to empirical laboratory tests.

- biochemical oxygen demand (BOD)
- total suspended solids (TSS)
- fecal coliform bacteria
- pH
- oil and grease

Group II. Toxic ("Priority") Pollutants

As part of the amendments to the CWA in 1977, Congress issued a list of 65 priority pollutants and classes of pollutants considered to be "toxic" [Section 307(a) of the CWA, Committee Print 95-100]. This number was later expanded by the Agency into 129 specific toxic-priority pollutants, three of which since have been deleted. This list includes 111 organics, 13 metals, plus cyanide and asbestos. During the development of the categorical effluent regulations, the wastewater must be analyzed to determine whether any of the priority pollutants are present.

Group III. Nonconventional Pollutants

Generically, the nonconventional group of parameters represents any pollutant or pollutant parameter that is not identified as either conventional or toxic.

Economic Impact Analysis

The economic analyses conducted in the development of effluent guidelines and standards compares the annualized costs of compliance on a national industry-wide basis to the impacts incurred by industry in meeting the limitations. This analysis evaluates the effects of compliance costs on operating companies, individual facilities, future product production, employment, Federal and State revenues, balance of trade considerations, and inflation. An analysis is also performed to determine whether impacts on small firms are disproportionate to those on large firms.

An additional part of an economic analysis involves the performance of a cost-effectiveness evaluation on alternative treatment options to compare the incremental costs versus toxics pollutant removal. Results of a cost-effectiveness analysis are expressed as the costs of an option to the incremental toxic pound-equivalents removed by that option.

Table 1. Pollutant concentrations for BPT treated produced water effluent

Pollutant	Settling Effluent Concentration (ug/l)
CONVENTIONAL AND NON-CONVENTIONAL POLLUTANTS	
Total Recoverable Oil and Grease	52,956.00
Total Suspended Solids	133,063.00
Ammonia	65,773.00
Chlorides	65,111,000.00
Total Dissolved Solids	84,036,000.00
Total Phenols	2,030.00
PRIORITY POLLUTANT METALS	
Antimony	166.00
Arsenic	10.80
Beryllium	5.56
Cadmium	22.80
Chromium	128.00
Copper	180.00
Lead	515.00
Mercury	0.58
Nickel	109.00
Selenium	250.00
Silver	252.00
Thallium	180.00
Zinc	329.00
OTHER METALS	
Aluminum	1,072.00
Barium	52,573.00
Boron	20,244.00
Calcium	2,501,000.00
Cobalt	83.60
Iron	15,492.00
Magnesium	615,699.00
Manganese	1,301.00
Molybdenum	86.90
Strontium	205,500.00
Sulfur	9,683.00
Tin	305.00
Titanium	32.40
Vanadium	96.60
Yttrium	25.00
PRIORITY POLLUTANT VOLATILE ORGANICS	
Benzene	4,285.00
Ethylbenzene	115.00
Methylene chloride	170.00
Toluene	3,370.00
Trichlorofluoromethane	294.00
OTHER VOLATILE ORGANICS	
Carbon Disulfide	8.48
Chloromethane	28.60
m-Xylene	136.00
o+p Xylene	86.10
Vinyl Acetate	29.40
2-Butanone	122.00
2-Hexanone	35.80
2-Propanone	913.00
PRIORITY POLLUTANT SEMI-VOLATILE ORGANICS	
Bis(2-ethylhexl)phthalate	46.00
di-n-Butyl Phthalate	46.00
Naphthalene	144.00
Phenol	553.00
OTHER SEMI-VOLATILE ORGANICS	
Benzoic Acid	3,813.00
Benzyl Alcohol	49.50
Hexanoic Acid	790.00
n-Decane	139.00
n-Docosane	38.00
n-Dodecane	225.00
n-Eicosane	68.00
n-Hexosane	36.10
n-Hexadecane	283.00
n-Octacosane	35.20
n-Octadecane	82.90
n-Tetracosane	38.20
n-Tetradecane	119.00
n-Triacontane	35.00
o-Cresol	121.00
p-Cresol	149.00
1,2:3,4-Diepoxybutane	71.10
2-Methylnaphthalene	67.20
2,4-Dimethylphenol	117.00
RADIONUCLIDES	
Gross alpha (pCi/l)	383.54
Gross beta (pCi/l)	312.63
Lead 210 (pCi/l)	64.28
Radium 226 (pCi/l)	172.18
Radium 228 (pCi/l)	228.40

Benefits Analysis

Executive Order 12291 requires that, under certain circumstances, a Regulatory Impact Analysis be performed for major rulemakings. A "major" regulation is defined as one having "an annual effect on the economy of $100 million or more, [inducing] a major increase in costs and prices..., or significant adverse effects on competition, employment, investment, [and] productivity....". The oil and gas industry rulemakings are among those considered to be major rules. A Regulatory Impact Analysis compares the environmental, human health, and certain socioeconomic benefits of implementing a regulation to the costs of the regulation compliance.

Public Participation

Throughout the regulatory development process, there is opportunity for the public and industry to provide comments and input on EPA's actions and decisions. Industry participation is invited during the development of effluent limitations in several ways. EPA usually arranges to meet with industry and their trade associations early in the process to make contacts with experts, to learn about industrial trends, and to educate industry on the regulation development plans. During the wastestream sampling process, EPA conducts on-site visits to industrial facilities where information is exchanged. EPA makes it a point to attend pertinent conferences and seminars especially as they relate to wastestream pollution technologies to ensure that they are cognizant of the most recent developments.

State regulatory authorities are also contacted for their local expertise of industrial operations and permitting activities. Consultation with environmental interest groups are conducted also for information and opinions regarding industrial practices and environmental concerns.

During the final phases of regulatory development, during which treatment control options are being formulated, EPA may hold public meetings to obtain comments on these options even before proposal of the regulation. State regulatory officials, environmental groups, public interest groups, and industry are all encouraged to attend.

EPA publishes the proposed regulation in the *Federal Register*. The regulation, as well as a detailed explanation of its development, is included in that publication. Public comment is also solicited on the proposal as part of the publication. Comments are received and the regulation is finalized. The final regulation is also published in the *Federal Register*. This publication includes a detailed explanation of EPA's responses to comments received after proposal, and any changes made from the proposed regulation. Both after proposal and promulgation of the regulations, public workshops are arranged to either receive additional comments (for a proposed rule) or to assist permit writers and industry in the enforcement of the regulations (for a final rule).

EFFLUENT GUIDELINES DEVELOPMENT FOR THE OIL AND GAS INDUSTRY*

Industry Description and Regulatory Background

For effluent limitations development purposes, EPA has divided the oil and gas industry into 5 subcategories. These are Onshore, Offshore, Coastal, Stripper Wells, and Agricultural and Wildlife Water Use. Offshore is defined as being seaward of the inner boundary of the territorial seas (essentially the low-water mark along the U.S. coastlines). The coastal subcategory is currently proposed as being on water bodies landward of the territorial seas. Onshore is basically anything that is not coastal or offshore. Stripper wells are defined as onshore wells that produce less than 10 barrels of oil per day (BPD). The Agricultural and Wildlife Water Use subcategory is defined as being west of the 98th

* The industry information presented in this section relating to effluent limitations development for the oil and gas industry comes from information included in two EPA publications: 1) *Development Document for Effluent Limitations Guidelines and New Source Performance Standards for the Offshore Subcategory of the Oil and Gas Extraction Point Source Category*, January 1993, EPA 821-R-93-003, and 2) *Development Document for Effluent Limitations Guidelines and Standards for the Coastal Subcategory of the Oil and Gas Extraction Point Source Category*, February 1995, EPA 821-R-95-009.)

meridian. These subcategories evolved in response to the geographical and topographical differences inherent in each which ultimately required, either for economic or technological reasons, different methods of wastestream management.

On April 13, 1979, EPA published final BPT limitations for all 5 of the subcategories. The regulation prohibited the discharge of all wastestreams from onshore facilities. Limitations for stripper wells were reserved until more detailed financial information about these size wells could be obtained to determine their capacity to afford pollution control technology. Discharges were allowed for produced waters for the remaining subcategories based on oil and grease limitations. Table 1 lists the produced water effluent limitations for all the oil and gas subcategories. The 1979 regulations have recently been updated to reflect BCT, BAT, and NSPS for the offshore and the coastal subcategories. This paper hereafter includes discussion focusing on produced water requirements for these two subcategories. On March 13, 1991, EPA published a notice proposing BAT, BCT, and NSPS limitations and standards for the offshore subcategory. After receipt and consideration of public comments, EPA promulgated the final rule on March 4, 1993. These final rules allow the discharge of produced waters based on oil and grease limitations (which are more stringent than the 1979 BPT limitations). EPA's recently proposed effluent limitations for the coastal subcategory (on February 17, 1995), require zero discharge of produced water for all coastal operations except those located in the Cook Inlet of Alaska. For Cook Inlet, the limitations reflect those required by the offshore rule. After considering and responding to public comments, EPA plans to promulgate these limitations by 7/96.

Offshore oil and gas activity largely exists in the Gulf of Mexico where, at the time of the rulemaking (1991), EPA estimated that a total of 2,233 and 284 structures existed in federal and state waters respectively. Thirty-two structures were identified in the Pacific, one in Alaskan offshore waters, and none in the Atlantic. (A structure is a platform that may have anywhere from 1 - 60 producing wellheads). These existing structures accommodated approximately 13,000 producing wells. Future projections were that 759 wells per year would be drilled (from 1993 until 2007), and that over this 15 year period, 759 new platforms would be installed in offshore waters (the fact that the estimated average annual number of wells is the same as the number of platforms installed is coincidental).

Coastal oil and gas activities are located on water bodies inland of the inner boundary of the territorial seas. These water bodies include inland lakes, bays and sounds, as well as saline, brackish, and freshwater wetland areas. Although the definition includes water bodies even in all inland U.S. states, EPA knows of no existing operations other than those in certain states bordering the coast. Thus, at this time, the coastal oil and gas operations are located only in coastal states.

As of 1992, coastal oil and gas activity consists of approximately 7,600 producing wells existing along the Gulf of Mexico coastal states of Texas, Louisiana, Alabama and Florida, in San Pedro Bay, California and also in Alaska's Cook Inlet and the North Slope areas. The majority of Gulf Coast activity takes place in Texas and Louisiana. There, EPA estimates approximately 853 coastal production facilities (or gathering centers which separate, treat, and dispose of produced water from one or more wells) exist in a number of topographical situations including bays, sounds, lakes, and wetlands. Coastal oil and gas activity in Alabama consists of several platforms located in Mobile Bay; and a small number of wells are also located in wetlands along the west coast of Florida.

Coastal oil and gas activity in California exists behind the barrier island that forms San Pedro Bay (in Long Beach Harbor). There, four man-made islands, housing approximately 586 productive wells, have been constructed solely for the purpose of oil and gas extraction.

Roughly one third of all the coastal oil and gas production activity exists in Alaska. Fifteen deep water platforms (13 of which are currently producing) exist in the northern part of Cook Inlet. In addition, approximately 2085 wells, resembling onshore operations (individual wellheads sporadically located above a producing formation, as opposed to deep water platforms where wellheads are centrally located) are situated on the tundra wetlands of Alaska's North Slope.

Produced Water Characteristics

By far the largest volume wastestream generated from the extraction of oil and gas is that of produced water. Produced water is brought to the surface during the oil and gas extraction process as part of a complex hydrocarbon mixture of oils and gases. The produced water itself includes: formation water extracted along with oil and gas; injection water used for secondary oil recovery that has broken through the formation and mixed with the extracted hydrocarbons; and various well treatment chemicals added during the production and oil/water separation processes. Depending on the age of a well and site-specific formation characteristics, the produced water can constitute between 2 percent and 98 percent of the gross fluid production at a particular well. Generally, in the early production phase of a well the produced water volume is relatively small and the hydrocarbon production makes up the bulk of the fluid. Over time, the formation approaches hydrocarbon depletion and the produced water volume usually exceeds the hydrocarbon production.

Offshore produced volumes per structure average 1,800 barrels per day (BPD), and may range from 2 to 150,000 BPD per structure. EPA estimates are that a total of 1.1 billion barrels per year (BPY) of produced water is discharged to surface waters from the offshore industry. EPA estimates that average produced water discharge rates from a coastal well ranges from approximately 1180 BPD in Cook Inlet to 270 BPD in the Gulf Coast. EPA estimates that 228 million BPY of produced water is discharged to surface waters by the coastal oil and gas industry.

As part of the offshore and coastal rulemakings, EPA has conducted extensive sampling and analysis programs to characterize the chemical constituents of produced water. While results were slightly different for each subcategory, generally produced water contains a variety of toxic organic pollutants and all 13 of the toxic metals on EPA's priority pollutant metals list. (The 13 toxic metals are antimony, arsenic, beryllium, cadmium, chromium, copper, lead, mercury, nickel, selenium, silver, thallium, and zinc). Studies analyzing BPT treated produced water (generally oil and water separation technology) , show the priority organics most often present in significant amounts were benzene, naphthalene, phenol, toluene, 2-Butanone, ethylbenzene, xylene, and n-Alkanes. In addition to the priority pollutants, EPA identified total suspended solids, oil and grease, and a number of nonconventional pollutants including barium, chlorides, ammonia, magnesium, strontium and iron present in produced water. Table 2 presents the characteristics of BPT treated produced water obtained from a sampling and analytical activity performed by EPA on coastal facilities in 1992.

While the pollutant characteristics of produced water are complex, the effluent limitations for it are comparatively simple in that oil and grease is the only pollutant parameter specified for regulation. EPA determined that it is not only infeasible to regulate each pollutant individually, but unnecessary as well since it has been shown that many of the organic pollutants are removed in produced water treatment technology along with the oil and grease. Because the organics are constituents of oil, they are directly controlled by the oil and grease limitation. Thus, oil and grease serves as an indicator for the removal of other toxic organics.

Produced Water Treatment Technology

The BPT regulations established for the onshore, coastal, and offshore subcategories of the oil and gas industry in 1979 are largely based on the performance capability of oil

Table 2. U.S. EPA oil and gas produced water effluent limitations

Subcategory	Year Promulgated	Level of Limitation	Effluent Limitation (Numerical limitations are for oil and grease)
Onshore	1979	BPT	No Discharge
Ag/Wildlife Water Use	1979	BPT	35 mg/l daily max*
Stripper Wells	1979	BPT	Reserved
Offshore	1979	BPT	48 mg/l 30-day ave*, 72 mg/l daily max
	1993	BAT/NSPS	29 mg/l 30-day ave, 42 mg/l daily max
Coastal	1979	BPT	48 mg/l 30-day ave, 72 mg/l daily max
	1995 (Proposed Rule)	BAT	No discharge for all coastal areas except Cook Inlet For Cook Inlet: 29 mg/l 30 day ave, 42 mg/l daily max
	1995 (Proposed Rule)	NSPS	No discharge

*The term "daily max" means "daily maximum", and "30-day ave" means "monthly average".

water gravity separation technologies, with chemical addition to enhance separation if necessary, followed by gas flotation to remove oil and grease. EPA's effluent limitations based on these technologies limit the concentration of oil and grease in the effluent to 48 mg/l monthly average, and 72 mg/l daily maximum.

Improved performance of gas flotation technologies (preceded by gravity separation and chemical addition) was investigated during EPA's development of BAT and NSPS effluent limitations for the offshore subcategory. Based on data from 4 intensive sampling and analysis studies directed at determining the performance capabilities of improved gas flotation technologies, (2 of which were EPA sponsored studies, and 2 of which were performed by industry) it was determined that reduced oil and grease effluent levels could be achieved from that of BPT technology. In the final offshore effluent limitations promulgated in March 1993, the daily maximum limitation for oil and grease was established as 42 mg/l and the monthly average limitation was set at 29 mg/l. The difference in the performance of BPT technology and that of BAT is due to closer supervision of the gas flotation units by operators, additional monitoring of operating parameters, proper sizing of the unit to improve hydraulic loading, additional maintenance of the process equipment, and addition and/or proper usage of flocculation enhancement chemicals.

EPA also considered improved gas flotation during its development of BAT and NSPS for the coastal subcategory. However, the extensive practice of produced water disposal by underground injection (as approved and regulated by EPA's Underground Injection Control (UIC) Program) drove EPA to consider this as the best available method of treatment and disposal where appropriate. In fact, no coastal oil and gas facilities are discharging produced water in Alabama, Florida, California or in Alaska's extensive oil and gas activities on the North Slope because of produced water disposal via underground injection. Underground injection is practiced due to a combination of factors including operational preference, waterflooding, and/or state requirements. Injection technology for produced water consists of injecting it, under pressure, into Class II UIC wells into underground formations. This option results in no discharge of produced water to surface waters.

Treatment of the produced water prior to injection is usually necessary, and such treatment often includes removal of oil and suspended matter by BPT oil separation technology followed by filtration technology. The removal of suspended matter prior to injection is required to prevent pressure build-up and plugging of the receiving formation and/or to protect injection pumps from damage.

Simple cartridge filtration was found to be commonly used among coastal operators as a pretreatment technology prior to injection, thus, it is included as part of the basis for the injection technology option for coastal operations. EPA determined from information gathered on site visits in the Gulf Coast area, as well as from industry contacts, that cartridge filtration is generally used following BPT oil/water separation technologies at injecting facilities accessible by water only. For facilities accessible by land, it was determined that rather than pretreat produced water using filtration, it is more cost effective to perform periodic well workovers on the injection well to remove clogged material from the wellbore. However, for those few land-based facilities treating produced water flows greater than 64,000 bpd, EPA determined that it would be more appropriate to employ granular filtration after BPT separation technology because it is more cost effective to use this technology for higher flows rather than cartridge filtration. Thus, in its proposed effluent limitations for coastal operations, EPA determined that zero discharge is the appropriate requirement for all facilities except those located in Cook Inlet.

Specific alternatives have been developed for Cook Inlet to account for the different operational practices, and geological situations that exist at these platforms. Injection of produced waters is not practiced in Cook Inlet because, where waterflooding is occurring, treated seawater is injected instead. Industry claims that injection of seawater other than produced water for enhanced recovery is practiced primarily because injection of produced water would cause formation fouling. Industry asserts that fouling would occur due to bacteria and scale formation in produced water, and otherwise not present in seawater. EPA has determined that formation fouling problems associated with produced water injection are not insurmountable because filtration and anti-fouling chemicals can be added prior to injection, and periodic downhole workovers can be performed to reopen clogged formation surfaces.

An additional problem with injecting produced waters in Cook Inlet is that no other formations exist there that can accommodate the large volumes of produced water other than the producing formation. Cook Inlet operators would experience significant additional cost associated with piping produced water if zero discharge was required from where it is currently treated to where it could be injected. Of the 13 producing platforms in the Inlet, 9 of them currently direct their extracted hydrocarbon fluids to one of 3 land-based separation and treatment facilities. These land-based facilities separate the hydrocarbons from the produced water, treat the produced water and then discharge it in accordance with EPA's NPDES general permit requirements. The Alaska Oil and Gas Conservation Commission has confirmed that no geological formations exist beneath the land-based facilities that are large enough to accept the approximately 100,000 barrels per day (bpd) of produced water generated from these facilities. Thus, produced water would be piped back to the platforms for injection if produced water discharges were prohibited. The costs for such piping would comprise 74 percent of the total costs for injection. This would be a major cost factor for the Inlet operations overall since the volume of produced water being discharged from these 3 land-based facilities amounts to approximately 99 percent of that discharged from all 13 platforms.

EPA determined, therefore, that due to economic unachievability of a zero discharge requirement in the Cook Inlet, discharge limitations would be allowed based on the same technologies as in the offshore regulations: improved performance of gas flotation. Thus,

Cook Inlet would be required to comply with the oil and grease limitations of 29 mg/l monthly average, and 42 mg/l daily maximum.

In addition to oil and gas extraction, produced water separation, treatment and disposal generally occurs at the production platform in offshore operations. The configuration of coastal produced water separation and treatment facilities is considerably more varied in that situations range from individual deep water multi-well platforms in large bays, to, more often, gathering centers located on land, barges, or shallow platforms servicing multiple satellite wells in shallow wetlands. The differences in location are principal reasons why different effluent limitations for these subcategories exist: Zero discharge, for example, was not selected as the final regulatory option for the offshore subcategory because of unacceptable geological conditions in some cases, the increase in air emissions associated with the injection technology, and the high costs.

Compliance Costs of the Effluent Limitations

For the offshore subcategory, costs are determined based on a 15 year period from 1993 until 2007. The combined BAT/NSPS capitol costs are estimated to be $527 million. The combined annualized operating and maintenance costs for the selected BAT and NSPS options is $52 million. (Costs are in 1991 dollars).

The total capital costs for gas flotation were complicated to determine because many operators currently use the gas flotation technology to comply with the current BPT regulations. To avoid over-costing by assigning capital costs to all platforms, EPA predicted the number of existing platforms that currently have gas flotation systems and the number of platforms that will have to install new flotation systems. The following paragraphs explain EPA's basis for its estimate of the total offshore industry costs for the gas flotation option.

EPA determined that to achieve BAT oil and grease limitations based on improved performance of gas flotation technology, operators who currently have gas flotation treatment systems would continue to use the same treatment units, although some changes to those systems or their manner of operation might be necessary. For existing platforms that do not currently have gas flotation systems and can not meet the limitations of the final rule with their currently installed treatment systems, some form of add-on treatment would be necessary. For costing purposes, EPA assumed that all facilities currently without gas flotation systems are unable to meet the BAT (or for new sources, NSPS) limitations, and that flotation units would need to be installed. This assumption does not take into consideration the fact that other treatment technologies currently used by the operators are available for use, such as parallel plate coalescers, corrugated plate interceptors, hydrocyclones, or filtration, and their use may enable operators to meet the requirements of this rule without requiring installation of flotation units. (The establishment of an effluent limitation based on a given technology does not require use of the treatment technology upon which the limitation is based.) EPA collected information from the Minerals Management Service and various industry sources to update projections of existing gas flotation systems, and learned that approximately thirty-five percent (35%) of the offshore facilities in Gulf of Mexico are now using gas flotation systems for produced water treatment. The distribution of gas flotation systems, in conjunction with capital and annual cost data compiled for flotation systems was used to estimate total compliance costs.

EPA calculated the BAT annual operating and maintenance (O&M) costs using the projections of existing gas flotation systems discussed above. For those platforms that already have gas flotation units installed, the annual O&M costs of complying with BAT limitations based on improved performance of gas flotation are estimated to be higher than their current annual O&M costs because of modifications and enhancements needed to improve system performance. Enhanced removals of oil and grease can be achieved by existing gas flotation systems through closer supervision of the units by the platform operators, additional monitoring of operating parameters, proper sizing of the unit to improve hydraulic loading, additional maintenance of the process equipment, and addition and/or proper usage of flocculation enhancement chemicals. These costs are incremental to the current annual O&M costs. EPA estimates that the additional labor and other improvements necessary to achieve compliance with BAT limits will approximately double the annual O&M costs for existing flotation systems currently achieving BPT quality effluent.

Existing facilities needing to install a gas flotation unit (or other technology) to comply with the limit are projected to incur costs to design and select a treatment system to meet the BAT oil and grease limit. Total annual O&M costs for existing platforms that will need to install gas flotation were determined to be approximately ten percent (10%) of the capital costs of the new flotation system.

A report by the Offshore Operators Committee (OOC) stated that even in the absence of produced water limitations more stringent than BPT, eighty percent (80%) of new development and production platforms would be designed with gas flotation systems for treatment of produced water. Thus, EPA based compliance cost projections on the assumption that in the absence of NSPS limitations, twenty percent (20%) of new platforms would not include a flotation unit in their treatment system design. This twenty percent (20%) of new platforms are considered to incur an incremental cost to comply with NSPS limitations.

In estimating NSPS capital costs, EPA assumed that it was necessary for the operator to add a flotation system to the produced water treatment system. The entire costs for adding on such a system were used in EPA's economic impact analyses.

For those eighty percent (80%) of new platforms that are expected to include gas flotation in the original design, capital costs consist only of an engineering redesign cost and not a new unit cost. It is assumed that the gas flotation units included in the existing design of the new platforms were able, at a minimum, to achieve the current BPT oil and grease limitations. For these systems to achieve the more stringent limitations of this final rule, EPA assumed that there may be an additional cost to upgrade the system. A design upgrade could consist of increasing the system's retention time through increasing the cell size or the addition of another cell, maximizing separation efficiency through properly sizing rotors, gas dispersers, and chemical injection equipment, and optimizing the system's performance through the addition of state-of-the-art instrumentation and controls. This system redesign cost was estimated at fifteen percent (15%) of the NSPS flotation system capital cost.

The NSPS annual O&M costs for new platforms were calculated using the same NSPS profile used in the development of the NSPS capital costs. For the twenty percent (20%) of new platforms assumed to not already include flotation systems in the design, the annual costs are ten percent (10%) of the capital costs. However, for those new platforms that already have flotation systems in the design plans of the facility, there are no incremental annual costs for compliance with the NSPS limitations. EPA assumed that since there is a flotation system in the design of a facility, there are also annual costs associated with operation of that system in the financial projection of that project. The improved performance of that system has been accounted for by improving the design parameters of the flotation system.

Combined BAT and NSPS costs (annualized capital plus yearly O&M) for zero discharge for all coastal operations, except Cook Inlet where oil and grease limitations apply, are estimated to be approximately $35 million per year (based on 1992 dollars). In developing the costs of zero discharge for this option, EPA determined, based on Texas and Louisiana state permit data, the number and volume of produced water discharges that would be discharging by the time the final rule is scheduled to be signed in July 1996. This investigation identified, by operator and oil and gas field, 216 produced water separation/treatment facilities that would be discharging approximately 180 million barrels per year (bpy) in Texas and Louisiana as of July 1996. Costs are calculated without taking into account the regulatory effects of the zero discharge requirement imposed by the EPA Region VI General Permits (currently covering Texas and Louisiana).

Based on extensively detailed site visits to operators practicing zero discharge, as well as information obtained through industry contacts, EPA was able to specifically identify the costs associated with produced water injection. Some of the more influential cost factors used in the costing analyses are presented below:

General

The only areas that will incur compliance costs are Cook Inlet in Alaska, Texas, and parts of Louisiana since all other coastal areas that have oil and gas activities currently practice zero discharge.

For Texas and Louisiana

Produced water would be injected into Class II UIC injection wells. The capacity of each Class II injection well is 5,000 BPD.

Ninety percent of the injection wells would be converted from previously producing wells or dry holes.

If a discharge is greater than 108 bpd (for water-based facilities) and 71 bpd (for land-based facilities), then the produced water would be injected onsite; if the discharge is less than those flows then it would be more cost effective to send the produced water offsite to a commercial facility for injection. (EPA's data from Texas and Louisiana coastal permits show that 77 percent of the produced water discharges would inject on-site).

For purposes of estimation, all Texas separation/treatment facilities are located on land and all Louisiana separation/treatment facilities are located over water. EPA is aware that this is not entirely the case, *i.e.* some facilities in Louisiana are located over land and some Texas facilities are located over water. In the absence of specific location information on all of the 216 discharging facilities, EPA determined this to be a good approximation since the coastal topography of Louisiana consists of more extensive wetlands than that of Texas. (Location is an important factor when determining the cost of drilling an injection well, and the cost of produced water transportation. EPA's state permit data base shows that 24 percent of the produced water discharges are in Texas and the separation/treatment facilities are therefore considered to be on land).

No pretreatment beyond BPT technology is required prior to injection for land-based facilities because it is more cost effective to perform downhole well workovers twice a year. Pretreatment beyond BPT treatment prior to injection consists of cartridge filtration for water-based facilities. For flows greater than 64,000 bpd, granular filtration is used as pretreatment.

Capital costs are based on sizing equipment to accommodate future produced water volume, estimated to be approximately 1.5 times current flow.

Where more than one produced water discharge location exists from one or more production facilities owned by the same operator in the same field, EPA combined the discharges to be injected into a single injection system. By combining discharges a savings would result due to installation of fewer injection wells.

For Cook Inlet

No geological formations are available for produced water injection except the producing formations.

No geological formations are available near or below the existing on land separation/treatment facilities. Thus, the produced waters would be required to be piped back to the platforms for injection.

Pretreatment prior to injection consists of gas flotation and multimedia filtration. However, operators will use existing equipment where it currently exists, and no costs would be incurred for such existing equipment.

During the development of this proposal, industry provided EPA with information on reservoir plugging and souring that may result from injecting produced water in the Cook Inlet. EPA, in its cost analysis, included costs for the addition of chemicals that would be added to the produced water being injected to alleviate the scaling and hydrogen sulfide (H2S) formation problems associated with injection in this area. Such chemicals include biocides and scale inhibitors. Annual workovers must also be performed on the injection wells.

EPA determined that no costs would be attributed to zero discharge for California, Florida, Alabama, certain parts of Louisiana, and the North Slope of Alaska because operators in these areas are already practicing zero discharge of all produced waters.

For improved gas flotation, costs were estimated based on an evaluation of this technology during development of the Offshore Guidelines (58 FR 12463). Improved performance of gas flotation units includes improved operation and maintenance of gas flotation treatment systems and chemical pretreatment to enhance system effectiveness. Costs are based on vendor-supplied data, industry information, cost analyses conducted by the Department of Energy, and EPA projections. Capital and O & M costs were applied specifically to the coastal oil and gas operations using nine modeled flows for land-and water-access production facilities. From these nine modeled flows, EPA conducted regression analyses to derive cost equations that would vary based on flow. These equations were then applied to the actual 216 discharging facilities to estimate costs on a site specific basis. Capital costs include equipment purchase, installation, and platform or concrete pad (for land based operations) retrofit. Operation and maintenance costs are estimated to be 10 percent of capital costs.

While it was determined that zero discharge was not the appropriate option for existing Cook Inlet facilities based on the costs to retrofit existing produced water transport and treatment operations, new facilities, on the other hand, can incorporate injection processes into the design of the facility's operation. Thus, zero discharge is proposed for Cook Inlet facilities for NSPS, as it is for all other coastal areas. However, the cost issue for new sources may be mute for Cook Inlet in that no new sources are expected in this area.

In addition to the cost evaluations performed on the selected regulatory technologies, EPA also conducted analyses to determine the limitations' economic achievability on a national level. The economic analyses evaluated the production and employment losses, federal and state revenue losses, and facility closures that may occur as a result of compliance with these regulations. While room does not allow a presentation of the results here, the effluent limitations promulgated for the offshore subcategory, and those proposed for the coastal subcategory have been determined to be economically achievable for the oil and gas industry.

Section II

CHEMISTRY AND TOXICITY ISSUES

5

SESSION SUMMARY

Alexis Steen

American Petroleum Institute

Five papers were presented which are summarized below. In addition, this section of the proceedings includes several papers which appeared in the seminar poster session. The presenters were also members of a panel for a brief plenary session. A synopsis of the oral session is provided below.

- **Estimating the acute toxicity of produced waters to marine organisms using predictive toxicity models - David Pillard/ENSR consulting (US).** The findings from common ion imbalance showed the importance of knowing the suitability of PW constituents for test species' health in contrast to acceptable seawater or freshwater constituent concentration. Data was used to develop a model to predict the differences in observed PW toxicity due to ion-imbalance toxicity versus chemical toxicity.
- **Approaches in toxicity identification evaluations of low and high-total dissolved solid produced water effluents - Theodor Sauer/Battelle (US).** Findings from Phase 1 Toxicity Identification Evaluation (TIE) trials were described,with a focus on methods modifications for PWs. Many potential acute toxicants were identified. Frequently, more than one toxicant per PW was observed. The Phase 1 TIE manipulations do not address oxygen scavengers or ion imbalances.
- **A comparison between the toxicity of il and condensate - F. Foekema, TNO (Netherlands).** The relative acute toxicity of "topped oil" and condensate from the North Sea and sunflower oil indicated that toxicity was a function of areal discharge. The main observation was that"toxicity" was due to physical effects, not chemical. Toxicity from all three oils was similar. This study used the CHARM model to evaluate relative risks.
- **Produced water composition, toxicity and fate: A review of recent BP North Sea studies - Steve Flynn/BP Exploration (Scotland).** An evaluation of PWs from plattforms in the North Sea found that toxicity was generally low, except during a production upset. I appears that toxicity may be related to the concentration of phenol in PW.
- **The chemistry and toxicity of oilfield chemicals after partitioning between oil and water phases - Peter Johnson Institute of Offshore Engineering (Scotland).** Oil:water partitioning is dependent on salinity and temperature. Measurements of PW acute toxicicty were made to examine the degree of sensitivity to salinity or temperature. Different PWs showed different sensitivities. Conse-

Produced Water2: Environmental Issues and Mitigation Technologies
Edited by M. Reed and S. Johnsen, Plenum Press, New York, 1996

quently, toxicity information should be accompanied by data on salinity, Kow, and temperature.

Field evaluation of toxicity effects and dispersion of produced water discharges from North Sea oil platforms: implications for monitoring acute impacts in the environment - Ron Stagg/SOAFD Marine Laboratory (Scotland).

Scotland evaluated the dispersion of PW plumes and PW toxicity. Plumes were difficult to track. Toxicity was low and mostly associated with concentrations of total hydrocarbons(not specific compounds). Toxicity associated with salinity was not observed.

SUMMARY OF QUESTIONS AND COMMENTS FROM THE AUDIENCE

1. Jim Ray/SHELL Oil to Ron Stagg/Scottish Office Agriculture and Fisheries Department (SOAFD). *Please explain why PAH concentrations would increase with depth and clarify the depths of PW discharges. Lastly, were any model predictions used to determine the point at which PW plume depth would stabilize?* The discharge depths were slightly variable (at, near, or just below the sea surface) in the North Sea. Samples were taken at the depth predicted from the ZENECA model that the plume would stabilize.
2. John Vandermuelen/Bedford Institute of Oceanography to Ron Stagg/ (SOAFD): *Why were no air:sea interface samples taken? While data on PAH concentrations in estuary water columns from the US are similar to those presented, air:sea interface concentrations have been measured at levels 2-3 orders of magnitude higher.* While the microlayer is very interesting, the relevance of this interface for water column organisms is unknown. For plume detection, microlayer measurement is difficult because the rapid spreading and mixing around the discharge area is great. So, the focus has been predominantly on the water column.
3. John Vandermuelen/Bedford Institute of Oceanography to Theodor Sauer/Battelle: The large variation in PW toxicity data and the need for replication was noted. *Care should be taken to avoid pseudo-replication in the conduct TIEs.* Clues of PW toxicants can be contradictory in Phase I TIEs, so duplicate samples and tests are preferred to increase the certaintly of results (rather than just single samples, without replication). In Phase 1 fractionation schemes, the purpose is simply to obtain basic information on what portion of the PW contained the toxicants.
4. Ted Sauer/ Battelle commented that it seems PW toxicity sources appear to be different between Europe (phenols, alkylated phenols, hydrocarbons) and US (many potential toxicants) Perhaps this is an artifact of the different evaluations yielding different results. If so, findings from similar test designs (e.g. TIEs) using PWs from US and Europe would be able to illuminate this.
5. Mark Reed/IKU offered a summary of the session and requested feedback from the panel: *What has been learned is that PW acute toxicity levels (based on known pollution rates) are not a great environmental problem, and should not be a focus for future research.*

   - Jerry Neff/Battelle commented that non-persistence fractions (volatile or complexed) are likely PW toxicants. *Given the rapid PW dilution in seawater from discharges, how useful are the current laboratory toxicity test designs to predict impacts in the field?* This question was not answered.

- Steven Flynn/BP noted that dilution model results predict low concentration in the field. Consequently, environmental effects should be commensurately low, and research effort could be better spent on examining longer-term effects and risks.
- Dave Pillard/ENSR remarked that some recent work has been done on modelling chronic toxicity based on ion imbalances using sublethal effects data longer-term LC50 data. This should enable prelimiary examination of chronic effects soon.

Research results have increased our understanding of the composition of PW and is aquatic toxicity. Most toxicity tests has been conducted to determine the short-term, acute toxicity of produced waters discharged to marine waters.The focus of this Session was on what levels of acute toxicity were observed and what behavior or constituents were obsered to have caused that toxicity. Information is relatively consistent on the variability in composition of individual produced waters; however research findings indicate that causative toxicants can be identified. From studies on US PWs, non-persistent compounds appear to be the most frequent toxicants. For European PWs (primarly from the North Sea), phenols and total hydrocarbon concentrations appear to be the dominant toxicants. Whether these distinctions were solely due to differences in experimental design is not known. Additionally, some claims of PW "toxicity" resulted from physical damage, not chemical toxicity.

The degree of acute toxicity observed from PW discharges is generally low. With the rapid dilution available in deep or fast-flowing waters, concerns of the research community with acute toxicty have lessened.Consensus from this Session was that future efforts on PW toxicity research may be better spent examining the more subtle effects from long-term, chronic exposure. Such tests are more Sophisticated, use lower exposure concentrations, and should begin to approach field conditions. Researchers still need to consider how to design tests that can mimic field conditions well enough to provide useful information about environmental effects.

6

ESTIMATING THE ACUTE TOXICITY OF PRODUCED WATERS TO MARINE ORGANISMS USING PREDICTIVE TOXICITY MODELS

David A. Pillard,[1] Joseph E. Tietge,[*1] and James M. Evans[2]

[1] ENSR Consulting, Engineering, and Remediation
4413 West LaPorte Avenue
Fort Collins, Colorado 80521
[2] Gas Research Institute
8600 West Bryn Mawr Ave.
Chicago, Illinois 60631

INTRODUCTION

Common ions in effluents can be toxic to aquatic organisms over a wide range of salinities. Although the toxicity of high salinity (hypersaline) solutions and low salinity (hyposaline) solutions may be due to osmotic conditions, toxicity may also be the result of specific ion concentrations. Therefore, even in effluents of normal freshwater or marine salinity, common ion toxicity to aquatic organisms may occur in the absence of other toxicants. Little historical data exist on the toxicity of specific ions to freshwater or marine test organisms. Because of the paucity of data on the toxicity of common ions in produced waters, the Gas Research Institute (GRI) sponsored a research program to examine the toxicity of major ions to aquatic test organisms. The goal of this research is to quantify the relationships between common ion concentrations and organism response (mortality), and then to use these empirical relationships to develop models that would allow the prediction of organism response based upon known ion concentrations.

The first phase of this research program quantified the toxicity of seven common ions to three freshwater organisms (*Ceriodaphnia dubia*, *Daphnia magna*, and *Pimephales promelas*) (Mount and Gulley, 1992; Tietge et al., 1995). It was demonstrated that common ions (e.g., potassium, bicarbonate) in produced waters are toxic to freshwater organisms (Gulley et al., 1992). This research resulted in a computer program (the ***GRI-FW STR***

* Current Address: USEPA, ERL-Duluth, 6201 Congdon Blvd. , Duluth, MN 55804

Program™) that predicts the acute toxicity of an effluent using ion concentrations as input variables (Tietge et al., 1995).

Because produced waters are also generated and released in marine environments, additional research was initiated to determine the relationship between common ions and toxicity to marine species. In general, produced waters have total dissolved solids (TDS) concentrations higher than most fresh and marine surface waters, although individual TDS values may vary widely (2,000 to 200,000 mg/L TDS; natural seawater is about 32,000 mg/L). Typically, the predominant cation and anion in produced waters are sodium and chloride. Produced waters can vary greatly in composition and salinity, depending on the type of production operation, geologic source of the water, and the treatment of the water once brought to the surface. For example, although chloride is usually the predominant anion and can sometimes reach concentrations of 200,000 mg/L in produced waters (Tibbetts et al., 1992), bicarbonate (>9,000 mg/L) was reported as the dominant anion in coalbed methane-produced water from Colorado in the United States (Simmons, 1992).

TDS, or salinity, in produced waters can cause acute toxicity to aquatic marine organisms due to an excess or deficiency of common ions (e.g., Ca, Mg, SO_4). In addition to these common ions, other chemicals may also be present in produced waters creating a complex mixture (Fillo et al., 1987; Wesolowski et al., 1987, O'Neil et al., 1989) including: low molecular weight aromatic compounds (e.g., benzene and toluene), higher molecular weight aliphatic compounds (e.g., C_{10} to C_{20} alkanes), and metals. Metals common to produced waters include iron, strontium, and barium. Heavy metals such as zinc, copper, and lead may also be present in significant (i.e., toxic) concentrations.

Since produced waters are complex mixtures, it is often difficult to separate and quantify toxicity due to common ions and toxicity due to other chemicals. Toxicity Identification Evaluation (TIE) studies have been developed to systematically identify the causative toxicants present in effluents and other complex mixtures (USEPA, 1991). Standard TIE procedures are usually effective in discriminating toxicity due to TDS, but they are not very effective in characterizing ion-specific toxicity. In addition, there are few studies in the literature which systematically examine common ion toxicity to the species used in biomonitoring of produced waters.

The determination of ion/acute toxicity relationships in three marine organisms constitutes the second phase of the research sponsored by GRI. As with the freshwater study, this research was conducted in three steps: 1) empirical quantification of the acute toxicity of the eight most common ions in seawater to three common marine test organisms, 2) development and validation of toxicity models which predict toxicity based on ion concentration alone, and 3) development of a computer program, based on the models, to facilitate application of these data to solve effluent management problems. This paper focusses on the total salinity/toxicity relationships for three species: mysid shrimp (*Mysidopsis bahia*), sheepshead minnow (*Cyprinodon variegatus*), and inland silverside (*Menidia beryllina*). It also discusses the preliminary results of single-ion models developed for the mysid shrimp.

MATERIALS AND METHODS

Preparation of Test Solutions

Test solutions used in this study were based on the GP2 artificial seawater formulation as recommended by USEPA (Weber 1993). GP2 consists of nine different salts with a final concentration of approximately 31 g/L (ppt) (Table 1).

The salt composition of the GP2 formulation used in this study was altered by substituting NaBr for KBr, while maintaining constant ion concentrations, such that all salts

Table 1. Modified GP2 synthetic seawater used in performing salinity toxicity studies. Deficient or excess solutions were prepared by modifying this basic formula[a]

Compound	Concentration (g/L)[b]	Molarity
NaCl	21.03	0.3598
Na_2SO_4	3.52	0.02478
KCl[c]	0.665	0.00892
NaBr[d]	0.076	0.000739
$Na_2B_4O_7$	0.018	0.000089
$MgCl_2$	4.456	0.0468
$CaCl_2$	0.997	0.00898
$SrCl_2$	0.012	0.0000757
$NaHCO_3$	0.17	0.00202
Total salinity	30.9	

[a] From Weber (1993)
[b] Unhydrated concentrations
[c] The concentration of KCl was increased to compensate for substituting NaBr for Kbr.
[d] In this study NaBr was substituted for KBr so all compounds were sodium or chloride based.

were either sodium or chloride based. This modification allows individual ion manipulations accompanied only by changes in the two counter ions, Na^+ and Cl^-. All solutions were prepared using reagent-grade salts (Mallinckrodt® or Sigma®).

In the studies to determine the toxic effects of total salinity, all compounds in GP2 were changed proportionately. A concentrated GP2 solution was prepared and diluted with standard GP2 or deionized water for the high salinity test solutions to create a dilution series. GP2 was diluted with deionized water for the low salinity test solutions.

In the studies to determine the toxic effects of specific ions, the concentration of each ion was manipulated by changing the concentration of the salt containing that ion in the artificial seawater. Salinity in these test solutions was maintained at 31 ppt by adjusting the concentration of sodium and chloride. The specific ions tested were potassium (K^+), magnesium (Mg^{++}), calcium (Ca^{++}), strontium (Sr^{++}), sulfate (SO_4^-), bicarbonate (HCO_3^-), bromide (Br^-), and borate ($B_4O_7^-$). To test the effects of increasing ion concentration, a salt concentration was increased in a modified GP2 formula and diluted with unmodified GP2. The toxicity of sulfate excess, for example, was determined by increasing the concentration of Na_2SO_4 to 6.9 times its normal concentration in seawater in a modified GP2 and then diluting the modified GP2 with unmodified GP2. Unlike typical toxicants, where acute toxicity only occurs with increasing chemical concentrations, some common ions may be essential to organism survival, even under acute testing conditions. Therefore, acute toxicity tests were conducted on solutions deficient in the salt of a specific ion in a modified GP2 solution diluted with unmodified GP2. Test solutions used to study the toxicity of sulfate deficiency, for example, were prepared by deleting all Na_2SO_4 from a GP2 solution and then diluting with unmodified GP2.

Test Procedures

Acute toxicity tests were performed using three common marine test organisms: the mysid shrimp (*Mysidopsis bahia*), the sheepshead minnow (*Cyprinodon variegatus*), and the inland silverside (*Menidia beryllina*). Organisms were from either ENSR's in-house cultures at the Houston Environmental Toxicology Laboratory (HETL) or from commercial suppliers. Tests were performed in both the HETL and the Fort Collins Environmental Toxicology Laboratory (FCETL). At test initiation, mysid shrimp were 2 to 6 days old, sheepshead minnows were 1 to 2 days old, and silversides were 9 to 11 days old. Studies ran for 48 hours under static conditions at 25° C. Test chambers were 120 ml plastic beakers containing 50 ml of test solution. Five organisms were placed in each chamber. Usually, five test concentrations and a control were tested. The mysid shrimp were fed brine shrimp (*Artemia*) three times daily to minimize cannibalism among the mysids. The two fish species were not fed during the study. Observations of mortality were made daily. A minimum of three successful toxicity tests of a single condition were conducted.

Chemical Analysis

Concentrations of major ions were determined analytically in all stock solutions used in testing. Cation concentrations were determined using inductively coupled plasma emission spectroscopy (ICP - USEPA method 200.0). Chloride and sulfate concentrations were determined by anion chromatography using USEPA method 300.0). Bicarbonate concentrations were determined indirectly by the measurement of phenolphthalein alkalinity according to American Public Health Association methods (APHA 1989). As bicarbonate is the predominate carbonate species present in the pH range of interest (pH 6.5 to 9.0), all alkalinity was assumed to be present as bicarbonate. Salinity was measured in all stock solutions using a Reichert Jung refractometer. Measurements of pH and dissolved oxygen (DO) were made on selected test solutions during actual toxicity testing, primarily on solutions near the threshold for acute toxicity. Measurements of pH were completed using an Orion model 720A pH meter; DO was measured using a YSI model 57 DO meter.

Data Analysis

All toxicity data were entered into a computer database using Lotus® 1-2-3® version 4.01 for Windows®. The database contained all relevant information about the tests. All models were developed from this database.

Total Salinity Models. Salinity toxicity data for all three species was analyzed using logistic regression (Hosmer and Lemeshow, 1989; Brown, 1991). Since the simple logistic models can only deal with one-tailed results, the salinity data were divided into low and high salinity data sets and analyzed independently. After successful modelling of the subsets, the models were combined in a multiplicative (product) model (Callahan, 1995). For all three species, there were a few aberrant data points that did correspond well with other data. For example, in the *M. bahia* data set, a few tests showed 80% survival at approximately 30 ppt salinity, while there was 100% survival in all other replicate tests. A difference of 20% often means little in toxicity testing (ie., one mortality in an initial exposure population of five). While these data did not substantially change the shape of the model, they did reduce the significance of the model and increase deviation. Since it is well-established that mysid shrimp survive at a salinity of 30 ppt (Weber, 1993), these data points were dropped from the data set for the purpose of model development.

Single-Ion Models for Mysidopsis bahia. Single ion models were also developed using logistic regression. Similar to the approach taken for the salinity models, high and low single ion data were analyzed separately, then combined in a multiplicative model, if appropriate.

Table 2. Predicted $LC_{50}s$[a] based on salinity (ppt) for the three species tested

Test Type	*Mysidopsis bahia*	*Cyprinodon variegatus*	*Menidia beryllina*
Ion Deficiency	3.9	NA[b]	NA[b]
Ion Excess	43.0	71.8	35.6

[a] LC_{50} = concentration where 50 percent of the test organisms die.
[b] Both species of fish had good survival at low salinity; no mortality is predicted by the model.

RESULTS

Total Salinity Studies

The total salinity investigations produced unique results for each of the three species tested. Actual survival of test organisms at different salinities are discussed in the following sections. The $LC_{50}s$ for each of the three species tested are presented in Table 2.

Mysidopsis bahia. Mysid shrimp survival was reduced in solutions with deficient salinity as well as excessive salinity. At the end of the 48 hour study, some mysid shrimp were still alive at salinities as low as approximately 4 ppt, although partial mortality generally occurred when salinity dropped below 7 ppt. Mortality also occurred when salinity exceeded approximately 40 ppt. The resulting survival curve for mysid shrimp shows 100 percent survival in salinities ranging from 7 to 40 ppt, with steep mortality curves at either end (Figure 1). The model that best explained the pattern of survival at 48 hours was a double logistic model with two constants and slopes for both the deficient and excess salinities (Figure 1). The pattern of survival at 24 hours was similar to the 48-hour curve although some organisms were still alive at 24 hours and the resulting model had a greater deviance.

Cyprinodon variegatus. Sheepshead minnow survival was high over a wide range of salinities, showing 100 percent survival in salinities ranging from as low as 1 ppt to as high as 60 ppt. Partial and complete mortality occurred when salinities approached zero; survival ranged from 0 to 60% after 48 hours in water with salinity near 0. When salinity exceeded approximately 75 ppt, *C. variegatus* showed 100 percent mortality (Figure 2).

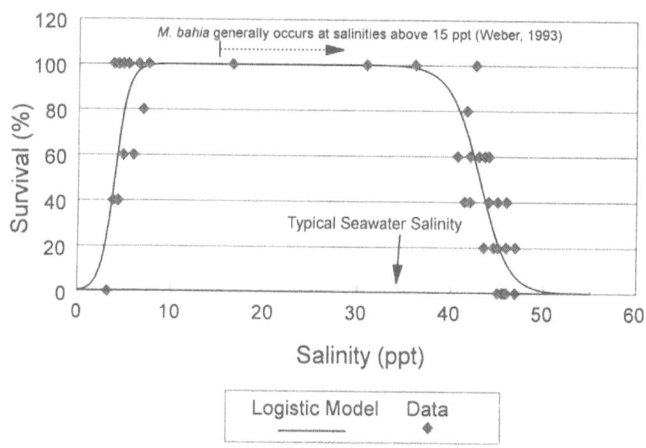

Figure 1. Survival of *Mysidopsis bahia* at different salinities and the logistic model predicting survival.

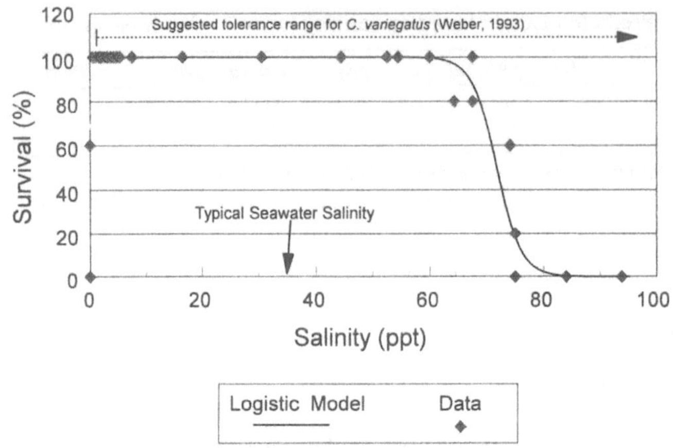

Figure 2. Survival of *Cyprinodon variegatus* at different salinities and the logistic model predicting survival.

Because there was incomplete and inconsistent mortality at low salinities, it was not possible to produce a meaningful model for this organism in the low salinity range. The resulting salinity/toxicity model was a "one-tailed" logistic model with a constant and slope representing organism response at high salinities (Figure 2).

Menidia beryllina. Inland silverside survival was high over a narrower range of salinities, relative to sheepshead minnow (Figure 3). Although survival was generally high at or below the salinity of seawater (approximately 31 ppt), it dropped substantially above that level, with 100% mortality at any salinities above 50 ppt. This low survival at high salinities is not, however, surprising. *Menidia* sp. is an estuarine taxon and generally occupies areas with salinities lower than would be found in the open ocean (Weber, 1993). Like sheepshead minnow, however, inland silverside survival was high even at very low salinities. The resulting salinity/toxicity model was a "one-tailed" logistic model with a constant and slope representing organism response at high salinities (Figure 3).

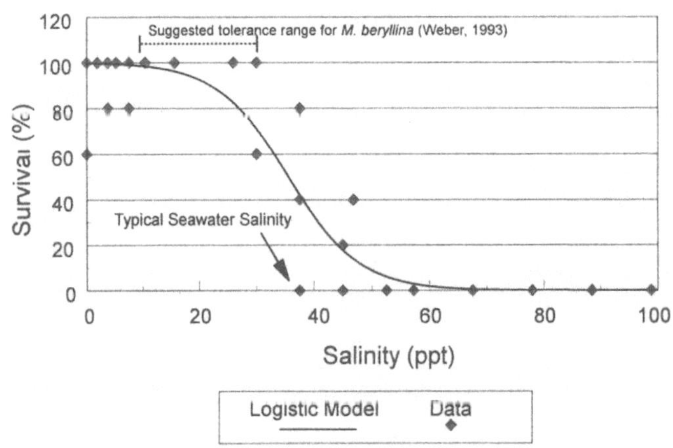

Figure 3. Survival of *Menidia beryllina* at different salinities and the logistic model predicting survival.

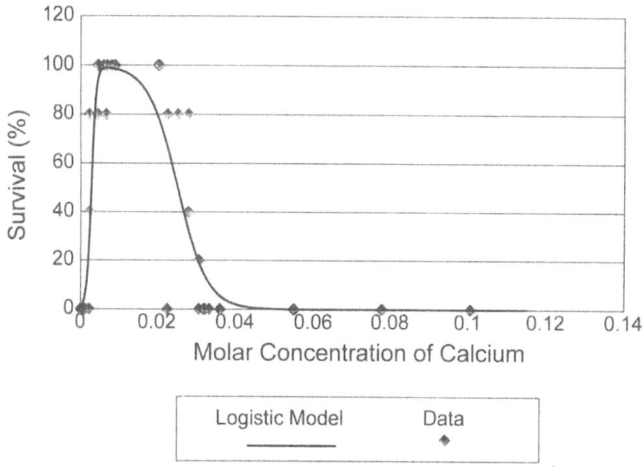

Figure 4. Survival of *Mysidopsis bahia* at different molar concentrations of calcium and the logistic model predicting survival.

Single Ion Studies with *Mysidopsis Bahia*

The eight ions tested produced substantially different patterns of toxicity in mysid shrimp. In general, three patterns of ion-specific acute toxicity occurred: 1) distinct toxicity occurring both at ion excess as well as deficiency, 2) distinct toxicity occurring only when an ion is in excess, and 3) no toxicity over the range of ion concentrations tested. The following discussion is separated into these three groups. Because of the limitations of this paper, results cannot be presented graphically for all ions tested, however, representative figures are shown.

Toxicity of Ion Excess and Deficiency. Both calcium and potassium caused toxicity at high and low concentrations. In excess, calcium was toxic when molar concentrations exceeded approximately 0.022 M. At concentrations greater than 0.03 M, mysid shrimp showed 100% mortality (Figure 4). Good survival of shrimp was observed at calcium concentrations as low as 0.003 M. At very low concentrations of calcium 100 percent mortality occurred. The resulting model that best described the pattern of mortality was a "two-tailed" product model which combined two logistic models, one describing toxicity of calcium in excess, the other describing toxicity of calcium in deficiency.

The pattern of toxicity for potassium was similar to that of calcium. The molar concentration range in which organisms survived was also similar, ranging from approximately 0.003 to 0.01. The potassium model was similar to that of calcium.

Toxicity of Ion Excess. Four ions, magnesium, bicarbonate, borate, and sulfate caused toxicity only when present in excessive amounts. Although slight toxicity may have occurred for the first three ions at low concentrations, survival was still high, even at concentrations near zero. Figure 5 shows the pattern of survival for magnesium. In all cases the model that best described the pattern of survival was a single logistic model. Toxicity studies of sulfate indicated some toxicity at the high end of sulfate concentrations. In one study, for example, 100% mortality occurred at approximately 0.177 M. However, it is possible that apparent toxicity due to sulfate may, in fact, be an artifact of the precipitation of calcium in association with the high sulfate concentration. In freshwater, sulfate was the

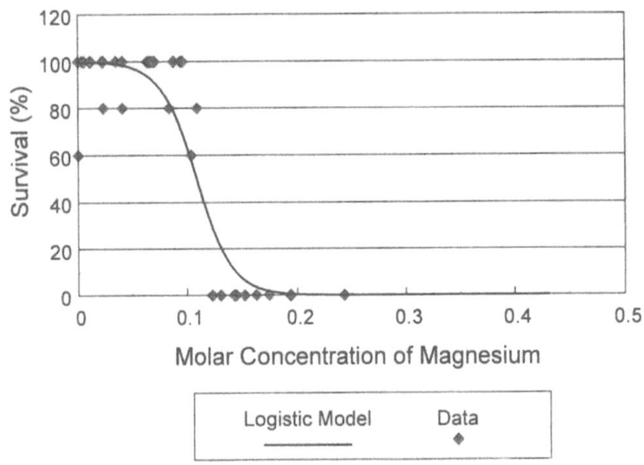

Figure 5. Survival of *Mysidopsis bahia* at different molar concentrations of magnesium and the logistic model predicting survival.

least toxic of the ions investigated during the earlier stages of this research (Gulley et al. 1992).

No Toxicity. Both strontium and bromide had essentially no effect on survival of mysid shrimp within the concentration ranges tested. In the bromide tests some partial mortality (≥60% survival) occurred at approximately 0.028 M and at 0 M. However, in other tests 100% survival occurred at these concentrations. In some of the strontium tests partial mortality (≥80% survival) occurred at the low range of ion concentrations. Once again, however, most studies showed 100% survival at those concentrations. No predictive models were developed for these two ions.

DISCUSSION

Salinity Studies and Models

The three marine organisms used in this study are reasonably tolerant of a wide range of salinities. The invertebrate, *M. bahia*, is less tolerant of low salinities and shows significant mortality at salinities below 4 ppt. The fish, on the other hand, can tolerate, at least on a short-term basis, salinities near 0. The intolerance of *M. beryllina* to high salinities is not unexpected considering the habitats in which it is frequently found. While *C. variegatus* is found in more open water areas, *M. beryllina* is more commonly found in estuaries where the salinity may be lower than in open water (Weber, 1993).

It is possible, and even likely, that survival of *C. variegatus* and *M. beryllina* may not be consistently high at very low salinities in toxicity tests of longer duration. Near-freshwater conditions are likely to stress the organisms and in some situations mortality would occur. However, since the salinity of produced waters generally is much higher than that of freshwater, it is unlikely that the salinities of produced waters being tested would be low enough to cause mortality to fish in short-term acute studies.

Single-Ion Study

Despite the tolerance of these three standard test organisms to a wide range of salinities, toxicity due to total salinity or to the individual ions that comprise the salinity can occur. Some produced waters have extremely high salinity and thus may cause toxicity which can be predicted by the salinity models. However, the single-ion studies confirm that standard test organisms react very differently to different ions. Because any of these ions, when present in excess amounts, can contribute to increased salinity, higher salinity in a produced water may not necessarily result in toxicity as predicted by the salinity models. It is important, therefore, that individual ion concentrations be determined in an effluent if accurate estimations of toxicity are to be determined. For example, an effluent might have high salinity which, based on the salinity models, might be predicted to be toxic. However, if the higher salinity is due to strontium or bromide, then the effluent may, in fact, not be toxic.

When results of the sulfate studies indicated that toxicity occurred at high sulfate concentrations, it was suspected that the manipulation of ion levels may be causing unexpected changes in solution concentrations of ions (other than those being manipulated). In order to address these unanticipated shifts in ionic composition, a Chemical Equilibration Model (CEM) was developed by modifying existing models (Oddo and Tomson, 1994) and applying them to the chemistry data from this project. This CEM was written directly into a spreadsheet that can be used to modify the existing data set to arrive at activity coefficients. Research is continuing to determine if the modified data set improves the models. Current activities indicate that application of the CEM may decrease the deviance, and increase the significance, of the models. Results of the CEM modeling will be discussed in future papers.

SUMMARY

It has been well-established through freshwater and marine studies that high (and low) salinity can cause toxicity to laboratory test organisms. Despite this toxicity, however, salinity has characteristics that set it apart from "traditional" toxicants (Gulley et al., 1992). An important characteristic of high or low salinity discharges is that the toxicity can be rapidly reduced by simple dilution with seawater. Since common ions do not bioaccumulate or biomagnify, there is no concern over food web effects. Furthermore, most or all of the common ions are important to normal physiological functions and are, in some cases, essential to life.

For many effluents containing high concentrations of common ions, including produced waters, the identification of the causative agent of toxicity has often been difficult. Separation of toxicity due to high salinity or elevated ion concentrations and that due to other toxicants (metals or organics) frequently presents a challenge. Because common ion concentrations cannot generally be reduced like some other toxicants (e.g., biodegradation of organics) without expensive treatment systems, it is important to differentiate between salinity-associated toxicity and toxicity caused by other materials. These studies demonstrate that common test organisms show a consistent, predictable pattern of survival when exposed to a range of salinities or to a range of concentrations of some ions. This consistent pattern can be accurately represented in the form of a mathematical model.

Application of the predictive toxicity models can be a useful way of interpreting the observed toxicity of a produced water effluent. If the model over-predicts the toxicity of an effluent or other mixture (i.e., the LC_{50} of the prediction is substantially lower than the actual LC_{50}), then there may be components in the effluent which somehow reduce toxicity. This type of difference between observed and predicted toxicity may not be very problematic since the actual toxicity is less than expected from the model prediction. If the model under-predicts the toxicity (i.e., the LC_{50} of the prediction is substantially higher than the actual LC_{50}), then there may be other components in the produced water which increase toxicity (e.g., metals or organic compounds). To determine if the model is accurately predicting toxicity, a "mock" effluent (made in the laboratory with deionized water and reagent-grade salts) that mimics ion concentrations and salinity can be used (Tietge et al., 1995). If the toxicity of the mock effluent is the same as is predicted by the model, then it is likely that other toxicants are present in the effluent or mixture.

Although only single-ion toxicity data for *M. bahia* are presented here, additional studies have been initiated and modeling has begun on other marine test organisms (i.e., *C. variegatus* and *M. beryllina*). When all modeling has been successfully completed, a computer program will be constructed. Like the ***GRI-FW STR Program***™, the marine STR program should provide a method to quickly estimate potential toxicity due to common ions in complex mixtures and to discriminate the ion related toxicity from toxicity caused by other constituents in the effluent.

ACKNOWLEDGMENTS

This research is funded by the Gas Research Institute (GRI) through Contract Number 5091-253-2160 with ENSR Consulting, Engineering, and Remediation. We would like to thank John Fillo and Darryl Brandon of the ENSR Pittsburgh office for their contributions to the study design. The authors wish to thank all of the technicians from the Houston and Fort Collins ENSR laboratories who conducted these tests used to develop the models. Salinity and single-ion modeling was completed by Dan Thomas of Lincoln Research. The Chemical Equilibration Model was developed by Dan Caudle of Caudle Consulting. Mark Brady of the ENSR Fort Collins Toxicology office was instrumental in preparing the figures for this paper.

REFERENCES

APHA, 1989. Standard methods for the examination of water and wastewater. American Public Health Association, Washington, DC.

Brown, R., 1991. All peaks aren't Gausian. *Personal Engineer. & Instrument. News*, pp. 51-54, July.

Callahan, E., 1995. Estimating dose-response curves when organisms respond at both high and low doses, (paper in preparation).

Fillo, J.P., Hamaker, T.A., Koraido, S.M., Stefenacci, R.G., Tallon, J.T. and Wesolowski, D., 1987. Production waters associated with the production, processing, transmission, and storage of natural gas: a literature survey. Gas Research Institute, Chicago, Illinois, GRI-87/0126

Gulley, D.D., Mount, D.R., Hockett, J.R., and Bergman, H.L., 1992. A statistical model to predict toxicity of saline produced waters to freshwater organisms, pp. 89-96, In: Ray, J.P. and Engelhardt, F.R. (Eds.), *Produced Water Technological/Environmental Issues and Solutions.* Plenum Press, New York.

Hosmer, D.W. and Lemeshow, S., 1989. Applied logistic regression, John Wiley & Sons, New York. 307 pp.

Mount, D.R. and Gulley, D., 1992. Development of a salinity toxicity relationship to predict acute toxicity of saline waters to freshwater organisms. Gas Research Institute, Chicago, Illinois, GRI-92/0301.

Oddo, J.E. and Tomson, M.B., 1994. Why scale forms and how to predict it. *SPE Production and Facilities*, SPE 21710:47-54.

O'Neil, P.E., harris, S.C., Mettee, M.F., Drottar, K.R., Mount, D.R., and Fillo, J.P., 1989. Biomonitoring of a produced water discharge from Cedar Cover Degasification Field, Alabama. Gas Research Institute, Chicago, Illinois, GRI-89/0073.

Simmons, B.F., 1992. Treatment and disposal of wastewaters produced with coalbed methane by reverse osmosis, pp. 593-604, In: Ray, J.P. and Engelhardt, F.R. (Eds.), *Produced Water Technological/Environmental Issues and Solutions.* Plenum Press, New York.

Tibbetts, P.J.C., Buchanan, I.T., Gawel, L.J., and Large, R., 1992. A comprehensive determination of produced water composition, pp. 97-112, In: Ray, J.P. and Engelhardt, F.R. (Eds.), *Produced Water Technological/Environmental Issues and Solutions.* Plenum Press, New York.

Tietge, J.E., Hockett, J.R., and Evans, J.M., 1995. Discrimination of ion toxicity to six produced waters using the freshwater salinity toxicity relationships and TIE procedures. Society of Petroleum Engineers, SPE 2970, PP. 393-402.

USEPA., 1991. Methods for aquatic toxicity identification evaluations: phase I toxicity characterization procedures. United States Environmental Protection Agency, EPA/600/6-91/003.

Weber, C.I., (Ed.), 1993. Methods for measuring the acute toxicity of effluents and receiving waters to freshwater and marine organisms. United States Environmental Protection Agency, EPA/600/4-90/027F.

Wesolowski, D., Broughton, A., Hansotte, C.A., Koraido, S.M., and Fillo, J.P., 1987. Characterization of produced waters from natural gas production operations. Gas Research Institute, Chicago, Illinois, GRI-87/0335.1.

7

A COMPARISON BETWEEN THE TOXICITY OF PRODUCED OIL AND GAS CONDENSATE USING A NEW SEDIMENT BIOASSAY FOR DEPOSITED OIL

E. M. Foekema,[1] H. P. M. Schobben,[1] J. Marquenie,[2] and M. C. Th. Scholten[1]

[1] TNO Institute of Environmental Sciences
Energy Research and Process Innovation
and Department of Ecological Risk Studies
P.O. Box 57, 1780 AB Den Helder, The Netherlands
[2] The Netherlands Oil Company (NAM)
P.O. Box 28000, 9400 HH Assen, The Netherlands

INTRODUCTION

Between 1981 and 1986, the effects of oil pollution and the applicability of different oil spill combat methods in intertidal coastal waters were studied in large scale ($20m^2$) experimental intertidal mesocosms ("OPEX"- project Kuiper *et al.*, 1983; Kuiper *et al.*, 1986; Scholten *et al.*, 1987). In these experiments North Sea oil (Forties) and fuel oil were tested, in various dosages. The most obvious effects were due to deposition of oil at the sediments during low tide. Small ostracods, together with the amphipods (mudshrimps and scuds) appeared to be the most sensitive species regarding oil pollution, followed by planktonic copepods, benthic bivalves (cockles and balthic tellins) and polychaete worms (*Heteromastes sp.* and *Capitella sp.*). The ranking of sensitivity of the exposed benthic species was not related to the oil type used. Moreover, the effect was clearly related to the dosage in $ml.m^{-2}$.

On the basis of these observations, it was concluded that the mudshrimp is a useful species for sedimentary oil toxicity testing. Mudshrimps are among the most sensitive benthic species and are relatively easy to collect and to handle, especially when compared with the small ostracods.

The most realistic way to assess the effect of stranded oil on a benthic community of a tidal flat is to work in a (semi) field situation. However, this kind of research is very expensive, influenced by weather conditions and carries with it the impossibility to perform short term experiments. Therefore, there is a need to develop smaller scale bioassays under controlled laboratory conditions. The objective of this study was to design a convenient

Produced Water2: Environmental Issues and Mitigation Technologies
Edited by M. Reed and S. Johnsen. Plenum Press, New York, 1996

experimental set-up to assess the specific effects of deposited oil on the benthic fauna of a tidal flat.

Most laboratory toxicity experiments for oil are designed to test the effect of the water soluble oil or the water accommodated fraction (Barre *et al*, 1994; Sophia & Balasubramianian, 1992). However, the physical and chemical reality of oil deposition on sediments is remote from such tests. When testing the effect of oil polluted sediments on benthic species the oil is usually either mixed with the sediment before exposure, or polluted sediments collected in the field are being used (van den Hurk *et al*, 1992; Chapman *et al*, 1992; Roddie *et al*, 1994). Although the introduction of sediment in the test is a realistic improvement, the actual exposure of organisms during an oil spill on a tidal flat is more dramatic and related to a combination of effects from the oil in water and sediment resulting from the deposition of the oil on the sediment surface.

In this paper a new bioassay for the assessment of the effects of an oil spill on benthic species is presented. The method is based on the deposition of oil in a more or less realistic way on the sediment surface of a small experimental unit with benthic test species (in this case mudshrimps). The experimental design is tested for a wide variety of oil types: topped forties (as a reference to the OPEX mesocosm studies), topped gas condensate, original gas condensate and sunflower oil.

The topped fraction is the residue of the oil after destillation up to 200°C, which can be considered as the fraction of the oil that actually will be deposited on a tidal flat in case of an oil spill after evaporation of the lighter components.

MATERIALS AND METHODS

The Mudshrimp (*Corophium volutator*)

The mudshrimp (*Corophium volutator*) is a marine amphipod with a maximal body length of about 1 cm, living in the toplayer of fine sediments in tidal areas, where they can reach very high densities of more than 10.000 individuals per m^2. Mudshrimps are capable of moving around by walking and swimming, but spend most of their life time in burrows, collecting food (organic detritus and associated micro-organisms) by use of their very long antennae (Green, 1968). The mudshrimp is known as one of the most oil sensitive benthic species of mudflats (e.a. Bernhem, 1982; Scholten *et al*, 1987; Bonsdorff *et al*, 1990).

The adult mudshrimps used in these experiments were collected from a mudflat area called the 'Mokbaai' located on the south-western part of the Wadden Sea island of Texel (the Netherlands). The animals were collected one day before introduction into the experimental units and were kept overnight at 15°C, with continuous water supply and without sediment. Individuals that were harmed during collection and/or transportation could after this period be distinguished from healthy ones and were excluded from experimentation.

Tested Oil Types

Within the scope of this study two experiments were conducted. A first preliminary experiment was performed to test the experimental design and to find a powerful dose range and test duration. In this experiment topped "Forties," topped condensate and original condensate were used. "Forties" is a light oil produced in the North Sea, whereas the condensate is produced with gas in the Wadden Sea.

The second actual experiment was optimised on the basis of the results of the preliminary test with the aim to quantify the toxicity of the three oil types used in the first experiment, also in comparison to a commercial blend of sunflower oil.

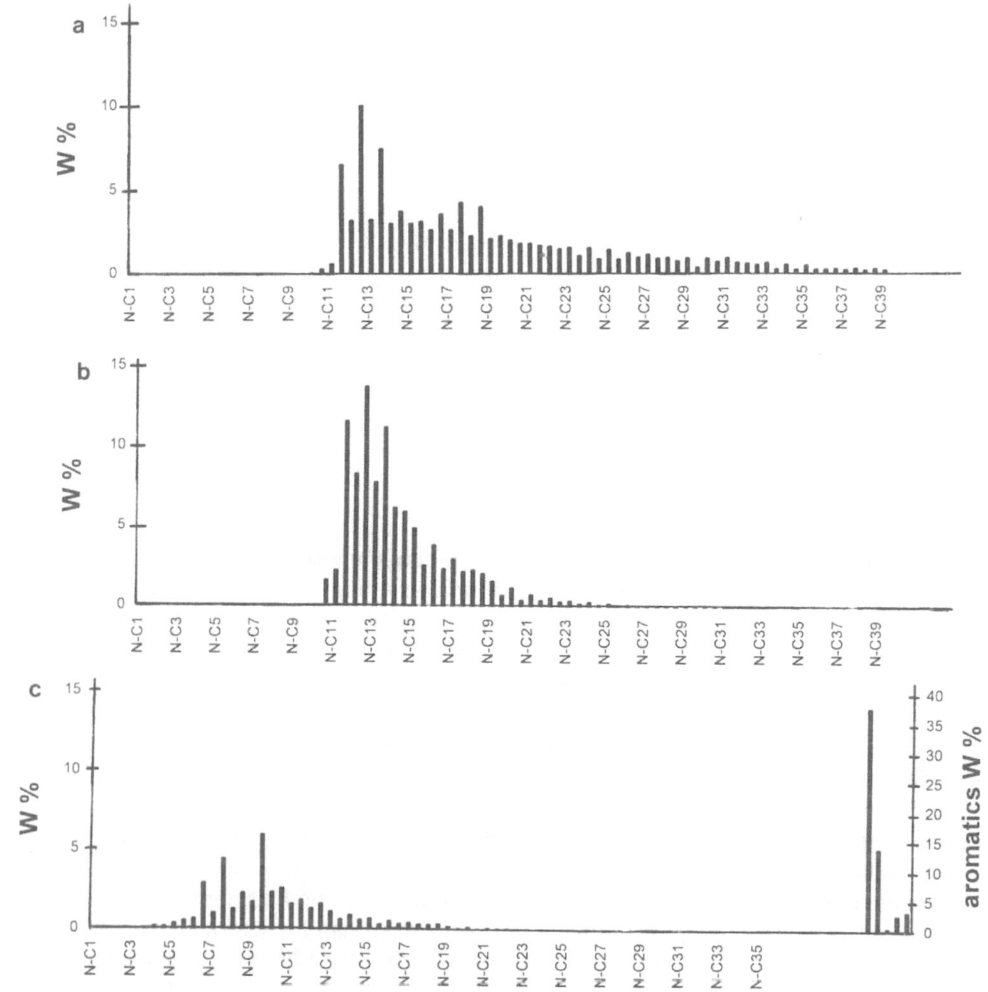

Figure 1. Composition of the applied topped forties (a), topped condensate (b) and condensate (c), based on gaschromatographic analyses. The five bars on the right of figure 1c (condensate) represent from left to right benzene. toluene, ethylbenzene. M=P-xylene and O-xylene.

Only the original condensate contained aromatics (about 50% on weight basis) and light weight (<12 C) hydrocarbons (Figures 1a,b,c). These compounds are absent in the topped fractions of condensate and "Forties" oil. Topping of the "Forties" oil and condensate resulted in a weight loss of about 53% and 87% respectively. Both topped forties and topped condensate are composed of linear hydrocarbons with more than 12 C-atoms. The relative concentration of larger molecules (>20 C) is higher in topped forties. Sunflower oil consists of about 90% of unsaturated and about 10% of saturated fatty acids.

Experimental Design

Both experiments were performed in a temperature controlled room with air temperatures around 15°C. Glass tanks with a surface of 0.05 m^2 (28*18 cm) and a height of 25 cm were used as experimental units, and were filled with a 10 cm thick layer of organic poor

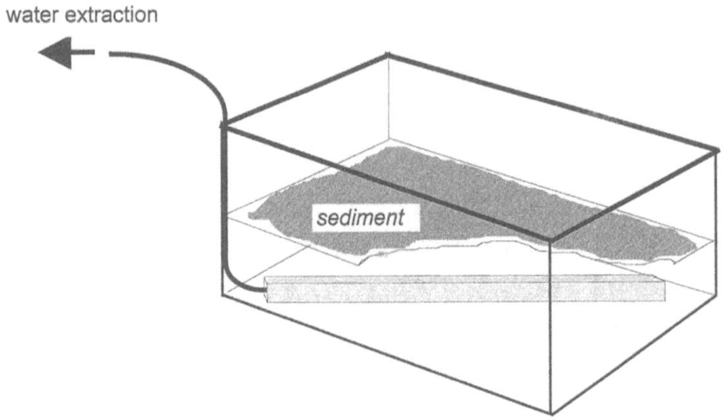

Figure 2. Experimental set-up.

sandy sediment collected from the same place as where the test organisms originated from. In the preliminary experiment the time consuming sieving of the sediment prior to the introduction was omitted. The results of this test showed however that sieving before starting an experiment was necessary. Therefore, in the following definite experiment only sieved (0.5 mm) sediment was used.

Before the sediment was added, an air diffusor (as used in fish-tanks) with a length of 27 cm and 1 cm square was placed diagonally on the bottom of each tank. To each of these air diffusors a tube of 35 cm was connected, which was led out of the sediment and attached in vertical position in the corner of the tank, leaving approximately 15 cm above the water surface (Figure 2). By extracting water from this tube low tide could be simulated in each individual tank. Due to the air diffusor the sediment could be drained equally and extraction of sediment during drainage was prevented. After application of the sediment the tanks were simultaneously filled by drops (40 litre.24 hr^{-1}) of filtered natural seawater. The final water level was controlled at 10 cm above the sediment surface by a hole in the short side of the tank, opposite to where the water entered. Except for the simulated low tide period during the application of the test compounds the refreshment of the watercolumn was continuous (33% hr^{-1}) throughout the duration of the experiment.

One day after the sediment introduction, groups of 100 (85 in first experiment) mudshrimps were distributed randomly over the experimental units. The animals were allowed two days to settle into the sediment and to acclimatise to the experimental conditions before the application of the oil took place.

Prior to the application of the different oil-types, the water refreshment was stopped and the water level in each tank was lowered till 1 cm above the sediment surface. By using a calibrated Finn pipette the oil was placed on the water surface. As soon as the oil spot visually stopped spreading (max. 10 seconds), the remaining water was drained off until a depth of 1 cm below the sediment surface was reached. This situation was extended for 5 hours (low tide), after which period the water refreshment was started again. The applied dosages of the oil are: 0.1; 0.3; 1; 3; 10; 30; and 100 $g.m^{-2}$. This range was based on the results of the OPEX study, in which 95-100% mudshrimp mortality was observed after the application of about 100 g oil per m^2. Since no effects were expected at low dosages of sunflower oil applications of less than 3 $g.m^{-2}$ of this oil type were omitted. After the experimental period all sediments were sieved gently through a 0.5 mm sieve and living mudshrimps in the residue were counted. The results of the experiments were analysed using the sigmoidal dose-response (variable slope) option incorporated in the software Graphpad Prism, version 1.0 (Graphpad Software, Inc).

RESULTS

Preliminary Range Finding Test

Remarkable differences in the reaction of the mudshrimps to the different oil-types were observed during the oil application. At the moment of draining of the reference (not treated) units only a few animals crawled out of the sediment, wandered around for a short period and dug themselves back in again. At the higher dosage (>3g·m^{-2}) of both condensate types, a large part of the test population showed a strong escape reaction. At the moment the oil touched the sediment surface the animals came rapidly out of the sediment and stayed on the dry surface. Application of topped forties did not produce this escape reaction to the sediment surface.

After the waterflow was started again a substantial part of the applied amount of all oil types came lose from the sediment and formed a floating layer on the water surface.

A number of mudshrimp in the treated units started floating and were unable to reach the sediment, for which reason their changes for survival were minimized.

Two and three days after the oil dosage numbers of dead mudshrimps on the sediment or water surface were counted. Floating individuals were counted as dead. Three days after application substantial mortality was recorded in tanks dosed with equivalents of 10 grams·m^{-2} topped forties or topped condensate and 30 grams·m^{-2} original condensate. Sieving of the sediments 7 days after oil application confirmed these observations. The test results however were hampered by the relative high mortality under reference conditions (caused by the presence of a predator *Nereis sp.*) and the fact that in some cases the numbers of living individuals at the end of the experiment exceeded the introduced numbers, due to the presence of a resident population in the introduced sediments. Both problems were solved in the next experiments by using sieved sediments. The results of the preliminary experiment are summerised in Table 1.

Actual Toxicity Test

The duration of the actual toxicity test was shortened to 3 days after oil application. Flight reactions were observed after addition of both condensate types and were absent when

Table 1. Number of observed dead mudshrimps 2 and 3 days after oil application, and number of living individuals 7 days after oil application (bold) in the preliminary test. Per experimental unit 85 individuals were introduced

	Topped Forties			Topped Condensate			Condensate		
Doses g.m^{-2}	2d	3d	**7d**	2d	3d	**7d**	2d	3d	**7d**
0 (reference)	0	3	**55**				3	4	**47**
0.1	2	3	**78**	3	6	**75**	5	5	**71**
0.3	3	3	**60**	7	7	**78**	2	4	**77**
1	4	5	**38**	2	3	**79**	3	4	**53**
3							1	3	**88**
10	8	32	**10**	32	22	**6**	4	3	**44**
31							39	39	**7**
100	52	47	**1**	65	80	**2**	35	28	**0**

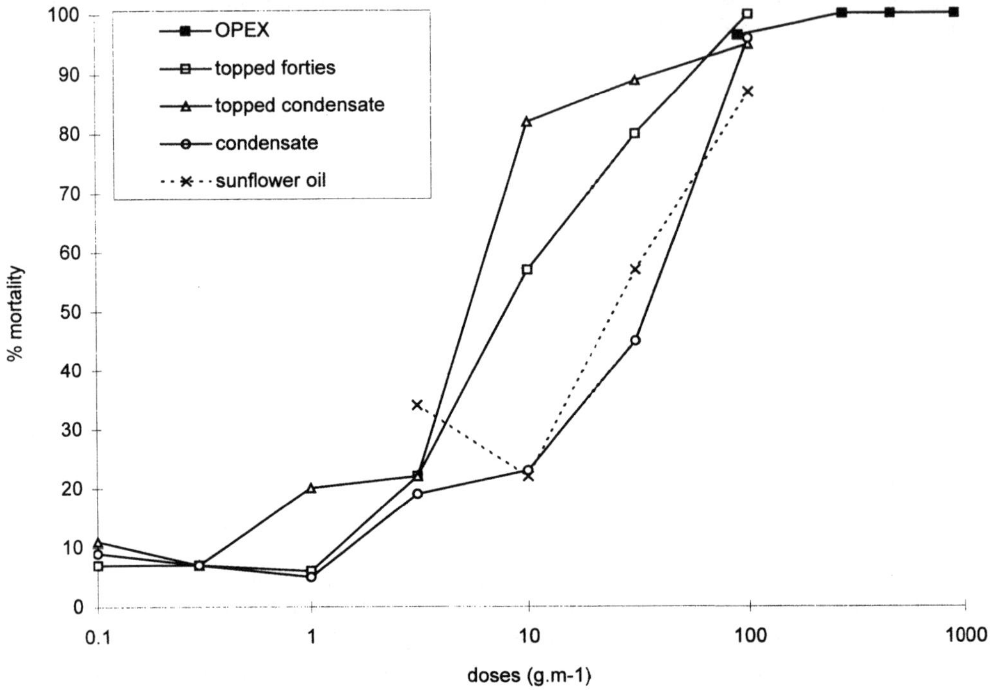

Figure 3. Mortality of mudshrimps exposed to oil in actual laboratory experiment (open markers) and in OPEX mesocosm study (closed markers).

topped forties was used, confirming the observations in the preliminary experiment. Sunflower oil did not cause an enhanced activity by the mudshrimps at the moment it reached the sediment.

Three days after the oil application, the surviving mudshrimps were collected from the sediments (for % mortality see Figure 3). A comparison of the LC50-values (Table 2) shows that no difference was found between topped forties and topped condensate (LC50 ≈ 6 $g.m^{-2}$), while the toxicity of original condensate is significantly lower (LC50 = 25.7 $g.m^{-2}$). The observations are consistent with those of the preliminary test. Due to the lack of a clear dose-related survival of the test organisms exposed to respectively 3 and 10 grams of sunflower oil the confidence intervals of the LC50 for this oil type are very broad, hampering the determination of significant differences with the other oil types tested. However, the observed mortality and the LC50-value suggest that this value is about 4 times higher than that of topped forties and topped condensate.

Table 2. Calculated $LC50_{(72h)}$-values of the different tested oil types

	$LC50_{(72h)}$ ($g.m^{-2}$)	95% Confidence intervals	R^2
Topped forties	6.9	4.9 - 9.7	0.98
Topped condensate	4.8	2.9 - 7.8	0.96
Condensate	25.7	13.3 - 49.7	0.91
Sunflower oil	19.6	4.8 - 79.6	0.84

DISCUSSION AND CONCLUSIONS

The observed mortality rate of the mudshrimps, three days after the application of topped forties and topped condensate in our experiments are comparable with that observed during the OPEX-mesocosm study (Figure 3). At the lowest dosed levels of 60 and 75 gram per m^2 (respectively in 1986 and 1985) the mortality of the mudshrimps one week after oil application occurred between 90 and 100% (Kuiper *et al.*,1986; Scholten *et al*, 1987). The experimental set-up presented in this paper can therefore be regarded as a useful test to assess the effects of deposited oil on benthic species.

Van Bernhem (1982) reported flight reactions of mudshrimps during exposure to different types of crude oil on sediments and suggested that this response was initiated by the "low boiling compounds" of the applied oil. We observed similar escape reactions when either condensate or topped condensate were applied, but not when mudshrimps were exposed to topped forties or sunflower oil. Gaschromatographic analysis confirmed the absence of aromatics and alkanes with less than 12 C-atoms in the topped condensate and forties and the substantial contribution of these compounds in the original condensate. The observed flight response to topped condensate should therefore be related to the relatively high concentration of C12-C20 compounds in condensate, compared with topped forties.

The lethal effects of topped forties and topped condensate on the mudshrimp were comparable. Original condensate, however, proved to be less toxic, even though only this oil type contained toxic (aromatic) compounds. The observed difference between the toxicity of topped and (less toxic) pure condensate can even be explained as the result of evaporation of the low molecular compounds from the pure condensate, before any acute harm was done. During the "topping" procedure only 13% of the pure condensate is retained in the residue (being topped condensate), the other 87% is evaporated. The LC50-value of topped condensate is 19% of the LC50 of the pure condensate. The combination of these observations indicates that the adverse effect on mudshrimps being exposed to pure condensate was only due to the higher molecular aliphatics. It is, therefore, concluded that the observed mortality was primarily caused by physical effects of the mixture of hydrophobic, long chained aliphatics, rather than by chemical intoxication due to insoluble aromatics.

This is in line with the findings of Van Bernhem (1982) who observed severe mudshrimp mortality after exposure to various mineral oil types only on individuals that had been "trapped" in the oil.

Without question the mortality caused by sunflower oil was completely the result of physical effects. Even though the confidence intervals are very broad the experimental results indicate a toxicity 4 times less than those of topped forties and topped condensate. This relative low toxicity of sunflower oil can be explained from the high surface tension (hampering the spreading of the oil spot) compared with the other oil types tested. In a small experiment it was confirmed that the surfaces of the oil spots produced by forties oil and both condensate types were of comparable size, while the addition of the same amount of sunflower oil resulted in an oil spot with $^1/_3$ to $^1/_4$ of this surface. Because physical effects can only occur when contact between animal and solid oil is made, the surface of the oil spot must be directly correlated with the intensity of the effects that occur.

In case of birds (Jenssen, 1994) and mammals (Engelhardt, 1983) it has already been recognised that physical contact with oil leading to staining of the body is an important cause of harmful effects. For smaller organisms little is known about this specific mode of action, mainly because effects of oil on these organisms are being studied in traditional laboratory experiments as mentioned in the introduction of this paper, or in (semi) field situations, where observations are of a population dynamical character and restricted to the resultant of the pollution effects on the density of the species.

Even though the mudshrimp is one of the most sensitive benthic species to oil pollution it seems obvious that other benthic species are affected by the physical effects of oil as well. In the OPEX mesocosm experiments, the other sensitive benthic species are either living at the sediment-water interface or active filter- or deposit-feeders and thus have an intensive contact with deposited oil. Evaluation of toxicity data regarding oil in water shows that filter-feeders are amongst the most sensitive species, probably due to contact with dispersed oil (Scholten, in prep).

Our experiments show that the acute impact of deposited oil on the benthic fauna of a tidal flat can be assessed based on the fraction of C12 and higher aliphatics in combination with physical characteristics responsible for the spreading of the oil on the water surface. This includes vegetable oils as well. As a consequence, the environmental hazard of non-mineral oils should be considered in the same way as mineral oil, for instance in the Marpol resolutions for the protection of the marine environment.

ACKNOWLEDGEMENTS

The authors thank Gerrit Hoornsman, Dennis van der Veen, Henk van het Groenewoud, Pim de Kock (all TNO), Henny Reerink (NAM) and Albert Bos (SERVO) for their contribution in experimentation, evaluation of the results or correction of the manuscript.

REFERENCES

Barre J.S., T.M. Bert & E.S. van Vleet (1994): Toxicity of the watersoluble fraction of diesel fuel to postsettlement juvenile Stone crabs (*Menippe mercenaria*). *Bull. Mar. Sci.* 55:240246.

Bernhem K.H. van (1982): Effect of experimental crude oil contamination on abundance, mortality and resettlement of representative mud flat organisms in the mesohaline area of the Elbe estuary. *Neth. J. Sea Res.* 16:538546.

Bonsdorff E., T. Bakke & A. Pedersen (1990): Colonization of amphipods and polychaetes to sediments experimentally exposed to oil hydrocarbons. *Mar. Pollut. Bull.* 21:355358.

Chapman P.M., R.C. Swartz, B. Roddie, H.L. Phelps, P. van den Hurk & R. Butler (1992): An international comparison of sediment toxicity tests in the North Sea. *Mar. Ecol. Prog. Ser.* 91:253264.

Engelhardt F.R. (1983): Petroleum effects on marine mammals. *Aquat. Toxicol.* 4:199217.

Green, J. (1968): The biology of estuarine animals. Biology series 21s. University of London at Westfield College

Hurk P. van den, P.M. Chapman, B. Roddie & R.C. Swartz (1992): A comparison of North American and West European infaunal amphipod species in a toxicity test on North Sea sediments. *Mar. Ecol. Prog. Ser.* 91:237243.

Jenssen, (1994): Review article: Effects of oil pollution, chemically treated, and cleaning on the thermal balance of birds. *Environ. Pollut.* 86:207215.

Kuiper J., H. van het Groenewoud en G. Hoornsman (1983): A study of marine oil pollution in outdoor model ecosystems representing a tidal flat (opex) - final report -. *TNO Report* R83/014.

Kuiper J. (1986): Opex 1984 - 1987: Het lot en de effecten van al dan niet bestreden olie in modelwadecosystemen. *TNO Report* R86/068

Roddie B., T. Kedwards, R. AshbyCrane & M. Crane (1994): The toxicity to *Corophium volutator* (pallas) of beach sand contaminated by a spillage of crude oil. *Chemosphere* 29:719727.

Scholten M.C.Th., T. Bowmer en H. van het Groenewoud (1987): Effects of a light crude oil and of two selected oil combat methods in experimental tidal flat ecosystems. Final report Oil Pollution Experiments 1986. *TNO-report* R87/348

Scholten M.C.Th. (in prep): A new appraisal of the environmental hazard of mineral oils, based on analysis of aquatic toxicity data.

Sophia A.J.A. & T. Balasubramanian (1992): Changes in the physical condition of Meretrix casta exposed to watersoluble fractions of refined and crude oil. *Arch. Environ. Contam. Toxicol.* 22:471474.

8

PRODUCED WATER COMPOSITION, TOXICITY, AND FATE

A Review of Recent BP North Sea Studies

Steven A. Flynn,[1] Ed J. Butler,[2] and Ian Vance[2]

[1] BP Exploration Operating Company Ltd.
Farburn Industrial Estate
Dyce, Aberdeen AB1 9TL, United Kingdom
[2] BP International Ltd.
Group Research and Engineering Centre
Sunbury-on-Thames, Middlesex TW16 7LN, United Kingdom

INTRODUCTION

BP Exploration currently operates 18 producing fields in the North Sea and has equity interests in a further 13. BP operated facilities produce about 500,000 barrels of oil per day and 900 million cubic feet of gas. The proportion of water produced with the oil from these fields is increasing as they mature and in 1994 the water production was similar to oil production.

In order to enhance the understanding of the possible environmental significance of produced water discharges, a series research studies have been carried out on fields operated by BP. Five different platforms were involved at various stages of the programme (Figure 1) and the broad aims were to:

1. Quantify more fully the chemical composition and physical characteristics of produced water at the point of discharge;
2. Establish and compare the ecotoxicity of discharges from different fields;
3. Understand which components may be more significant from the environmental point of view;
4. Study the influence of dilution and biodegradation processes in mitigating any possible impacts.

The first phase of the study looked in detail at the chemical, physical and ecotoxicological properties of the produced water at the point of discharge for four BP platforms. Magnus was sampled twice, first during normal operations and secondly, under upset conditions to evaluate worse case effluent quality. Platforms studied were:

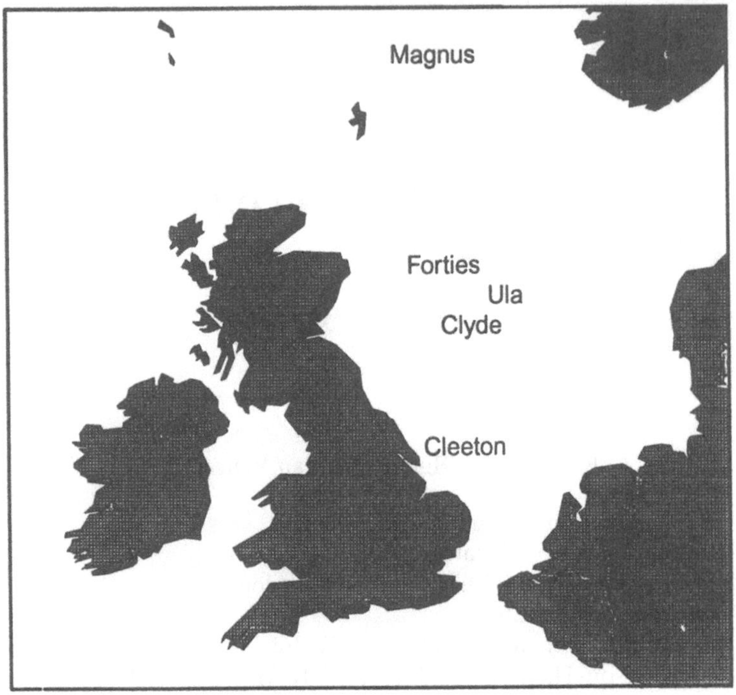

Figure 1. Locations of the platforms studied.

- Magnus (oil field)
- Forties (oil field)
- Ula (oil field)
- Cleeton (gas condensate)

The fields studied are located in the region from the southern North Sea gas fields to the northern edge of the East Shetland basin. The chosen fields included a range of reservoir conditions, water compositions, and stages of operational maturity.

The detailed chemical analysis was carried out by M-Scan Ltd and ecotoxicological tests were performed by Environment and Resource Technology Ltd. For this work, samples of whole produced water were prepared on the platforms, transported to the laboratories onshore within 24 hours and stored under carefully controlled conditions. Supplementary compositional data were collected for the Clyde field and detailed studies were carried out on the fate of produced water after discharge. This included biodegradation studies and computer modelling of dispersion characteristics which were carried out by the Brixham Environmental Laboratory.

The most recent studies involving direct sampling of the sea around the Clyde and Forties platforms, were lead by UK Government scientists from the Scottish Office Agriculture and Fisheries Department (SOAFD) and the Ministry of Agriculture Fisheries and Food (MAFF). This provided field confirmation of very low values of acute toxicity in samples of the water column taken from around the platform (Stagg et al. 1996) which was predicted by the present study.

Table 1. Summary of installations included in studies

	Magnus I + II [3]	Forties Delta	Ula	Cleeton	Clyde	Forties Charlie	N. Sea Platforms
Platform details							
Typical water discharge (mbd)	6-16	38	13	0.05	50	32	0.01-250 [1]
Approximate depth (m)	180	120	70	50	80	120	20-180 [2]
Bulk parameters							
Sample oil content (mg/l)	8-360	68	37	58	34	50	2-220 [1]
Total organic carbon (mg/l)	140-160	340	71	290	45	250	na
Salinity (mg/l Cl^-)	17-18,000	44,400	40,440	31	72,700	34,200	44,630 [1]
Temperature ° Centigrade	59-68	76	68	14.3	72	76	n/a
Specific Gravity gcm^{-3} @ 20° C	1.020	1.049	1.048	1.000	1.080	1.040	n/a

na = not available
[1] E&P Forum Report No. 2.62/204, May 1994.
[2] BP in-house data
[3] Denotes the range of Magnus data

PRODUCED WATER COMPOSITION AND TOXICITY

Details of the platforms are shown in Table 1 together with some of the bulk parameters for the discharged produced water.

Samples were collected from the final discharge point of the platform according to a predefined protocol. Effluent quality was monitored prior to sampling to ensure that the samples were representative of normal platform operation. Care was taken to avoid the possibility of sample contamination resulting from drilling operations. Where appropriate, the samples were chemically conditioned and preserved offshore to minimise any changes

Table 2. Summary of produced water characterisation parameters

Parameter	Technique
Bulk parameters	
Total Organic Carbon	Combustion and infrared absorption
Chemical Oxygen Demand	Chromic acid oxidation
Organic constituents	
Total/soluble oil	Extraction and infrared absorption before and after filtration
Volatile fatty acids	Gas chromatography
Acid fraction	Solvent extraction at pH 2 base neutrals have been removed, gas chromatography-mass spectroscopy (GC-MS) of trimethylsilyl derivatives
Volatiles	Purge and trap followed by thermal desorption and GC-MS
Base/neutral fraction	Dichloromethane extraction at pH 12 and GC-MS
Bioassays	
Photobacterium phosphoreum (bacteria)	15 min EC_{50} by 'Microtox 500' analyser system utilising luminescent bacteria
Acartia tonsa (copepod)	48 hr LC_{50} by ISO / Parcom ring test method
Skeletonema costatum (algae)	72 hr EC_{50} for growth rate reduction by ISO / Parcom ring test method

in composition that might occur during transit. Samples were quickly cooled and transported in refrigerated containers by helicopter and courier to the onshore laboratories, within 24 hours, for subsequent analyses. Volatile and aromatic compounds were analysed by carrying out the purge and trap part of the analysis using the methods described by Tibbetts et al (1992) in the platform laboratory, in order to minimize the losses of these compounds during transportation.

Composition

Some of the analytical techniques employed during the programme are summarised in Table 2.

A detailed analysis of inorganic components was also carried out, but the present review focuses mainly on organic components.

Salinity of the produced water samples varied widely:

- almost fresh on the Cleeton gas platform (31 mg l^{-1} Cl^{-})
- similar to seawater on Magnus (17,450 mg l^{-1} Cl^{-})
- approximately twice that of seawater on Ula and Forties (ca 40-44,000 mg l^{-1} Cl^{-}).
- approximately four times that of seawater on Clyde (ca 72,000 mgl^{-1} Cl^{-})

Trace metals and anions were all within the ranges generally found on oil and gas production platforms reported by E&P Forum (1994) and Tibbetts et al (1992).

The distribution of organic compounds between different broad categories is shown in Table 3.

Volatile fatty acids were found to be the major class of organics present, with acetic acid being the main constituent. Monocyclic aromatic compounds (BTEX + C3 and C4 Benzenes) were present in the range 10 to about 40mg/l, additionally, levels of phenols observed ranged from 0.1 to 30 mg/l. Between 0.5 and 7 mg/l of naphthalene and polyaromatic hydrocarbons (PAH) were detected in the water (Figure 2).

Naphthalenes were by far the most abundant group in this class with individual higher PAH compounds present at less than 0.5 mg/l.

Attempts were made to identify individual production chemicals in the produced water samples using Fast Atom Bombardment Mass Spectroscopy (FABMS). Although

Table 3. Main organic constituents of produced water

Organic constituents (mg/l)	Magnus I + II [3]	Forties Delta	Ula	Cleeton	Clyde	Forties Charlie	N. Sea Platforms
Volatile fatty acids	220-240	380	13	52	25	300	80-930 [1]
Aliphatics	2.2-87	20	14	27	n/a	2.2	n/a
Phenols	7.5-15	10.2	0.09	29.5	8.8	5.5	2-23 [2]
MAH	23.1-29.8	9.9	12.4	40.7	4.8	19.6	0.5-670 [1]
Total bases	0.26-0.47	0.06	0.09	0.29	n/a	0.005	n/a
PAH	1.4-6.9	3.0	3.0	0.5	1.5	0.66	0.0005-3.0 [3]

na = not available
MAH = Monocyclic aromatic hydrocarbons (BTEX + C_3, C_4 benzenes)
PAH = Polycyclic aromatic hydrocarbons
[1] E&P Forum Report (1994)
[2] Tibbetts. PJC, et al, (1992)
[3] M-Scan Ltd data
[4] Denotes the range of Magnus data

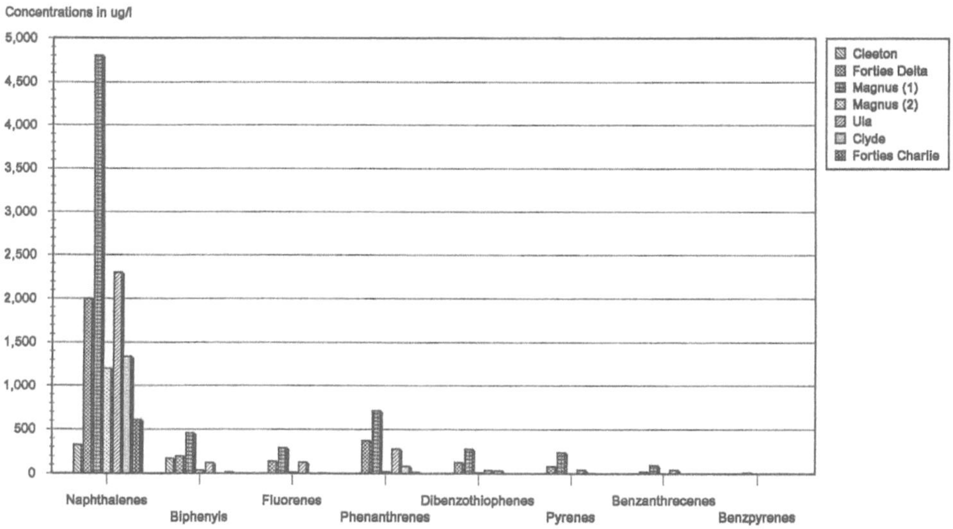

specific functional groups were detected, it was not possible to quantify or to assign these to individual production chemicals.

Toxicity

Three types of bioassays were carried out, as indicated in Table 2 and the results are given in Table 3. The acute toxicity (EC_{50}) of produced water from oil production platforms was in the range 5-25% by volume (Table 4).

The toxicity was higher (2% by volume) for the low volume of produced water discharged from the gas platform. Overall, the results lie within the range reported for other North Sea Platforms (Figure 3).

The sensitivity of the bacteria and copepod tests to produced water toxicants correlated very well (98% confidence level). However, interestingly the results of the Skeletonema toxicity tests showed an inverse relationship to the Microtox™ data.

Table 4. Toxicity of produced water by a range of bioassays

Toxicity (%)	Magnus I + II [3]	Forties Delta	Ula	Cleeton	Clyde	Forties Charlie	N. Sea Platforms
Microtox™ 15 mins EC_{50}	9.8-10.7	11.0	24.4	2.4	8.6	7.4	5.0-6.0 [1]
Acartia tonsa 48 hr LC_{50}	14.2-18.0	8.0	24.8	2.0	na	na	0.3-5.0 [2]
Skeletonema costatum 72 hr EC_{50}	10.0-35.1	42.5	7.9	1.0	na	na	4.5-67 [2]

na = not available
[1] Jacobs. RPWM, et al. International Produced Water Symposium (1992) San Diego
[2] E&P Forum Report No. 2.62/204
[3] Denotes the range of Magnus data

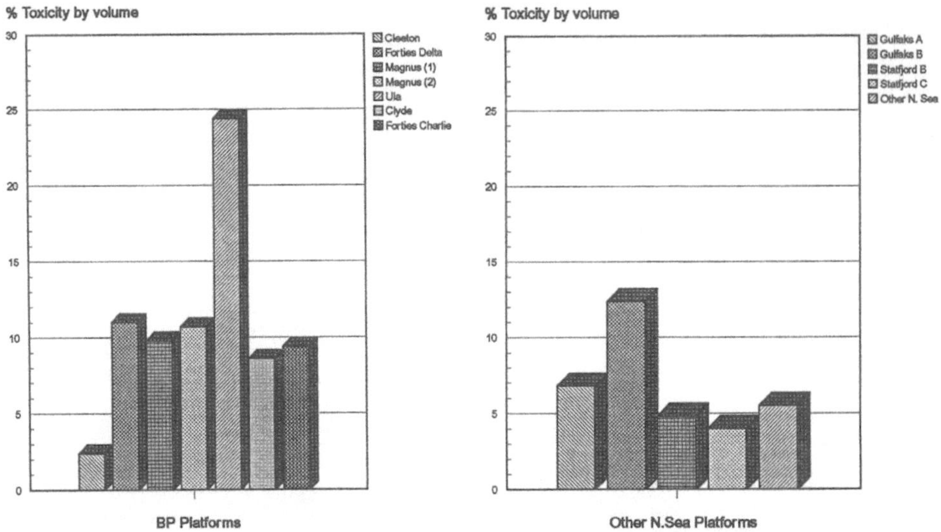

Figure 3. Comparison of toxicity results for various North Sea produced waters.

Good correlations, at the 98% confidence levels, were also found between both the Acartia and Microtox™ tests and total phenols (Figure 4), suggesting that phenols may be a significant source of toxicity.

There was no significant correlation between BTEX concentration and toxicity to Acartia and Microtox™.

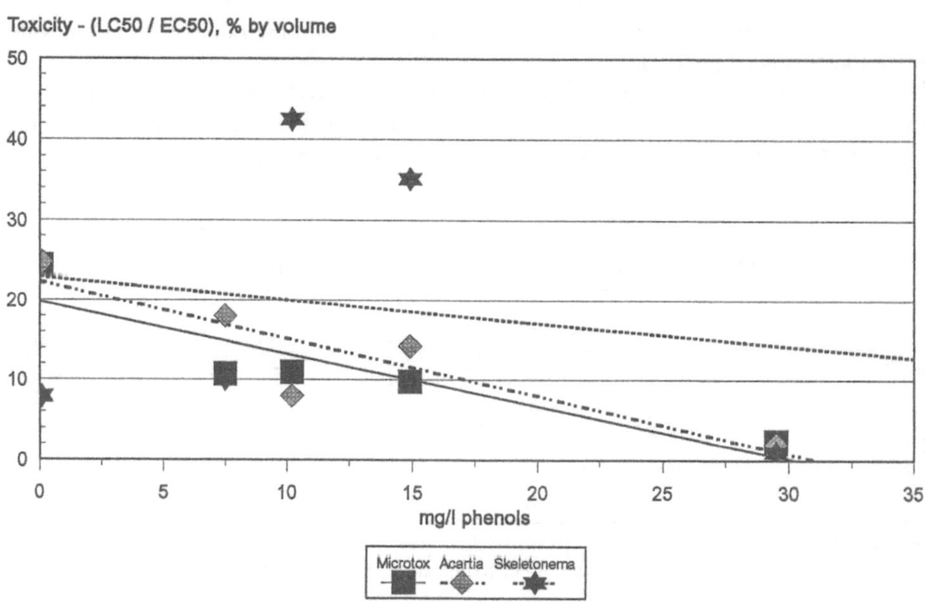

Figure 4. Correlation of phenol with Acartia tonsa and Microtox™ toxicity results.

PRODUCED WATER FATE

Once water has been discharged, dilution will have the most immediate effect in mitigating potential environmental impacts of produced water. Volatilisation and biodegradation are also important mechanisms in mitigating the environmental impacts of organic constituents. To gain some insight into the relative significance of these mechanisms two further phases of study were carried out for the Clyde platform. These involved:

- Laboratory biodegradation and volatilisation experiments on samples of produced water.
- Computer modelling of the plume dispersion.

The laboratory work quantified changes in chemical composition and relative toxicity with time and at different temperatures during biodegradation and volatilisation experiments. The results of the biodegradation study were then incorporated into the mathematical model in order to establish the relative importance of this mechanism for eliminating organic compounds from the plume.

Biodegradation

As before, produced water samples were rapidly transported by helicopter and overnight courier to the onshore laboratory, packed in cooled containers and then stored at 4°C. Chemical composition was determined using the methods described by Tibbetts et al (1992). However, for this phase of the study volatiles were analysed onshore, rather than carrying out the purge and trap procedure offshore. Toxicity testing was carried out using the Microtox assay technique.

Two types of biodegradation tests were used:

- Closed bottle biodegradation tests were used to provide samples for chemical and bioassay after fixed times. Tests were carried out in triplicate at 20°C and 10°C according to OECD Guideline 301/d, 1993.
- Manometric respirometry was used to follow the biodegradation profile with time. Tests were carried out on 30% and 40% dilutions of produced water according to OECD Guideline 301/F, 1993.

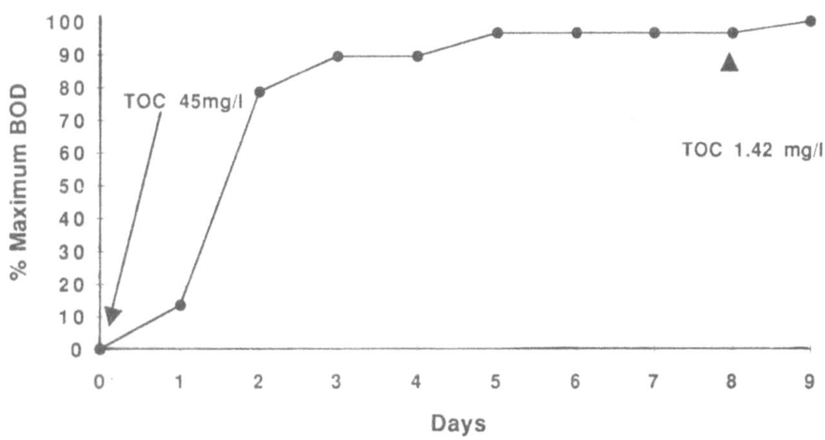

Figure 5. Clyde produced water BOD.

Table 5. Effect of biodegradation on selected organics

Time (Days)	TOC (mg/l)	VFA's (mg/l)	Phenols (μg/l)	BTEX (μg/l)	Total B/N (μg/l)*	PAH (μg/l)
0	45	25	8810	4800	9210	1460
8 days	1.42	6	4.7	<0.5	610	99
28 days	0.52	4	na	<0.5	630	4.6

na = not available
* Total Bases/Neutral Fraction, BTEX + non-volatile base neutrals

Inocula for biodegradation tests were acclimated to sea water and Clyde produced water, as rate measurements were carried out on relatively concentrated samples.

The majority of organic materials in Clyde produced water were readily biodegradable. The ultimate BOD was reached in less than 9 days (Fig 5).

Measurements of total organic carbon (TOC) showed that 90% of the organics had been biodegraded after 8 days and after 28 days the proportion had risen to 96%. The addition of inorganic nutrients to the BOD tests made little difference to the results, suggesting that the nutrient status of produced water is sufficient to support substantial biodegradation. The chemical composition of the water before and after biodegradation is shown in Table 5.

Over 99% of phenols and PAH were degraded, and BTEX was reduced below the limit of detection (< 0.5ppb). The Microtox toxicity was initially 8.6% (EC_{50}), but was undetectable after 8 days, indicating that the components responsible for the initial toxicity are readily biodegradable. These results are similar to the findings of Brendehaug et al (1992) where rapid biodegradation of organics in produced water resulted in up to a 10 fold reduction in toxicity.

Volatilisation

Volatilisation was studied since some produced water outfalls are above the sea surface and volatilisation could make a contribution to toxicity reduction. Also, gas stripping has been proposed as a possible method for removing organics from some discharges, particularly for gas platforms. The effect of volatilisation on Clyde produced water was determined by sparging with nitrogen at 1.25 v/v/min at 20°C and 10°C for 7 hours. The sample sparged at 20°C was subsequently subjected to a 28 day closed bottle biodegradation test.

The results of the volatilisation experiments are shown in Table 6. Sparging was found to:

1. have little effect on volatile fatty acids
2. remove some of the Phenols (< 50%)
3. remove over 90% of BTEX and PAH (presumably the more volatile naphthalenes, which represent the majority of compounds in this class)

Toxicity measured by the Microtox assay fell from an EC50 value of 8.6% to 16% at 10°C and 25% at 20°C. This would suggest that the volatile compounds are not necessarily the most important compounds in determining the overall acute toxicity of the produced water.

Dispersion and Dilution

A particle tracking model, as described by Staples et al (1993), was used to predict the dilution of the plume after discharge. A random walk component was superimposed to

Table 6. Effect of volatilisation on selected organics

Sparging Time	VFA's (mg/l)	Phenols (μg/l)	BTEX (μg/l)	Total B/N (μg/l)*	PAH (μg/l)
0 hours	25	8810	4800	9210	1460
7 hours @ 10° C	28	5030	17	1820	28
7 hours @ 20° C	26	5190	5	680	2.9
7 hours @ 20° C+ 28 days Biodegradation	1.9	1.8	<0.5	420	5.4

* Total Base/Neutral Fraction, BTEX + non-volatile base neutrals

simulate the effect of dispersion. This model has the advantage of simulating the effect of tidal oscillations and transverse harmonic components. The model also incorporated the effects of plume buoyancy and biodegradation of produced water components.

The model details and results of this study are described by Murray-Smith et al, (1996). The model predicted that the centre of the discharge plume would sink about 10m due to its high salinity and initial momentum. As the plume entrains more seawater it becomes neutrally buoyant and the centre line of the plume stabilises in the water column. The model has given a fascinating insight into the complex nature of the effluent plume in the near field and how its location changes during the tidal cycle. The plume is relatively long and narrow (approximately 50m x 4m in cross section at 100m from the platform) and actually rotates around the platform with time. The greatest proportion of the dilution effect is manifest in the first 100m from the discharge point (Murray-Smith et al, 1996). At 500m from the platform the model predicts that the minimum dilution, at any point in the plume, will approach a value of 775 for a short period during the tidal cycle (Figure 6).

However, the volume of water affected is relatively small and for most of the tidal cycle the dilutions are much greater.

The minimum dilution was predicted to reach about 9,000 at 20 km. Over this distance, however, biodegradation would begin to have a significant impact. At the average winter sea temperature of 6°C, the effective minimum dilution of the organic material would be 13,000, at 20 km from the platform. Higher predicted biodegradation rates at summer sea

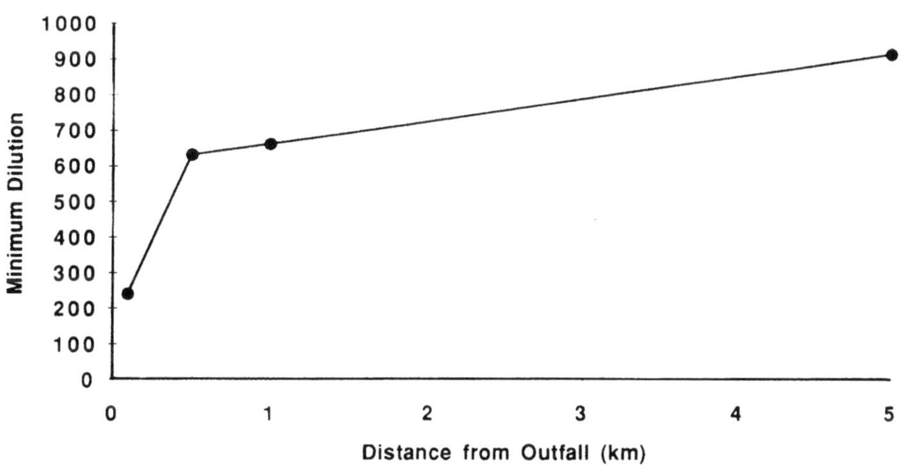

Figure 6. Minimum dilution with distance in the near field.

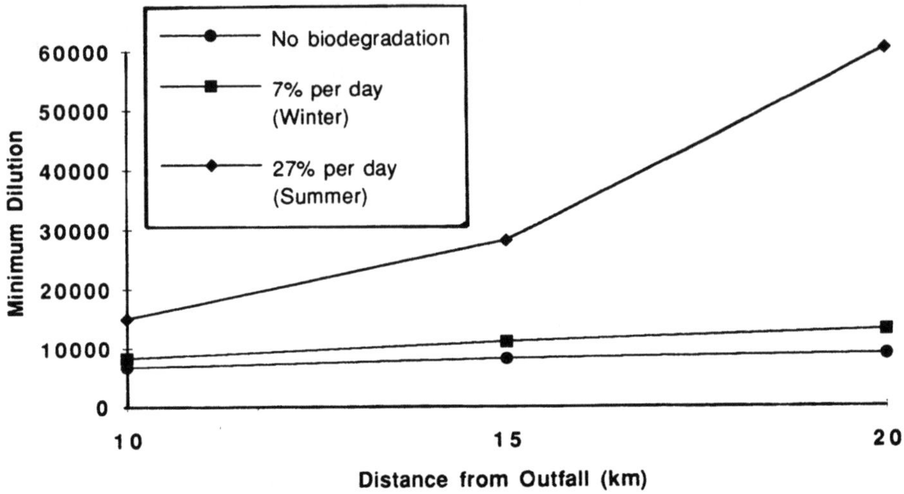

Figure 7. Effect of biodegradation on minimum dilution in far field.

temperatures of 15°C, result in a minimum dilution of about 60,000, again at 20 km from the platform (Figure 7).

Sampling of the produced water discharge plume was carried out by Government scientists during the North Sea "Dana cruise" in 1994, which included BP's Forties and Clyde fields, along with the Shell Brent and Marathon's Brae field. Full details of the sampling methods and results are given by Stagg et al, 1996. These field observations indicated that the model provided a reasonable qualitative description of the behaviour of the plume under actual conditions. Measurements and samples taken within the plume suggest that the model is conservative and the dilutions observed in practice were greater than the minimum values predicted from the simulation. Further discussion of these points appears in Stagg et al, 1996 and Murray-Smith et al, 1996.

DISCUSSION

BP studies described here are consistent with previous observations that soluble organics, particularly phenols, have an important role in determining acute toxicity of produced water. It is also encouraging that biodegradation occurs for those components that make a significant contribution to produced water toxicity. This is consistent with the previous studies reported by Brendehaug et al (1992).

The results of the programme to date suggest the acute toxicity effects are unlikely outside the immediate vicinity of the platform. The rapid dilution that occurs within the first hundred metres of the platform is probably the most important mechanism for mitigating any environmental effects. Further confirmation of the actual levels of dilution together with a broader base of toxicity studies would appear to be the most fruitful areas for future work.

CONCLUSIONS

The composition, toxicity and fate of produced water discharges have been studied since 1992 in a programme involving five BP operated fields in the North Sea . This paper reviews some of the techniques used and the main findings of the programme. In particular it focuses on the relationship between composition and toxicity and the likely fate of organic compounds present in produced water. Naturally occurring, short chain carboxylic acids have been found to represent the major proportion of the organic material in the produced water in most cases. Of more interest from the environmental point of view are the variations in trace amounts of more toxic aromatic components benzene, toluene, ethylbenzene, xylene (BTEX), phenols, and polyaromatic hydrocarbons (PAH) in different produced waters. The presence of naturally occurring organic constituents, particularly phenols and other aromatics, appeared to have a strong influence on the ecotoxicity of produced waters. The acute toxicity (EC_{50}) of produced water from oil production platforms was in the range 5-25% by volume. Discharges from gas platforms had a somewhat greater acute toxicity (EC_{50} 2% by volume).

Dispersion modelling of produced water discharges has indicated that the minimum dilution of the plume is likely to exceed 700 fold within 500 m of the platform. A minimum dilution of over 10,000 times was predicted by the time the plume has travelled about 20 km. A high proportion of the organic material in produced water samples has been observed to be highly biodegradable. Over 99% of phenols and PAH were broken down in standard laboratory biodegradation tests and BTEX was reduced below the limit of detection (<0.5ppb). A corresponding decrease in toxicity was observed. The most recent studies have involved attempts to validate the dilution model in the field.

1. Volatile fatty acids usually represented the major proportion of organic material in produced water from the fields studied, being present in the range of 13 - 380 mg/l.
2. The concentration of phenols correlated well with the acute toxicity of the produced waters, (98% confidence).
3. In general, the organic constituents of produced water were readily biodegradable. Over 99% of the concentration of total phenols, for example, was biodegraded within 8 days.
4. Dilution of discharged produced water occurs most rapidly within 100m of the platform. The behaviour of the discharge plume and the extent of dilution is influenced by a number of factors, including the rate of discharge, the density of the produced water as well as tide and wind conditions.

ACKNOWLEDGMENTS

The authors would like to thank BP Exploration for their enthusiastic support of the environmental programme and their kind permission to publish this paper.

REFERENCES

Brendehaug, J., Johnsen, S., Byrne, K.H., Gjøse, A.L., Eide, T.H. and Aamot, E., 1992 Toxicity Testing and Chemical Characterization of Produced Water - a Preliminary Study. In, J.P. Ray and F.R. Engelhardt (eds.), Produced Water. Technological/Environmental Issues and Solutions. Environmental Science Research Volume 46. Proceedings of the 1992 International Produced Water Symposium, Feb. 4-7, 1992 San Diego, California. Plenum Press, pp 245 - 266.

Jacobs, R.P.W.M., Grant, R.O.H., Kwant, J., Marquenie, J.M., 1992. The Composition of Produced water from Shell Operated Oil and Gas Production in the North Sea. In, J.P. Ray and F.R. Engelhardt (eds.), Produced Water. Technological/Environmental Issues and Solutions. Environmental Science Research Volume 46. Proceedings of the 1992 International Produced Water Symposium, Feb. 4-7, 1992 San Diego, California. Plenum Press, pp 13 - 21.

Murray-Smith, R., Gore,D., Flynn, S.A., Vance, I. & Stagg, R., 1996. Development and Appraisal of a Particle Tracking Model for the Dispersion of Produced Water Discharged from an Oil Production Platform in the North Sea. [This symposium]

Stagg, R., Gore, D.J., Whale, G.F., Kirby, M.K., Blackburn, M. Bifield, Vance, I., Flynn, S.A. & Grogan, W.C. 1996 Field Evaluation of Acute Toxic Effects of Produced Water Discharges from North Sea Oil Production Platforms. [This symposium]

Staples, T.B., Lewis, R.E., Owens, N.J.P., Woodward, E.M.S., Rees, A.P. and Plummer, D.H., 1993. A Modelling Study of the Contribution of Riverine and Anthropogenic Inputs of Ammonium to a North Sea Coastal Area. Estuarine, Coastal and Shelf Science 37, 497 - 507.

The Oil Industry International Exploration and Production Forum (E&P Forum), 1994. North Sea Produced Water: Fate and Effects in the Marine Environment. Report No. 2.62/204, May 1994. E&P Forum, 25-28 Old Burlington Street, London W1X 1LB.

Tibbetts, P.J.C., Buchanan, I.T., Gawel, L.J. and Large, R., 1992. A Comprehensive Determination of Produced Water Composition. In, J.P. Ray and F.R. Engelhardt (eds.), Produced Water. Technological/Environmental Issues and Solutions. Environmental Science Research Volume 46. Proceedings of the 1992 International Produced Water Symposium, Feb. 4-7, 1992 San Diego, California. Plenum Press, pp 97 - 112.

FIELD EVALUATION OF TOXIC EFFECTS AND DISPERSION OF PRODUCED WATER DISCHARGES FROM NORTH SEA OIL PLATFORMS

Implications for Monitoring Acute Impacts in the Environment

R. Stagg,[1] D. J. Gore,[2] G. F. Whale,[3] M. F. Kirby,[2] M. Blackburn,[2] S. Bifield,[2] A. D. McIntosh,[1] I. Vance,[4] S. A. Flynn,[5] and A. Foster[6]

[1] Scottish Office Agriculture and Fisheries Department
Aberdeen AB9 8DB
[2] Ministry of Agriculture, Fisheries and Food
Burnham-on-Crouch, Essex CMO 8HA
[3] Shell Research Limited
Sittingbourne Research Centre, Kent ME9 8AG
[4] BP Research and Engineering
Sunbury-on-Thames, Middlesex TW16 7LN
[5] BP Exploration Europe
Farburn Industrial Estate, Dyce, Aberdeen AB2 OPB
[6] Marathon Oil UK Limited
Anderson Drive, Aberdeen AB2 4AZ

INTRODUCTION

Produced water is co-produced with oil or gas and is usually separated on the platform and discharged as a waste stream into the sea. It originates from water trapped in the geological structure from which the oil/gas is being recovered and also from water injected into the formation to maintain reservoir pressure. In 1993 148 million tonnes of produced water were discharged from oil production platforms in the U.K. sector of the North Sea (HMSO, 1994). The amount of produced water being discharged on the UK continental shelf rose steadily up until 1991/2 (Figure 1). Although the amount of oil and water discharged has reached a plateau since 1990 both are projected to rise as a function of the increased number of production units and because during the life of a producing oil field, the volume of produced water will increase. This occurs mainly as a consequence of the injection of increasing volumes of seawater to maintain reservoir pressure in older fields and the subsequent break-through of this water to the production wells. The volume of produced

Produced Water2: Environmental Issues and Mitigation Technologies
Edited by M. Reed and S. Johnsen, Plenum Press, New York, 1996

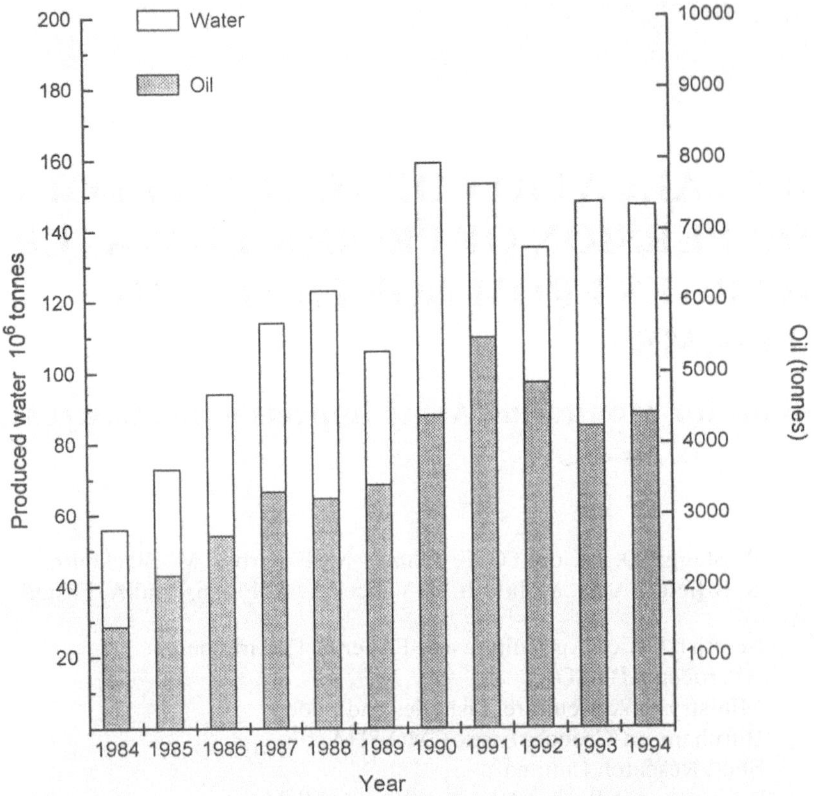

Figure 1. Oil and produced water discharges on the UK continental shelf.

water may exceed that of the oil by as much as ten times and during the later stages of production, the produced water may account for as much as 98% of the total extracted fluids.

In the United Kingdom the discharge of oil-contaminated produced water from offshore installations is permitted by an exemption granted under the Prevention of Oil Pollution Act 1971. A condition of this exemption is that the oil content of the effluent discharged should not exceed a monthly average concentration of 40 parts per million, with no more than 4% of samples exceeding 100 ppm. However, because of the rise in the total volume of produced water the amount of oil discharged has been rising in parallel with the volumes of water discharged (Figure 1). Similar increases in produced water discharges (up to 90 million tonnes) and associated oil are also expected to occur in the Norwegian sector of the North Sea (Brandehaug *et al.*, 1992).

The oil discharged in produced water will comprise a variety of aliphatic, monoaromatic and polyaromatic hydrocarbons (PAH's). In addition to the oil, produced water is usually rich in inorganic constituents, which arise mainly as a result of the contact between the formation water and the geological structure, and also contamination from the chemicals used in the production process. Very few measurements of the concentrations of process chemicals in produced water have been made due mainly to the fact that methods are not developed sufficiently (Tibbets *et al.*, 1992). Measurement of these chemicals in seawater is likely to be even more difficult because of the lower detection limits required. However, several studies have quantified the concentrations of organic acids, BTEX (benzene, toluene, ethylbenzene and xylene) and the hydrocarbon concentrations in produced water (see review by van Hattum *et al.*, 1992).

The acutely toxic effect of produced water to marine species has been widely reported (Somerville et al., 1987; Girling and Streatfield, 1988; Brandehaug *et al.* 1992; Brown *et al*, 1992, Jacobs *et al*, 1992; Strømgren *et al.*, 1995) and planktonic species may be the most sensitive (Slager *et al.*, 1992). On the basis of detailed composition (van Hattum *et al.*, 1992), and the known toxicity of more than 40 of the components in produced water, Slager *et al.* (1992) predicted that aliphatic oil, monoaromatics and heavy metals account for most of the toxicity observed in produced water on the Dutch continental shelf. However the authors acknowledged that the information on the toxicity of some chemicals such as phenols, PAH's, cresols, xylenols and production chemicals was not sufficient to allow a good estimation of the acute effects of these components.

The regulation of discharges from the offshore E&P industry in the Paris Convention area is currently managed by a number of national control schemes. At the third North Sea Conference (The Hague, 7-8 March 1990) PARCOM was requested to develop and adopt a harmonized mandatory control system for the discharge and use of chemicals used offshore. It has also been proposed that this might be achieved by the adoption of a chemical hazard assessment and risk management (CHARM) model to predict the likely environmental risks posed by chemicals used offshore (TNO, 1994). The basis of this model is the comparison of predicted environmental concentration (PEC) with the perceived no effect concentration (NEC) derived through toxicity tests. The model predicts the discharge and environmental partitioning of chemicals from the amount of substance used in the production process and the octanol-water partition coefficient. The steps in the model from initial use to environmental fate is largely based on a series of untested assumptions about the behaviour of chemicals. Output is further confounded by the difficulties in accurately determining the octanol-water partition coefficient for many of the chemical products. It is therefore important that the model is field validated before it is implemented as a regulatory tool and a vital part of this validation will require the determination of the concentrations and toxicity of components of produced water in the waste stream and in the receiving waters around platforms.

The objective of this study was to assess the feasibility of undertaking field validation of the CHARM model for production chemicals. In 1994 a field survey was undertaken at four North Sea oil production platforms in the East Shetland Basin and the Central North Sea to assess the acute toxicity and the concentrations of selected organic chemical components of produced water, at source, and in the water column. Chemicals, from production process, are thought to enter the sea via produced water, however, because of the methodological problems associated with measuring individual production chemicals the concentrations of aliphatic, mono- and poly-aromatic hydrocarbons co-discharged in the effluent were monitored as tracers to measure dispersion of the effluent plume. These were determined in discrete water samples collected at individual sampling locations at various distances along transects of the predicted produced water plumes and also with a fluorimeter (Law, 1984) deployed *in situ*. The toxicity of undiluted produced water and seawater transect samples was assessed using a 48 hour growth inhibition test with a marine unicellular alga, *Skeletonema costatum* and an acute 48 hour immobilisation test with the marine calanoid copepod, *Acartia tonsa*, which are both PARCOM specified toxicity tests (OSPAR, 1995). In addition, an acute 48 hour immobilisation test using the harpacticoid marine copepod, *Tisbe battagliai* and a rapid 15 minute luminescence inhibition test based on the marine bacterium, *Photobacterium phosphoreum,* (Microtox^{TM}) were conducted. *T. battagliai* bioassays were also done on solvent-extracted concentrates of seawater.

The results from this study are discussed in the context of the implications for verifying the CHARM model.

Table 1. Operational conditions at the time of sampling (April 1994)

	Brent	Brae	Forties	Clyde
Latitude	61° 07.57'N	58° 41.33'N	57° 41.33'N	56° 43.38'N
Longitude	01° 44.10'E	01° 16.54'E	00° 50.50'E	02° 17.19'E
Density g ml^{-1}	1.018	1.039	1.039	1.080
Salinity ppt	24	30	34	71
Temperature °C	40	69	76	72
Water Discharge m^3 day^{-1}	17222	4335	8583	9062
Oil Production m^3 day^{-1}	10800	3791	4284	4134
Oil in water mg l^{-1}	7 - 8	68	50	40

MATERIALS AND METHODS

Sampling Programme

The studies were conducted from R.V. DANA in April 1994 around Brent Delta, Brae Alpha, Forties Charlie and Clyde oil platforms on the UK continental shelf. The platforms selected are considered to be representative of different areas of the Northern and Central North Sea and of the different types of produced water discharged. The location of these platforms and the chemical and physical characteristics of the produced waters discharged at the time of the survey are given in Table 1. Produced water and seawater samples were collected around each platform and a fluorimeter was deployed from the vessel to take *in situ* measurements of hydrocarbon fluorescence.

Produced Water

Neat produced water samples were taken from the usual U.K. Department of Trade and Industry (DTI) sampling point on the overboard discharge line. The sample point was flushed with produced water for 5 minutes prior to sampling into 2.5 litre glass safebreak bottles and EPA vials.

In situ Continuous Sampling

To define and characterise the effluent plumes at each location, continuous *in situ* measurements of hydrocarbon fluorescence were obtained over the full water depth by undulating a fluorimeter between the sea surface and bottom whilst the ship was on station. Sampling stations were positioned at distances of 100, 250, 500 and 1000 metres from the platform along transects of the predicted produced water plumes (Figure 2). The fluorimeter was also towed at 4 knots on an undulating track behind the vessel across the presumed position of the plumes. Fluorescence (λEx 310 nm λEm 360 nm) was measured using a Chelsea Mark I Aquatraka connected to a battery pack and a modified Chelsea Instruments Aquapack CTD data logger mounted in an instrument frame. The fluorimeter was calibrated with a water soluble fraction (WSF) of Forties crude oil. The oil content of the WSF was determined by extraction of 500 ml of sample by 2 x 50 ml aliquots of dichloromethane. The dichloromethane layer(s) were collected, pooled, dried by the addition of anhydrous sodium sulphate, made up to 100 ml and the hydrocarbon content of the extract determined using a Perkin Elmer LS5 Luminescence Spectrometer itself calibrated against reference Forties crude oil at λEx 310 nm λEm 360 nm. The remaining water soluble fraction was then diluted as appropriate in artificial seawater to give calibrant solutions between 0.5 and 100 $\mu g\ l^{-1}$.

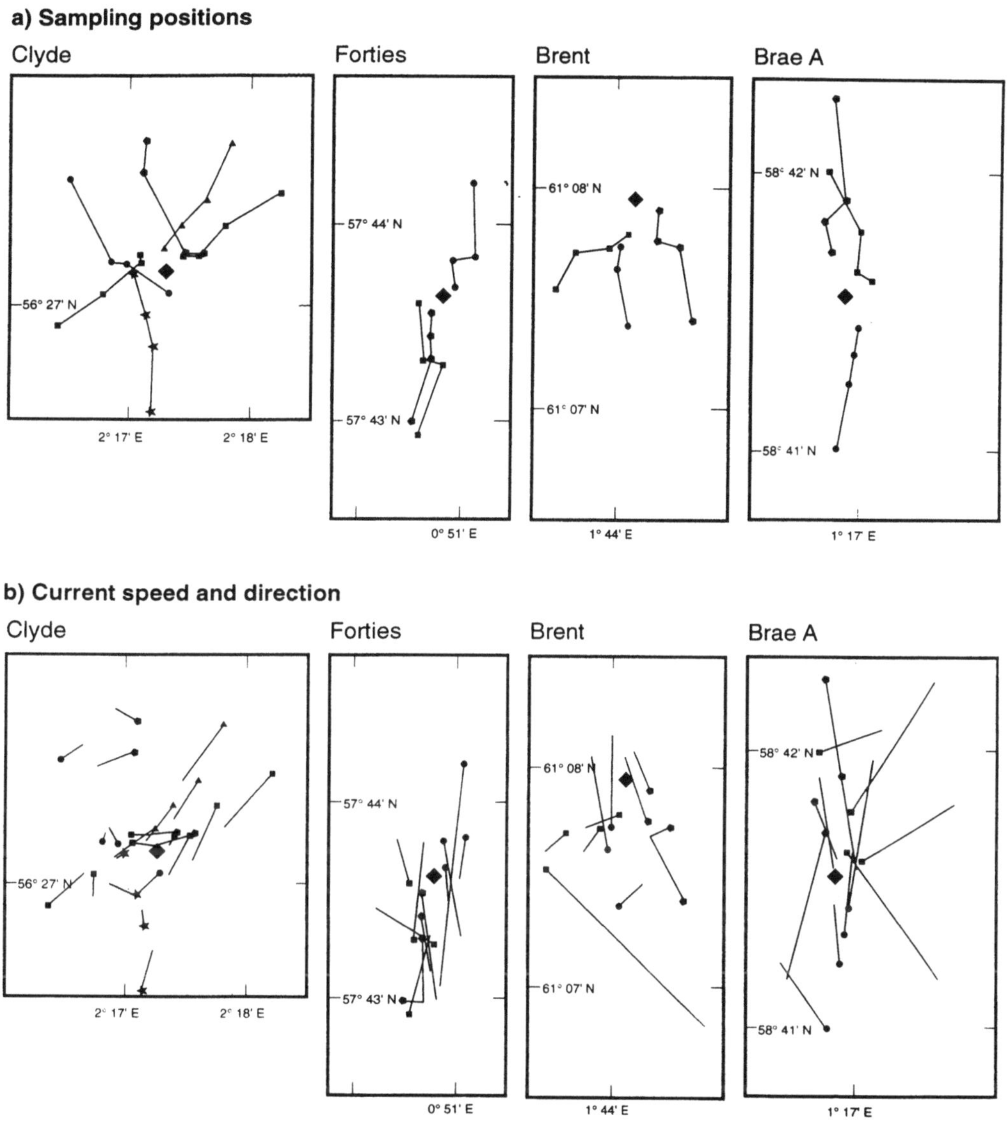

Figure 2. a) Sampling transects around the Clyde, Forties, Brent and Brae platforms. b) The location of the sampling stations in relation to the direction of the tidal stream at the time of sampling. The lines point to the position of the water mass sampled one hour before the sample was taken as determined by the resolution of vectors from the ADCP at the same depth as the sampler.

These were applied to the Chelsea Aquatracka, in the dark in a 100 ml quartz cuvette placed on the excitation window. The limit of detection of this instrument is 0.1 $\mu g\ l^{-1}$. The hydrocarbon data generated by the fluorimeter was used to assess the existence of hydrocarbon gradients around the platform and the existence of discrete plumes. The position of the plume at each platform was predicted using a particle dilution model (Murray-Smith *et al.*, 1996; this volume) and chart tidal data. The orientation of the sampling points in relation to the discharge point was subsequently compared to the speed and direction of the tidal streams calculated from *in situ* measurements of current using an acoustic doppler current profiler

(ADCP). Dilution effects were calculated on the basis of the hydrocarbon concentration measured in the produced water discharge.

Discrete Water Sampling

Seawater samples were taken, along transects of the predicted produced water plumes (see Figure 2) at distances of 100, 250, 500 and 1000 metres from the platform, via a clean diaphragm pump with the inlet held 2-3 m from the side of the vessel. The samples were taken at either 5 or 15 metres to coincide with the centre of the predicted effluent plume (Murray-Smith *et al.*, 1996; this volume). With the exception of the solvent extraction studies, transect samples for toxicity assessment were stored in hexane rinsed 2.5 l amber glass Winchester bottles. Reference 'pristine' seawater samples were also collected from the location 61° 44.6' N and 00° 34.8' E, to serve as field blanks for background chemical concentration levels.

Analytical Chemistry

Analytical grade solvents, purchased from Rathburn Technical Ltd, were used for all analytical work.

With the exception of the Brae produced water samples, which were stored cold (approximately 4 °C) for 14 days prior to processing, the produced water samples and bulk seawater samples were extracted on board the ship and stored at -20°C for up to 60 days prior to analysis. Samples for determination of volatiles were stored in sealed ampoules, intact at 4°C prior to analysis.

Total hydrocarbon levels were determined by extraction of 2 l of sample by 2 x 50 ml aliquots of dichloromethane in a separating funnel, the dichloromethane layer(s) were collected, pooled, dried by the addition of anhydrous sodium sulphate and made up to 100 ml. The hydrocarbon content of the extract was determined using a Perkin Elmer LS5 Luminescence Spectrometer calibrated against reference Forties crude oil at λEx 310 nm λEm 360 nm.

Phenol and methylphenol, methyl naphthalenes, benzenes, toluenes, ethylbenzene, xylene (BTEX), and the alkanes (pentadecane (C15), eicosane (C20) and docosane (C22)) were extracted from aqueous samples as described in Law *et al.* (1991). One litre samples of neat produced water and 2.5 litre samples of seawater were extracted with 2x50 ml aliquots of dichloromethane, after the addition of internal standards (deuterated xylene for the neat produced water samples and xylene for the transect samples). The produced water extracts were then reduced in volume to 10 ml and the seawater extracts were reduced in volume to 1 ml. For determination of methyl napthalenes, phenols and alkanes 1 μl samples were injected onto a Finnigan MAT Incos 50 quadrupole mass spectrometer interfaced to a Hewlett Packard HP5890 Gas Chromatograph (GC/MS) for quantitative analysis, relative to the external standards naphthalene and xylene, by full scan screening. The limit of detection of this instrument was 0.01-0.05 $\mu g\ l^{-1}$ and reproducibility of the GC/MS was found to be within 5% standard deviation for 5 replicates of a phenolic mixture. The concentration of only three alkanes was reported - pentadecane, eicosane and docosane because the profile of these on the aliphatic GC trace from the produced water and the seawater samples was identical. Thirteen volatile organic compounds (most notably benzenes, toluenes, ethylbenzene and xylene) were measured and quantified by purge and trap GC/MS using a standard method described in Dawes and Waldock (1994). This method uses similar instrumentation to that above except that the GC was fitted with a Chrompack purge and trap autosampler. The limit of detection for this analysis was 0.1$\mu g\ l^{-1}$ and 5 replicates of each standard of BTEX compounds gave a coefficient of variance of between 0.9 and 21%.

The concentrations of 6 PAHs were measured in the neat produced water samples from the Clyde, Forties and Brent platforms. Dichloromethane extracts were reduced in volume to 200 µl and reconstituted in 1 ml acetone. 50 µl were analysed by HPLC using a Chrompack PAH analyser equipped with a fluorescence detector and quantified relative to a deuterated anthracene internal standard. The limit of detection for this instrument was 100 ng l^{-1} with a coefficient of variance for each PAH (n=5) of between 0.9 and 5%.

Procedural and solvent blanks were run alongside each batch of samples and were always below the detection limits of the instruments.

Bioassay

Bioassays were conducted on board within 48 hours of collection except for those samples from the Brae platform which were stored at approximately 4 °C for between 7 and 14 days prior to testing. All bioassays were conducted using standard methods and met the validity criteria specified in the test protocols. Dilutions were made using filtered (10µm) reference seawater obtained from the River Crouch (MAFF, Burnham-on-Crouch).

The *Skeletonema costatum* growth inhibition tests were conducted in accordance with the method approved by OSPAR and based on the ISO 147/SC5/WG5N protocol. Dilutions of produced water were made in nutrient enriched, reference seawater with inhibitory effects on average algal growth rate being measured over 48 hours.

Acartia tonsa immobilisation tests were conducted following the general principles of the method approved by OSPAR and based on the ISO TC147/SC5/WG2 protocol. On board bioassays were conducted in 20 ml sealed perspex pots with a nominal 10 animals per container and a minimum of two replicates per dilution or seawater sample. On board the cultures were maintained semi-statically in 10 l glass culture vessels with replenishment of approximately half the seawater containing algal food every 48 hours.

Microtox™ (*Photobacterium phosphoreum*) were supplied as small freeze dried batches of reagent by Microbics (U.K.) Ltd., Hitchin, Hertfordshire. The Microtox™ tests were conducted on board following the guidelines given in the instruction manual supplied with the Microbics Model 500 Microtox™ toxicity analyser (Microbics 1991).

Tisbe battagliai tests were an adaptation of that described by Hutchinson and Williams (1989). First and second stage (4-6 day old) *Tisbe* copepodites were used in the test. At each concentration five copepodites were exposed in each of four 5 ml replicates set out in 12-well sterile culture dishes. Mortality was checked microscopically after 24 and 48 hours and organisms were classified as being dead if no appendage movement could be detected. Bulk water samples were solvent extracted resulting in a 1000 fold nominal concentration of the contaminants. The procedure was chosen because it is robust enough to be conducted at sea (Thain and Kirby, 1994). 40 l of seawater were placed into a modified alloy milk churn; 500 ml hexane was added and the container sealed with a lid incorporating a P.T.F.E. venting valve. The churn was shaken at approximately 100 rpm for 20 minutes, allowed to stand for 15-20 minutes and the hexane extract was removed by displacement with seawater. The extract was dried by the addition of anhydrous sodium sulphate, combined with two hexane washes of the residual sodium sulphate, reduced in volume to 5-10 ml by rotary evaporation at 40°C and stored at -20°C until bioassays were conducted. The hexane was substituted with 4 ml of acetone and the samples reduced to 40 µl before being re-constituted in 40 ml of clean, filtered (10 µm) seawater. 20 ml of this stock solution was used as the highest concentration and a range of dilutions prepared by serial dilution with clean seawater. The results of these bioassays were calculated as 48 hour LC_{50} (median lethal concentration) and expressed as the concentration factor of the original seawater sample.

For the produced water samples and where the data permitted the median effect concentration (EC_{50}) or LC_{50} values and 95% confidence limits were calculated using probit analysis (Finney, 1971; Litchfield and Wilcoxon, 1959). Where there was no intermediate

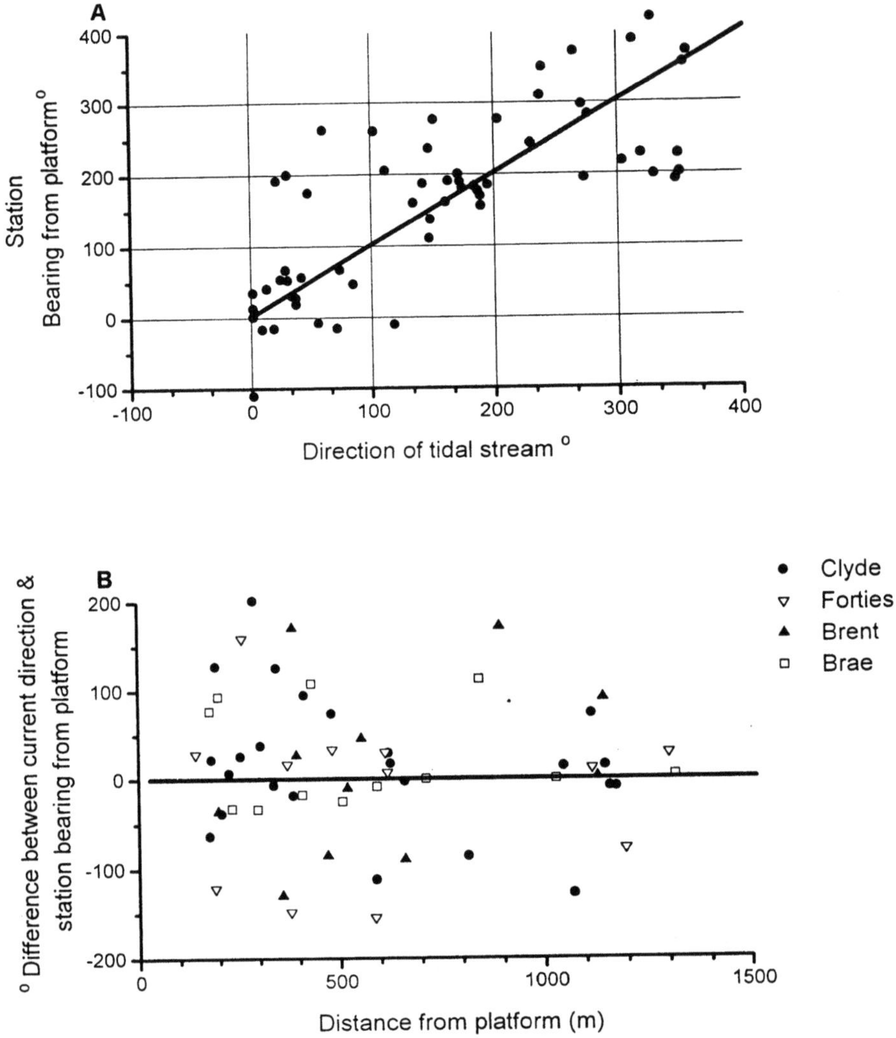

Figure 3. a) Orientation of sample position to each platform compared to the direction of the tidal stream. b) Difference between the current direction and the station bearing from the platform in relation to the distance of the sample point from the platform.

response values the EC_{50} or LC_{50} were calculated using moving average angle analysis (U.S. Environmental Protection Agency (EPA), 1985). The results of the seawater bioassays were expressed as the total adverse effect for each sample.

RESULTS

Sampling Positions

The transect sampling positions occupied in this study where chosen to coincide with the most stable tidal plumes predicted by the particle tracking models developed at the

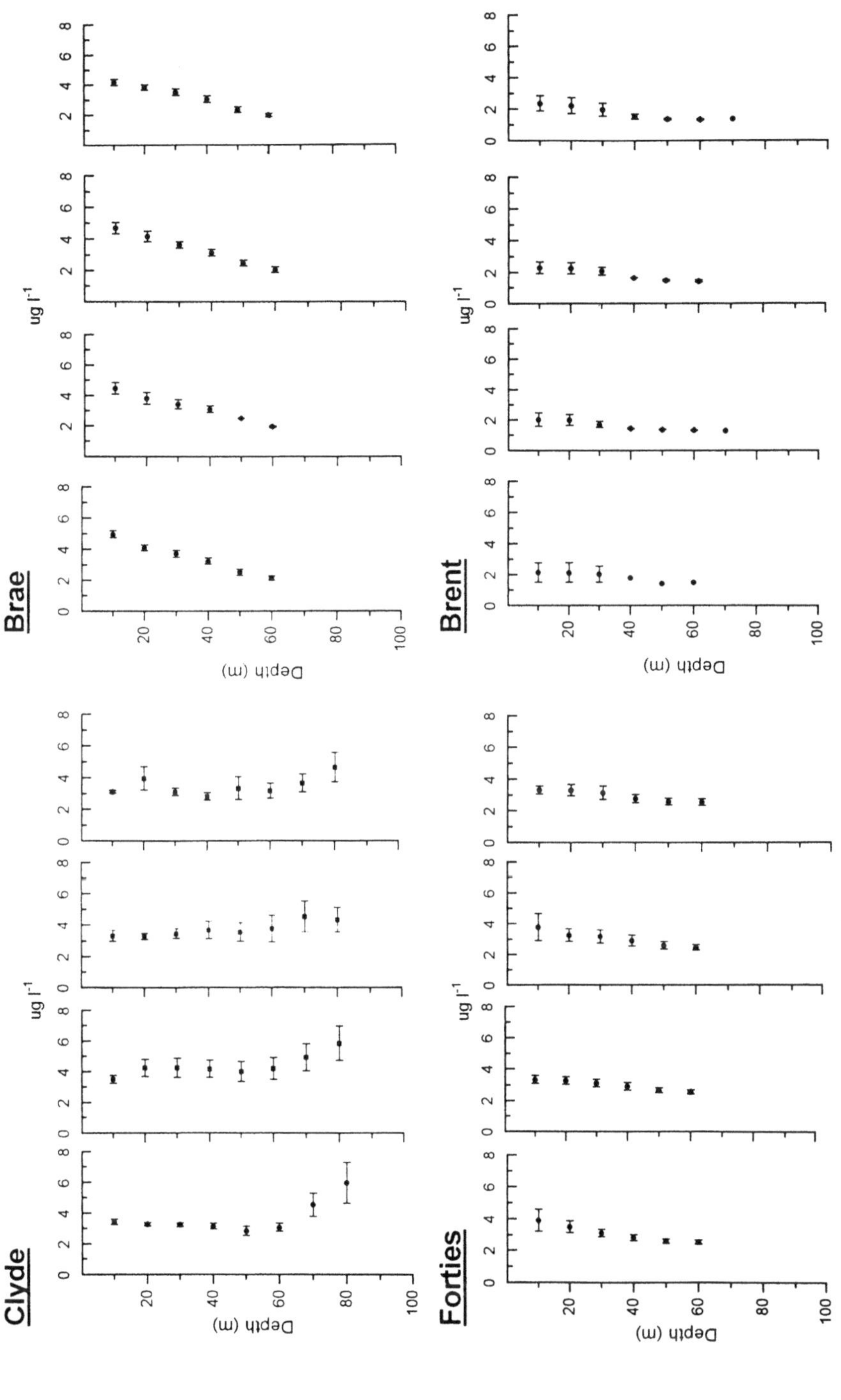

Figure 4. Levels of oil on transects from the Clyde, Forties, Brent and Brae platforms in the North sea determined by *in situ* fluorimetery.

Zeneca Laboratory (Murray-Smith *et al.*, 1995). The origin of the water mass sampled at each station is indicated in Figure 2b and shows that on many occasions the water sampled would not have passed close to the platform in the hour preceding the sample being taken. Figure 3a compares the orientation of these stations with the actual current direction determined by the ADCP at the time of sampling and shows that although there is an excellent correlation between the bearing of the sampling points to the platform and the direction of the current flow there is a considerable variability in the data. Expression of these data as the difference between the observed current direction and the orientation of the sample points to the platform (Figure 3b) showed that only a small proportion of the sampling stations were within ± 5° of the true current stream from the platform particularly in those stations closest to the platforms.

In situ Continuous Sampling

Hydrocarbon levels of between 1 and 8 μg l^{-1} were measured around the oil platforms studied using a fluorimeter either deployed *in situ* at fixed stations or towed from the ship The results for the fixed station vertical deployments on a transect away from each platform in the direction of the predicted effluent plume are shown in Figure 4. The data showed no hydrocarbon gradient between 100 and 1000 m although at Brent, Brae and Forties there were higher levels at the surface compared to deeper waters. In contrast at Clyde which has the most dense produced water (Table 1) there is some indication of higher and more variable hydrocarbon levels closer to the bottom. The towed fluorimeter detected elevated hydrocarbons only occasionally as peaks, and this was most obvious at Brae and Forties where there is a narrow tidal ellipse and thus a narrower less dispersed effluent plume. Some typical traces are given for the Forties platform in Figure 5 and show that effluent plumes exist where the hydrocarbon concentration may reach double the background concentration. However, these peaks were seen to represent very narrow bands (50-100 m) and were only found within 300 m of the effluent discharge.

Chemical Analyses

The levels of oil measured in the produced water samples (Table 1) were 40, 50, 7, and 68 mg l^{-1} for Clyde, Forties, Brent and Brae respectively. Detectable concentrations of phenols, methyl phenols, methyl naphthalenes, alkanes (Σ pentadecane, eicosane and docosane) and the volatile compounds benzene, toluene, ethyl benzene, xylene and 1,2,4-trimethylbenzene were determined in the neat produced water by GC/MS (Table 2). The concentrations of these hydrocarbons within the four produced water streams were surprisingly similar with the biggest differences being noted for alkanes and the benzenes. The concentrations of PAH's in the produced waters, determined by HPLC, were also similar and reflect the spectrum of aromatic compounds found in fresh, un-degraded oil in that lighter, more water soluble compounds, such as naphthalene and phenanthrene dominated.

Measurement of hydrocarbon fluorescence in discrete water samples show very variable results. Generally, sample concentrations were less than 5 μg l^{-1} (ppb) with occasional samples being greater than 10 μg l^{-1}; at none of the platforms was there a measurable gradient away from the platform (Figure 6). The concentrations of phenol, methylphenol and all volatile organic chemicals, including BTEX, in the transect samples at each platform were below the detection limits (0.01-0.05 μg l^{-1} for phenol and methyl phenol and 0.1 μg l^{-1} for BTEX). However, trace levels (<1 μg l^{-1}) of methylnaphthalenes and alkanes were detected in the seawater samples from around the platforms. Methylnaphthalene concentrations were not significantly above background levels but alkane concentrations in several of the samples were an order of magnitude above background. The

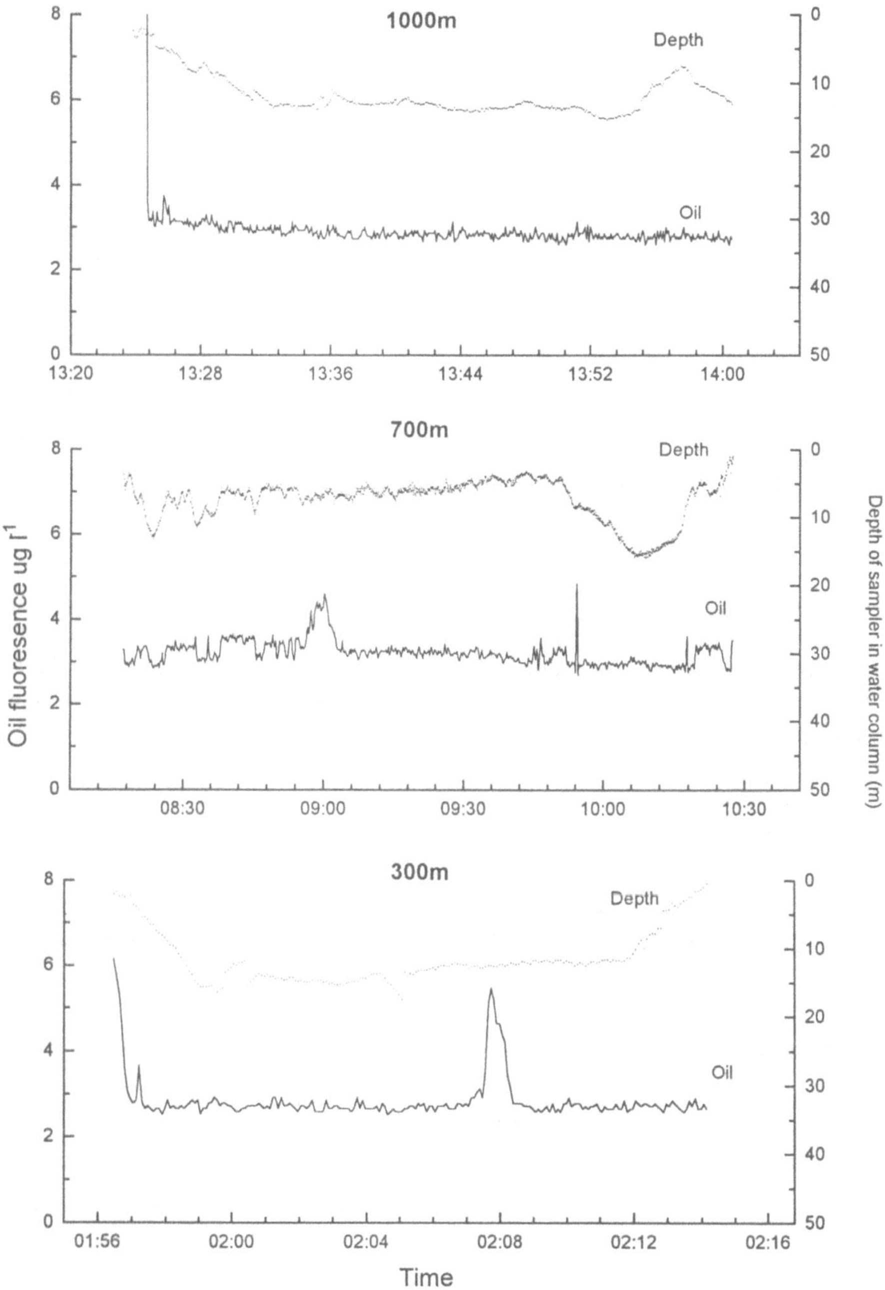

Figure 5. Typical traces of hydrocarbon fluoresence (ug l^{-1} of Forties equivalents) in seawater measured with a towed *in situ* fluorimeter at the Forties platform together with the depth of the sampler in the water column.

Table 2. Analytical measurements (μg l^{-1}) of neat produced water

	Platform			
Analyses	Clyde	Forties Charlie	Brent Delta	Brae Alpha
Phenol	1240	1440	1470	1270
Methyl Phenols	890	1560	1160	2530
Methyl Naphthalenes	1510	1700	770	1210
1,2,4-Trimethylbenzene	741	360	717	607
Benzene	4154	1417	6853	5291
Toluene	2772	2174	2947	2391
Ethyl Benzene	902	425	1013	1503
Xylene	2926	675	1791	3411
Alkanes[1]	2170	540	920	1440
Naphthalene[2]	161	169	172	ns
Phenanthrene[2]	44	15	50	ns
Fluorene[2]	34	14	51	ns
Acenapthene[2]	4.8	2.4	7.8	ns
Anthracene[2]	0.2	0.5	0.7	ns
Fluoranthene[2]	30	10	4.9	ns

ns: no sample
[1] Sum of pentadecane, eicosane and docosane alkanes
[2] Analysed by HPLC, all other determinands analysed by GC or GC-MS

worst case concentrations for these two analytes are given in (Table 3) for each sampling location. Pentadecane, eicosane and docosane were selected for aliphatic analysis because the profile of these three alkanes in the seawater samples most closely matched those in the produced water samples. Despite this the alkane concentrations in the water samples taken along each platform transect did not show a clear gradient of concentration but the distribution was patchy and discontinuous.

Bioassays

The results of the neat produced water toxicity assessments expressed as percentage of produced water (i.e. equivalent to g 100 ml^{-1}) are summarised in Figure 7. Undiluted produced water samples from each production platform were found to be acutely toxic to

Table 3. Worst case analytical measurements (ng l^{-1}) of seawater sampled along transects of the predicted produced water plume

Analyses	Sample	Clyde	Forties Charlie	Brent Delta	Brae Alpha
Methylnaphthalenes	Neat PW	1510	1700	770	1210
	100 m	<0.05	<0.04	0.05	0.02
	250 m	0.09	0.11	0.02	0.12
	500 m	0.03	<0.1	<0.04	<0.05
	1000 m	0.05	<0.04	0.12	0.04
Alkanes[1]	Neat PW	2170	540	920	1440
	100 m	<0.10	0.03	0.27	0.06
	250 m	0.18	0.09	0.15	0.09
	500 m	0.44	0.05	0.12	0.09
	1000 m	0.24	0.05	0.22	0.63

[1] Sum of pentadecane, eicosane and docosane.

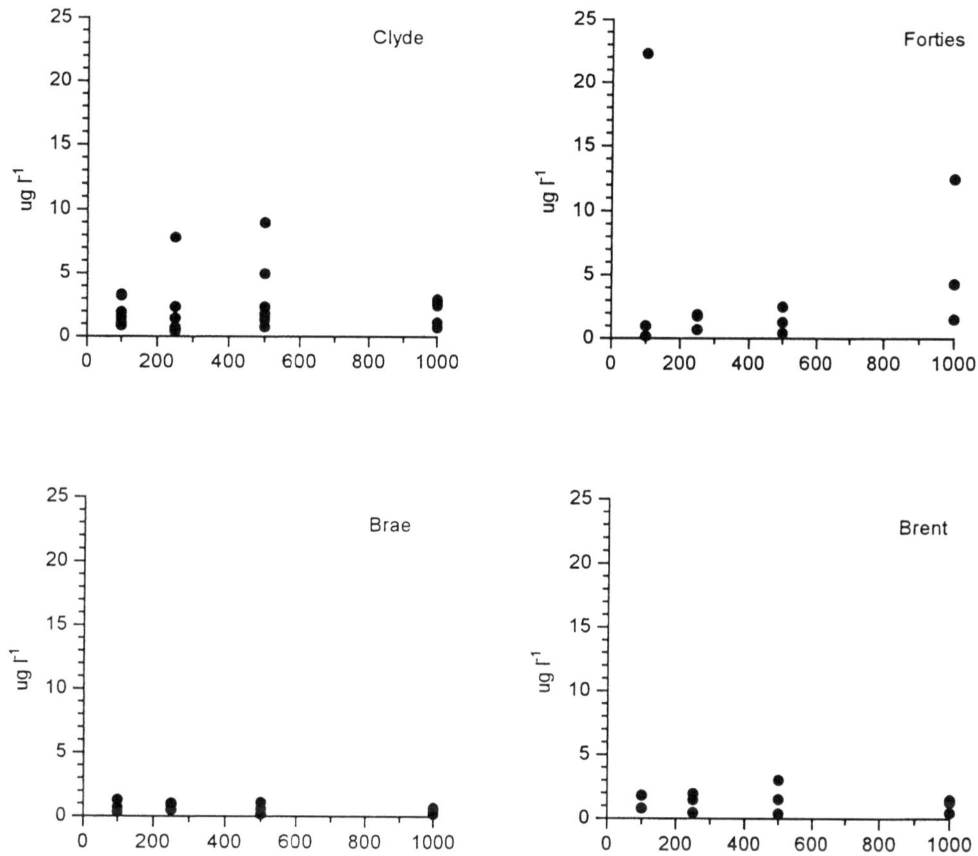

Figure 6. Hydrocarbon concentrations determined by UV fluorescence in discrete water samples from around the Clyde, Forties, Brent and Brae platforms in the North Sea.

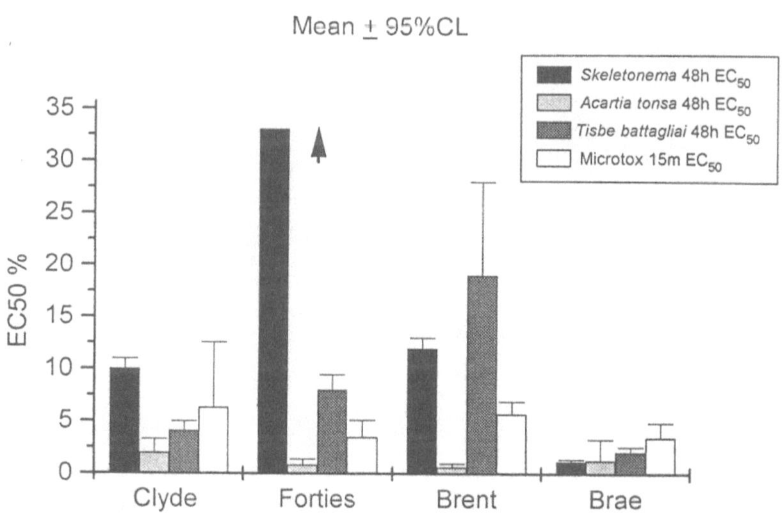

Figure 7. Comparison of the acute toxicity of neat produced water from Clyde, Forties, Brent and Brae platforms to four marine species.

S. costatum, *A. tonsa*, *T. battagliai* and Microtox^{TM}. The toxicity to *A. tonsa* varied by less than a factor of 3.2 ranging from 2% for Clyde to 0.63% for Brent Delta. For Microtox^{TM} the variation in toxicity between the platforms was also low and ranged from 6.3% for Clyde to 3.5% for Brae Alpha and Forties Charlie. Greater differences between the platforms was seen with the *T. battagliai* and *S. costatum* tests but there was no clear pattern of toxicity between the platforms although produced water from the Brae Alpha platform was the most toxic to both (Figure 7). The greatest variance was seen with *S. costatum* where a factor of > 27 was observed in the toxicity of the produced water ranging from >33 % for the Forties Charlie platform to 1.2% for samples from the Brae Alpha platform. For *T. battagliai* the toxicity varied by a factor of 9 ranging from 2.1% for Brae Alpha to 19% for Brent Delta produced water. The produced water samples from each platform did not exhibit a consistent pattern of toxicity to the range of test species used. For example, the produced water from the Brent Delta platform showed the greatest toxicity to *A. tonsa* but the smallest toxicity to

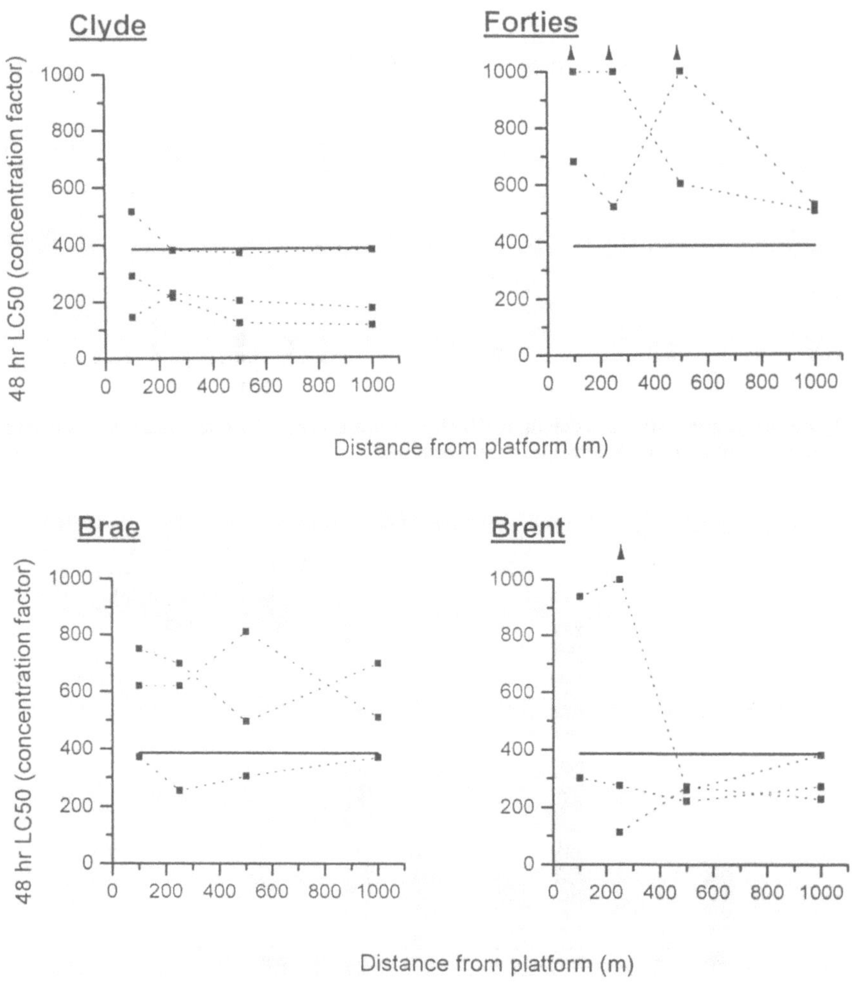

Figure 8. Acute toxicity of concentrated extracts of water samples from around the Clyde, Forties, Brent and Brae platforms to *Tisbe battagliai*. Seawater was concentrated 1000 fold by extraction with hexane and redissolving the extract in a small volume of seawater.

T. battagliai. Similarly, the Forties Charlie produced water showed the second highest toxicity to *A. tonsa* but the lowest toxicity to *S. costatum*.

In the seawater samples taken from around the platform no significant growth inhibitory effects where found with *S. costatum*. Some mortality was observed in *T. battagliai* exposed to un-extracted seawater although, this could not be correlated to distance from platform nor to the toxicity subsequently observed in solvent-extracted samples tested against the same species. Some toxicity was also found to *A. tonsa* and Microtox™ but this was subsequently attributed to hexane contamination of the sample bottles. Experiments showed that low levels of hexane (approximately 0.07 ml l^{-1}) in toxicity tests will result in false positive toxicity being recorded. The *A. tonsa* and Microtox™ tests were the only species demonstrating this sensitivity because of the test designs. Microtox™ was conducted rapidly with little time for the solvent to evaporate whereas *A. tonsa* tests were conducted in sealed vessels which acted as solvent traps. Taking these effects into account no distance-related or consistent toxic effects were seen with any of the transect bioassays with the seawater samples.

The results of the *T.battagliai* bulk solvent-extraction tests are summarised in Figure 8 and also show that no significant distance related toxicity was observed. In some cases the results obtained from the transect samples were less toxic than the background level determined for the reference seawater samples. However, transects 2 and 3 at both the Clyde and Brent platforms appear to show elevated toxicity, albeit, not distance related. It should also be noted that the 48 hour EC50 of 390 obtained for the reference site implies that there was significant toxicity associated with this sample. Chemical analysis of the seawater sample obtained at this site showed the presence of trace amounts of methyl-naphthalene and alkanes which would indicate chemical contamination of the sample.

DISCUSSION

The characterisation of the chemical nature of the produced waters investigated in this study was restricted to hydrocarbon analysis. The results showed that the aromatic hydrocarbon residues are dominated by monoaromatic (BTEX and phenols) and lighter polyaromatic residues. These are considered to be the most acutely toxic components of the organic contaminants in produced waters (Slager *et al.*, 1992) although heavy metals may also contribute to the observed toxicity (Slager *et al.*, 1992; Strømgren *et al.*, 1995). Although the total hydrocarbon concentrations (Table 1) of the various produced waters varied by an order of magnitude there was little difference in the specific aromatic or aliphatic components measured (Table 2).

In this study the acute toxicity of produced water from four oil production platforms has been measured using four marine species in laboratory toxicity tests. The results show that the whole neat produced water samples are acutely toxic to *S. costata*, *A. tonsa, T. battagliai* and Microtox™ and that the LC_{50} of the produced waters used was between 30 and 0.63% depending on the species used in the test. A recent review on the acute toxicity of produced waters (Slager *et al.*, 1992) reported that acute toxicity measured as either the LC_{50} or EC_{50} may vary over three orders of magnitude depending on the test species and the produced water used. *A. tonsa* (48 h LC_{50} 2.0 - 0.63%) was the most sensitive of the species used in this study and the toxicities were not dissimilar to those reported by Girling (1989) of between 4.5 and 2.6% using Dunlin A produced water. One factor which may have made this test appear to be more sensitive than the others is that the *A. tonsa* tests were conducted in sealed pots. As a result the volatile organic chemicals, such as BTEX, which are present in the neat produced water samples may have been retained to contribute to the toxic effect on the test organisms. With the exception of the Brae Alpha produced water, *S. costatum* was the least sensitive of all the test species. The *S. costatum* test also produced the most variable results (48 h LC_{50} >33 - 1.2%) and is similar to that reported previously (Brandehaug *et al.*, 1992; Strømgren *et al.*, 1995). Likewise the results for the Microtox^{TM} tests are comparable, both in terms of actual toxicity and variability, to those found by Brandehaug *et al.* (1992) and Somerville *et al.* (1987) in their studies of produced water from the Gullfaks A and the Brent platform respectively. For example, the 15 min EC_{50} values for Microtox^{tm} for the Brent Delta production water were between 6.2 and 4.3% compared with values of between 5 and 6% reported by Somerville *et al.* (1987) for the same platform. The consistency of Microtox^{TM} results combined with relative sensitivity in comparison with other test methods supports the views of Whale (1994) that this test could be used to monitor the toxicity of produced water discharges.

It was not possible to identify the primary chemical constituent responsible for the observed toxicity. The composition of produced water is unique to each production system (Stephenson, 1991) and is affected by the type of oil, the geology and geochemistry of the reservoir structure and the individual chemical treatment. In the light of this it was perhaps surprising to find such a uniform spectrum of contaminants in the produced waters measured, especially given the observed differences in toxicity. A broad range of toxicants are present in produced water at low levels, including heavy metals which were not measured in this study and which have been targeted as contributing to acute toxicity (Slager *et al.*, 1992; Strømgren *et al.*, 1995). It is possible that this may explain the apparent contradiction between the observed differences in toxicity and the similarity in the observed concentrations of the contaminants.

The fluorimetry data showed that background levels of hydrocarbons around the oil platforms studied were significantly higher than the levels reported in pristine areas remote from anthropogenic influence. In a separate study of the northern North Sea (Stagg *et al.*, 1996) hydrocarbon concentrations of between 0.1 and 0.5 $\mu g\ l^{-1}$ were measured to the north of the East Shetland Basin in areas remote from oil exploration and development using identical instrumentation. These levels contrast with the concentrations reported here of between 2 and 4 $\mu g\ l^{-1}$ either in vertical profiles at fixed stations around the platforms or in undulating tows behind the ship. The source of this elevated background concentration is not clear at present (Stagg *et al.*, 1996). The towed fluorimeter detected elevated hydrocarbons only occasionally as peaks, and this was most obvious at Brae and Forties where there is a narrow tidal ellipse and thus a narrower less dispersed effluent plume. Some typical traces are given for the Forties platform in Figure 5 and show that effluent plumes exist where the hydrocarbon concentration may reach double the background concentration. However, these peaks were seen to represent very narrow bands (50-100 m) and were only found within 300 m of the effluent discharge. Generally the data shows that plumes are difficult to track, they are often fragmented into smaller discrete ribbons rather than a discrete plume, and are difficult to distinguish from the elevated background concentrations.

The fixed station sampling programme was designed to obtain discrete water samples from within the effluent plumes as they are discharged from the platform. Models of the discharges (Somerville *et al.*, 1987; Murray-Smith *et al.*, 1996) show the progressive dilution of plumes as they extend from the platform. However, the stability of plumes is primarily dependant on the tidal flows and in this study, model predictions were used to focus the sampling positions and times at the point were the plumes were most stable. Retrospective analysis of the speed and direction of the tidal flow at the time of sampling (Figure 2b and 3) obtained from ADCP data showed that on most occasions the samples would not have been taken from within the plume but that some of the samples were within ± 5° of the target. Furthermore, Murray-Smith *et al.* (1996) showed that plumes are a transient occurence at any particular location and may only occur at a fixed station for between 20 and 30 minutes. This again makes sampling extremely difficult because of the timescale required to position a ship and deploy sampling equipment. Therefore, the data indicate that there are many technical difficulties associated with sampling the produced water plume and that it is extremely difficult to locate the plume and to sample within it with accuracy and within the timescale imposed by the tidal regime.

The data on specific chemicals measured in seawater showed that the levels of phenol, methyl phenols, methyl napthalenes, alkanes and all volatile organic compounds (all less than 0.5 $\mu g\ l^{-1}$) approached the detection limits of the methods and were present at much lower concentrations than in the produced water effluents. At none of the platforms was there any difference in concentrations between the 100 and 1000 m stations. Likewise no acute toxicity could be detected in seawater samples taken from around the platforms using *S. costata, T. battagliai, A. tonsa* and Microtox™ bioassays. The lack of observed acutely toxic effects in the majority of the seawater transect and solvent extracted samples implies (depending upon the accuracy of sampling within the produced water plume) that a rapid dilution /dispersion of those chemical constituents responsible for the toxicological response in neat produced water occurs once it is released into the marine environment. However the absence of acute effects does not necessarily indicate that there will be no chronic or sublethal impacts from produced water discharges.

Implications for CHARM Validation

Validation of the CHARM model could not be achieved with the results of this study. Due to analytical constraints it was not possible to obtain qualitative or quantitative measurements of the production chemicals known to have been used on the day of sampling in the waste stream. Nor was it possible to conduct a CHARM hazard or risk assessment for the individual chemicals used at each platform because of the lack of ecotoxicological and appropriate chemico-physical (Log P_{ow}) data essential for running the model. This emphasises the need for all new and existing offshore chemicals to comply with the data requirements specified by OSPAR. Nevertheless there were valuable lessons to be learnt about the feasibility of sampling the water column around oil platforms and the methods to be used in any subsequent validation study.

It is unlikely that the use of discrete water sampling devices allied to the current analytical capability will yield interpretable data on the concentrations of chemicals in the marine environment around oil platforms. Water samples were not able to distinguish gradients in concentration for any of the chemicals measured in this study. This indicates both a patchy or discontinuous distribution in the water column and that many of the water samples were not taken within the main produced water plume. Even using continuous *in situ* measurements it was difficult to locate and maintain station in the plume. The sizes of the plumes were also small in relation to the length of vessel used and it is unlikely that a smaller vessel could be used in the areas of the North Sea sampled. Deployment of a towed sampler which covered a much larger sampling area than sampling at fixed stations was able to detect plumes at some locations (e.g. Forties Fig 5). Since the discharge level was 40 mg $l^{-1}$and the maximum concentration observed was of the order of 8 $\mu g\ l^{-1}$ this represents a dilution of about 9-10,000. Likewise the dilution of alkanes and methylnapthalene at 500m from the platforms, calculated from the data in Table 3, were 5,000 - 16,000 and 17,000 - 50,000 respectively. These values contrast with the dilution of 1000 at 500 m which has been built into the CHARM model.

The bioassay data were similarly constrained by the fact that they are reliant on discrete water sampling. In addition there is much evidence which suggests that the acute toxicity of produced water is dominated by the concentrations of aliphatic oil, monoaromatics and heavy metals (Slager *et al.*, 1992; Strømgren *et al.*, 1995) and effects of chemicals in acute studies will therefore be difficult to detect.

SUMMARY AND CONCLUSIONS

A field survey was undertaken in the East Shetland Basin and the Central North Sea to assess the feasibility of sampling and measuring the toxicity and dispersion of produced water. Specific objectives of the survey were to measure the concentrations of selected components and acute biological effects in an attempt to provide field validation for the CHARM (Chemical Hazard Assessment and Risk Management) model which has been developed to evaluate the potential environmental risks posed by chemicals used offshore. Extensive work was undertaken close to platforms and in the wider field using hydrocarbon concentrations, determined with an *in situ* fluorimeter as a marker for effluent plumes and as an indicator of regional dispersion. The position of the effluent plume at each platform studied was predicted using a particle dilution model, tidal data and subsequently compared with the true direction of the tidal plume by *in situ* measurements of current using an on board acoustic doppler current profiler (ADCP). The hydrocarbon data generated was used to assess the existence of hydrocarbon gradients around the platform and the existence of discrete plumes. Dilution effects were calculated on the basis of the hydrocarbon concentrations measured in the produced water discharge. The toxicity of produced water and seawater and analytical characterisation of the hydrocarbons in samples obtained at various distances along transects of the predicted produced water plumes was assessed. Four marine organisms (*Skeletonema costatum, Acartia tonsa*, *Tisbe battagliai* and Microtox^{TM}) were used and no toxicity or chemical gradient was found using either neat or concentrated seawater samples. The results emphasise the difficulty in adequately sampling the water column environment around platforms but nevertheless suggest that the effluents are rapidly diluted by tidal movements to levels which are not acutely toxic to marine organisms. The possibility of chronic long term effects cannot be excluded.

Sampling produced water plumes around oil platforms in this study was difficult because of the characteristics of the produced water plumes (narrow, transient and unstable), the patchiness of the water column environment, and the length of time taken to deploy a sampler. From measurements of current at the time of sampling it would appear that most samples were not taken from a point downstream of the tidal flow from the platform.

Neither *in situ* measurements nor point source water samples were able to detect contaminant gradients from the platform. Towed instrumentation detected plumes at only one platform (Forties) with a narrow tidal ellipse.

Elevated levels of alkanes and hydrocarbon fluorescence compared to local background concentrations were measured at some points around platforms. These data indicate that the dilution of produced water streams is of the order of 5 - 10,000 in the immediate vicinity (100 - 1000m) of the platform.

Although acute toxicity was measured in the produced water no acute toxicity was found in water samples obtained from around the platform. Using an organic extraction method water samples concentrated a thousand fold did show some toxicity to *Tisbe battagliai* but there was no clear gradient.

ACKNOWLEDGEMENTS

The authors would like to thank BP Exploration Operating Company Limited, Shell (U.K.) Exploration and Production particularly Ali Onder, Marathon Oil, U.K. Ltd and the Brae Field partners for their cooperation and permission to publish the data. Ed Butler for his great help in organising the sampling from BP platforms and devising a sampling programme They would also like to thank Robin Law, Judith Barr and Vanessa Dawes of

MAFF for their analytical endeavours and expertise. In addition, they are grateful to Dick Rycroft, Julie Osborne and Jan Rusin for their general assistance in the preparation of the paper. Finally, the authors would like to thank the crew of the Danish research vessel, RV 'Dana' for their helpful assistance with the collection of field samples.

REFERENCES

Brandehaug, J., Johnsen, S., Bryne, K.H., Gjose,A.L., Eide, T.H. and Aamot, E., 1992. Toxicity testing and chemical characterisation of produced water a preliminary study. In: Produced Water, Eds. Ray, J.P. and Engelhardt, F.R., Plenum Press, New York, pp.245256.

Brown, J.S., Sauer, Jr., T.C., Wade, M.J. and Neff, J.M., 1992. Chemical and toxicological characterisation of produced water freon extracts. In: "Produced Water, Technical/Environmental Issues and Solutions", James P. Ray and F. Rainer Engelhardt (eds), Environmental Science Research 46: 113131.

Dawes, V.J. and Waldock, M.J., 1994. Measurements of volatile organic compounds at UK National Monitoring Plan stations. Mar. Pollut. Bull. Vol. 28:291298.

Finney, D.J., 1971 . Probit Analysis. Cambridge University Press, 3rd Edition, p. 333.

Girling, A.E., 1989. Acute and chronic toxicity of produced water from a North Sea oil production platform to the calanoid copepod Acartia tonsa. Bull. Environ. Contam. Toxicol. 43: 280286.

Girling, A.E., and Streatfield, C.M., 1988. An assessment of the acute and chronic toxicity of produced water from a North Sea oil platform based upon laboratory bioassays with a calanoid copepod*Acartia tonsa* (Dana) In: "Environmental Protection of the North Sea", P.J. Newman and A.R. Agg (eds), Heinemann Professional Publishing, Oxford, pp. 789791.

HMSO, 1994. The Energy Report, Oil and Gas Resources of the United Kingdom, Department of Trade and Industry.

Hutchinson, T. H. and Williams, T. D., 1989. The use of sheepshead minnow (*Cyprinodon variegatus*) and a benthic copepod (*Tisbe battagliai*) in shortterm tests for estimating the chronic toxicity of industrial effluents. Hydrobiologica 188/189: 567572.

Jacobs, R.P.W.M., Grant, R.O.H., Kwant, J., Marquenie, J.S. and Mentzer, E., The composition of produced water from shell operated oil and gas production in the North Sea, 1992, in: "Produced Water, Technical/Environmental Issues and Solutions", James P. Ray and F. Rainer Engelhardt (eds), Environmental Science Research, Volume 46, Plenum, pp. 1321.

Law, R. J., 1984. The continuous measurement of low concentrations of hydrocarbons in water using a towed fluorimeter. ICES Marine Environmental Quality Committee CM 1984/E:8.

Law, R.J., Fileman, T.W. and Matthiessen, P., 1991. Pthalate esters and other industrial organic chemicals in the North and Irish Seas. Wat. Sci. Technol. 24: 127134.

Litchfield, J.T. and Wilcoxon, F., 1949. A simplified method of evaluating doseeffect experiments. J. Pharm. Exp. Ther. 96: 99113.

Microbics, 1991. Microtox Manual. Preliminary Release 53091. Microbics Corporation, Carlsbad, California, U.S.A.

Murray-Smith, R., Gore, D., Flynn, S.A., Vance, I. & Stagg, R., 1996 Development and appraisal of a particle tracking model for the dispersion of produced water discharged from oil production platforms in the North Sea. This volume

OSPAR, 1995. PARCOM Protocols on Methods for the Testing of Chemicals Used in the Offshore Oil Industry.

Slager, L.K., van Hattum, B., Meesters, M.T., Scholten, M.C.Th., Kaag, N.H.B.M., Cofino, W.P. and Feenestra, J.F., 1992. Environmental aspects of produced water discharges from oil and gas production on the Dutch Continental Shelf. Part III. Environmental effects. Commissioned by NOGEPA, pp103.

Somerville, H.J., D. Bennet, J.N. Davenport, M.S. Holt, A. Lynes,A. Mahieu, B. McCourt, J.G. Parker, R.R. Stephenson, R.J. Watkinson and T.G. Wilkinson, 1987. Environmental effect of produced water from North Sea oil operations. Mar. Pollut. Bull. 18: 549558.

Stagg R.M., Gillibrand P.A., McIntosh, A.M. & Turrell,W.R., 1996. The effects of produced water on hydrocarbon levels and monooxygenase activity in fish larvae in the northern North Sea. This volume.

Stephenson, M.T., 1991. Components of produced water: A compilation of results from several industry studies. In proceedings of the first international conference on health, safety, and environment in oil and gas exploration and production, 1114 November 1991, The Hague, Netherlands, Vol 2, pp 2538. SPE 23313, Society of Petroleum Engineers Inc.

Strømgren, A., Sørstrøm, S.E., Schou, L., Kaarstad, I., Aunaas, T., Brakstad, O.G. & Johansen, Ø., 1995. Acute toxic effects of produced water in relation to chemical composition and dispersion. Mar. Environ. Res. 40: 147-169.

Thain, J. E. and Kirby, M. F., 1994 . Improving the sensitivity of biological water quality measurements in marine waters. Quality status report, Conference of the North Sea, Ebeltoft. (Proceedings in press).

Tibbets, P.J.C., Buchanan, I.T., Gawel, L.J. and Large, R., 1992 A comprehensive determination of produced water composition. Environmental Science Research Series 46: 97 - 112.

TNO, 1994. CHARM 2.1: Chemical Hazard Assessment and Risk Management of offshore exploration and production chemicals. TNO Report: TNOMW94/315.

U.S. Environmental Protection Agency, 1985. Methods for Measuring the Acute Toxicity of Effluents to Freshwater and Marine Organisms. (Third Edition). EPA/600/485/013.

Van Hattum, B., Cofino, W.P. & Feenstra, J.F., 1992 Review of characteristics of produced water from offshore platforms. Commissioned by NOGEPA, pp40.

Whale, G.F., 1994. Potential applications of the Microtox toxicity test within the offshore oil and gas industry. Society of Petroleum Engineers 27176.

10

SUB-LETHAL EFFECTS OF OIL-PRODUCED WATER ON THE EARLY LIFE STAGES OF TURBOT

Susan M. Stephens,[1] J. Anne Brown,[1] and Martin A. Ferguson[2]

[1] Department of Biological Science
Hatherly Laboratories, University of Exeter
Exeter, EX4 4PS, United Kingdom
[2] Amerada Hess
33 Grosvenor Place
London, SW1X 7H7, United Kingdom

INTRODUCTION

Oil extraction generates produced water, a complex mixture of formation and injection waters contaminated with oil, metals and various treatment chemicals (Fillo et al 1992; Tibbetts et al 1992). Although the amount of oil entering the North Sea declined by approximately one third between 1988 and 1993, produced water release increased by a similar proportion over this period (Taylor 1994). In 1990, approximately 160 x 10^6 tons of produced water were discharged into the UK sector of the North Sea (HMSO 1991). With produced water volumes of this magnitude, disposal is an important issue for both operators and regulatory authorities.

To date, most studies of the effects of oil-produced water on fish have been concerned with establishing the concentrations which induce mortalities (Somerville et al 1987; Schiff et al 1992). As the early life stages of fish generally appear to be more sensitive to petroleum hydrocarbons than adult fish (Carls 1987; Overton et al 1994) fish larvae might be expected to be particularly vulnerable to produced waters. The sub-lethal physiological disturbances of fish larvae exposed to petroleum hydrocarbons have hardly been examined, but will ultimately determine growth and development, and thus in turn affect the ability of larvae to capture prey, escape predation and develop into adult fish.

We have investigated the effects of oil-produced water on larval survival, their activity and various indices of sub-lethal physiological stress. Heart rates of fish exposed to produced waters were monitored, to provide an index of adrenergic ('flight or fight') responses (Mazeaud et al 1977) or evidence of any narcosis. Gill structure, which has been shown to be affected by exposure to petroleum hydrocarbons (Correa and Garcia 1990; Prasad 1991), was examined by scanning electron microscopy. Adult fish exposed to petroleum hydrocarbons have elevated plasma cortisol concentrations (LeLoup-Hatey and

Hardy 1985; Thomas and Rice 1987), reflecting an activation of generalised endocrine stress responses (Pickering 1992). Cortisol responses were investigated in turbot larvae exposed to oil-produced waters by measurement of their whole body cortisol content.

METHODS

Oil-Produced Water

Water produced by the Ivanhoe and Rob Roy reservoirs are processed on the AH001 platform. The AH001 currently processes approximately 39,000 Barrels of Water per Day (BWPD). The process is treated using hydrocyclones to achieve statutory discharge oil-in-water quality limits. The process has recently been upgraded by retrofitting high efficiency 'K-Liner' hydrocyclones in addition to the existing hydrocyclones to improve separation efficiency and throughput. The produced water was collected in 2.5 litre, amber, acid-rinsed glass bottles from the sample point immediately prior to discharge. An initial batch of produced water (Batch 1) was taken prior to the system upgrade whilst Batch 2 was taken shortly after the upgrade. Samples were kept cool prior to transportation, and upon arrival in Exeter (8-9 days later) were stored in the dark at 12°C.

Source and Maintenance of Turbot

Turbot *Scophthalmus maximus* (L.) larvae, approximately 30 days old, (500°C days) were obtained from Mannin Sea Farm, Isle of Man and maintained under a natural photoperiod in aerated seawater at 15°C, fed with live *Artemia salina* until experimentation.

Experimental Exposure of Larvae to OilProduced Water

Experimental exposures to oilproduced water dilutions were carried out between 10.00am and 10.00pm at 15°C. Larvae were transferred from the stock supply of larvae to dishes containing 0% (control), 0.001%, 0.01%, 0.1%, 1%, 5% and 10% oilproduced water.

Larvae were observed at 15 minute intervals and any dead fish were removed. Mortalities, counts of the number of fish swimming and their orientation were recorded. At the end of the 12 hour exposure period, remaining larvae were killed, rapidly rinsed in distilled water, blotted dry and frozen in liquid nitrogen. All larvae were stored at 60°C until analyses.

Oil-Produced Water Analyses

Deuterated naphthalene (175µg), an internal standard, was added to 1 litre aliquots of produced water in glass bottles, and 25ml pentane was added. After 2 hours rolling (Luckman multimix MM1) the contents were poured into a stoppered separatory funnel and left to settle for 1 hour. The pentane fraction was collected and cleaned with sulphuric acid and glass wool prior to GCMS analyses (Hewlett Packard 5890 series II plus gas chromatograph with Hewlett Packard Ultra 2 column (25m, 0.25mm id containing 5% methyl phenyl), linked to a 5972 series mass selective detector. The temperature was initially raised by 2.71°C/min to 40°C, and then raised at 7°C/min to 270°C, at which it was held for 15 min).

Fluorescent spectroscopy was used to monitor the loss of hydrocarbons from the 10%, 5% and 1% produced water in static exposure systems (excitation wavelength 310nm, emission wavelength 360nm; Perkin Elmer LS5 luminescent spectrometer). Fluorescence spectroscopy was found to be unreliable for measurement of hydrocarbons in 0.1%0.001% produced water.

The osmolarity of the stock produced water was determined by freezing point depression (Roebling osmometer).

Heart Rate of Turbot Larvae

The heart of the turbot larvae was visible to the naked eye. The heart rates of 8 individual larvae, exposed to each batch of produced water at a the range of exposure concentrations were determined 2 hours after transfer to the exposure media, and again after 12 hours of exposure.

Whole Body Cortisol Content of Turbot Larvae

Cortisol content of turbot larvae exposed to batch 2 produced water was measured by homogenising individual larvae in phosphosaline buffer (61.2mM NaH_2PO_4; 154mM NaCl; 15.4mM sodium azide; 39.1mM $Na_2H_2PO_4$; 1g/L gelatin; pH 7.6; 5 times volume by weight) and doubleextraction of 250 μl aliquots of homogenate with 1ml of ether (de Jesus et al, 1991). Ether extracts were combined, dried, and then resolubilised in 800 μl phosphosaline buffer prior to radioimmunoassay as previously described (Brown et al, 1989).

Scanning Electron Microscopy of Larval Gills

Larvae exposed to a 10% solution of batch 2 produced water were killed and the gills were rapidly excised, rinsed in distilled water and fixed in 3% glutaraldehyde for 2 days. After washing, gills were postfixed in 1% osmium tetraoxide overnight (4°C) before dehydration in graded acetones. Gills were then critical point dried, sputtercoated with gold and examined by scanning electron microscopy (Cambridge Stereoscan 100).

Statistical Analyses

Swimming activity was statistically analysed using one-way analysis of variance. Larval heart rates were analysed by one and two-way analysis of variance. The whole body cortisol content was log transformed prior to one-way analysis of variance and regression analysis of cortisol content against produced water concentration. Where F-values from ANOVA were significant, *a priori* comparisons of values in control larvae and values in animals exposed to each of the concentrations of produced waters were made by analyses of least significant differences.

RESULTS

Produced Water Composition

GCMS showed that batch 1 of oilproduced water contained a not unexpectedly complex mix of petroleumderived aromatic and aliphatic hydrocarbons (Figure 1). Batch 2 of the oilproduced water was relatively rich in light aromatic hydrocarbons, but had considerably reduced quantities of the same range of aliphatic hydrocarbons and polycyclic aromatic hydrocarbons (PAHs) as found in batch 1. No treatment chemicals were detected in either batch of produced waters. Fluorescence spectroscopy readings estimated that over the 12 hour exposure period, approximately 16% of hydrocarbons were lost from batch 1

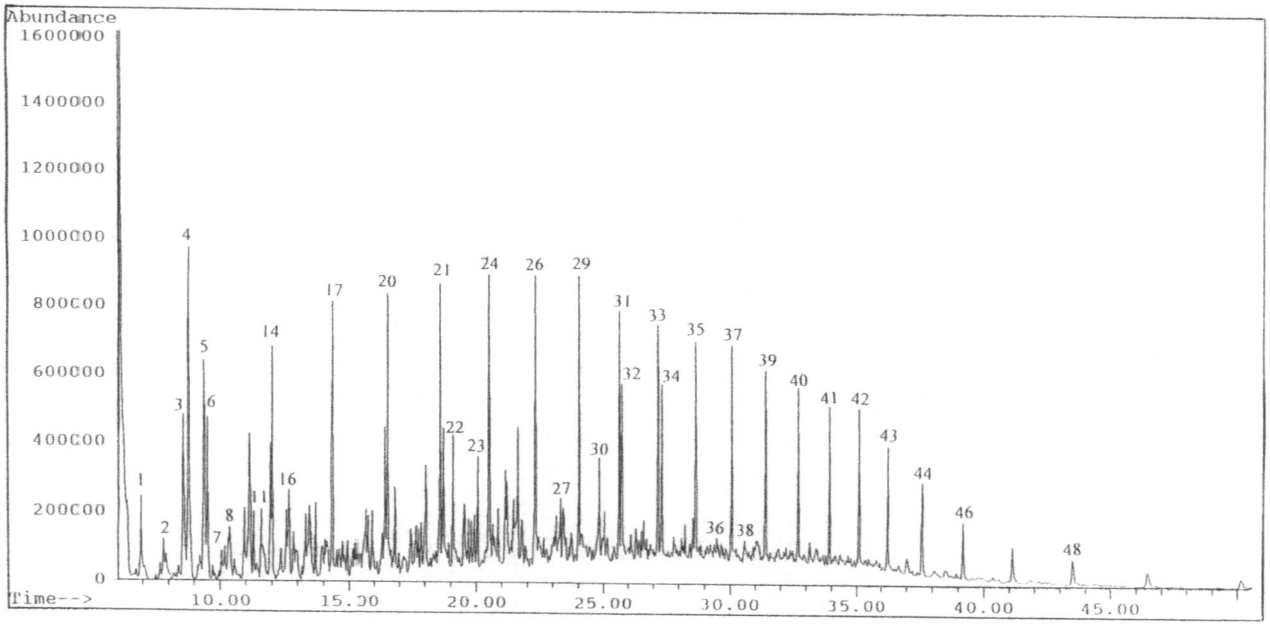

1. octane
2. cyclohexane, ethyl-
3. benzene, ethyl-
4. benzene, 1,4-dimethyl
5. benzene, 1,3-dimethyl-
6. nonane
7. benzene, (1-methylethyl)-
8. cyclohexane, propyl-
9. benzene, propyl-
10. benzene, 1-ethyl-2-methyl-
11. benzene, 1-ethyl-4-methyl-
12. benzene, 1,3,4-trimethyl-
13. benzene, 1,2,4-trimethyl-
14. decane
15. benzene, 1-methyl-3-propyl-
16. cyclohexane, butyl-
17. undecane
18. naphthalene, 1,2,3,4-tetrahydro-
19. azulene
20. dodecane
21. tridecane
22. naphthalene, 1-methyl-
23. hexadecane, 2,6,10,14-tetramethyl-
24. tetradecane

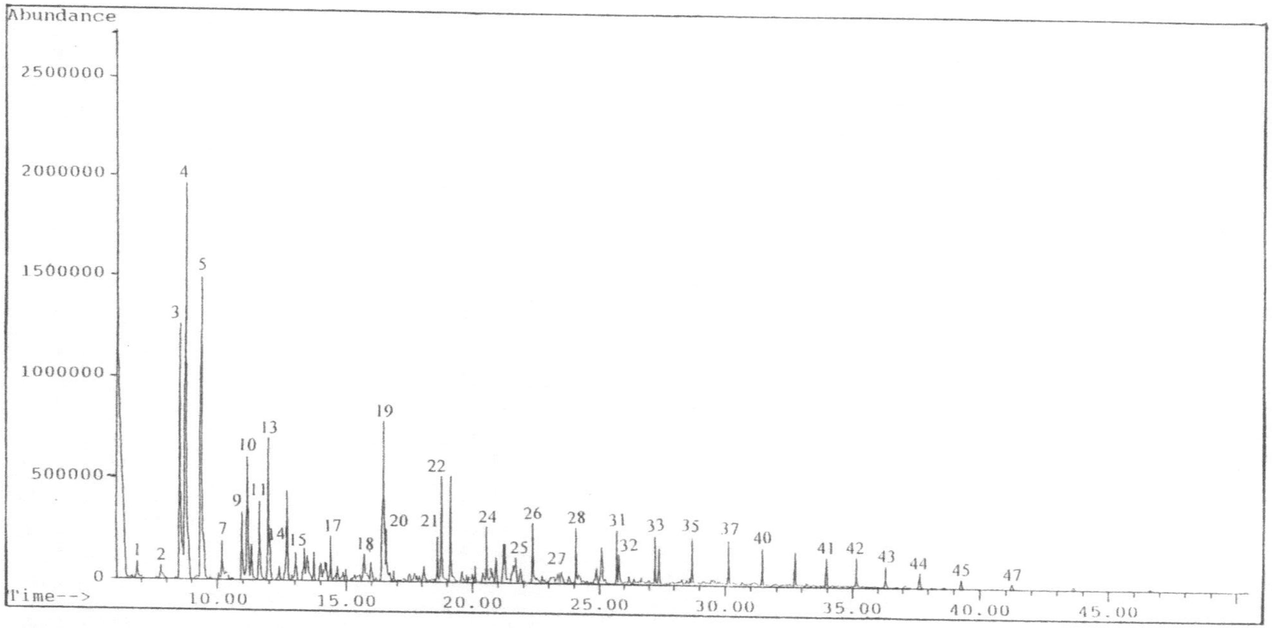

25. naphthalene, 2, 3-dimethyl
26. pentadecane
27. naphthalene, 1,4,6-trimethyl-
28. naphthalene, 2,3,6-trimethyl
29. hexadecane
30. decane, 2-methyl-
31. heptadecane
32. pentadecane, 2,6,10,14-tetramethyl-
33. octadecane
34. hexadecane, 2,6,10,14-tetramethyl-
35. nonadecane
36. phenanthrene, 4-methyl
37. eicosane
38. anthracene, 9,10-dimethyl-
39. heptadecane, 2,6,10,15-tetramethyl-
40. docosane
41. tricosane
42. heptacosane
43. pentacosane
44. hexatriacontane
45. nonacosane
46. docosane, 11-decyl-
47. dotriacontane
48. tetratriacontane

Figure 1. Gas Chromatography - Mass Spectroscopy analyses of batch 1 (A) and batch 2 (B) oil-produced waters from the AH001 (Amerada Hess) platform. The major compounds identified with 90% or more certainty are listed.

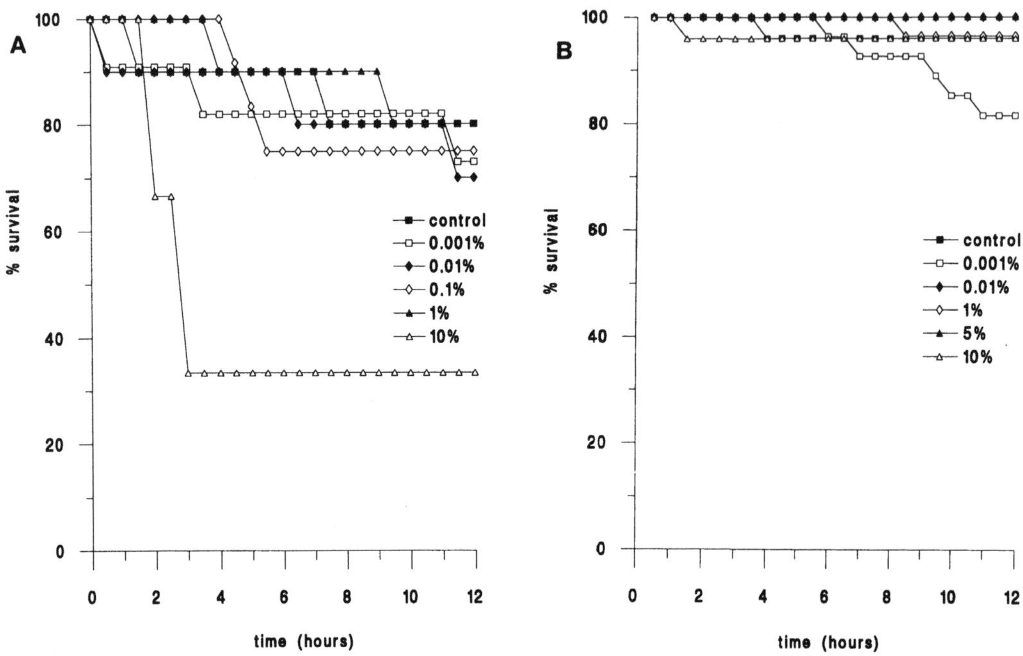

Figure 2. Survival of turbot larvae exposed for 12 hours to dilutions of batch 1 (A) and batch 2 (B) oil-produced waters (0.001% to 10%) from the AH001 (Amerada Hess). Values for control larvae held in seawater are also shown.

and 22% were lost from batch 2. The produced water batches were similar in osmolarity (2282 and 2202 mOsmolar for batch 1 and 2 respectively).

Larval Mortality

There were significant larval mortalities of turbot exposed to a 10% solution of batch 1 oil-produced water. In this solution, only 30% of larvae survived the 12 hour exposure (Figure 2A). With a greater dilution of this produced water, mortalities did not exceed those of the control group. The cleaner, batch 2 produced water did not induce any significant mortalities (Figure 2B).

Swimming Activity and Orientation

Swimming activity was significantly increased in larvae exposed to 1% batch 1 produced water compared to larvae held in clean seawater. However, in 10% produced water swimming activity was significantly depressed. Larval swimming activity was significantly reduced in all dilutions of batch 2 produced water (Figure 3).

A maximum of 16% of the larvae were upside down during exposure to batch 1 produced water, but there was no clear relationship between the concentration of produced water and the proportion of disorientated larvae at any time. Fewer larvae were disorientated during exposure to the cleaner batch 2 produced water (maximum 8%), but again no trends were apparent between the proportion of disorientated larvae and the concentration of produced water.

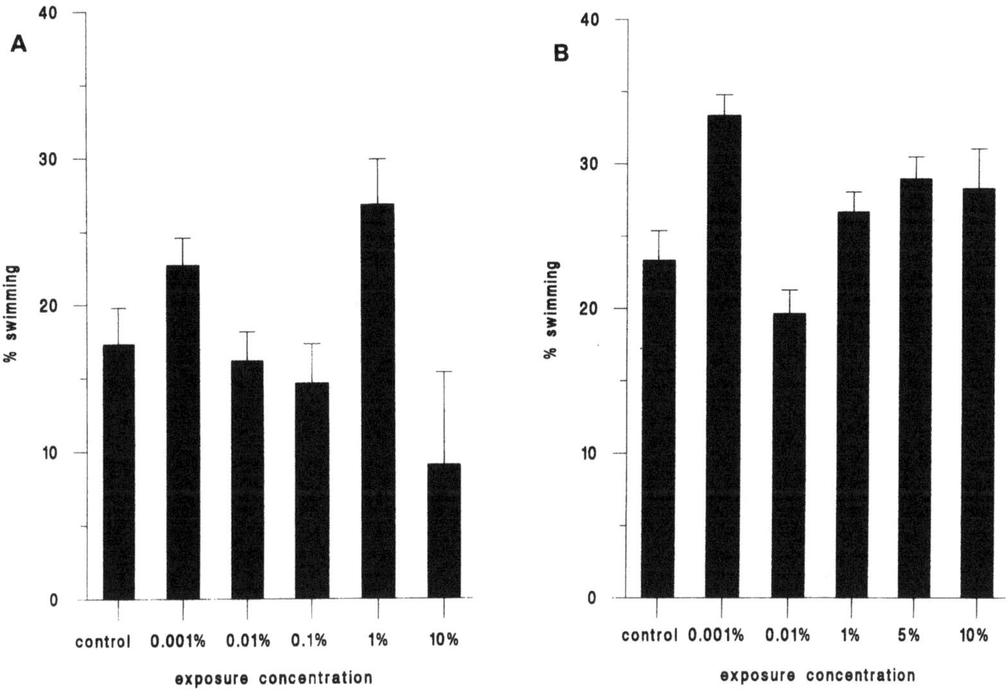

Figure 3. Percentage of larvae swimming during exposure to batch 1 (A) and batch 2 (B) oil-produced waters (0.001% to 10%) from the Ivanhoe/Rob Roy AH001 (Amerada Hess). Values for control larvae held in seawater are also shown. All values are means ±SE.

Heart Rate of Turbot Larvae

Table 1 presents the mean heart rates of larvae exposed to both batches of produced waters Two-way ANOVA of heart rates of larvae exposed to batch 1 showed a significant effect of time ($P< 0.005$). In larvae exposed to 10% produced water, heart rate was significantly depressed at 12 hours compared to the heart rate at 2 hours ($P<0.001$), but lower concentrations of produced water had no significant effect on heart rate. In larvae

Table 1. Heart rates of turbot larvae exposed for 12 hours to oil-produced waters (0.001% to 10%) from the AH001 (Amerada Hess) platform. Values for control larvae held in seawater are also given. All values are means ±SE

	Batch 1		Batch 2	
	2 hours	12 hours	2 hours	12 hours
Control	110.6±2.7	110.0±2.4	122.1±1.6	116.3±1.8
0.001%PW	109.3±0.9	107.2±2.9	130.9±3.7	117.5±3.6
0.01%PW	107.7±1.2	106.9±1.5	118.8±3.7	112.5±1.6
0.1%PW	112.6±1.7	107.8±3.5	-	-
1%PW	111.0±1.3	105.8±1.9	114.9±3.6	113.1±3.2
5%PW	-	-	120.6±3.8	98.4±17.7
10%PW	118.0±1.2	106.4±0.9	123.7±2.8	113.6±2.0

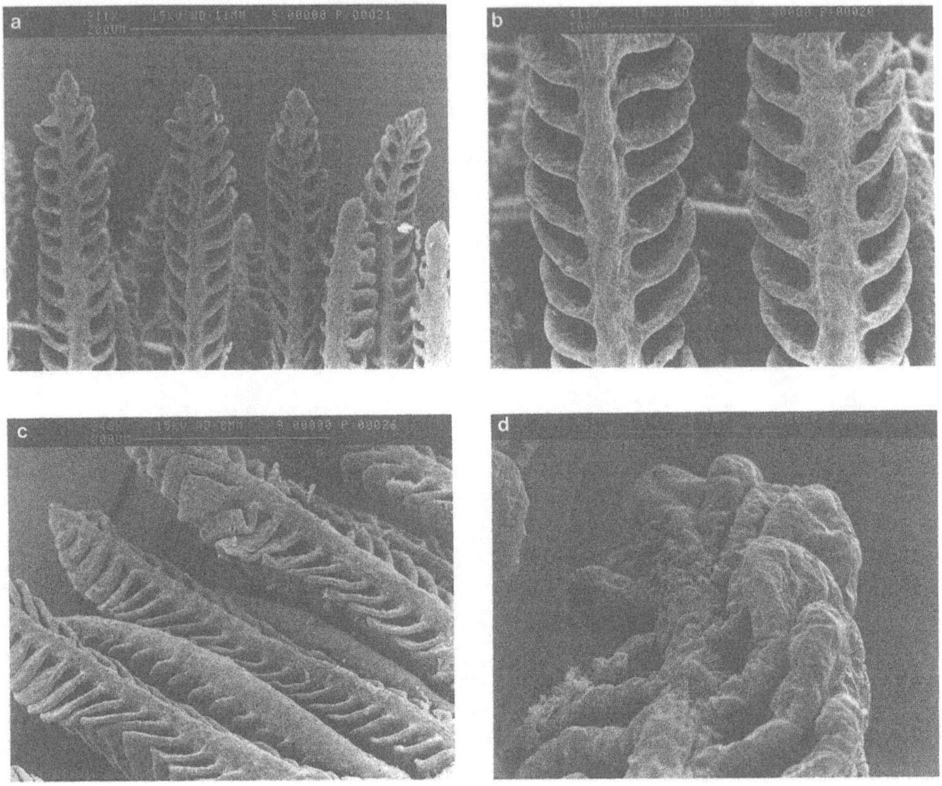

Figure 4. Scanning electron micrographs of gills from control turbot (Figure 4a & 4b) and turbot larvae exposed to 10% oil-produced water (batch 2) from the Ivanhoe/Rob Roy AH001 (Amerada Hess) platform (Figure 4c & 4d). Low magnification photomicrographs (4a, 4c) show several gill filaments with the regularly-arranged series of respiratory lamellae. The lamellae are examined more closely at higher magnification (4b & 4d).

exposed to batch 2 oil-produced waters, 2 way ANOVA indicated both a time effect and a produced water concentration effect ($P< 0.001$ and 0.025 respectively). Subsequent one way ANOVA indicated a significant produced water concentration effect at 2 hours ($P<0.04$) although this was not striking, and was not confirmed or localised by comparisons of heart rate in control animals and larvae in each produced water group; there was no apparent produced water concentration effect at 12 hours (one way ANOVA). Analyses of the time-related effect on heart rate indicated, as with batch 1 exposed larvae, that heart rate was significantly depressed after 12 hours exposure to 10% produced waters ($P<0.05$), but not in larvae exposed to more diluted produced water.

Gill Morphology

Exposure of turbot larvae to 10% oilproduced water (batch 2) resulted in some swelling and folding of the gill lamellae (Figure 4).

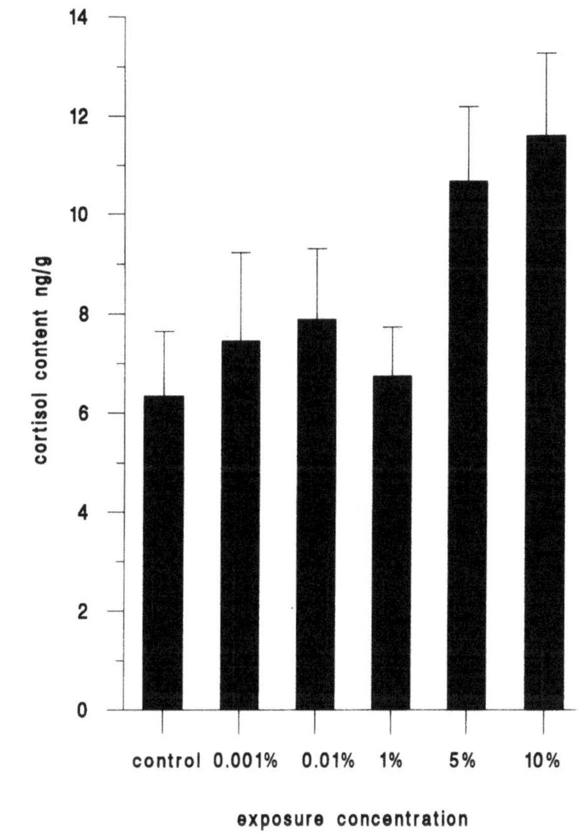

Figure 5. Whole body cortisol content of turbot larvae exposed for 12 hours to batch 2 oil-produced water (0.001% to 10%) from the Ivanhoe/Rob Roy AH001 (Amerada Hess) platform. Values for control larvae held in seawater are also shown. All values are means ±SE.

Whole Body Cortisol Content of Turbot Larvae

There was a significant relationship between whole body cortisol content of turbot larvae and produced water concentration (correlation coefficient 0.42, $p<0.01$). Cortisol content of larvae exposed to a 10% solution of produced water was significantly increased ($P< 0.05$; Figure 5), but there was no apparent increase in cortisol content in larvae exposed to more diluted produced waters.

DISCUSSION

Oil produced water contains a complex mixture of petroleum hydrocarbons, metals, salts and treatment chemicals (Fillo et al 1992; Tibbetts et al 1992). In the North Sea, after clean up and removal of oil (Taylor 1994) this water is discharged from the platforms to the marine environment, but so far there have been few published studies which have investigated the effects of these discharged oil-produced waters on fish. What studies there are have been concerned with establishing the concentrations which induce mortalities, measuring LC_{50} for example (Gamble 1987; Schiff et al 1992).

The early life stages of fish are generally more sensitive to petroleum hydrocarbons than adult fish (Rice 1985; Carls 1987; Overton et al 1994). The high sensitivity of larvae may reflect their relatively high surface area relative to their volume and thus increased uptake of hydrocarbons. The intensity of ongoing developmental processes may also predispose fish larvae to pollutant stress. In the wild, the limited mobility of larvae is likely to make them more vulnerable when exposed to any chemical pollutants. With extremely high natural losses of the early life stages of fish, even relatively small additional losses could have considerable impact on fish populations.

Our studies indicate only a very low acute toxicity of oil-produced water from the Rob Roy/Ivanhoe AH001 platform, and furthermore that the recently improved clean up procedures, with fitting of improved hydrocyclones, eliminated the acute toxicity of this produced water. Prior to this, 10% produced water caused 70% mortality in 12 hour static exposure of turbot larvae, but there was insignificant mortality in larvae exposed to more diluted produced water. These results are comparable to those of Gamble et al (1987) who found no acute toxicity to cod and herring larvae of 0.25-0.125% produced water from the Auk and Forties platforms.

Produced water from the AH001 platform (batch 1), in the dilution range of 0.001% to 0.1%, did not produced consistent effects on larval swimming activity. Activity was significantly increased by exposure to 1% produced water, but significantly reduced by exposure to 10% produced water. However, larvae exposed to dilutions of batch 2 produced water consistently showed reduced swimming activity. Increased activity may reflect irritation and in the wild, despite the limited mobility of fish larvae, may be sufficient to enable avoidance of discharged produced waters. Decreased swimming activity in larvae exposed to 10% batch 1 produced water and to all dilutions of batch 2 produced water probably reflects non-specific and reversible narcotic effects of the produced water (Hermens 1986). The reduction in swimming activity of larvae exposed to batch 2 produced water, seems likely to reflect the high concentrations of light aromatic hydrocarbons.

After 12 hours exposure to 10% oil-produced water the heart rate of turbot larvae was significantly depressed which seems likely to reflect narcotic action. More diluted produced water did not significantly depress heart rate. There was no evidence of activation of adrenergic stress responses which might have elevated heart rates (Mazeaud et al 1977), although these responses may not be fully developed in premetamorphic larvae.

Despite the recently improved clean up of the oil-produced water discharged from the AHOO1 platform, slight gill damage was noted in fish exposed to the 10% solution of batch 2 oil-produced waters. The swelling of gill lamellae was similar in nature to the swelling of respiratory lamellae reported in adult fish exposed to petroleum hydrocarbons (Prasad 1991; Correa and Garcia 1990; AlKindi 1995). However, the damage after exposure to produced water for 12 hours was rather slight. Much more extensive gill abnormalities are evident when adult fish are exposed to crude oil (Prasad 1991) or the water soluble fraction of crude oil (AlKindi 1995), and here severe respiratory stress results (AlKindi 1995; AlKindi et al 1996). By contrast, respiratory stress of the turbot larvae seems very unlikely even during exposure to the 10% solution of batch 2 produced water.

Larvae exposed to oil-produced water showed an activation of an endocrine stress response (Pickering 1992), with cortisol content of larvae exposed to oil-produced waters related to the concentration of oil-produced water and significantly elevated in larvae exposed to 10% produced water. The degree of elevation of whole body cortisol content of these larvae was, however, less than the change in whole body cortisol content of turbot larvae of a similar age during exposure to the water soluble fraction of crude oil (Stephens et al 1996), and, the increase in cortisol content of herring larvae exposed to drilling muds (Ramsay 1991). Whole body cortisol was not significantly affected in turbot exposed to less than 10% produced water, which implies that there was little acute physiological sublethal stress in these larvae.

In the light of our studies it is important to consider whether a 10% solution of produced water is environmentally relevant. A major factor in assessing the field effect of discharges of produced water is the dilution by dispersion, as well as the loss of volatile components (Davies and Kingston 1992). Read and Blackman (1980) predicted a 400 to 800-fold dilution by 0.5-1km from the discharge point. Thus a concentration of 10% produced water might only occur *very* locally. Nevertheless, the surface discharge of oil produced waters will render planktonic animals, including fish larvae vulnerable to exposure to the initially high concentrations at the discharge point.

It is not possible to be confident that oil-produced water from other fields and platforms has a similarly low acute toxicity as appears to be true of water currently discharged from the Rob Roy/Ivanhoe platform. The well known variation in produced water composition from field to field and within one field depending upon its life and operation procedures (Davies and Kingston 1992; Tibbetts 1992) demands assessment of the effects of a range of produced waters on fish before any generalisations can be arrived at.

In summary, oil-produced water from the AHOO1 platform had little apparent acute toxicity to turbot larvae. Only the 10% solution (a concentration likely to be present only near to the discharge point) caused any significant sub-lethal stress. The recent fitting of improved hydrocyclones appears to have significantly improved the clean up of the produced water prior to discharge and reduced its acute toxicity. However, in view of the prevalence of many PAHs in the produced waters, and of the potential cocktail of treatment chemicals, the effects of chronic exposures should now be investigated.

ACKNOWLEDGMENTS

We are grateful to Greenpeace Analytical, Exeter, for help and assistance in use of their GC-MS, and to Gavin Wakley for advice on the use of the SEM and printing facilities.

REFERENCES

AlKindi, A.Y.A. (1995). The effects of the water-soluble fraction of crude oil on flounders and larvae of turbot. PhD thesis, University of Exeter.

AlKindi, A.Y.A., Brown, J.A., Waring, C. P. and Collins, J.E. (1996). Endocrine responses of flounder *Platichthys flesus* physiologically disturbed by exposure to the water soluble fraction of crude oil. Submitted to Gen. Comp. Endocrinol.

Brown, J.A., Edwards, D. and Whitehead, C. (1989). Cortisol and thyroid hormone responses to acid stress in the brown trout, Salmo trutta L. J. Fish. Biol. 35: 7384.

Carls, M.G. (1987). Effects of dietary and waterborne oil exposure on larval pacific herring, Clupea harengus pallasi. Comp. Biochem. Physiol. 58C: 4147.

Correa, M. and Garcia, H.I. (1990). Physiological responses of juvenile white mullet, *Mugil curema*, exposed to benzene. Bull. Environ. Contam. Toxicol. 44: 428-434.

Davies, J.M. and Kingston, P.F. (1992). Sources of environmental disturbance associated with offshore oil and gas developments. In: North Sea Oil and the Environment. Developing oil and gas resources, environmental impacts and resources. Ed. W.J. Cairns. Elsevier Applied Science. pp. 417-439.

de Jesus, E.G., Hirano, T. and Inui, Y. (1991). Changes in cortisol concentration during early development and metamorphosis in Japanese flounder, Paralichthys olivacceus. Gen. Comp. Endocrinol. 82: 369376.

Fillo, J.P., Koraido, S.M. and Evans, J.M. (1992). Sources, characteristics, and management of produced waters from natural gas production and storage operations. In: Produced Water. Eds. J.P. Ray and F.R. Engelhart. Plenum Press, New York. pp. 151-161.

Gamble, J.C., Davies, J.M., Hay, S.J. and Dow, F.K. (1992). Mesocosm experiments on the effects of produced water discharges from offshore oil platforms in the northern North Sea. Sarsia 72: 383-386.

Hermens, J. (1986). Quantitative structure - activity relationships in aquatic toxicology. Pestic. Sci. 17: 287-296.

HMSO. (1991). Development of the oil and gas resources of the United Kingdom, HMSO, ISBN 0 11 413705 6.

Leloup-Hatey, J. and Hardy, A. (1985). Effet d'une pollution par les hydrocarbures sur la fonction de l'anguille (*Anguilla anguilla* L.). Ichtyophysiol. Acta 9: 39-50.

Mazeaud, M.M., Mazeaud, F. and Donaldson, E.M. (1977). Primary and secondary effects of stress in fish: Some new data with a general review. Trans. Am. Fish. Soc. Vol. 106: 201-212.

Overton, E.B., Sharp, W.B. and Roberts, P. (1994). Toxicity of petroleum. In: Basic environmental toxicology. Eds. L.G. Cockerham and B. Shane. U.S. CRC Press. pp. 133156.

Pickering, A.D. (1992). Rainbow trout husbandry: management of the stress response. Aquaculture 100: 125-139.

Prasad, M.S. (1991). SEM study of the effects of crude oil on the gills and air breathing organs of climbing perch *Anabas testudineus* Bull. Env. Contam. Toxicol. 47: 882-889.

Ramsay, N.C. (1991). The effects of pollutants on osmotic and ionic regulation of herring, *Clupea harengus*, Ph.D. thesis. University of Aberdeen.

Read, A.D. and Blackman, R.A. (1980). Oily water discharges from offshore North sea installations. A perspective. Mar. Pollut. Bull. 11: 44-47.

Rice, S.D. (1985). Effects of oil on fish. In 'Petroleum effects in the Arctic Environment' (Ed F.R. Engelhardt). pp 157-182, Elsevier.

Schiff, K.C., Reish, D.J., Anderson, J.W. and Bay, S.M. (1992). A comparative evaluation of produced water toxicity. In: Produced Water. Eds. J.P. Ray and F.R. Engelhart. Plenum Press, New York. pp. 199-207.

Somerville, H.J., Bennett, D., Davenport, J.N., Holt, M.S., Lynes, A, Mahieu, A., McCourt, B., Stephenson, R.R., Watkinson, R. J., Wilkinson, T.G. (1987). Environmental effects of produced water from North Sea oil operations. Mar. Pollut. Bull. 18: 549.

Stephens, S.M., AlKindi A.Y.A., Waring, C.P., and Brown J.A. (1996). Corticosteroid and thyroid responses of larval and juvenile turbot, *Scophthalmus maximus* L. exposed to the water soluble fraction of crude oil. Submitted to Gen. Comp. Endocrinol.

Taylor, B.G.S. (1994). Minimising the environmental impact offshore. Scottish Environtec. pp. 30.

Thomas, P. and Rice, S.D. (1987). Effects of water soluble fraction of cook inlet crude oil on swimming performance and plasma cortisol in juvenile coho salmon, *Oncorhynchus kisutch*. Comp. Biochem. Physiol. 87C: 177-180.

Tibbetts, P.J.C., Buchanan, I.T., Gawel, L.J. and Large, R. (1992). In: Produced Water. Eds. J.P. Ray and F.R. Engelhart. Plenum Press, New York. pp. 97-112.

Tilseth, S., Solberg, T.S. and Westrheim, K. (1984). Sublethal effects of the WSF of Ekofisk crude oil on the early life stages of cod, Gadus morhua (L.). Mar. Envir. Res. 11: 116.

EXPOSURE OF COD LARVAE (*Gadus morhua L*) TO PHENOLS IN PRODUCED WATER DURING THE CRITICAL PERIOD OF FIRST FEEDING

R. Farestveit,* S. Johnsen, H. Riksheim, and M. Østensen

STATOIL Research Centre
Postuttak, 7005 Trondheim, Norway

The discharge of produced water rapidly increases as an oil field ages, eventually reaching a level where the production of injection and formation water exceeds the production of oil. The processing capacity of water and gas is normally the limiting factor for a mature oil field (Ynnesdal and Furuholt, 1994). The composition of produced water includes dissolved salts, dissolved and colloidal hydrocarbons, trace metals, residual chemicals added in the petroleum production processes and suspended solids.

While the total discharged amount of produced water in the Norwegian sector of the North Sea barely exceeded 16 million m3 in 1990, the volume is expected to reach 90 million m3 within the year 2000 (Anon. 1994). Hydrocyclones are the preferred technology for treating large amount of produced water, normally bringing the mean concentration of oil in water down to a level in the range of 10 - 30 mg/l. Traditional oil/water separation systems act to remove particulate dispersed oil, but are not effective in removal of dissolved petroleum hydrocarbons.

The amount of dissolved hydrocarbons in produced water depends on the fractional composition of the petroleum and is inversely related to the molecular - weight distribution of the hydrocarbon fractions (Anon. 1993b). Low molecular weight organic components, organic acids, phenols and BTX , are the largest groups of dissolved organic carbons (Middleditch 1984). The concentration range for phenols found in the produced water from the Statfjord and the Gullfaks fields is 1-10 ppm, and the aromatic compounds range from 1 to 5 ppm (Brendehaug et al. 1993; Sørstrøm et al. 1993).

The toxicity of North Sea produced water is generally low.The EC50 values for the Microtox bacteria are 5 - 20% of produced water in sea water, algae (Skeletonema costatum) 2-55%, copepod (Acartia tonsa) 2-8%, LC50 fish larva (Clupea harengus) 1-10% (Sommerville et al. 1987; Serigstad and Klungsøyr 1992; Brendehaug et al, 1993. Sørstrøm et al. 1993).

---

* Present address: Ocean Climate, Thormøhlens gt 55, 5000 Bergen, Norway.

Produced Water2: Environmental Issues and Mitigation Technologies
Edited by M. Reed and S. Johnsen, Plenum Press, New York, 1996

Previous work has demonstrated that phenols and aromatics represent the major sources of acute toxicity observed by the Microtox bacteria test (Johnsen et al. 1994). Aromatics, especially PAH's, and alkylated phenols show potential for bioaccumulation (Tollefsen et al. 1995), even though clear evidence of accumulation in organisms near the oil fields has not been presented.

The present experiment was established as a first step in order to identify possible long term effects of phenols in produced water on the reproduction process in fish. The early development of a newly hatched fish larvae is probably the most sensitive developmental stage in the life cycle of fish and the investigation of possible toxic effects of exposure to produced water phenols in this period was the focus of this work. Cod has been shown to be the most sensitive among several fish species from the North Sea in toxicity tests with different oils and oil products (Falk-Pettersen and Kjørsvik 1987). The metabolic activity of fish larvae depends on a sufficient supply of oxygen from the ambient sea water to the mitochondria in the cells, where the energy production for synthesis, regulatory processes and locomotory activity takes place (Serigstad 1987). Oxygen uptake is a direct measure of the aerobic energy production of the fish larvae, and a quantitative expression for the rate of consumption of its energy stores. It is assumed that the development of the larvae is optimised so that a major part of the energy production is directed for synthesis and growth, while a minor part goes to locomotory activities and regulatory processes. The use of oxygen as a physiological stress indicator is based on the assumption that normal oxygen uptake of the fish larva is optimised for its growth and development, and that any deviation from these normal conditions will have a serious impact on the survival of the animal. If the cod larvae do not get food just after the yolk sac is absorbed, they will starve to death through a period with decreasing oxygen uptake (Ellertsen et al. 1980; Serigstad 1987). On the other hand if the larvae are supplied with adequate food they will grow rapidly and the oxygen uptake will increase (Folkvord et al. 1985; Serigstad 1987).

Cod larvae showed reduced oxygen uptake in a total of 7 different experiments, when continuously exposed in a flow-through biotest system to the water soluble fraction of Statfjord crude oil in the range of 50 - 264 ppb, measured as BTEX content (Serigstad 1987). The accumulated exposure dose (time multiplied by concentration) is relatively high using a flow-through system, compared to the expected dose in nature were the larvae will drift through the discharge plume in a period of typically a few hours. This experiment was designed to evaluate possible effects of a realistic exposure regime (5 to 2400 ppb - hours) on startfeeding success, survival and growth for cod larvae.

Preliminary tests with exposure of phenols in a continuous flow system indicated that effects could be found for concentrations down to 25- 100 ppb over several days of exposure in a flow-through system (Smith 1993). The exposure dose in these experiments ranged from 600 - 12 000 ppb hours.

MATERIALS AND METHODS

Experiments were conducted at the SINTEF Aquaculture Research Laboratory in Trondheim. Eggs from cod (Gadus morhua L.), fertilised 17. May 1994, were obtained from the Marine Biological Station in Tromsø, transported by air to Trondheim and incubated at 8°C in an up-stream automated system designed for incubation of pelagic fish eggs (Jørgensen 1987). The eggs were hatching during the period from 30. - 31. of May, and the exposure of the eggs started 1. June 06.00 am. The composition of phenols in the experiment was based on analytical results obtained from the Statfjord and Gullfaks fields in the North Sea (Brendehaug et al. 1993, Sørstrøm et al. 1993). The composition of the phenol mixture is shown in Table 1.The exposure regime was developed by a factorial experimental design

Table 1. Composition of the phenol mixture in the experiment

Phenols	Initial composition	Percent
- phenol	600.0 mg	60.4
-o-cresol	101.4 mg	10.2
-p-cresol	104.8 mg	10.5
-m-cresol	101.5 mg	10.2
-2,5-dimethylphenol	24.4 mg	2.5
-m-ethylphenol	20.4 mg	2.1
-p-ethylphenol	29.6 mg	3.0
-o-iso propylphenol	2.5 mg	0.25
-o-n-propylphenol	2.9 mg	0.29
-p-n-propylphenol	2.6 mg	0.26
-2,3,5 trimethylphenol	3.7 mg	0.37
Total	993.8 mg	100

with time and concentration as the two variables. Figure 1 gives an overview over the design and the actual exposure doses for each group of larvae in the study.

The egg material was divided into 6 equal batches and distributed into 6 incubators, each with a volume of 45 liters (Each exposure vessel corresponds to a point on the X—axis in Figure 1). The phenol mixture was dissolved in distilled water (100mg/l) and added to the exposure tanks. The temperature was kept constant at 8°C during all parts of the experiments. Unhatched egg and dead larvae were removed from the incubators and the water flow was stopped before the introduction of the phenol mixture. During the exposure period, the phenol concentrations in the different systems were monitored by GC/MS analysis, according to the method described by Brendehaug et al. (1993). After a specific exposure time (according to the experimental design, Figure 1) the larvae were transferred from the exposure vessels to 431 tanks containing clean sea water and the evaluation period was initiated. The polluted transport water were exchanged with clean sea water to remove any remaining traces of phenols associated with the larvae. The number of tanks in the evaluation

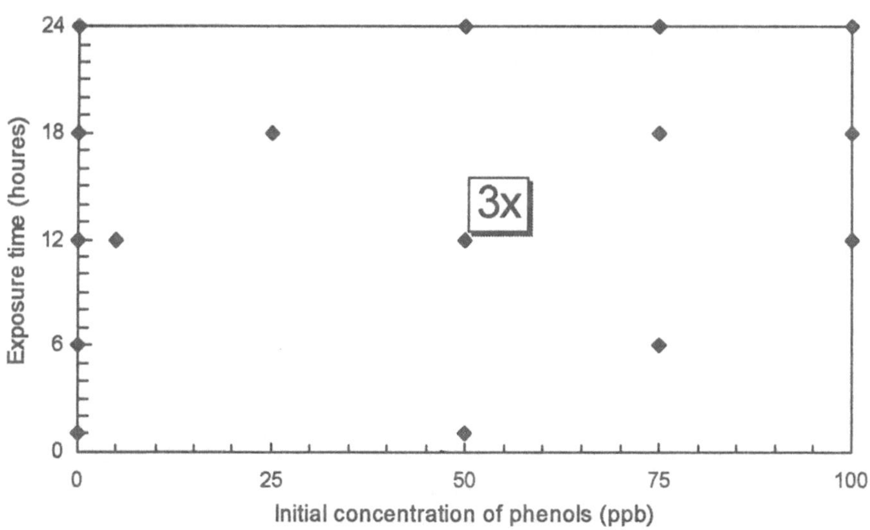

Figure 1. Experimental design showing exposure regime. X-axis represents initial concentrations of phenols added, while y-axis shows exposure time, ranging from 1 to 24 hours.

period was corresponding to the number of points in the two-dimensional experimental design (Figure 1).

Immediately after the transfer of larvae to clean sea water, 2x5 larvae were placed in respirometers with a stirring devise in a temperature- regulated (8°C) water bath, following a method adapted from Serigstad (1987). The initial oxygen tension was the tension of fully air saturated sea water, and the oxygen tension was measured using a polarographic oxygen electrode (Strathkelvin model 1302, 1994) mounted in a temperature regulated cuvette. The oxygen uptake was calculated from the decrease in oxygen partial pressure as a function of the incubation time in a gastight glass syringe. The oxygen uptake (V02 expressed as ml ind.-1 h-1) was calculated using the formula:

$$V0_2 = dpO_2 \text{ x } aO_2 \text{ x } VH20 / N \text{ x } dt \qquad (1)$$

where dpO_2 was the difference in oxygen tension reading between the respirometer water of the control and the experimental groups. The solubility of oxygen (aO2) was established from literature (6,63 ml /litre at 8,0 °C and salinity of 34,5 ppt). VH20 is the respirometer water volume, N is the number of larvae and dt is the elapsed time in the respirometer.

The experiment was designed to find the dose - response relationship for phenols to newly hatched cod larvae (Figure 1). The centre point of the experiment (12 hours - 50 ppm) was a triplicate, intended to describe the variance in the system.

Total carbon and nitrogen were analysed by using a Carlo Erba Elemental Analyser, model 1106. At sampling, each individual larva was rinsed in distilled water, placed in a tin capsule and dried at 60 °C for a minimum of 72hours. The sample was combusted at 1016 °C in a quartz tube were nitrogen and carbon are converted to NO_2 and CO_2. The combustion gases passes a reduction reactor to remove excess of oxygen, and to reduce NO_2 to nitrogen gas (N_2). CO_2 and N_2 are separated on a chromatographic column. Acetanilide was used as an external standard.

The day after finishing the exposure period (approximately 2 days after hatching) the larvae were added an algae culture (Tetraselmis sp.), 700 ml to 43 l of sea water), and 24 hours later the rotifer Brachionus plicatilis (500 ml with a concentration of 200 ind. pr. ml) was added as live feed organism for the fish larvae. Every second day 20% of the water was replaced, using a Reheim aquarium pump removing dead larva and particles from the bottom of the bucket. Rotifers were added to the system every second day to keep a constant density of the live feed organism. The light regime was a normal cyclic regime with daylight period as found in April west of Lofoten in Norway. The density of the larva in the tanks were counted at the time of start-feeding. The water in the bucket was gently mixed and the total number and larva in a 5-800 ml sample was counted, (Table 2).

RESULTS AND DISCUSSION

No acute lethal effects were observed by visual inspection during the 24 hours exposure of newly hatched cod larvae to phenols. However, the concentration of phenols was observed to decrease during the 24 hours of exposure. In general, the residual phenol concentration in the exposure vessels after 24 hours was in the range 32-39 percent of the initial. Phenol concentration was the most rapidly decreasing, while the concentration of C1-alkylated phenols remained relatively constant during the exposure period. The C2- and C3 alkylated phenols decreased rapidly initially and remained constant during the remaining part of the exposure period. The observed changes can most likely be explained by accumulation/absorbtion in organic material, evaporation, biodegradation and adsorption to active surfaces.As a consequence, the true exposure doses in ppb x hours were lower than

Table 2. Density and reduction (mortality and elimination by sampling) of larvae during the experiment

Exposure	Initial density, larvae pr. litre (June 3)	Density larvae pr. litre at the end of the experiment (June 17)	Reduction during the experiment (14 days) (%)
Control, 1 hr.	82	25	69
50 ppb	93	37	61
Control, 6 hrs.	41	21	50
25 ppb	149	38	75
75 ppb.	82	56	32
Control, 12 hrs.	33	11	66
5 ppb.	176	38	79
50 ppb. (I)	162	20	88
50 ppb. (II)	19	8	56
50 ppb. (III)	59	6	90
100 ppb.	100	26	74
Control, 18 hrs.	80	23	72
25 ppb.	44	17	63
75 ppb.	73	26	64
100 ppb.	34	11	68
Control 24 hrs. (a)	34	9	75
Control 24 hrs. (b)	111	12	89
50 ppb.	75	33	57
75 ppb.	71	15	79
100 ppb.	110	27	76
Range (average)	19-176 (81)	6-56 (23)	32-90 (69)

intended from the experimental design (effective dose 3 to 1000 ppb - hours). However, based on results from dilution modelling and field studies at the Statfjord and Gullfaks fields, the actual exposure doses in the present experiment fall inside the realistic exposure range for organisms in the discharge influence area in the recipient (Furuholt 1995, Rye et al. 1995).

During the two weeks of evaluation following the exposure period, the density of larvae in the tanks decreased from an average of 81 to 23 larvae pr. litre. The average reduction, involving mortality and reduction by sampling, was 69 percent, as shown in Table 2. The density of larvae in the tanks at the start of the experiment was positively correlated with the density 14 days later at the end of the experiment (Spearmann Rank Correlation coefficient equal to 0.65, $p = 0.02$). The result indicate that there was no lethal effect of the chemicals during the experimental period.

The density of fish larvae was not expected to be critical for growth and survival within the range selected, due to the fact that the food organisms (algae and rotifers) were added in excess. During early start-feeding cod larvae have a positive buoyancy and most of the larvae were observed in the upper part of the water column. If 90% of the larvae was located in the upper 10 cm of the water column as an average, the result would be a larval density in the range of 73 - 681 larvae pr. litre. The minimum prey density required for growth and survival of cod larvae lies between 0.2 and 1.0 ind. / ml, using collected live zooplankton and a density of 26 cod larvae pr. litre in a total volume of 38 litre (Buckley 1979). In our experiment the initial density of rotifers were approximately 2 ind./ ml.

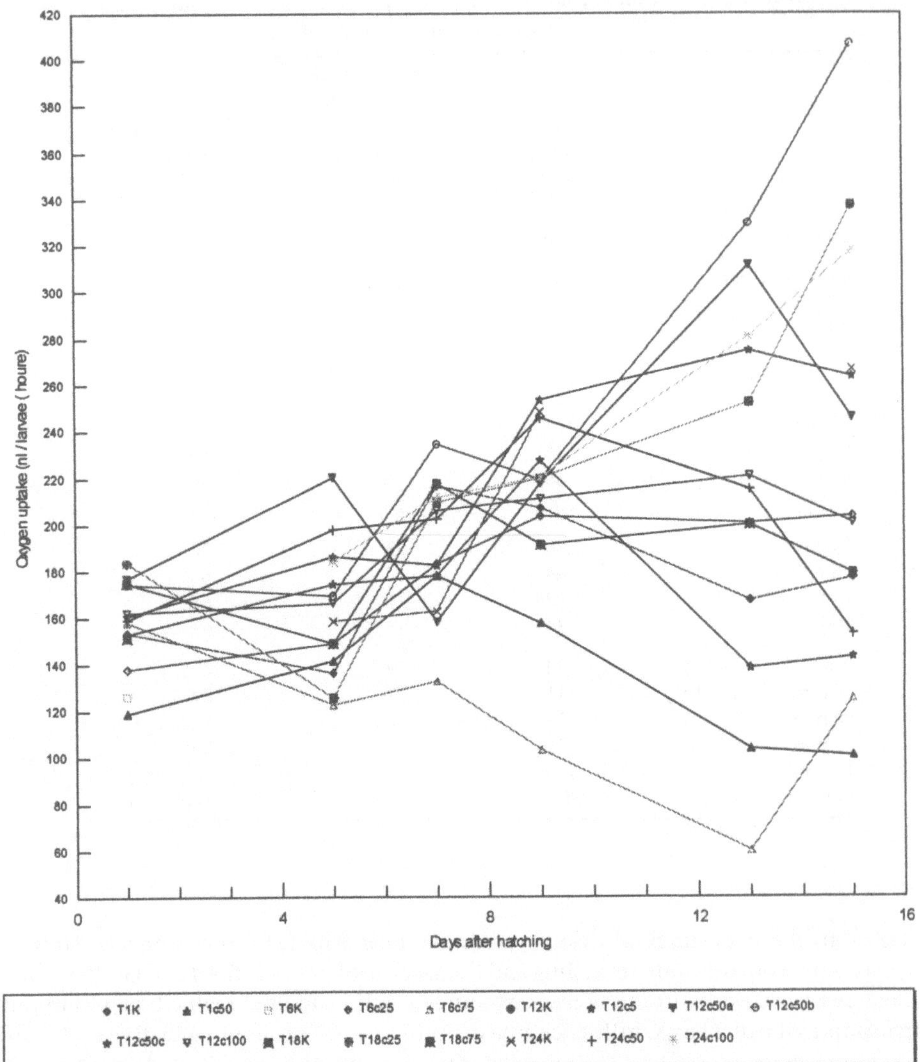

Figure 2. Oxygen uptake in developing larvae of cod. T— is the exposure time in hours, c—is the concentration (ppb) of the phenol mixture and K is the control. Example:T1K is the control transferred from the egg incubator 1 hour after starting the experiment, T1c50 represents the larvae that were exposed 1 hour in 50 ppb phenol mixture and so on.

There was no correlation (Spearmann, rank correlation) between the density of larva, and the reduction of larvae in the tanks during the experiment, indicating that the density of larvae did not affect the observed mortality in the experiment. No correlation was observed between the density of larva at the end of the experiment and exposure dose (time vs. concentration), indicating that the exposure up to a dose of 2400 ppb - hr did not affect the mortality.

The uptake of oxygen, shown in Figure 2, increased from approximately 160 to 220 nl per larva per hour during the experiment. The group with the highest oxygen uptake at the end of the experiment was one of three parallels in the centre point, 50 ppb, 12h (II).

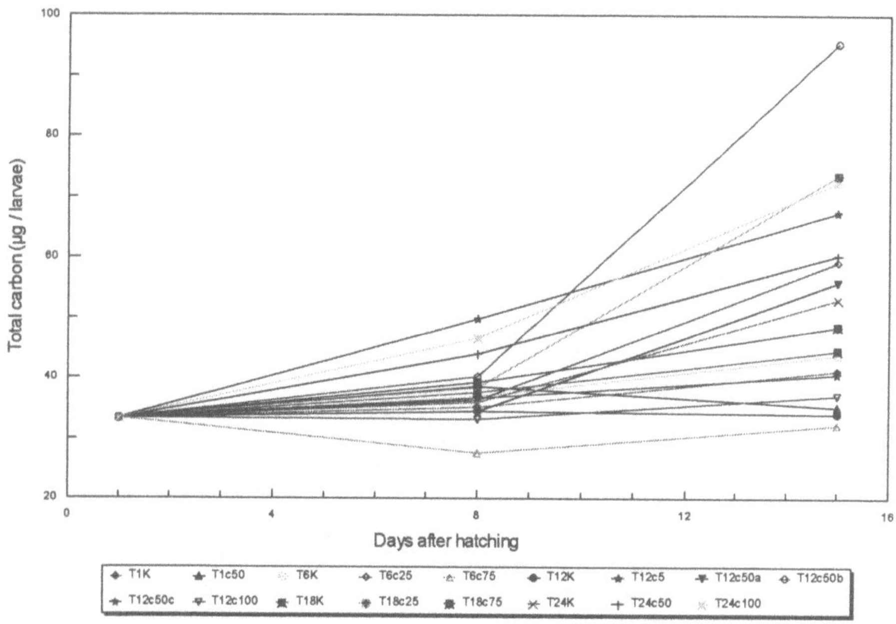

Figure 3. Total carbon content in developing larvae of cod. T— is the exposure time in hours, c—is the concentration (ppb) of the phenol mixture and K is the control. Example: T1K is the control transferred from the egg incubator 1 hour after starting the experiment, T1c50 represents the larvae who were exposed 1 hour in 50 ppb phenol mixture and so on.

This was larvae from the bucket with the lowest initial density in the experiment, 19 larvae /litre and low final density of 8 larvae /litre. The density of larvae at the end of the experiment was negatively correlated with the measured oxygen consumption (corr. coeff. -0.74, p=0.009) indicating higher oxygen consumption in the tanks with a low density of larvae (Figure 2). This observation was supported by the negative correlation between larval density at the end of the experiment and total content of carbon (corr. coeff. - 0.58, p= 0.037) and total content of nitrogen (corr. coeff. -0.58, p=0.04) indicating bigger larvae in the tanks with fewer larvae (Figure 3).

The total carbon and nitrogen status is a well established method in the study of fish eggs and larvae, and in the search for optimal physical condition for embryonic and larval development (Jørgensen, 1985). Total carbon is linearly related to the dry weight of the larva, and total nitrogen is an adequate approximation for the body protein.

The total content of carbon increased for all groups during the 16 days following hatching, as shown in Figure 3. The variation in the total carbon content between the groups increased during the period, indicating different rate of growth in the different tanks. The group with highest content of total carbon was (50 ppb, 12h,b), the group with the highest oxygen uptake and the lowest initial density. Figure 4 shows the variation in content of total nitrogen, which followed the same pattern as observed for total carbon during the experiment, the Speermann rank correlation was equal to 1, indicating at the parameters are parallel ways to describe larval growth. Close dependency between the measure of total carbon and total nitrogen can be expected, since the result are generated from analysis of the same larva. The ratio between total carbon and total nitrogen remained fairly constant close to 4 during the experiment.

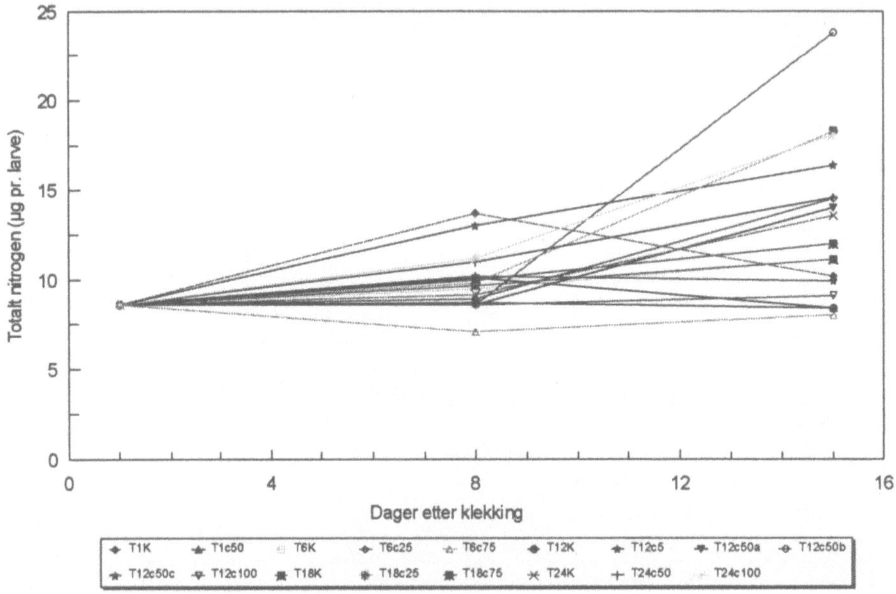

Figure 4. Total nitrogen content in developing larvae of cod. T— is the exposure time in hours, c—is the concentration (ppb) of the phenol mixture and K is the control. Example :T1K is the control transferred from the egg incubator 1 hour after starting the experiment, T1c50 represents the larvae who were exposed 1 hour in 50 ppb phenol mixture, etc.

CONCLUSIONS

Based on the results from the present experiment, it can be concluded that no lethal effects to newly hatched cod larvae were caused by exposure doses of produced water phenols in the range 3 to approximately 1000 ppb - hours. This range represents a realistic exposure regime for the Statfjord and Gullfaks fields in the North Sea. In addition, no effect on oxygen consumption or growth caused by the produced water phenols could be observed during the first 14 days of start feeding after the exposure period. On the other hand, high larval density resulted in decreased growth and oxygen consumption for some of the groups in the study.

REFERENCES

Anon. 1993a. Federal Register (USA), Vol. 58, no. 41, p. 12454.

Anon. 1993b. GESAMP (IMO/FAO/UNESCO/WHO/UN/UNEP Joint Group of Experts on the Scientific Aspects of Marine Pollution). Impact of oil and related chemicals on the marine environment. Rep. Stud. GESAMP (50): 180pp.

Anon. 1994. OLF, "Environmental programme, Discharges to Sea" Report phase 1, Part B. OLF, Stavanger Norway, 1991.

Brendehaug, J., Johnsen, S., Bryne, K.H., Gjøse, A.L., Eide, T.H. and Aamot, E. 1993. Toxicity testing and chemical characterization of produced water - a preliminary study. Produced Water. Tecnologies, Environmental Issues and Solutions. Plenum Press New York.ødtekst16.

Buckley, L.J. 1979. The relationships between RNA-DNA Ratio, Prey Density, and Growth Rate in Atlantic Cod (Gadus morhua) Larvae. J. Fish. Res. Board Can. 36: 1497-1502.

Eilertsen,, B., Morsness, E., Solemdal, D., Strømme, T., Tilseth, S., Westergard, T. and Øistad, V. 1980. Some biological aspects of cod larvae. Fiskeridirektoratets havundersøkelse : 17, p29-47.

Falck-Pettersen, I. B. and Kjørsvik, E. 1987. Acute toxicity tests of the effect of oil and dspersants on marine fish embryos and larvae - a review. Sarsia 72, p411-413.

Folkvord, A., Kvenseth, P.G, Petersen, T. and Øiestad, V. 1985. Mass production of juvenile atlantic cod in a pond. Results and new aplications in 1985. ICES C.M. 1985, F63, 12pp.

Furuholt, E. 1995. Environmental effects of discharges and reinjection of produced water. Presented at the 1995 International Produced Water Meeting, Trondheim, Norway.

Johnsen, S. A. Th. Smith, J. Brendehaug, H. Riksheim and A.L. Gj>se. 1994. Identification of sources of acute toxicity in produced water. SPE 27138. Djakarta.

Jørgensen, L. 1987. An automated system for incubation of pelagic fish eggs. MIC., p8:47-50.

Middleditch, B.S. 1984. Ecological effects of produced water effluents from offshore oil and gas production platforms. Ocean Management, 9, p191-316.

Rye, H., Reed, M., Johsen, S. and Melbye, A. 1995. An intercomparison between field measurements and three different calculation models for dilution factors. Presented at the 1995 International Produced Water Meeting, Trondheim, Norway.

Serigstad, B.1987. Respiratory studies on cod with special reference to effect sof oil exposure on eggs and larvae. Dr. Scient thesis, University of Bergen, Norway.

Serigstad, B. and Klungsøyr, J. 1992. Giftighetstester av utslippsvann fra Ekofisk-tanken. Effektstudier på silde-egg og -larvae. Marine Research Institute, Bergen, Norway, Report No. 4/1992/HSM.

Smith, A. Th. 1993. Unpublished results.

Sommerville, H.J., Bennet, D.,Davenport, J.M., Holt, M.S.,Lynes, A., Mathieu, A., McCort, J., Parker, J.B, Stephenson, R.R., Watkinson, R.J. and Wilkinson T.G. 1987. Environmental effect of produced water from North Sea oil operations, Marin Poll. Bull. 16.10:549.

Sørstrøm, S.E., Aunås, T., Bjerkem, S., Brakstad, O.G., Johamsem, Ø., Kaarstad, I., Mork, G., Strømgren, T. andThendrup, A. 1992. Produced water. Chemistry and toxicity study. Reort made for OLF, Continental shelf and Petroleim Research Institute No 22.2059 00/01/02

Tollefsen, K.E., Ingebrigtsen. K., Olsen, A. and Johnsen, S. 1996. Lethality and toxicokinetic of 4-heptylphenol in juvenile atlantic cod. Submited for publication to Envron. Toxicol. and Chem.

Ynnesdal, H. and Furuholt, E. 1994 Environmental effects of produced water from large fields in the North Sea. An overview. SPE paper 27150, Djakarta.

Section III

ENVIRONMENTAL EFFECTS

12

SESSION SUMMARY

James P. Ray

The basis for most of the discussion and controversy related to produced water discharges into the world's oceans centers on concerns related to possible negative environmental effects. For the past decade, the scientific community has been studying a broad diversity of habitats and a variety of marine invertebrates and vertebrates. Studies have ranged from the determination of the fate, using both field assessments and mathematical models, to acute and chronic toxicity.

Presentations at this conference advanced our understanding of produced waters in all of the above categories. Clelia Booman and Ron Stagg reviewed recent studies into the effects of water soluble fractions of crude and produced water on fish larvae and crustaceans, and the relationship of produced water in the North Sea relative to metabolic enzymes in fish larvae.

In contrast, Dr. Scott Carr of the U.S. National Biological Survey discussed results of studies related to long term produced water discharges in shallow estuarine waters. They utilized a multi-tier testing approach (Sediment Quality Triad) which incorporates sediment chemistry, sediment toxicity, and in situ effects). Somewhat in contrast, Dr. Mulino presented results from a recent wetlands study conducted in the state of Louisiana for the U.S. Department of Energy. One of the key questions for the future in the U.S. relates to the low term recovery of shallow water, long energy environments, once the produced waters cease. Data from this study is important to support future cleanup vs. no cleanup decisions.

Dr. Neff focused on the issue of bioaccumulation of produced water related components. He noted that many of the organics present in produced water would not/do not bioaccumulate because of their physical characteristics. A fairly good predictor of bioaccumulation potential is the use of the octanol/water partition coefficient (Kow) for each chemical entity. Those compounds with Kow's above 5.0 have the potential for bioaccumulation. From actually studies, it has been determined that observed Biological Concentration Factors (BCF's) are almost always lower than the predicted BCF's. Data such as presented by Dr. Neff is a good example of the potentially over-conservative estimates made by modeling biological/chemical interactions.

Dr. Stagg's presentation sparked considerable interest and discussion from the attendees. As pointed out by Dr. Glickman, only certain of the PAH's induce EROD activity in selected marine species. Many of the PAH's present in produced waters that might be EROD inducers are present in only low concentrations. Dr. Stagg pointed out that the five and six ring PAH's are the primary inducers. An attendee asked if there had been any documented changes in fish chemistry due to produced water discharges. Several people responded and were not aware of any such changes.

Another interesting point made in the discussion of Dr. Stagg's paper was related to the origin of the "aromatics" being measured by their towed fluorometer. The instrument

Produced Water2: Environmental Issues and Mitigation Technologies
Edited by M. Reed and S. Johnsen, Plenum Press, New York, 1996

was integrating information over long towed transects. Especially in the more southern parts of the North Sea, numerous other sources could be contributing to the measured aromatic levels, including nonpoint source runoff, vessel spills, and previously discharged oil based mud cuttings which used diesel fuel as a base material. More detailed analytical techniques will be required to determine sources for the low levels of aromatics being detected.

Questions were raised regarding previous consideration of effects on general fish stocks. In most cases, the year-to-year natural variability makes it almost impossible to discern cause and effect relationships related to produced water. This has been the case in both the North Sea and in the Gulf of Mexico.

In summary, investigators from both Europe and the U.S. are using a variety of approaches and tools to evaluate the environmental fate and effects of produced water discharges. Because of the low level of individual components present, the rapid dilution upon discharge, and the relatively low exposure time to marine organisms, it is very difficult to measure impacts in the field. Laboratory approaches are being pursued to try and determine low level, sublethal responses. Care will need to be taken in the future to ensure that laboratory exposure methods and species are realistic and truly representative of real world situations. Two major field studies have just been completed in the Gulf of Mexico and were not available for publication in this proceedings. These were the U.S. Minerals Management Service program called GOOMEX which was conducted by the Geochemical Environmental Research Group at Texas A&M University, and the offshore bioaccumulation study conducted by the Offshore Operators Committee (OOC). The GOOMEX study assessed field impacts from produced water discharges at several locations, and included sublethal responses including reproductive effects, EROD induction, pore water toxicity, and genetic effects. The OOC study assessed actual bioaccumulation in platform resident fish, molluscs, and crustaceans from twelve different high volume discharging platforms across the entire Gulf.

Likewise, modeling approaches must also be cautious about incorporating unrealistic, sometimes overly conservative assumptions. There was much interest, and some differing opinions, of the appropriateness of the North Sea CHARM model for predicting effects and selecting treatment chemicals for use in produced water treatment systems. The importance of field verification of model predictions was noted.

13

ECOTOXICOLOGICAL RISK OF PRODUCED WATER DISCHARGED FROM OIL PRODUCTION PLATFORMS IN THE STATFJORD AND GULLFAKS FIELD

Chris C. Karman,[1] Ståle Johnsen,[2] Henk P. M. Schobben,[1] and Martin C. Th. Scholten[1]

[1] TNO Institute of Environmental Sciences, Energy Research and Process Innovation
Environmental Quality and Safety Division
Department of Ecological Risk Studies
PO Box 57, 1780 AB Den Helder, The Netherlands
[2] STATOIL
Postuttak
7005 Trondheim, Norway

INTRODUCTION

The actual ecotoxicological risk of the discharge of produced water from an offshore oil or gas production platform is mainly based on the dilution processes around the platform. Risk analysis based on the chemical composition of the produced water may give some insight into the contribution of the individual components to the overall ecotoxicological risk, allowing specific measures to reduce the discharge of components associated with the highest risk.

The basic principle of ecotoxicological risk assessment is the comparison of the exposure of biota to an environmental concentration of a chemical with the sensitivity of this biota. The exposure concentration can be estimated by the Predicted Environmental Concentration (PEC) of a component in the water by multiplying the concentration in the production water with a dilution factor. The sensitivity of biota can be estimated by the PNEC (Predicted No-Effect Concentration). The basis of the calculation of a PNEC is the translation of laboratory toxicity data to the sensitivity of species in the field. The PEC:PNEC ratio is used as a measure of ecotoxicological risk (Basietto *et al.*, 1990; EC, 1993).

In this study a method will be demonstrated allowing the estimation of the ecotoxicological risk of the effluent (produced water) at various dilution rates on the basis of PEC:PNEC analysis of single constituents.

Produced Water2: Environmental Issues and Mitigation Technologies
Edited by M. Reed and S. Johnsen, Plenum Press, New York, 1996

Table 1. Concentration of the components of production water, discharged from oil platforms in the Gullfaks and Statfjord fields

Chemical	*Concentration ($\mu g.l^{-1}$)*	*Chemical*	*Concentration ($\mu g.l^{-1}$)*
Aliphates	15000	Cadmium	10
Naphtalene	942	Copper	2
Benzene	2000	Cr-VI	10
Toluene	1000	Lead	50
Xylene	1000	Zinc	6.8
Total Petroleum Hydrocarbons	*19942*	Butyric acid	39000
Acenaphtene	0.001	Caproic acid	9000
Benzo(a)antraceen	0.0193	Propionic acid	106000
Benzo(a)pyreen	0.001	Valeric acid	18000
Benzo(ghi)peryleen	0.001	Biocid process (MB544)	1200
Benzo(k)fluorantheen	0.0197	Biocid waterinjection (Kathon OM)	0.0001
Chrysene	0.226	Corrosion inhibitor (PK 6050)	1500
Fluorantheen	0.0854	Corrosion inhibitor(VN 6000 K)	300
Fluorene	12	Scale inhibitor process (SP 2500)	2100
Phenantrene	41.7	Scale inhibitor process (SP 2945)	200
Phenol	5100	Scale inhibitor well treatment (S 432)	6800
C1-Phenol	2500	Anti foam (AF 119)	100
C2-Phenol	400	Flocculant (ML 2317 W)	2400
C3-Phenol	130	Glycol	7700
C4-Phenol	26	Methanol	300
C5-Phenol	16		
C6-Phenol	13		
C7-Phenol	5		

METHODS

PEC

From measurements at the outlet of the specific platforms, the concentration of various components in produced water is known (Table 1). The concentration of each of the components in the ambient water is usually much lower than the concentration at the outlet of the platform due to dilution in the water surrounding the platform. The concentration of a component in the ambient water can be calculated by multiplying the concentration at the outlet with a factor representing the dilution at a specific distance from the platform. In this study the ecotoxicological risks are presented as a function of the dilution of produced water.

PNEC

The PNEC represents the highest concentration at which all (or, in practice, almost all) biota in the ecosystem are protected. A general ecosystem sensitivity value (PNEC) can be extrapolated from single species toxicity data obtained from laboratory tests. To derive a value that ensures protection of all species in the ecosystem, the toxicity data has to be corrected for the uncertainties related to the extrapolation of these test data toward the actual field situations. Among these uncertainties are differences between laboratory and field conditions, differences between test species and local biota and indirect effects due to species interactions.

Scholten & Schobben (in prep) developed a method to calculate a PNEC from all available toxicity data, on the basis of OECD (OECD 1993) and HC5 (Kooijman 1987; van

Straalen & Denneman 1989; Aldenberg & Slob 1991) principles. The geometric mean EC_{50}, summarising the toxicity information, is divided by a safetyfactor which is dependent on the amount of available data. The PNEC is thus defined as:

$$PNEC = \frac{GM}{1000 / \sqrt{n}}$$

in which: GM = geometric mean of all available EC_{50} values; n = number of species for which toxicity data is available for the chemical concerned.

In Table 2 the PNEC estimates of all components are presented. These estimates are based on toxicity data obtained from MEDUSA's Head (the database system of the TNO Department for Ecological Risk Studies), except for production chemicals for which toxicity data was provided by Statoil.

The PNEC of total petroleum hydrocarbons is extrapolated from toxicity data of various crude oils (Scholten *et al.*, 1993). Because of a lack of data for C2 and C3-alkylphenols, the PNEC for C4-alkylphenols is in this ecotoxicological risk analysis used for comparison with the sum of the PEC of C2, C3 and C4-alkylphenols. In fact this is a conservative approximation as the toxicity of alkyl-phenols decreases with decreasing C-number. The PNEC for C6-alkylphenols is used for comparison with the sum of the PEC's of C6 and C7-alkylphenols, which might give an underestimation of the risk.

Risk

The environmental risk of all components is first expressed as the PEC:PNEC ratio of the single components respectively. To calculate the risk of produced water the separate PEC:PNEC values of all components have to be combined into one single risk value. It has been demonstrated that that for a combination of more than five toxicants an additive

Table 2. Overview of the PNEC of the components of production water

Chemical	*PNEC* *(μg.l⁻¹)*	*Chemical*	*PNEC* *(μg.l⁻¹)*
Total petroleum hydrocarbons	19.9	Cadmium	5.9
Naphtalene	13.4	Copper	2.5
Benzene	248.4	Cr-VI	3.0
Toluene	62.4	Lead	2.6
Xylene	23.8	Zinc	2.9
Acenaphtene	2.2	Butyric acid	237.3
Benzo(a)antraceen	$6.0\ 10^{-3}$	Caproic acid	175.5
Benzo(a)pyreen	1.0	Propionic acid	152.8
Benzo(ghi)peryleen	$<1.0\ 10^{-3}$	Valeric acid	109.0
Benzo(k)fluorantheen	$1.0\ 10^{-3}$	Biocid process (MB544)	3.8
Chrysene	2.4	Biocid waterinjection (Kathon OM)	1.6
Fluorantheen	4.1	Corrosion inhibitor (PK 6050)	2.9
Fluorene	5.7	Corrosion inhibitor(VN 6000 K)	9627.2
Phenantrene	1.0	Scale inhibitor process (SP 2500)	1774.7
Phenol	71.6	Scale inhibitor process (SP 2945)	864.3
C1-alkylphenol	111.9	Scale inhibitor well treatment (S 432)	374.7
C4-alkylphenol	4.1	Anti foam (AF 119)	107.2
C5-alkylphenol	2.5	Flocculant (ML 2317 W)	3413.7
C6-alkylphenol	0.5	Glycol	5163.1
		Methanol	25942.2

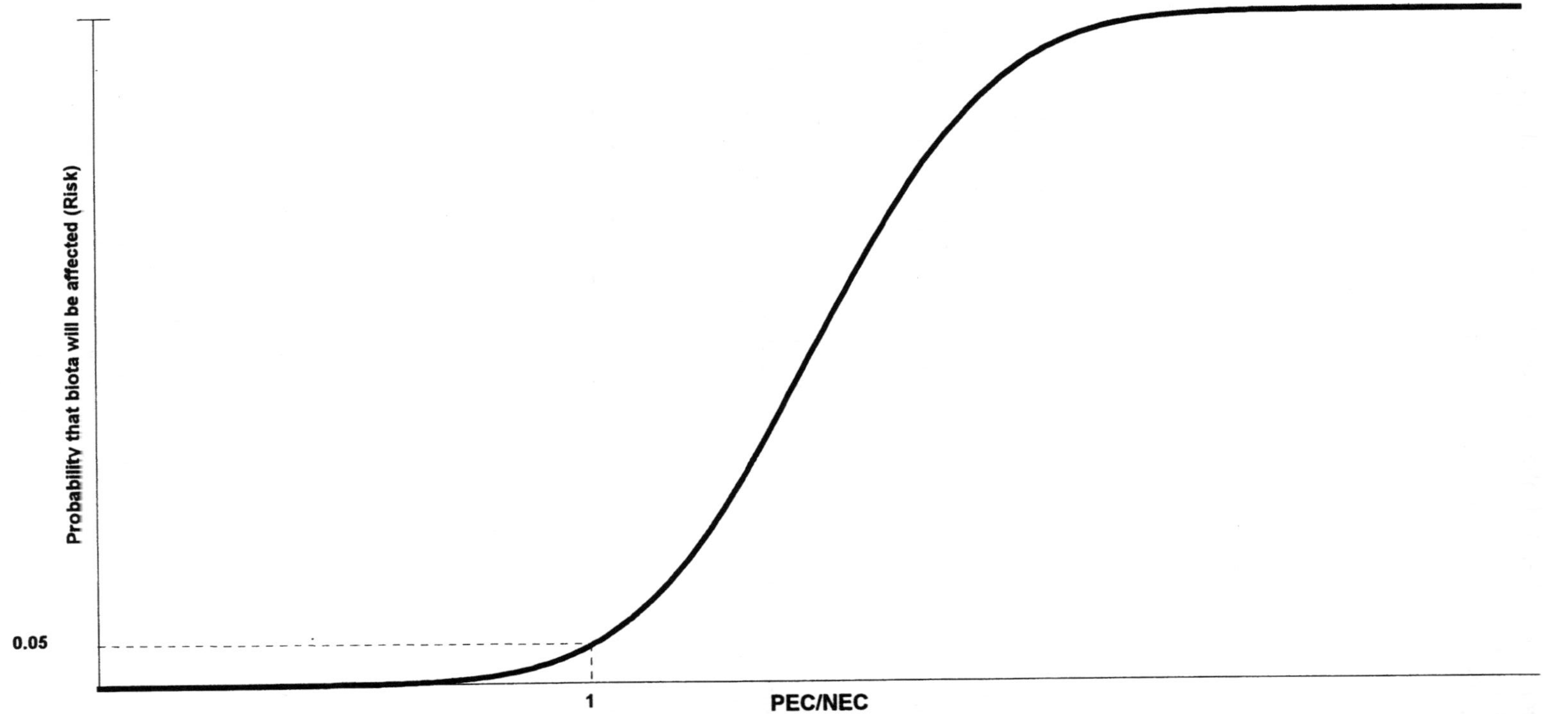

Figure 1. An example of the relationship between the PEC:PNEC ratio and the actual ecotoxicological risk. The shape of the curve is determined by the variation in sensitivity of biota for benzene. The relationship is calibrated so, that a PEC:PNEC ratio of 1 equals an ecotoxicological risk of 5%.

approach is valid, as synergistic and antagonistic effects are cancelling each other out. However, this does not mean that the PEC:PNEC ratios of the individual components can simply be added to derive a PEC:PNEC value for the produced water. The PEC:PNEC ratio's are first translated into a real risk estimate (i.e. the probability that biota are affected) on the basis of ecological variation in sensitivity amongst biota (see Figure 1). Subsequently, the set of probability estimates are combined into an overall probability estimate representing the environmental risk of produced water, using the rule for combining independent probabilities:

$$P(A+B) = P(A) + P(B) - P(A) * P(B)$$

in which:

P(A+B) = probability that biota will be affected by a combination of toxicants A and B.
P(A) = probability that biota will be affected by toxicant A.
P(B) = probability that biota will be affected by toxicant B.

RESULTS

In Figure 2 the estimated overall risk of produced water of the Statfjord and Gullfaks field is presented as a function of dilution. At low dilution rates (1:1 to 1:32) the probability that biota are affected by the produced water is 1. At a dilution varying between 1:100 and 1:1000 the risk of the discharge of produced water is reducing. At higher dilution rates the risk of the discharge of produced water is 0.

The relative proportions of the ecotoxicological risk of the components of produced water, presented in Figure 3, change at increasing dilution rates. When dilution is lower than 1:1000, the proportion of organic acids increases until it almost equals the proportions of production chemicals and petroleum hydrocarbons together (46%:25%:25%). At higher dilution rates the proportion of organic acids decreases, leading to a somewhat higher proportion of production chemicals and total petroleum hydrocarbons.

Although not presented in the figures, the influence of an increased concentration of oil (concentration of total aliphatics: 30 000 $\mu g.l^{-1}$) in the produced water was studied as well. It was shown that the environmental risk of produced water only slightly increases at higher concentrations of aliphatics. The proportion of total hydrocarbons relative to the other components is dominating when dilution reaches 1:100000.

DISCUSSION

Jacobs and Marquenie (1991) performed a series of toxicity tests with effluents of gas production platforms in the North Sea, and calculated median effect concentrations expressed as dilution rates. The EC_{50} of produced water appeared to be in the range of 1:4 and 1:1100. Effluents from oil production platforms are generally somewhat less toxic (EC_{50} in the range of 1:10 - 1:100; Slager *et* al, 1992). In this study the dilution giving a risk estimate of 0.5, which is comparable to the median effect concentration, is established between 1:100 and 1:320. This indicates that the method for estimation of the risk of produced water from the individual compounds generates results that are very comparable to measured values, although somewhat more conservative. This might be due to degradation or evaporation of chemicals during transfer of effluents into the test systems.

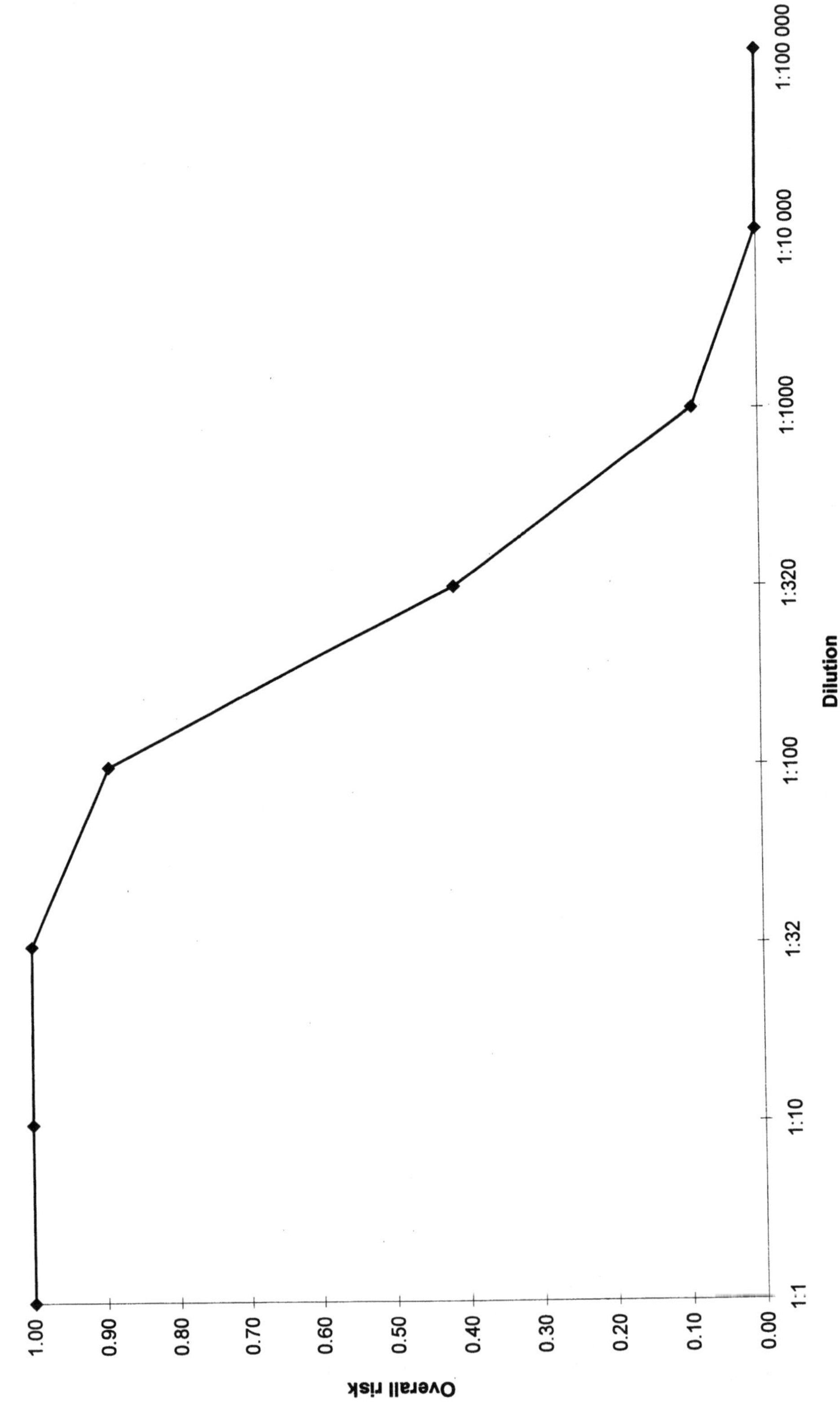

Figure 2. The overall risk of produced water, discharged from oil production platforms in the Gullfaks and Statfjord fields, as a function of dilution.

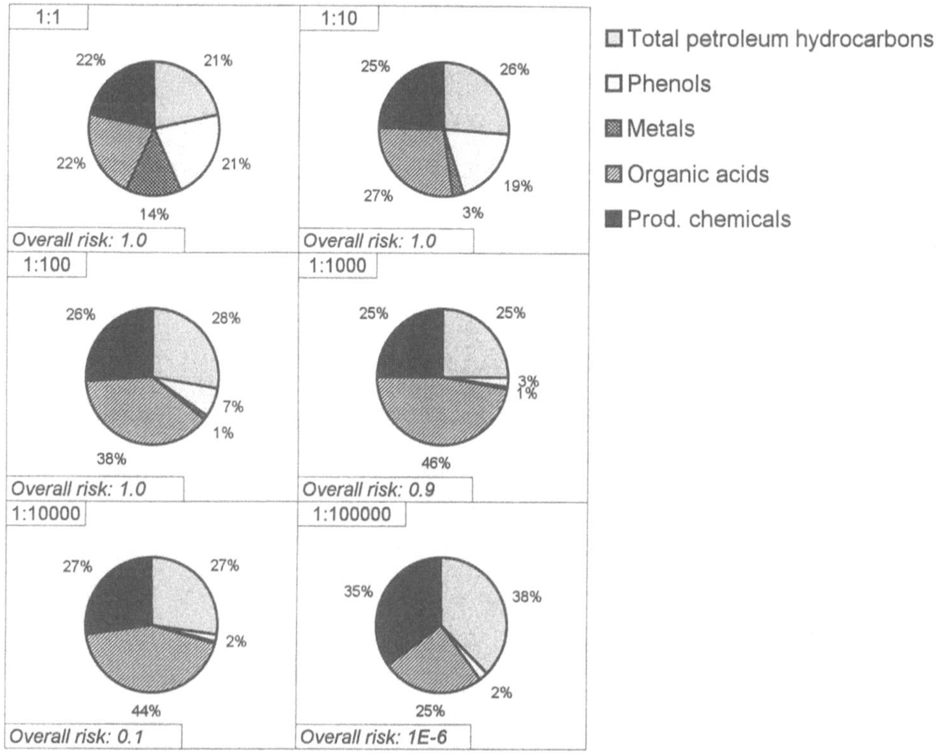

Figure 3. Relative proportions of the ecotoxicological risk of the components of produced water, presented for a range of dilution rates.

This study reveals that there is an asynchronous change of risk when dilution increases. This behaviour is due to differences in variation of sensitivity between biota for a specific chemical (Smit *et al.*, 1995).

At a dilution slightly higher than 1:1000 the risk of produced water reaches a level that is often regarded to be safe for the environment, because the probability that the ecosystem is adversely affected due the discharge of produced water is then lower than 5%. Although it seems that at this dilution rate there is no influence on the environment, it must be noted that the results of this analysis are based on acute toxicity data. An extensive risk analysis should also focus on long term, chronic, environmental effects. Discharges of lipofilic and persistent chemicals over a period of years might result in substantial bioaccumulation which potentially affect species at the end of the foodchain (such as carnivorous fish, seals and purpoises).

CONCLUSIONS

- This study demonstrates that ecological risk analysis can be a valuable tool for the evaluation of the environmental effects of produced water.
- The overall risk estimated from the risks of single components are consistent with observations from toxicity tests with complete effluents (produced water).

- Risk of produced water decreases rapidly with dilution, reaching an acceptable level (5%) when dilution is slightly higher then 1:1000.
- The overall risk mainly determined by organic acids, total petroleum hydrocarbons and production chemicals.

REFERENCES

Aldenberg T. & W. Slob (1993): Confidence limits for hazardous concentrations based on logistically distributed NOEC toxicity data. Ecotoxicol. Environ. Saf. 25:48-63.

Bassieto, J., D. Hinckley, J. Plafkin, M. Slimak (1990): Ecotoxicity and ecological risk assessment. Environmental Science & Technology, 24: 11-15.

EC (1993): Richtlijn 93/67/EEG van de commissie van 20 juli 1993 tot vaststelling van de beginselen die gelden bij de beoordeling van de risico's voor mens en milieu van stoffen die zijn aangegeven krachtens Richtlijn 67/548/EEG van de raad. Publikatieblad van de Europese Gemeenschappen Nr. L 227/9.

Jacobs R.P.W.M. & J.M. Marquenie (1991): Produced water discharges from gas/condensate platforms: Environmental considerations. Proc. 1st Int. Conf. on Health, Safety and Environment, The Hague, The Netherlands, 10-14 nov. 1991. Society of Petroleum Engineers. pp. 89-96.

Kooijman, S.A.L.M. (1987): A safety factor for LC_{50} values allowing for differences in sensitivity among species. Water Research, 21: 269-276.

OECD (1993): Guidance document for aquatic effects assessment. Report prepared for the OECD Hazard Assessment Advisory Body.

Scholten, M.C.Th., H.P.M. Schobben, C.C. Karman, R.G. Jak & H. van het Groenewoud (1993): De berekening van het maximaal toelaatbare risico van olie en oliecomponenten in water en sediment. TNO Report R93/028.

Slager, L.K., B. van Hattum, M. Tromp Meesters, M.C.Th. Scholten, N.H.B.M. Kaag, W.P. Cofino & J.F. Feenstra (1992): Environmental aspects of produced water discharges from oil and gas production on the Dutch continental Shelf. IVM Report E-92/19.

Smit, M.G.D., H.P.M. Schobben & C.C. Karman (1995): Hellingen van concentratie-effect relaties: Een verkennend onderzoek. TNO Report R95/135.

Straalen N.M. van & C.A.J. Denneman (1989): Ecotoxicological evaluation of soil quality criteria. Ecotoxicol. Environ. Saf. 18:241-251.

14

THE N-OCTANOL/WATER PARTITION COEFFICIENT

A Critical Parameter in Environmental Risk Assessment of Offshore E&P Chemicals

Eilen Arctander Vik,[1] Siri Bakke,[1] David R. Johnson,[2] and Johan Verburgh[3]

[1] Aquateam - Norwegian Water Technology Centre A/S
P.O. Box 6326 Etterstad, 0604 Oslo, Norway
[2] Conoco Norway Inc.
P.O. Box 488 Dusavik, 4001 Stavanger, Norway
[3] European Chemicals Bureau
JRC-ECB (TP280), I-21020 Ispra (VA), Italy

INTRODUCTION

The objective of the Oslo and Paris Commission's (OSPAR) work connected to offshore exploration and production (E&P) chemicals, is to reduce the use and discharge of the chemicals posing an unacceptable risk to the marine environment. Within OSPAR the required test procedure for offshore E&P chemicals have, to a large extent, been harmonized. The standard methods used are generally based on well established OECD Guidelines or proposed guidelines and require that that the tests are conducted according to OECD's GLP (Good Laboratory Practice) principles. The basic data set includes the following ecotoxicological tests (OSPAR, 1995):

Toxicity	Algae: *Skeletonema costatum* Herbivore: *Acartia tonsa* Sediment reworker: *Corophium volutator* Fish: *Scopthalmus maximus*
Biodegradation	A 28-day seawater test
Bioaccumulation	n-octanol/water partition coefficient, P_{ow}

It is generally accepted that environmental risk assessment needs to be performed to evaluate the potential hazard of chemicals. Risk assessment of chemicals is facilitated by the use of mathematical models describing partitioning, transport and transformation processes.

To assess environmental risks of chemicals and pollutants, quality criteria need to be based on an integrated approach including air, water/groundwater, sediment/soil and biota compartments. The rationale behind this is that the concentrations in surface water/seawater and sediments, or in groundwater and soil, are expected to approach thermodynamic equilibium, where partitioning plays an important role.

The Paris Commission (PARCOM) was in 1990 requested by the North Sea Ministers to develop a mandatory control system for the use and discharge of offshore E&P chemicals. In 1993, the Dutch and Norwegian authorities together with the offshore industry initiated the development of a decision supporting model, called CHARM (Chemical Hazard Assessment and Risk Management) model. OSPAR has recommended that all North Sea countries should accept and use the CHARM model as an internal control tool to evaluate and control the discharges of chemicals used offshore. The CHARM-model has been developed to predict the potential environmental risk of offshore chemicals (Øfjord et al., 1995). The model includes four modules:

1. Prescreening,
2. Hazard Assessment,
3. Risk Analysis, and
4. Risk Management.

The model is based on general hazard/risk assessment principles, where the predicted environmental concentration (PEC) is compared to the no-observed-effect concentration (NEC). The resulting PEC/NEC ratio is a central component in many hazard/risk assessment models. The hazard module is based on environmentally conservative principles. The environmental hazard assessment and risk analysis, include three ecological compartments i.e. the marine water column, sediments and food chain (biota). The highest PEC/NEC ratio of the three is designated as the $PEC/NEC_{ecosystem}$.

The n-octanol/water partition coefficient (P_{ow}) is for most chemicals a key environmental parameter, often serving as basic input data for the purpose of environmental exposure analyses (OECD, 1986). It has been shown by several authors (Beek et al., 1992) that there is a significant correlation between P_{ow} and the bioaccumulation potential e.g. fish (OECD, 1989). The P_{ow} also serves as basic input data for correlations with water solubility (Isnard and Lambert, 1989), soil and sediment adsorption (OECD, 1992), and with parameters predicting biological, biochemical and toxic effects. P_{ow} is also a useful parameter for establishing quantitative structure activity relationship (QSAR) for a wide variety of biological effects (OECD, 1989). Chemicals used offshore are often slighty water soluble, surface active and comprise complex mixtures of inadequately defined substances. The two OECD procedures recommended by OSPAR (OECD 107 and 117) have specific limitations with respect to analysis of these types of chemicals. Conoco Norway Inc. (CNI) initiated in 1995 and R&D project with the objective to find reliable methods for determination of log P_{ow} and validate the CHARM algoritms.

This paper describes the available methods to determine P_{ow}, and the weaknesses relating to typical offshore chemicals. It also describes the importance of P_{ow} in hazard/risk assessments, and the resulting uncertainty in such assessments. The paper also suggests a way forward for determination of reliable P_{ow} data for typical offshore chemicals.

METHODOLOGIES FOR DETERMINATION OF P_{OW}

The n-octanol/water partition coefficient for organic chemicals can be determined based on several methods, although OSPAR requires the use of OECD Guideline 107 and 117. The n-octanol/water partition coefficient (P_{ow}) is defined as:

$$P_{OW} = \frac{C_O}{C_W} \qquad (1)$$

where C_o and C_w are the equilibrium concentrations of the chemical in the octanol and water phases, respectively.

Shake Flask Method

This method is based on OECD test guideline 107 (1981). It was endorsed by the Council in 1981 and is now widely used as an internationally harmonized and recognized test method. The chemical to be tested is added to a flask containing equal volumes of octanol and water. The flask is mixed (~ 20 hours) and the two phases allowed to separate. The concentrations of the chemical in the octanol phase and in the water phase are then determined. The range of application of the method is limited by several factors, e.g. 1) the lack of chemical-specific analytical methods for certain chemicals, 2) adsorption of the chemical on equipment, and 3) the large volumes needed in the case of low solubilities. Increasingly, experience has shown that for this method the range of reliable results lie between log P_{ow} -2 and 4 (OECD, 1981). The method is especially applicable for single component chemicals which are water soluble, do not associate, dissociate, and are not surface active.

HPLC Technique

This technique is based on OECD test guideline 117 (1989). The test chemicals are injected into an analytical column wherein the different chemical structures are retained in proportion to their hydrocarbon - water partition coefficient. A set of suitable reference compounds with known log P_{ow} is used to calibrate the system. The log P_{ow} is determined from the retention times of the chemical in the column. For chemical mixtures exhibiting a fairly wide range of log P_{ow} values, the whole range should be reported. The use of "weighted average" log P_{ow} (based on peak area %) has therefore been recommended for use in some environmental hazard and risk evaluations of homologs, e.g. the CHARM-model (OSPAR, 1995). The HPLC-method is a rapid and inexpensive method which enables log P_{ow} determination in the range of 0-6. The method is not recommended for use with surface active chemicals (OECD, 1989).

Slow-Stirring Method

For highly lipophilic substances which solubilise only slowly in water, the "slow-stirring" method has been recommended (de Bruijn et al., 1989). The test method is similar to OECD 107, but the stirring time is longer (2-7 days) and the stirring intensity lower. De Bruijn et al. (1989) recommend 2-3 days for chemicals with log $P_{ow} < 4$, and from 3-4 days for chemicals with higher log P_{ow}. A good correlation is seen between results from this method and the Shake Flask method when log $P_{ow} < 4$.

QSAR Calculation Methods

Where no experimentally determined data are available, data derived from reliable and internationally recognized QSAR (Quantitative Structure Activity Relationship) models, based on molecular descriptors like fragmental constants, water solubility etc. as e.g. the ClogP-program (Meylan and Howard, 1994) can be used. Limitations of relationship has

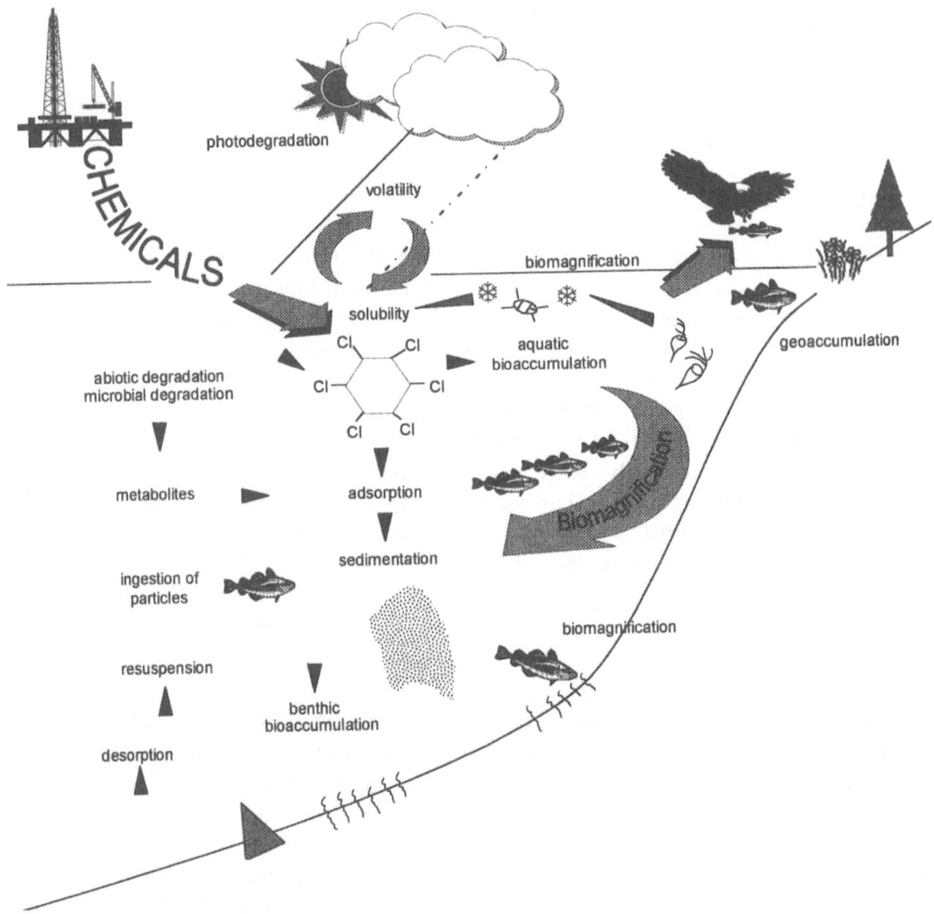

Figure 1. Offshore chemicals discharged with produced water go through several adsorption/desorption, sedimentation, degradation and bioconcentration processes.

been suggested for chemicals with log P_{ow} < 0 and > 5, but Pedersen et al. (1994) show a good relationship even at higher log P_{ow} values. For chelating compounds and surfactants, the program is stated to have limited reliability. The predictions are also considered to be unreliable for chemicals containing structures which are not included in the program. It is necessary to know the structure and the formulae of the chemicals, to find reliable P_{ow} data. This is seldom the case for typical offshore chemicals.

PARTITIONING OF OFFSHORE CHEMICALS

Offshore E&P chemicals discharged with produced water will undergo a number of partitioning processes. The potential complexity is illustrated in Figure 1. In general, the produced water is discharged below the sea surface, and distributed into the water column.

In the CHARM model the partitioning of E&P chemicals occur in the following steps (see illustration in Figure 2):

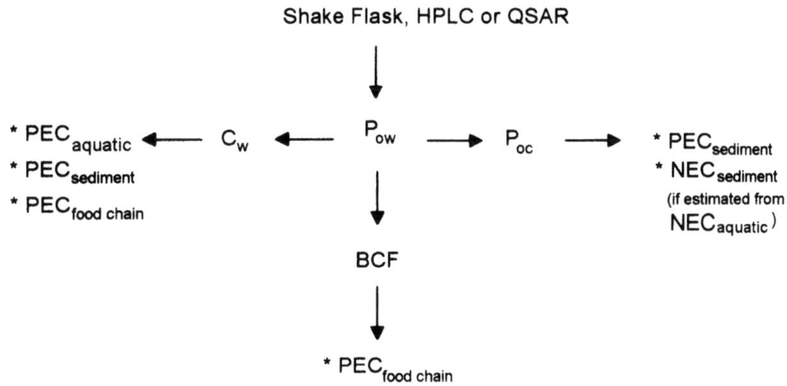

Figure 2. P_{ow} - an important parameter in the CHARM project, where

P_{ow} = octanol/water partition coefficient
C_w = chemical concentration in discharged/produced water
P_{oc} = sediment/water partition coefficient
BCF = bioconcentration factor
PEC = Predicted Environmental Concentration
NEC = No observed Effect Concentration

- Before discharges to the sea i.e. the partitioning of the chemicals between the oil and the water phases;
- After discharges to the sea i.e. the partitioning of the chemicals between the water column and the sediments, and organisms (bioconcentration).

The partitioning between water and air is for the typical offshore chemicals, not considered to be of significant importance.

Oil/Water Partitioning

Offshore chemicals, e.g. corrosion and scale inhibitors, are added to the production stream which typically consists of a mixture of oil and water. The added chemicals will partition between the oil and water phases. The oil is separated from the water and the produced water subsequently discharged. It is assumed that the concentration of the chemical in the produced water is dependant on the partitioning between the oil and water phases. In a study Bakke et al. (1993) defined the oil/water distribution factor (OWDF) as the oil/water partitioning equilibrium concentrations in seawater (C_w) and a "model" oil (Exxol D100S) (C_{oil}):

$$OWDF = \frac{C_{oil}}{C_W} \tag{2}$$

The Shake Flask method was modified to determine the OWDF for production chemicals, and a linear relationship was found between OWDF and P_{ow} (see Figure 3). In a mass balance study of offshore chemicals conducted by OLF (the Norwegian Oil Industry Association) (Hansen et al., 1995), a similar relationship between OWDF and P_{ow} was demonstrated for a system using crude oil and filtered produced water as the oil and water phases, respectively.

Some reviewers were, however, concerned that equilibrium conditions were not reached between the oil and water phases, in offshore processes, before the produced water

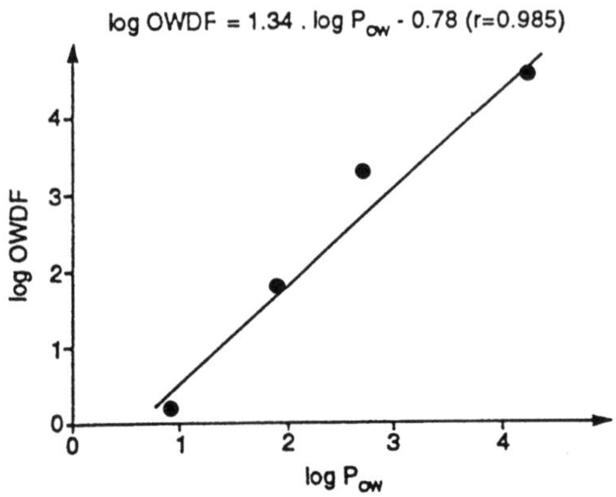

Figure 3. Correlation (in log scale) between the oil/water distribution factor (OWDF) and the n-octanol/water partition coefficient (P_{ow}) for a standard set of chemicals (Bakke et al., 1993).

was discharged. Accordingly, the partition coefficients were determined in the laboratory for different mixing times using a modified Shake Flask test (Hansen et al., 1995) (see Figure 4).

From the results in Figure 4 the following general conclusions can be drawn:

- There exist a relationship between OWDF and P_{ow}.
- More than 6 hours mixing are needed for laboratory tests to reach equilibrium for the chemical substances tested.
- The log OWDF varies from experiment to experiment.

After one hour mixing time the log OWDF varies from 0.8 to 1.9, and after 18 hours from 1.7 to 3.4 (Figure 4). The average log OWDF reaches 2.7 after 18 hours.

To compare the test results obtained in the laboratory with real offshore conditions the produced water was analyzed for the content of imidazoline (active ingredient in the corrosion inhibitor). The log OWDF values were found between 3.1 and 3.2 (Hansen et al., 1995). Comparing these results to the results seen in Figure 4, the oil/water partitioning was equal to those reached after 18 hours mixing in the laboratory. The field analysis of produced

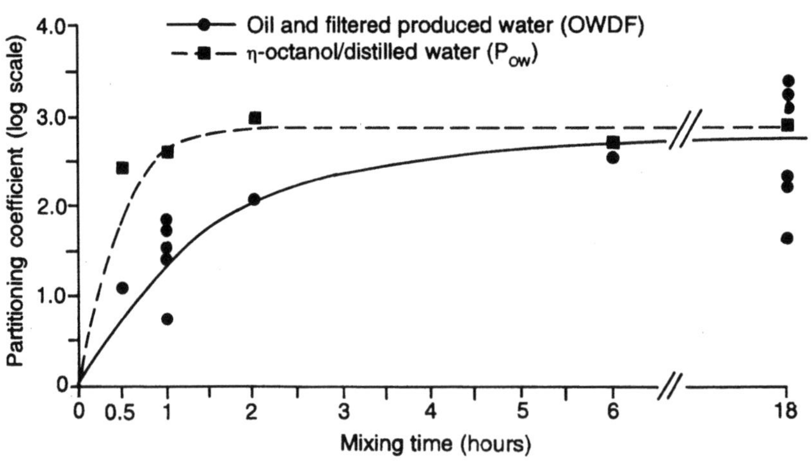

Figure 4. The influence of mixing time, on the organic phase/water partitioning, log OWDF and log P_{ow} for imidazoline (the active ingredient of the corrosion inhibitor). Data is based on Hansen et al., 1995.

water reported in the OLF study indicated that equilibrium conditions are reached offshore in spite of the short mixing times. The authors suggested that this was due to the high mixing intensity typically experienced in offshore production streams.

The potential influence of particulate matter and high temperatures were also studied in the laboratory, showing no effect. The influence of high solvent content in the produced water was also tested in the laboratory, but no consistent conclusions could be drawn (Hansen et al., 1995).

Based on the available test results, the CHARM project Steering Committee agreed on using the following equation to determine the concentration of a chemical discharged with produced water:

$$C_W = C_t \cdot f + 0.1 \cdot C_t \text{ (as long as } C_W \leq C_t) \quad (3)$$

where C_t is the chemical dosage, f is the fraction released of the chemical, and 0.1 • C_t is a safety factor to account for chemical dispersed in the water or adsorbed to particulate matter. The safety factor is intended to account for the variability in operating conditions. The fraction released is inversely proportional to P_{ow}.

Sediment/Water Partitioning

Surface active chemicals and chemicals with low water solubility are easily adsorbed to particles and to sediment. The sediment partitioning coefficient (P_{oc}) is defined as the ratio of the equilibrium concentration of a chemical in the water and the sediment phases. For non-polar organics, it has been established that partitioning is primarily dependant on the organic carbon content of sediment (OECD, 1992). For polar organics, it has been shown that organic carbon and cationic exchange capacity are important factors (Orth et al., 1994). In CHARM, the following relationship is used:

$$P_{OC} = f_{OC} \cdot P_{OW} \quad (4)$$

where f_{oc} = fraction of organic carbon in the sediments. In the hazard module of CHARM, f_{oc} has been fixed at 0.04, assuming 4% organic carbon in the sediments.

Air/Water Partioning

The water/air partioning can be based on Henry law's constant. In the CHARM model, it is, however, assumed that the concentration of production chemicals in air will be so small that it need not be taken into consideration.

Bioconcentration Factor (BCF)

It has generally been agreed that there is a linear relationship between P_{ow} and BCF up to a log P_{ow} of 6-7 for organic chemicals. The relationship between P_{ow} and bioaccumulation in organisms/fish has been demonstrated by several authors, (Veith, 1979; Mackay, 1982). Above log P_{ow} of 6 the polynominal equation of Connel and Hawkee (1988) or Buntein (1993) can be used. These equations show a breakdown of linearity above log P_{ow} of 6. In CHARM, the BCF is important factor in predicting the environmental concentration in biota, according to the following relationship:

$$BCF = f_{lip} \cdot P_{OW} \quad (5)$$

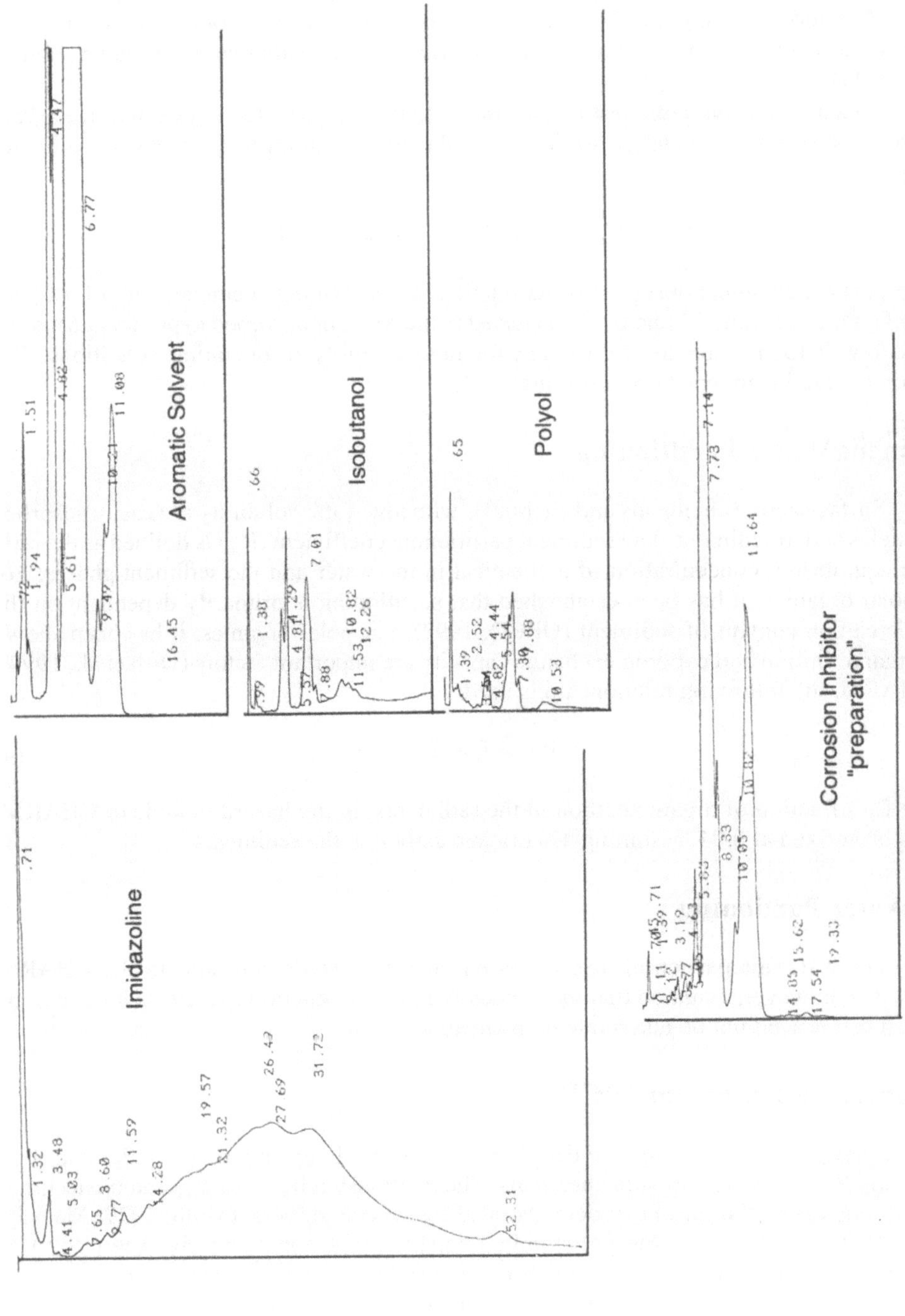

Figure 5. HPLC chromatogram of the corrosion inhibitor and its four substances.

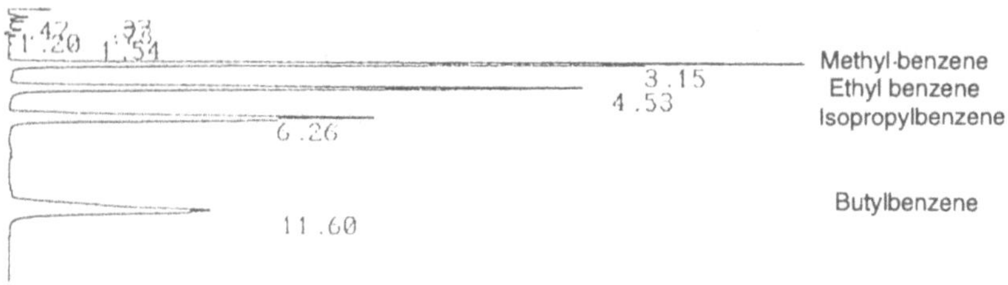

Figure 6. HPLC chromatogram of the prepåred mixture of benzene homologs.

The lipid content (f_{lip}) in biota is an obvious variable. In the development of CHARM, it was agreed to fix the lipid content at a relatively high level of 10%, (f_{lip} = 0.1) for puposes of calculating environmental hazard.

RESULTS AND DISCUSSION

For the purpose of this presentation log P_{ow} has been determined for two different chemical preparations. These are:

- A corrosion inhibitor typically used in North Sea oil operations, and
- A prepared mixture of well characterized homologs (containing relevant chemical substances used offshore).

Log P_{ow} have been determined using different methods. The two preparations have been selected to illustrate the difficulties connected to determination of partitioning of typical offshore chemicals.

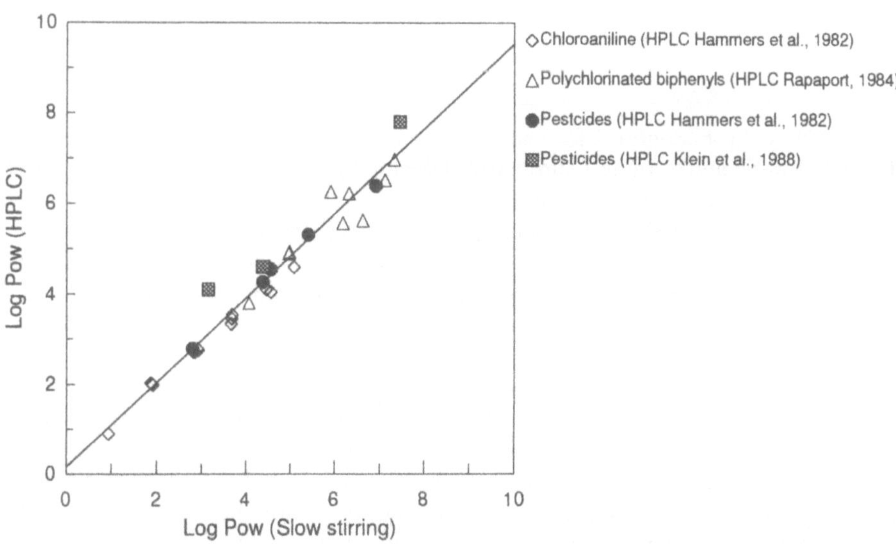

Figure 7. Comparison of log P_{ow} test results obtained from hydrophobic chemicals with the de Bruijn et al's (1988) slow-stirring technique and the HPLC technique. [Regression Output: Log Pow (HPLC)=0.92530*Log Pow (Slow Stirring)+0.2019 r2=0.9581, Std. err. Coef.=0.0347, n=33, Deg. of Freedom 31].

Table 1. Log P_{ow} data found in literature, determined from laboratory tests, in the literature and through QSAR calculations for the two chemical preparations studied

		log P_{ow}-data							
		HPLC		Shake Flask			QSAR and literature		
Chemical preparation	Chemical substances	Range	Weighted average	(1)	(2)	(3)	(4)	(5)	(6)
1. Corrosion Inhibitor	Imidazoline	0-6.1	5.7	3.0			7.0-8.5	3.1	6.5-6.6
	Aromatic solvent	3.5-4.7	4.1				3.2-3.7		
	Isobutanol	0-4.8	4.1				2.5-4.7		
	Polyol	0-4.6	4.1						
	Chemical preparation	3.7-4.8	4.5[7)]		1.3				
	Chemical preparation	0-6.1	5.1[8)]						
2. Prepared mixture	Methyl benzene	3.0	3.0			2.7	2.6		
	Ethyl benzene	3.5	3.5			3.2	3.2		
	Isopropyl benzene	4.0	4.0			3.7	3.6		
	Butyl benzene	4.7	4.7			4.6	4.2		
	Chemical preparation	3.0-4.7	4.2						

(1) Hansen et al., 1995
(2) Bakke et al., 1993
(3) OECD 117 data
(4) C log P (Leo and Weninger, 1988)
(5) Ghose et al., 1989
(6) Moreau et al., 1984
(7) Measured on preparation
(8) Calculated from weighted average of substances and composition of preparation
* Calculated log P_{ow} value from the known composition of corrosion inhibitor is 5.1

The HPLC chromatograms for each chemical substance in the two preparations are presented in Figures 5 and 6. Table 1 summarizes the log P_{ow} data, determined using different experimental techniques and QSAR-models.

Each substance in the corrosion inhibitor has been extensively characterized and the chemical composition has been determined in so far as possible with the application of relatively advanced test methods, i.e. GC/MS (gas chromatography/mass spectroscopy). The information of the structure of each substance was used to perform QSAR calculations to determine the log P_{ow} values.

Analyses showed that imidazoline, consisted of eleven different similar structures covering the mass formula $C_{20}H_{38}N_2$ to $C_{22}H_{41}N_3$. A model compound for the imidazoline, $C_{22}H_{45}N_3$, was used to find data through QSAR. Log P_{ow} was also estimated based on Moreau et al. (1984) to 6.5-6.6. A contributing factor to the high hydrophobicity determined was the long (C_{17}) alkyl chain. A lower log P_{ow} value may be reasonable because of a hydrophilic amino group at the end of the C_2 chain. For organic compounds, log P_{ow} >6 indicates "high hydrophobicity". When the method of Ghose et al. (1989) was used, log P_{ow} was found equal to 3.1, which is three orders of magnitudes lower. Using Leo and Weningers Clog P approach (1988) log P_{ow} is determined in the range of 7.0 to 8.5.

Log P_{ow} has also been determined by HPLC for each of the four substances of the corrosion inhibitor (see Figure 5). The individual homologs within each substance are subsequently calculated into a single weighted average value by integration of the area under the various peaks exhibited in the chromatogram. As seen from Table 1, QSAR calculated log P_{ow} data varies between 3.1 and 8.5 for the imidazoline, as compared to experimental HPLC data varying over the range 0-6.1 with a weighted average calculated to 5.7. Hansen et al. (1995) reported Shake Flask data to be 3.0 for a similar substance. The imidazoline is a typical surfactant, and according to OECD (1989), neither the Shake Flask nor the HPLC technique is recommended used. De Bruijn et al., 1989 recommend, however, the slow-stirring method for such chemicals, apparently on the basis of the deviations seen between the HPLC determined results and those obtained with the slow stirring technique. However, as shown in Figure 7 comparison of datasets generated by HPLC and the slow stirring techniques show a good correlation (n = 33 and r2 = 0.96). The largest deviations seen are at high log P_{ow}-values e.g. ± 0.3 log P_{ow} units, which is in fully acceptable agreement with results obtained by other methods (see Table 1). Log P_{ow} of 8.3 is the highest determined.

The characterization of the aromatic solvent showed that it consisted of a group of benzene isomers with the molecular formula C_9H_{12}. The QSAR calculations gave log P_{ow} values between 3.2 and 3.7, while the log P_{ow} values determined with HPLC were in the range of 3.5-4.7. The weighted average log P_{ow} was determined to 4.0. No Shake Flask data are available. The HPLC chromatogram shows a slightly higher weighted average log P_{ow} than the QSAR data.

The characterization of the isobutanol showed that this substance consisted of mass fragments of ethylmethyl benzenes homologs. The QSAR calculations indicated log P_{ow} data varying between 2.5 and 4.7 where the HPLC results varied between 0 and 4.8, with a weighted average log P_{ow} of 4.3.

The structure of the polyol was not sufficiently determined to permit the use of QSAR for log P_{ow} determination. Neither have any literature data been discovered regarding prospective structure. Accordingly, the only data available are these determined by the HPLC. These data show that log P_{ow} varies between 0 and 4.6 with a weighted average of 3.6.

The corrosion inhibitor used in this study has in fact been used in the industry for many years. It performs well with respect to corrosion inhibition. There is, however, limited information available regarding either the chemical structure or the behaviour of the preparation in the environment. It has also proven difficult to detect the chemical in produced water. The surface active properties and the poor water solubility make it more difficult to handle in existing laboratory tests, and there is uncertainty connected to the reliability of the log P_{ow} data determined in the laboratory. Log P_{ow} has also been determined for the preparation using HPLC, showing a log P_{ow} range of 3.0 to 4.7 with a weighted average of 4.5. The Shake Flask experiments showed a weighted average log P_{ow} of 1.3. A weighted average log P_{ow} of 5.1, was calculated from the known composition and the weighted average of each substance. This is significantly higher than the 4.5 found as the weighted average from the chromatogram of the whole preparation, which consists of several significantly different substances.

In contrast to the commercial corrosion inhibitor, Chemical Preparation no. 2 is an "ideal" mixture consisting of four benzene homologs. It consists of four pure substances used as reference compounds in OECD Guideline 117. For this preparation, only limited variability is seen between the experimental data from log P_{ow} determined with HPLC, Shake Flask, literature data as well as QSAR (see Table 1). This result was expected and was verified.

The CHARM model has been used to determine the ecosystem hazard quotient (PEC/NEC ratio) for each of the four substances in the corrosion inhibitor, covering the range

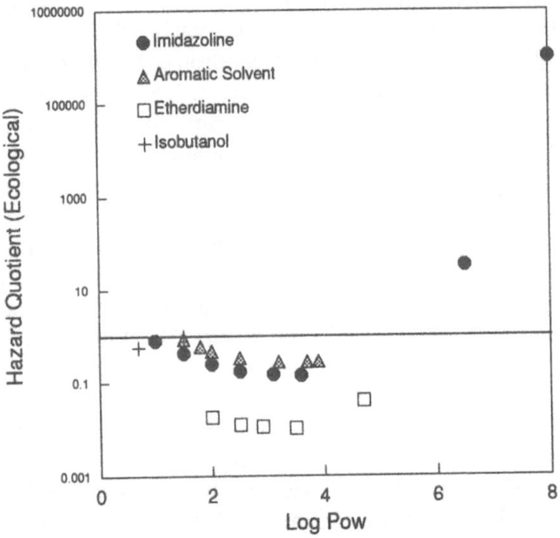

Figure 8. The influence of the variability of reported log P_{ow} -values on the PEC/NEC ratios determined in the hazard module of CHARM for the four substances of the corrosion inhibitor.

of log P_{ow}-values found in the literature and determined in the laboratory. Figure 8 presents the PEC/NEC ratio as a function of log P_{ow}. The resulting PEC/NEC calculated ratio for a chemical is dependent on the log P_{ow}-value, as seen in Figure 8. At low log P_{ow}-values the majority of the respective chemicals is released with the produced water, i.e. the fraction released is high (see equation 3). In these cases the pelagic ecological compartment is the one primarily affected. At the high log P_{ow}-values, the benthic ecological community is the primarily affected, and as shown in Figure 8, the hazard quotient reaches extremely high and presumably unacceptable values.

The hazard quotients for the ecosystem was determined for each chemical substance as well as for the two chemical preparations using the weighted average log P_{ow}-values. The results are presented in Table 2.

Table 2. Hazard quotients for the ecosystem ($HQ_{ecosystem}$) determined for the corrosion inhibitor and the prepared mixture, as well as for each substance

Chemical preparation	Chemical substances	Log P_{ow}-value used	$HQ_{ecosystem}$
Corrosion inhibitor	Imidazoline	5.7	4.72
	Aromatic solvent	4.1	0.25
	Isobutanol	4.1	0.10
	Polyol	4.1	0.02
	Chemical preparation	4.5[1)]	1.27
	Chemical preparation	5.1[2)]	5.07
Prepared mixture	Methyl benzene	2.7	0.15
	Ethyl benzene	3.2	0.13
	Isopropyl benzene	3.7	0.13
	Butyl benzene	4.6	0.38
	Chemial preparation	3.8	0.51

[1)] Measured on preparation
[2)] Calculated from weighted average of substances and composition of preparation

CONCLUSIONS

The following conclusions can be drawn from this work:

- The n-octanol/water partition coefficient (P_{ow}) is for most chemicals a key parameter serving as basic input data for environmental exposure analysis. In the CHARM model P_{ow} is essential for calculating the fraction released of these chemicals, the bioconcentration factor in the biota, and the chemical concentration in the sediments, and is therefore one of the most critical factors in assessing the environmental risk of these chemicals.
- P_{ow} can be determined by; the Shake Flask test (OECD 107), the HPLC test (OECD 117), and QSAR calculations. Emulsion breakers, corrosion inhibitors, flocculants and pseudo oil based drilling fluids are typical offshore E&P products falling into the category of surfactants, complex mixtures and chemicals with low water solubility. All the above methods reports limitations when testing this cathegory of chemicals.
- For a preparation of pure, single substance chemicals, the Shake Flask and HPLC technique can be used to determine log P_{ow}. For well characterized chemicals, reliable P_{ow} values can also be determined through QSAR or from the literature.
- The slow-stirring method which is a modification of OECD 107, has been recommended used for surfactants.
- Correlations of results from the slow-stirring method with HPLC test results, indicate that the HPLC method can give reliable results, within acceptable limits $\pm$ 0.3 log P_{ow} units, in the log P_{ow} range of 0 to 6. The HPLC technique has a great potential for surfactants and hydrophobic chemicals.
- QSAR calculations can only be performed for well characterized compounds, and several orders of magnitude difference can be obtained in P_{ow} values if the method is used for poorly characterized chemicals, like typical offshore chemicals.
- The Shake Flask method has limitations for non-water soluble and surface active chemicals. Three orders of magnitudes variations are seen between P_{ow} values determined by Shake Flask and determined by HPLC for typical offshore chemicals. This is demonstrated for imidazoline, the active ingredient in the corrosion inhibitor.
- The HPLC technique has a potential for being used for all typical offshore chemical. The method should be used to determine a weighted average log P_{ow} for each substance (set of homologs). The weighted average log P_{ow} of the preparation can be used in the CHARM model. This should be determined from the composition and the weighted average log P_{ow} of each substance.

REFERENCES

Aster/Aquire: Assessment tools for the evaluation of risk. US. EPA. Env. Research Lab., Duluth. MN 55804, Report No. EPA/DF/MT-93/050.

Bakke, S., Vik, E.A., Johnson, D.R., and Wright, N.H., 1993: Ranking Criteria for offshore Production chemicals based on environmental hazard assessment. Paper presented at the Fourth International Symposium on Oil Field Chemicals, Geilo, April 18-21.

Bintein, S.J., Devillers, J. and Karcher, W., 1993: Non-linear dependence of fish bioconcentration on n-octanol-water partition coefficient. SAR QSAR, Environ. Res. 1, 29-39.

Connel, D.W., and Hawker, D.W., 1988: Use of polynomial expressions to describe the bioconcentration of hydrophobic chemicals by fish. Ecotoxicol. Environ. Saf. 16, 242-257.

de Bruijn, J., Busser, F., Seinen, W., and Hermens, J., 1989: Determination of octanol/water partition coefficient for hydrophobic organic chemicals with the "slow-stirring" method. Environmental Toxicology Chemical, 8, 499-512.

Ghose et al., 1989: Four additional parameters for hydrophobic and dispersive interactions and their applications for an automated superposition of certain naturally occuring nucleosid antibiotics. Comput. Sci., 27, 163-172.

Hansen, N., Graversen, P., Christensen, F.B., and Linde, S.A., 1995: Massbalance of production chemicals - phase II. The Norwegian Oil Operators Association (OLF) Final Report.

Hammers, W.E, Mewis, G.J., and de Ligny, C.L., 1982: Correlations between liquid chromatographic capacity ratio data on Lichrosorb RP-18 and partition coefficients in the octanol/water system. J. Chromatogr. 247, 1-13.

Isnard, P., and Lambert, S., 1989: Aqueous solubility/n-octanol-water partitioning coefficient correlations. Chemosphere 18: 1837-1853.

Klein, W., Kördel, W., Weiz, M., and Poremski, H.J., 1988: Updating of the OECD test guideline 107. "Partition Coefficient n-octanol/water": OECD Laboratory Intercomparison Test of the HPLC Method. Chemosphere 17, 2, 361-386.

Leo, A.J. & Weninger, 1988: Clog P_{ow}. Computer program. Medchem Project. Pomona College, Claremont, CA.

Mackay, D., 1982: Correlation of bioconcentration factors. Environ. Sci. Technol. 16: 274-278.

Meylan, W., and Howard, P., 1994: User's guide for log K_{ow} octanol-water partition coefficient program, January, Syrancuse Research Cooperation Chemicals Hazard Chemistry Center, Syracuse, N.Y. 13210, USA.

Moreau, G.; Broto, P. and Vandycke, C., 1984: Molecular structures: perception, auto correlation descriptor and SAR studies. Eur. J. Chem. - Chim. Ter., 19, 71-78.

OECD, 1981: Partition Coefficient (N-Octanol/Water) Flask Shaking Method. OECD Guidelines for Testing of Chemicals, OECD 107, May 12.

OECD, 1986: Report of the OECD Workshop on Practical Approaches to the Assessment of Environmental Exposure; Chemicals Division, OECD, Paris, Report No. 9002H/KW/JdK/ 1/8/86.

OECD, 1989: Partition Coefficient (N-Octanol/Water) High Performance Liquid Chromatography (HPLC) Method, OECD Guidelines for testing of Chemicals. OECD 117, March 30.

OECD, 1992: Report of the OECD workshop on effect assessment of chemicals in the sediment. OECD Environment Monographs No. 60. OCDE/6D (92) 170.

Orth, R.G., Powell, R.L., Kutey, G., and Kimerle, R.A., 1995: Hazard and Risk Assessment. Impact of sediment partitioning methods on environmental safety assessment of surfactants. Environ. Toxicol. and Chem. 14, 2, 337-343.

OSPAR, 1995: Harmonized offshore chemical notification format. Oslo and Paris Commissions - Summary report of the meeting of the meeting of the programmes and measures committee (PRAM), Annex 10, Oviedo, 20-24 February.

Pedersen, F., Tyle, H., Niemelä, J.R., Guttmann, B., Lander, L. and Wedebrand, H., 1994: Environmental and hazard classification - Data collection and interpretation guide for substances to be evaluated for classification as dangerous to the environment. Tema Nord 1994, 589. The Nordic Council of Ministers.

Rapaport, R.A., and Eisenreich, S.J., 1984: Chromatographic determination of octanol-water partition coefficients (K_{ow}'s) for 58 polychlorinated biphenyl cogeners, Environ. Sci. Technol. 18, 163-170.

Sangster, J., 1989: Octanol-water partition coefficient of simple organic compounds. J. Phys. Chem. Ref. Data, 18, 3, 1111-1229.

van Leeuwen, C.J., van der Zandt, P.T.J., Aldenterg, A., Verhoor, H.J.M, Heunen, J.L.M., 1992: Hazard assessment application of QSARs, entrapolation and equilibium partitioning in aquatic effects assessments. I. Narcotic Industrial Pollutants. Environ. Toxicol. Chem. 11: 267-282.

Veith, G.D., DeFoe, D.L., and Bergstedt, B.V., 1979: Measuring and estimating the bioconcentration factor of chemicals in fish. J. Fish. Res. Board Can. 36: 1040-1048.

Øfjord, G.D., Vik, E.A. and Bakke, S., Johnson, D.R. and Edvardsson, T. (1995): CHARM - Chemical Hazard Assessment and Risk Management of offshore E&P Chemicals. Proceedings from the 6th International Symposium Oil Field Chemicals, Geilo, Norway, 19-22 March, No. 5570-7020.

15

EFFECTS OF THE WATER SOLUBLE FRACTION OF CRUDE OIL ON MARINE FISH LARVAE AND CRUSTACEANS

Clelia Booman and Lars Føyn

Institute of Marine Research
P.O. Box 1870, N-5024 Bergen, Norway

INTRODUCTION

The amount of produced water annually released in the Norwegian sector of the North Sea has been increasing dramatically, exceeding 20 million m^3 in 1992. Produced water has become the main contributor to oil discharges from Norwegian offshore operations, and over 2,000 tons of oil are expected to be released in the year 2000 (OLF, 1993). Oil separators are used to keep oil concentrations in discharged water well below the internationally regulated maximum value of 40 mg/liter. However, with the available technology it is not possible to efficiently recover the water soluble components of oil before the production water is released to the marine environment (OLF, 1993). The aromatic fraction, widely considered as the main contributor to both crude oil and produced water toxicity (Shiu et al., 1990, Johnsen et al., 1994) is thus enriched in produced water compared to concentrations in the original crude oil.

There has been an increasing concern about the possibility of negative effects of the off-shore oil exploration and production on the marine environment and particularly on fisheries and fish resources. Acute toxicity studies reported by Craddock (1977) indicate that fish are generally the marine organisms most susceptible to the WSF of crude oil. Adult cod can sense and avoid water masses containing concentrations down to 50 µg hydrocarbons from diesel oil per liter (Bøhle, 1986) but younger stages are considered to be more vulnerable. Except for capelin and herring eggs which develop on the sea bottom, most marine fish eggs and larvae in Norwegian waters are encountered in the upper 20-50 meter water layer (Fossum and Øiestad, 1992). These life stages are planktonic, as they drift in the water masses they would have limited or no ability to avoid a specific area even if they could sense the presence of pollutants.

Several studies have been conducted on the effects of exposure to oil and oil products on fish eggs and larvae. Toxicant concentrations in many of these studies (Carls, 1987; Falk-Pettersen and Kjørsvik, 1987 and others) were unrealistically high when compared to concentrations likely to occur in the field. Other studies concentrated on determining sublethal effects of exposure to lower concentrations (for example Fyhn and Tilseth, 1986;

Goksøyr et al., 1988 and Serigstad, 1991). Such effects could be coupled to delayed effects in later life stages that could have an impact at the population level.

In the present study, fish eggs and larvae were exposed to low concentrations of the water soluble fraction of crude oil (WSF) and to phenol to determine the effect concentrations on different biological parameters. Cod was chosen because among different fish species tested in the egg and larval stages, cod was the most sensitive to oil and oil products (Falk-Pettersen and Kjørsvik, 1987; Serigstad 1988, 1991) and because of its important role in Norwegian fisheries.

Oxygen consumption rates of marine fish eggs and larvae can be affected by exposure to oil and oil products (Eldridge et al., 1977; Davenport et al., 1979; Serigstad, 1987a) and we have used that method to evaluate the sensitivity of cod larvae to phenol.

Fish resources could be indirectly affected if the availability of their prey were reduced. Certain crustaceans that play a key role in the food chains in the North Atlantic and the Barents Sea ecosystems were chosen for this study. Acute toxicity tests and measurements of oxygen consumption were performed in field experiments with planktonic copepods, krill and Arctic amphipods.

Larvae of cod and most fishes in the North Atlantic continue their development on internal reserves deposited in the yolk-sac during a few days after hatching. In order to grow and survive, the larvae need to search and succeed in visualizing and capturing live zooplanktonic prey before the yolk is used up. The onset of exogenous feeding is thus a crucial milestone in a fish's life and requires an adequate function of the muscular and nervous systems. It could be assumed that a disfunction in one of these systems caused by exposure to xenobiotics would lead to impairment in capturing prey. One of the purposes of this study was to determine the effect concentrations of WSF on startfeeding success and on growth and survival after first feeding in cod larvae.

MATERIAL AND METHODS

The main characteristics of the ten laboratory and field experiments reported in this study are summarized in table 1.

Table 1. Overview over the experimental conditions

Exp. number	Type of experiment	Organisms tested	Compounds used	Concentration ($\mu g \cdot l^{-1}$)	Exposure time	Parameter
L1	laboratory	cod larvae	phenol	25-100-500	2 days	O_2 consumption
L2	"	"	"	25-100-500	5 days	"
L3	"	"	WSF	60	24 hours	startfeeding success
L4	"	cod eggs/larvae	"	85	24 hours	growth
L5	"	cod larvae	"	25	24 hours	"
F1	field	adult copepods	"	5 to 7200	24-48 hours	LC_{50}
F2	"	copepod nauplii	"	300 to 5600	24-48 hours	"
F3	"	krill	"	50-80	18 hours	O_2 consumption
F4	"	amphipods	"	70-140	12 hours	"
F5	"	"	"	1300	24 hours	mortality

Table 2. Composition of the phenol blend used in the experiments with cod larvae

Component	Percentage
phenol	45
o-, p- and m-cresol	35
2,5 dimethylphenol	5
m- and p-ethylphenol	10
o-isopropylphenol	1.5
o-n- and p-n- propylphenol	2
2,3,5 trimethylphenol	1

Exposure Set-Up

Phenol. Laboratory experiments (L1 and L2) were conducted with exposure of cod (*Gadus morhua*) eggs and larvae to phenol in a flow-through system allowing stable experimental conditions. Flows of a phenol stock solution and of pure seawater were accurately adjusted to provide a continuous supply of test solution to the experimental aquaria at the desired concentrations. However, during the last 24 hours of the exposure periods, the larvae were placed in closed chambers (see section on oxygen consumption later in this chapter). The aquaria were kept in a temperature regulated water bath (±0.2°C). The composition of the phenol mixture used to prepare the stock solution is presented in table 2.

Water Soluble Fraction of Crude Oil. All remaining experiments were conducted with exposure to the water soluble fraction (WSF) of North Sea crude oil (Veslefrikk). Benzene, toluene, xylenes and ethyl-benzene (BTX components) usually represent 80-90% of total hydrocarbons in the WSF (Westrheim and Palmork 1986). The test solutions were analyzed for BTX contents with gas chromatography after extraction with dichlormethane. The procedure from Westrheim and Palmork 1986 was followed, except for the use of an internal standard (n-propylbenzene) that was added to the samples before extraction. The remaining fraction, dominated by C_3 and C_4benzenes, naphthalene, C_1-to C_3 naphthalenes and phenols (Westrheim and Palmork, 1986), was not analyzed for. The WSF concentrations in the test solutions are thus presented in terms of µg (mg) BTX per liter seawater (µg or mg/L).

Three laboratory experiments (L3, L4 and L5) were conducted with exposure of cod eggs and larvae to WSF in a flow-through system. An emulsion-free stock solution of WSF in seawater was prepared continuously. Stock solution and seawater were accurately dosed into a blending chamber and temperated before flowing into the experimental aquaria. Water samples were taken daily from the aquaria and from the stock solution tank to monitor BTX concentrations.

Field experiments (F1 to F4) were conducted during a research cruise on board the research vessel "Lance" in the Northern Barents Sea in spring 1994. Different species of marine crustaceans were exposed to WSF in closed vessels or syringes. A stock solution of WSF was prepared daily by gentle stirring a volume of seawater, with a thin layer of crude oil on top, during 6 hours. Emulsification was avoided, as confirmed by the chromatograms. Samples from the stock solutions were extracted on board and later analyzed on land. The concentrations in the test solutions were then calculated according to the dilutions.

Table 3. Relative BTX composition in two samples of test solution prepared with Veslefrikk crude oil in the laboratory (flow-through system) and during the field experiments (closed system)

	Percentage	
Component	Flow-through	Closed system
benzene	68	28
toluene	18	43
ethyl-benzene	0	4
xylenes	14	25

Parameters Measured

Acute Toxicity Tests. LC_{50} tests were conducted with the planktonic copepods *Calanus finmarchicus* and *C. glacialis* (Copepoda, Crustacea) with exposure to WSF in closed systems. The LC_{50} (Lethal Concentration) tests express the concentration of the test chemical causing 50% mortality after exposure during a given amount of time (from 24 to 96 hours). Experiment F1 includes three tests carried out with adult stages (copepodite V and VI) from the same plankton haul. Experiment F2 includes three tests with nauplii, the first larval stages, from the same species obtained from a natural spawning in the laboratory on board. The main lines in the international standard described in Anon. (1993) were followed, but no groups were exposed longer than 48 hours.

Oxygen Consumption. Changes in the organisms oxygen consumption rates after exposure to the test solutions were determined by closed respirometry (Serigstad, 1987b). The organisms were transferred to closed respiration chambers (10 ml gas-tight glass syringes) together with test solution and placed in a gently rotating device in a temperated water bath. The oxygen partial pressures in the experimental chambers and in the blank chambers (four blanks per treatment) were determined after a 24-hour incubation period.

Oxygen partial pressures were determined with a Strathkelvin 781b oxygen meter equipped either with a Strathkelvin 1302 or a Radiometer E5046 electrode. The water samples were injected directly from the syringes into a Strathkelvin MC100 microcell containing the electrode. The microcell was connected to a water bath keeping the electrode and the sample at the same temperature as in the incubation bath.

In the experiments with cod (experiments L1 and L2), three larvae were introduced in each syringe in order to increase the resolution of the oxygen measurements. The larvae, originating from a one-night natural spawning of a capture broodstock (Holm and Andersen, 1991), were homogeneous in age and size and were assumed to have similar oxygen demands. At least four replicates were run per treatment. The oxygen consumption rates of cod larvae were calculated with equation 1 from Serigstad (1987b).

$$VO_2 = (\Delta pO_2 \cdot \propto O_2 \cdot V_k) \cdot (N \cdot T)^{-1} \quad (1)$$

where
ΔpO_2 = oxygen partial pressure difference (mmHg) between blank and experimental chambers
$\propto O_2$ = solubility coefficient of O_2 at the given temperature and salinity (nl $O_2 \cdot ml^{-1} \cdot mmHg^{-1}$)
V_k = respiration chamber volume (ml)
N = number of organisms in the chamber
and T = incubation time (minutes).

The crustaceans used in the oxygen consumption experiments (F3 to F5), were obtained from plankton hauls at sea and were therefore heterogenous as regards size, developmental stage and condition. Since they were also expected to have different oxygen demands, only one organism was placed in each respiration chamber. From 8 to 12 organisms were measured in each treatment. Immediately after the oxygen measurement each organism was observed in the microscope to determine the species and developmental stage and thereafter placed in a drying oven at 60°C during 24 hours. On land, the organisms were placed again in the oven during two hours prior to weighing in a Cahn 29 Automatic Electrobalance (±1 µg). Specific oxygen consumption rates of crustaceans were calculated with equation 2 and are expressed in µl O_2 . mg dry weight^{-1} hour^{-1}.

$$VO_2 = (\Delta pO_2 \cdot \propto O_2 \cdot V_k) \cdot (DW \cdot T)^{-1} \qquad (2)$$

where
$\propto O_2$ is expressed in µl $O_2 \cdot ml^{-1} \cdot mmHg^{-1}$
and DW = dry weight of the organism (mg).

Start Feeding Success

In experiment L3, cod yolk-sac larvae were exposed to WSF of crude oil at a concentration of 60 µg BTX/L during 24 hours starting day 2 after hatching. On day 5 the larvae were transferred from the flow- through aquaria to 6-liter buckets containing seawater and live prey (nauplii from a monoculture of the copepod *Acartia tonsa*) at similar concentrations (230 nauplii per liter). There were approximately 80 cod larvae in each of four replicates in the control and in the exposed group. After 6 hours the larvae were concentrated and preserved in formalin and later dissected under the microscope to determine the presence of prey in the digestive tract.

Growth

In experiment L4, cod eggs were exposed to WSF at a concentration of 85 µg BTX/L during 24 hours and kept in the flow-trough aquaria during hatching and until the first feeding stage. At this time the larvae were transferred to 70 liter outdoor tanks suspended in a circulating water bath where they were fed live wild zooplankton. Three replicates with 300 larvae were kept until the end of the experiment to calculate survival. A fourth tank for each group was stocked with 300 larvae and sampled twice a week for growth monitoring (n=10). At the end of the experiment, weight from all remaining larvae from all four tanks were accounted.

In experiment L5, cod were exposed at the more vulnerable yolk-sac stage to WSF at a concentration of 25 µg BTX/L during 24 hours starting day 3 after hatching. There were 250 larvae in each replicate. Immediately after the exposure period they were transferred to the rearing tanks placed in a water bath indoors. Wild zooplankton was added daily to the tanks until the end of the experiment. The rearing and sampling protocols were otherwise identical to those in L4.

RESULTS

Cod

Experiment L1: There was no significant difference between groups of cod yolk-sac larvae exposed to 25, 100 and 500 µg phenol /L during 2 days compared to the control group ($p > .05$ 1-way ANOVA, figure 1).

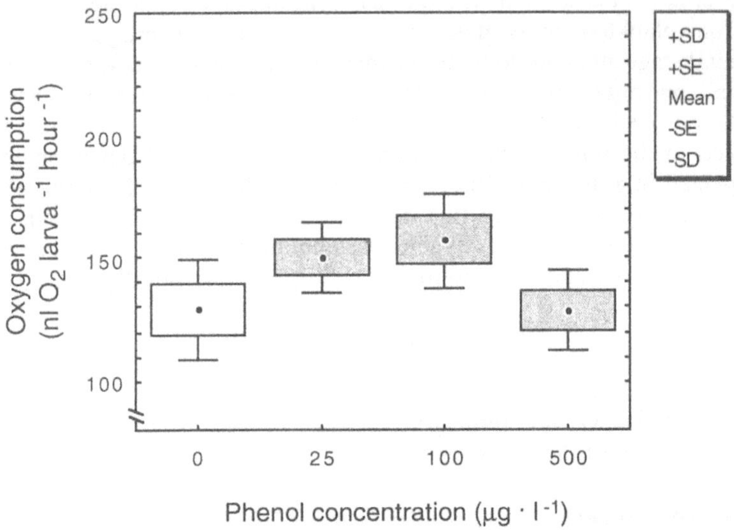

Figure 1. Oxygen consumption of Atlantic cod yolk-sac larvae exposed to different phenol concentrations during 2 days (shaded boxes) compared to the control group (empty box). Temperature: 5°C. n = 4 in each group.

Experiment L2: Groups of cod yolk-sac larvae from the same batch exposed to phenol during 5 days presented variations in oxygen consumption rates. There was a significant reduction in oxygen consumption in the groups exposed to 25 and 100 μg phenol/L compared to the control group ($p<.05$, Tukey HSD-test), while there was no significant difference between the group exposed to 500 μg phenol/L and the control group ($p>.05$, Tukey HSD-test, figure 2).

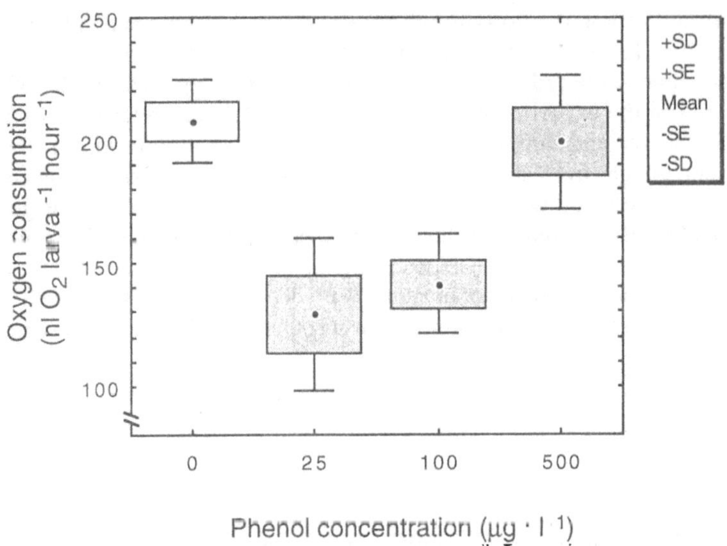

Figure 2. Oxygen consumption of cod yolk-sac larvae exposed to different phenol concentrations during 5 days (shaded boxes) compared to the control group (empty box). Temperature: 5°C. n = 4 in each group.

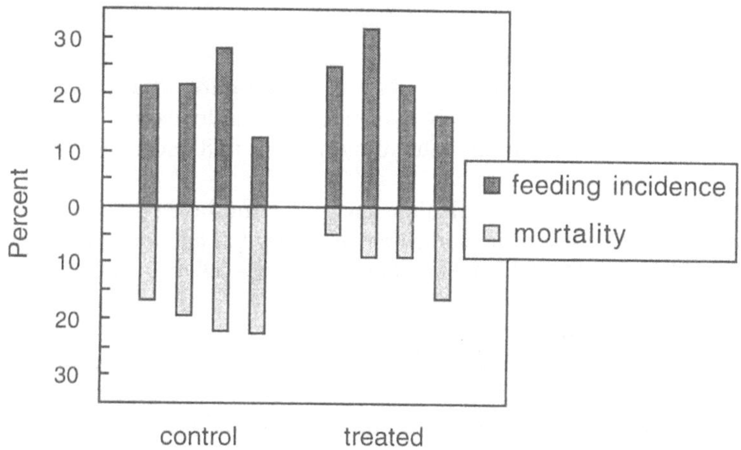

Figure 3. Startfeeding success and mortality of cod yolk-sac larvae exposed to WSF of crude oil at a concentration of 60 µg BTX/L during 24 hours starting two days after hatching. Feeding took place during 6 hours at day 5. Temperature: 5°C during the exposure period and 8°C during the feeding phase. n = approx. 80 in each of four replicates.

In experiment L3, cod larvae were exposed to WSF from day 2 to 3 posthatch and startfed 2 days later with live prey. There was no significant difference in feeding success (percentage larvae with prey in their digestive tracts) between the exposed larvae and those in the control group ($p>.05$, t-test, figure 3). Handling mortality, accounted from transfer to the feeding units, was lower in the treated group ($p<.05$, t-test, figure 3).

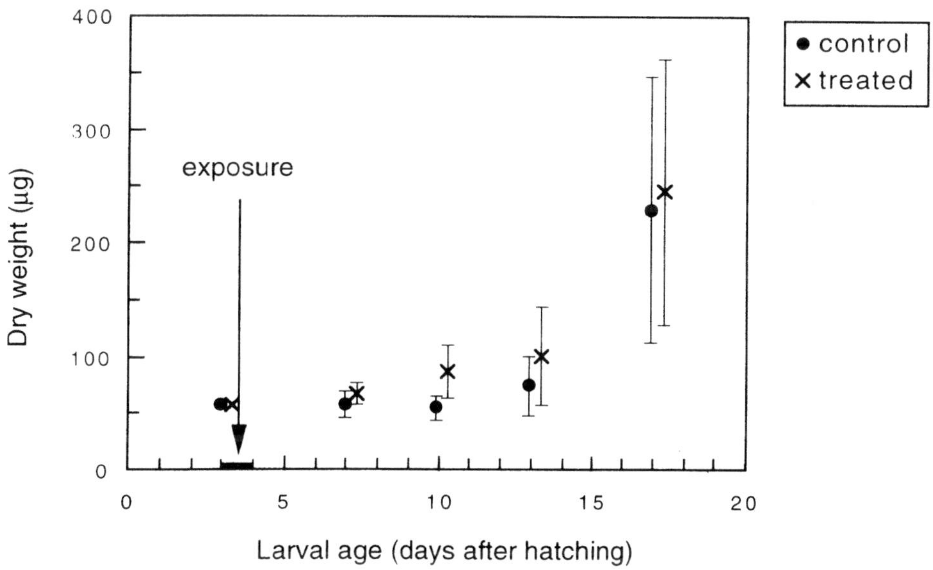

Figure 4. Dry weight of cod larvae exposed to WSF of crude oil at a concentration of 85 µg BTX/L during 24 hours in the egg phase compared to the control group (mean ±SD). Exposure took place 5 days before hatching. Temperature: 8°C during exposure and 9-10°C during the feeding period.

In experiment L4, exposure to WSF took place in the egg phase during 24 hours, and growth and survival after hatching were followed during 17 days after hatching. There was no significant difference in growth between the exposed and the control group (p>.05, ANCOVA, figure 4). Only the last 4 points plotted in the figure, after positive growth had started, were considered in the analysis. N = 10 in each sample, except for the last sample that includes all remaining larvae in the four subgroups (n = 58 and 92 for the exposed and control group respectively).

Three replicates were used to calculate survival from the egg phase until the end of the experiment. There was no significant difference in survival between the exposed and the control group (p>.05, 1-way ANOVA). Percent survival was 5.0, 5.0 and 6.0 in the exposed group, and 5.3, 7.3 and 9.0 in the control group.

In experiment L5, the cod were exposed during 24 hours as yolk-sac larvae to a lower concentration than the previous experiment. Growth of cod larvae exposed to WSF at a concentration of 25 μg BTX/L was higher than in the control group (p intercept <.05, figure 5). N = 10 in each sample, except for the last sample that includes all remaining larvae in the four subgroups (n = 54 and 44 for the exposed and the control group respectively).

There was no statistical difference in survival to the end of the experiment between the exposed and the control group (p>.05, 1-way ANOVA). Percent survival was 3.2, 4.4 and 8.8 in the control group and 4.8, 7.2 and 6.0 in the exposed group.

Crustaceans

Copepods. Experiment F1, adult *Calanus finmarchicus* and *C. glacialis* . The 24 hour LC_{50} was estimated at 7 mg BTX $\cdot l^{-1}$ (figure 6) while results for the 48 hour test are less consistent.

The same kind of test was conducted with nauplii (the first larval stages) from *Calanus spp* that had spawned in the laboratory on board. The data in figure 7 are corrected

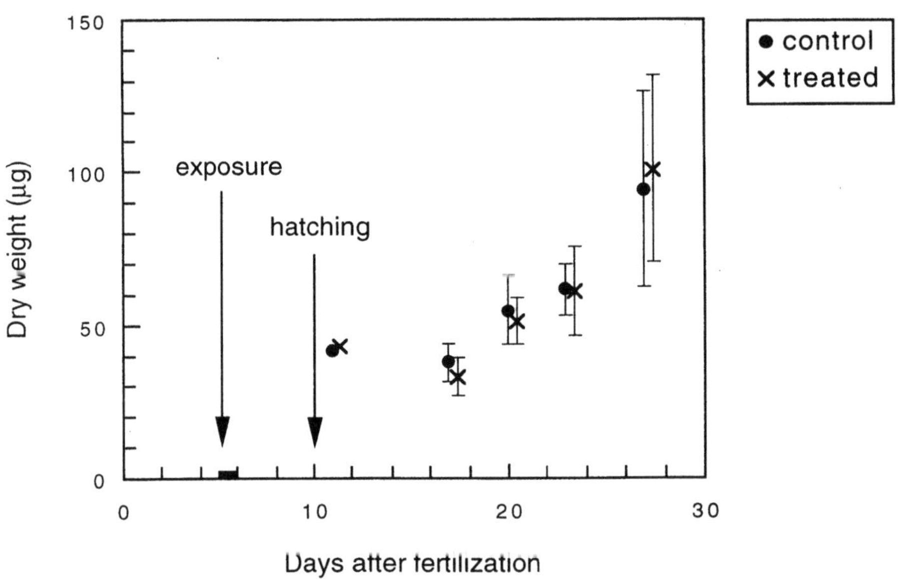

Figure 5. Dry weight of cod larvae exposed to WSF of crude oil at a concentration of 25 μg BTX/L during 24 hours compared to the control group (mean ±SD). Temperature: 8°C during the whole experimental period.

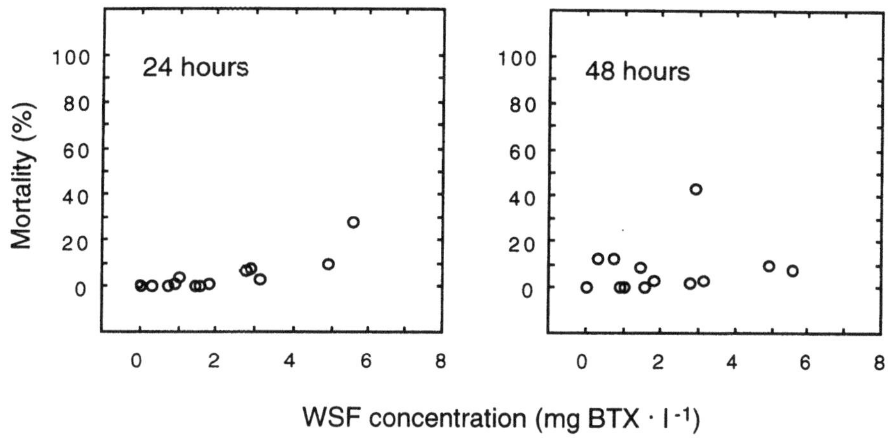

Figure 6. Mortality of *Calanus spp.* in the LC_{50} tests after 24 and 48 hours exposure to WSF of crude oil. n = 5 (occasionally 4 or 6) for each data point. Temperature: 4°C.

for mortality of up to 4 % that occurred in the control group of two of the three tests. Mortality was lower than 50% in all tests, even at the highest concentration tested (5.6 mg/L) and the 24 hour LC_{50} would likely lie between 6-7 mg/L . An overall reduction in swimming activity was observed at the highest concentrations.

Krill. Krill (*Meganychtiphanes norvegica*) and the other crustaceans were tested for changes in specific oxygen consumption after exposure to WSF in closed vessels.

There was no significant difference in specific oxygen consumption rates between krill exposed to WSF of crude oil at concentrations of 50 and 80 µg BTX/L during 18 hours and the control group (p>.05, ANCOVA). However, the plot in figure 8 corresponding to 80

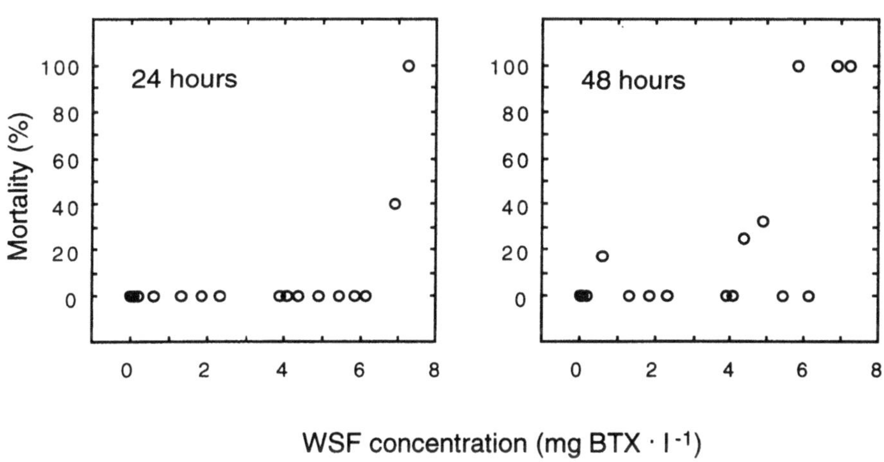

Figure 7. Mortality of *Calanus spp.* nauplii in the LC_{50} tests after 24 and 48 hours exposure to the WSF of Veslefrikk crude oil. Data from 3 different tests conducted with organisms from the same spawning batch. n = 40-130 for each data point. Temperature: 4°C.

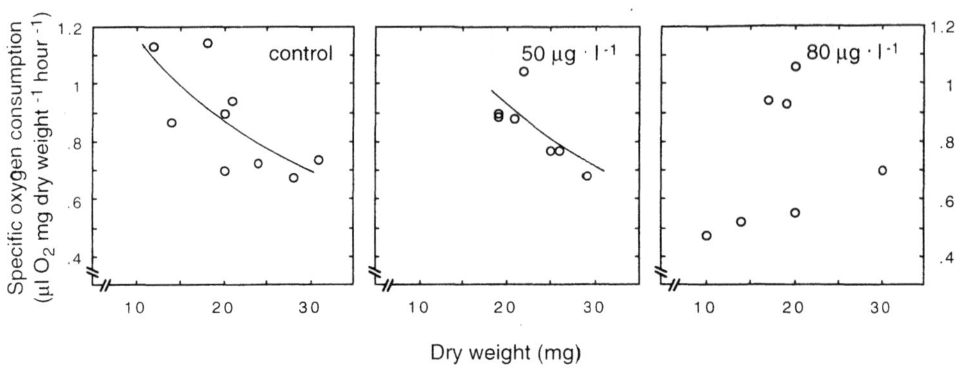

Figure 8. Oxygen consumption of krill, *Meganyctiphanes norvegica*, exposed to WSF of crude oil (50 and 80 µg BTX/L) compared to the control group. Temperature: 5°C.

µg BTX/L looks quite different from the control and the 50 µg/L-group, as the smallest individuals show a very low oxygen consumption rate in relation to the body weight.

Amphipods. There was a significant difference in specific oxygen consumption rates between organisms of the species *Themisto abyssorum* exposed WSF of crude oil at concentrations of 70 and 140 µg BTX/L compared to the control group ($p<.05$, ANCOVA, figure 9), the largest organisms being more affected.

Another experiment was conducted with *Themisto libellula*, a species related to *T. abyssorum*. After 24 hours exposure to WSF of crude oil (1300 µg BTX/L), most organisms were either dead (40%, n=22) or had a low activity level (40%). This indicates that the 24 hour LC_{50} value probably lies around 1300 µg BTX/L.

DISCUSSION

Changes in respiration rates have been observed in fish eggs and larvae exposed to benzene (Eldridge et al. 1977) or crude oil extracts containing high concentrations of aromatic and aliphatic components (Davenport et al. 1979). Changes in different anatomical

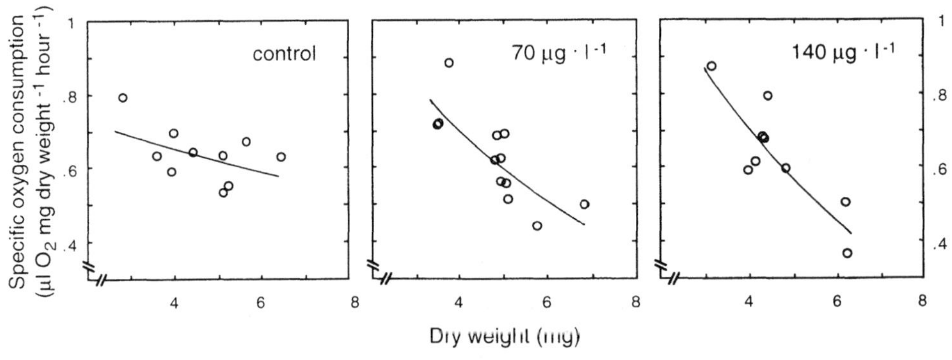

Figure 9. Oxygen consumption of the crustacean *Themisto abyssorum* (Amphipoda, Hyperidea) exposed to WSF of crude oil (70 and 140 µg BTX/L) during 12 hours compared to the control group. Temperature: 4°C.

characteristics and physiological parameters have been investigated in cod eggs and larvae exposed to the water soluble fraction of crude oil. Measurement of changes in oxygen consumption rates was the most sensitive of the methods tested and the one giving the most consistent results (Fyhn and Tilseth 1986, Serigstad 1986). The long term consequences of such physiological effects are yet unknown, but Serigstad (1987b) found that a decrease in oxygen consumption in oil exposed cod larvae was accompanied by a decrease in activity. He further argued that larvae affected in such a way could be impaired to catch their prey in the startfeeding phase.

In the present study, oxygen consumption rates of yolk-sac larvae were not affected by exposure to 25, 100 nor 500 μg phenol per liter during two days. Oxygen consumption rates were reduced in larvae exposed to 25 and 100 μg/L during 5 days, while larvae exposed to 500 μg/L had similar rates as those in the control group. These apparently contradictory results suggest there may be other physiological mechanisms triggering an increase in oxygen consumption at 500 μg phenol/L as compared to 100 μg/L. Some of these mechanisms could be: initiating of detoxification processes with higher metabolic cost, or increase in activity as an avoidance response. The phenolic fraction was identified as one of the main contributors to the acute toxicity of an artificial produced water, right after the aromatic fraction (Johnsen et al., 1994). Concentrations affecting the oxygen consumption of cod larvae can be encountered in the vicinity of oil platforms. Concentrations of 500 μg/L were measured in produced water simulated with reservoir water from the Njord field in the Norwegian Sea (Anon., 1991 reported by Anon., 1995) and produced water from oil and gas production in the Gulf of Mexico may contain up to 4 and 22 mg/L respectively (Stephenson, 1992). However, such high concentrations are rapidily diluted. Furuholt (1996) calculated a dilution factor of 1000 times at 50 and 250 from the discharge point, depending of the discharge rate. Riksheim and Johnsen (1994) calculated even higher dilution factors from analysis of water samples taken downstream of the Statfjord field.

Tilseth *et al.* (1984) observed reduced weight in cod larvae that had been continuously exposed to 50 and 150 μg WSF of crude oil per liter from the egg phase. The volatile aromatic compounds escape the water column fast and in case of an oil spill, it is unlikely that planktonic fish eggs or larvae would be exposed to these "high" concentrations of WSF for longer than one day. In the present study, exposure of yolk-sac cod larvae to 60 μg BTX during 24 hours did not affect their success in capturing live prey later in the startfeeding phase. Growth and survival rates of cod larvae that had been exposed to WSF at a concentration of 80 μg BTX/L in the egg phase were also unaffected. Ellingsen et al. (1992) reported a decrease in oxygen consumption rates of cod yolk-sac larvae exposed to the WSF of crude oil in similar conditions. Thus, the ecological significance of this physiological effect that we aimed to clarify still remains unclear.

In the experiment with exposure of yolk-sac larvae to WSF at a concentration of 25 μg BTX/L, the exposed larvae grew slightly better than the untreated larvae. This low toxicant concentration might have affected the bacterial flora in the test aquaria, giving the larvae a better growth potential. In both experimental and commercial rearing of marine fishes, eggs are often treated with bacteriostatics or fungicides. For example, the treating of Atlantic halibut eggs with 400 mg/L glutardialdehyde during 10 minutes has resulted in higher larval growth and survival in the startfeeding phase compared to a control group (Harboe *et al.* 1994).

Pelagic fish eggs and larvae drift with the water masses and may thereby be exposed to discharges from more than one oil field. Stephenson (1992) reported average concentrations of benzene, toluene and C_2-benzene in production waters from the Gulf of Mexico of 12 and 3 mg/L from gas and oil production respectively. Discharge water of a platform in the UK sector of the North Sea contained from 1 to 93 mg BTX / liter (Tibbetts et al., 1992) and simulated production water from the Njord field in the Norwegian Sea contained

100 mg/L (Anon., 1991 reported by Anon., 1995). These concentrations will rapidly be reduced due to the high volatility of these compounds together with high dilution rates in the open sea.

Our acute toxicity tests with the crustaceans *Calanus spp.* showed a 24-hour LC_{50} value of 7 mg BTX/L for the adult stages and over 5.6 mg/L for the naupliar stages of these species. These values suggest that these species are less vulnerable to the water soluble fraction of crude oil than marine fish eggs and larvae for which reported lethal concentrations lie between 0.1 and 1 mg of soluble petroleum hydrocarbons /L (Craddock, 1977). The data from the 48-hour exposure are more "noisy" than the 24-hour data, both for adults and nauplii and this was probably due to exposure taking place in closed vials, without renewal of the test solution. Fish eggs and larvae can accumulate and metabolize aromatic hydrocarbons (Carls and Rice, 1988 and Solbakken *et al.*, 1984). If this was the case also for *Calanus spp.*, significant changes in the relative composition of test water over time may have occurred in our 48-hour LC_{50} tests. This suggests that 96-hour LC_{50} values obtained by the internationally standardized procedure might be inadequate to test these kind of compounds.

Our results indicate that concentrations as low as 80 μg BTX/L may affect the respiration rates of krill juveniles. Considering also a 24-hour LC_{50} value close to 700 μg/L (Booman et al., 1995), we can regard krill as being rather vulnerable to the water soluble fraction of crude oil.

Themisto abyssorum and *T. libellula* (Amphipoda, Hyperiidea) are about 20 and 40 mm long respectively. They are associated to the ice and play a key role in the food chains from Arctic and Subarctic waters, being prayed upon by fish, seals and sea birds. Exposure to WSF in concentrations down to 70 μg/L during 12 hours affected oxygen consumption rates of *T. abyssorum*. Contrary to the tendency observed in fish larvae (Serigstad, 1987b) and in krill, the largest organisms seem to be more affected.

CONCLUSIONS

Based on data on concentration of phenol and BTX components at the discharge point and on calculated dilution rates of produced water discharges, it is unelikely that effects as those determined in our experiments would occurr in the field.

Crustaceans like krill and Arctic hyperidians can be placed in a category together with fish larvae, as organisms relatively vulnerable to the water soluble fraction of oil when compared to calanoid copepods. Marine fish larvae are probably more suitable for toxicity testing of chemicals in produced water discharges to the sea than the commonly used calanoid copepod *Acartia tonsa* .

ACKNOWLEDGMENTS

This work was partially supported by the Interministerial Working Group for Impact Assessment of Oil Activities (AKUP), under the Norwegian Ministry of Industry and Energy. We wish to thank the Norwegian Polar Institute for inviting us to two research cruises in the Northern Barents Sea and the crew of R/V "Lance" for their support on board. We also wish to thank Frank Midtøy at the IMR and Anita Smith at Statoil for their participation in the experimental work and Kjell Westrheim at the IMR for the GC-analyses. Statoil supplied us with crude oil and the phenol blend used in our experiments.

REFERENCES

Anon. 1995. Njord. Impact analysis. Norsk Hydro. Oslo. 66 pp. In Norwegian.

Anon. 1991. Njord reservoir formation water analysis. D.B. Robinson Research Ltd.

Anon. 1993. Water quality -Determination of the inhibition of the mobility of *Daphnia magna* Straus (Cladocera, Crustacea)- Acute toxicity test. International Organization for Standardization. ISO/TC 147/SC 5. 11 pp.(draft).

Booman, C., Midtøy, F., Smith, A. T., Westrheim, K. and Føyn, L. 1995. Effects of oil on marine organisms with particular reference to first feeding of fish larvae. Fisken og Havet 9. 142 pp. Institute of Marine Research. Bergen. In Norwegian.

Bøhle, B. 1986. Avoidance from petroleum hydrocarbons by the cod (*Gadus morhua*). Fisk Dir. Skr. Ser. Hav Unders.18: 97-112.

Carls, M. G. 1987. Effects of dietary and water-borne oil exposure on larval pacific herring (*Clupea harengus pallasi*). Marine Environmental Research 22: 253-270.

Carls, M. G. and Rice, S. D. 1988. Sensitivity differences between eggs and larvae of walleye pollack (*Theragra chalcogramma*) to Hydrocarbons. Marine Environmental Research 26: 285-297.

Craddock, D. R. 1977. Acute toxic effects of petroleum on arctic and subarctic marine organisms. In: Malins, D. C., (ed.) Effects of Petroleum on Arctic and Subarctic Marine Environments and Organisms. New York. Academic Press 2: 1-93.

Davenport, J., Lønning, S. and Sæthre, L. J. 1979. The effects of Ekofisk oil extract upon oxygen uptake in eggs and larvae of the cod *Gadus morhua* L. Astarte 12: 31-34.

Eldridge, M. B., Echeverria, T., and Whipple, J. A. 1977 Energetics of Pacific herring (*Clupea harengus pallasi*) embryos and larvae exposed to low concentrations of benzene, a monoaromatic component of crude oil. Trans. Amer. Fish. Soc. 106: 452-461.

Ellingsen, T., Vigrestad, A. and Føyn, L. 1992. Effects on cod eggs and larvae of low concentrations and short exposures to Veslefrikk and Gullfaks crude oils. HSMM 14: 26 pp. Institute of Marine Research. Bergen. In Norwegian

Falk-Pettersen, I. -B and Kjørsvik, E. 1987. Acute toxicity tests of the effects of oils and dispersants on marine fish embryos and larvae - A review. Sarsia 72: 411-413.

Fyhn, H. J. and Tilseth, S. 1986. Fish Larval Physiology and Anatomy - Basic Research and Effects of Oil. A cooperative research project between University of Bergen and Institute of Marine Research, supported by A.S. Norske Shell. University of Bergen.

Fossum, P. and Øiestad, V. 1992. The early life stages of fish and the threat from oil activities. Final report from the egg and larvae program - HELP (1985-1991). Institute of Marine Research. Bergen. In Norwegian.

Furuholt, E. 1996. Environmental effects of dishcarge and reinjection of produced water. *In:* Reed, M. and S. Johnsen, eds. Produced Water. Environmental Science Research Series. Plenum Press.

Goksøyr, A., Serigstad, B, Solberg, T. S. andStegeman, J. J. 1988. Response of cod (*Gadus morhua*) larvae and juveniles to oil exposure detected with anti-cod cytochrome P-450c IgG and anti-scup cytochrome P-450E MAb 1-12-3. Marine Environmental Research 24: 31-35.

Harboe, T., Huse, I. and Øie, G. 1994. Effects of egg disinfection on yolk sac and first feeding stages of halibut (*Hippoglossus hippoglossus* L.) larvae. Aquaculture 119:157-165.

Holm, J. C. and Andersen, E. 1991. Production of eggs and larvae. In: Holm, J. C., Svåsand, T. and Wennevik, V. (eds.) Manual on Cod Culture - Broodstock and Juvenile Production. Institute of Marine Research, Aquaculture Centre, Bergen: 29-42. In Norwegian.

Johnsen, S., Smith, A. T., Brendehaug, J., Riksheim, H. and Gjøse, A. L. 1994. Identification of sources of acute toxicity in produced water. International Conference on Health, Safety and Environment in Oil and Gas Exploration and Production. Jan 1994, Jakarta. (v. paper no. SPE 27138).

OLF. 1993. Environmental Program Phase 2. Summary report. Norwegian Oil Industry Association (OLF).134 pp. In Norwegian.

Riksheim, H. and Johnsen, S. 1994. Determination of produced water constituents in the vicinity of offshore production fields in the North Sea. International Conference on Health, Safety and Environment in Oil and Gas Exploration and Production. Jan 1994, Jakarta. (v. paper no. SPE 27151).

Serigstad, B. 1986. The effect of oil exposure on the oxygen uptake of eggs and larvae of the cod (*Gadus morhua* L.). In: Fhyn, H. J. and Tilseth, S. (eds.) Fish Larval Physiology and Anatomy. Basic Research and the Effects of Oil. University of Bergen: 203-252.

Serigstad, B. 1987a. Effect of oil-exposure on the oxygen uptake of cod (*Gadus morhua* L.) eggs and larvae. Sarsia 72: 401-403.

Serigstad, B. 1987b. Respiratory studies on cod (*Gadus morhua* L.) with special reference to effects of oil exposure on eggs and larvae. Dr. scient. thesis in Environmental Physiology. University of Bergen.

Serigstad, B. 1991. Effects of Gullfaks and Veslefrikk crude oils on fish eggs and larvae. HSMM 15. 52 pp. Institute of Marine Research. Bergen. In Norwegian.

Serigstad, B., Sværen, I., Ellingsen, T. and Føyn, L. 1988. Age dependent sensitivity of marine organisms to oil. BKO 8806: 37 pp. Institute of Marine Research. Bergen. In Norwegian.

Shiu, W. Y., Bobra, M., Bobra, A. M., Maijanen, A., Suntio, L. and Mackay, D. 1990. The water solubility of crude oils and petroleum products. Oil Chem. Pollut. 7: 57-84.

Solbakken, J. E., Tilseth, S. and Palmork, K. H. 1984. Uptake and elimination of aromatic hydrocarbons and a chlorinated biphenyl in eggs and larvae of cod (*Gadus morhua*). Marine Ecology Progress Series 16: 297-301.

Stephenson, M. T. 1992. A survey of produced water studies. In: Ray, J. P. and Engelhardt, F. R. (eds.) Produced Water. Technological/Environmental Issues and Solutions. New York, Plenum Press: 1-11.

Tibbetts, P. J. C., Buchanan, I. T., Gawel, L. J. and Large, R. 1992. A comprehensive determination of produced water composition. In: Ray, J. P. and Engelhardt, F. R. (eds.) Produced Water. Technological / Environmental Issues and Solutions. New York, Plenum Press: 97-112.

Tilseth, S., Solberg, T. S., Westrheim, K. 1984. Sublethal effects of the water-soluble fraction of ekofisk crude oil on the early larval stages of cod (*Gadus morhua* L.). Marine Environmental Research 11: 1-16.

Westrheim, K. and Palmork K. H. 1986. Chemical analyses performed under protect contract No. 83-524: "Effect of oil on the physiology and development of fish larvae" - 1983-1985. In: Fhyn, H. J. and Tilseth, S. (eds.) Fish Larval Physiology and Anatomy. Basic Research and the Effects of Oil. University of Bergen: 289-328.

16

AROMATIC HYDROCARBONS IN PRODUCED WATER

Bioaccumulation and Trophic Transfer in Marine Food Webs

Jerry M. Neff and Theodor C. Sauer, Jr.

Battelle Ocean Sciences Laboratory
397 Washington Street
Duxbury, Massachusetts 02332

INTRODUCTION

Objectives

Aromatic hydrocarbons are among the most abundant and toxic components of oil and gas well produced water. The objective of this paper is to present an overview of the bioaccumulation of aromatic hydrocarbons by marine animals, with emphasis on bioaccumulation of PAHs from produced water discharged to the ocean. The focus of this paper is on bioaccumulation data collected during recent monitoring studies in the U.S Gulf of Mexico.

Sources of Produced Water

During millions of years of geologic time, petroleum and natural gas may accumulate in porous sediments (e.g., sands) trapped between layers of impermeable rock deep within the earth (Neff et al., 1987). Water may be trapped with the oil and gas. This water may be derived from ancient sea water (connate water) or from ancient rain water or lake water (meteoric water) and often is as old as the fossil fuels in the reservoir (Collins, 1975). When the fossil fuel reservoir is tapped by an oil or gas well, the oil and gas pumped from the well may contain fossil water. In some oil fields, surface fresh water or sea water is pumped into the hydrocarbon-bearing formation through an injection well to enhance recovery of fossil fuels from the production wells. Sometimes, this injection water breaks through and mixes with the hydrocarbons in the reservoir and is pumped back to the surface with the fossil fuels. The fossil and modern water produced with oil and gas is called produced water, production water, formation water, or oilfield brine.

In many parts of the world, produced water from offshore production platforms is treated to remove oil and discharged to the ocean (Swan et al., 1994). The oil/gas/water

Produced Water2: Environmental Issues and Mitigation Technologies
Edited by M. Reed and S. Johnsen, Plenum Press, New York, 1996

mixture is processed through several separation devices, usually on the production platform, to separate the three phases from one another. The oil/water separation is not 100 percent efficient. Small amounts of oil, other natural ingredients, and additives to the original mixture remain in the produced water and are discharged with it. Current limits for the concentration of total oil and grease in produced water discharged to offfshore territorial waters of the United States are less than a daily maximum of 42 mg/L and a monthly average of 29 mg/L total oil and grease (57 FR 224:54642, November 19, 1992 and 58 FR 231:63964, December 3, 1993). Maximum permissible concentrations of oil and grease in produced water discharged to marine waters of Europe and Australia are in the range of 40 to 50 mg/L (Black et al., 1994). Other chemicals in produced water are not regulated.

Composition of Produced Water

Produced water contains a variety of chemicals that have been dissolved from the geologic formations in which the produced water resided for millions of years. These chemicals include inorganic salts (essentially the same salts that are found in seawater), several metals and metaloids, and a wide variety of organic chemicals.

Petroleum hydrocarbons are the organic components of greatest environmental concern in produced water. Petroleum hydrocarbons (measured as oil and grease by infrared spectrometry) account for 8.5 to 16 percent of the total organic carbon (TOC) in produced water samples from the Gulf of Mexico analyzed by Neff et al. (1989). Most of the remaining organic matter in produced water is low molecular weight organic acids, such as acetic acid. It should be pointed out that standard methods for analysis of oil and grease are not specific for petroleum hydrocarbons and measure several other ingredients in produced water (e.g., some organic acids) in addition to petroleum hydrocarbons (Brown et al., 1992).

Although the oil/water separator system on a modern offshore platform is able to remove most of the dispersed oil droplets, some finely dispersed oil and dissolved petroleum hydrocarbons may remain in the produced water and be discharged with in to the ocean. The solubility of petroleum hydrocarbons in sea water decreases as their size (molecular weight) increases (McAuliffe, 1966; Eastcott et al., 1988). Because the oil/water separator is efficient in removing oil droplets but not dissolved oil from the produced water, most of the petroleum hydrocarbons remaining in the produced water after treatment are low molecular weight aromatic and saturated hydrocarbons that are dissolved in the produced water.

The most abundant hydrocarbons in produced water are the one-ring aromatic hydrocarbons, benzene, toluene, ethylbenzene, and xylenes (the BTEX compounds). BTEX may be present in produced water from different sources at concentrations ranging from 68 to occasionally as high 600,000 μg/L. In the North Sea, and probably elsewhere, produced water from gas wells usually contains higher concentrations of BTEX than produced water from oil wells (Stephenson et al., 1994). Toluene often is the most abundant BTEX compound in produced water, followed by benzene. Ethylbenzene and the three xylene isomers usually are present at only a small fraction of the concentrations of benzene and toluene.

Polycyclic aromatic hydrocarbons (PAHs: also called polynuclear aromatic hydrocarbons), defined as hydrocarbons containing two or more fused aromatic rings, are the petroleum hydrocarbons of greatest environmental concern in produced water, because of their toxicity and persistence in the marine environment (Neff, 1987). Concentrations of total PAHs in produced water typically range from about 80 to 3,000 μg/L, with most produced waters containing less than about 1,000 μg/L. Naphthalene and occasionally phenanthrene and their alkyl homologues are the only PAHs that are sometimes present at higher than trace concentrations. These lower molecular weight PAHs often are present at higher concentrations in produced water from gas wells than in produced water from oil wells (Stephenson et al., 1994).

Higher molecular weight PAHs, such as benzo(a)pyrene, usually are present in produced water at concentrations near the quantification limit of the best available analytical techniques. Middleditch (1981) reported an average concentration of 1.2 μg/L benzo(a)pyrene in produced water from the Buccaneer oil and gas platform off Galveston, TX. Produced water from four platforms discharging to Bass Strait off the southeast coast of Australia contained 0.009 to 0.1 μg/L benzo(a)pyrene and up to about 18 ppb total five- and six-ring PAHs (Terrens and Tait, 1993). Produced water from North Sea gas and oil wells contained less than 0.1 μg/L and 2.0 μg/L, respectively, benzo(a)pyrene (Stephenson et al., 1994).

Bioaccumulation

One of the major concerns with discharges of nonpolar organic chemicals, such as petroleum hydrocarbons, to ocean waters is the potential for bioaccumulation by local organisms and possible biomagnification of the compounds in the marine food web to concentrations toxic to the food web organisms themselves or to human consumers of fishery products. Nonpolar (not electrically charged and therefore unable to engage in electrostatic interactions) organic chemicals in treated produced water effluents discharged to the ocean tend to become distributed among different environmental phases (air, water, sediments, biota) according to their affinities for the different phases (Connell and Hawker, 1986). Because of their low affinity for the water phase, they tend to partition out of the water and sorb to any available organic phase, such as tissues of aquatic organisms, organic coatings on sediment particles, and organic particles such as humic materials and organic detritus (Clayton et al., 1977; Karickhoff et al., 1979; Knezovich et al., 1987). They also have an affinity for sorption to or complexation with certain polar, dissolved organic chemicals, such as humic acids (McCarthy and Jiminez, 1985).

At least some of the external body surfaces of all water-breathing marine animals are permeable (e.g., the gills) and nonpolar organic contaminants can partition passively (both inward and outward) between the lipids of the animal and the surrounding water through these permeable surfaces (equilibrium partitioning). According to equilibrium partitioning theory (Davies and Dobbs, 1984; Bierman 1990), when an aquatic animal is exposed to a nonpolar contaminant dissolved in the ambient water, the contaminant partitions into the tissue lipids until an equilibrium, approximated by the octanol/water partition coefficient (K_{ow}) for the chemical, is reached. At equilibrium, the rates of absorption into and desorption from the lipid phase of the animal are equal. Because of the relationship between log K_{ow} and bioaccumulation, it is possible to predict the bioconcentration factor (concentration in tissues/concentration in ambient water: BCF) for a particular compound from its log K_{ow}, if all the underlying assumptions are met (Veith and Kosian, 1983; Davies and Dobbs, 1984).

Bioconcentration (defined as bioaccumulation from water alone) is a completely artificial parameter. Because nonpolar organic chemicals dissolved in water have a strong tendency to partition from the water onto or into all hydrophobic interfaces, they become distributed rapidly among different liquid and solid compartments of the environment according to various interphase partition coefficients. A marine animal in its natural environment is never exposed for an extended period of time to only the dissolved chemical in the ambient water. Food organisms that the animal consumes accumulate the chemical from the water and from their food. Animals living near or buried in bottom sediments are exposed to the chemicals bound to sediment particles and in sediment pore water. Thus, the animal is exposed to and may accumulate the chemical simultaneously from the water, food, and sediments in its environment. Accumulation of a bioavailable chemical from all possible sources is called bioaccumulation. As a general rule, nonpolar organic chemicals with log K_{ow}s less than 5.0 are accumulated primarily from the water; food becomes an important source of bioaccumulation for organic chemicals with log K_{ow}s greater than about 5.0 (Thomann, 1989).

The process of bioaccumulation of chemicals from food is called trophic transfer. Trophic transfer of contaminants in marine food webs involves many of the same physical and chemical processes that are involved in accumulation of contaminants from water and sediments. However, bioaccumulation occurs primarily or exclusively in the unique environment of the gut of the consumer (Gobas et al., 1993).

Contaminants may be spread throughout a marine food web by trophic transfer. If the trophic transfer is reasonably efficient and the consumers at each trophic level in the food web are inefficient in excreting the contaminants, the contaminants may biomagnify in the food web. Biomagnification is the process whereby a chemical, as it is passed through a food chain or food web by trophic transfer, becomes increasingly concentrated in the tissues of animals at each higher trophic level. There are two requirements for biomagnification. First, the efficiency of uptake of the chemical from the gut must be high. Second, once the chemical has been absorbed from the food by the consumer, it must be retained in or bound tightly to the consumer's tissues so that its rate of passive or active excretion is low. If these conditions prevail, and the food remains contaminated for a long time, the consumer will, over a long period of time eat several times its own body weight of contaminated food and may retain the chemical in its tissues at a concentration higher than the concentration in the food.

MATERIALS AND METHODS

Sample Collection

As part of an offshore production platform monitoring study sponsored by the U.S. Department of Energy, produced water and environmental samples were collected during the summer of 1993 from four oil production platforms on the continental shelf off Louisiana and Texas. Reference seawater samples were collected at four sites without platforms and produced water discharges at water depths and distances from shore similar to those of the study sites.

Treated produced water samples were collected directly from the effluent pipe just before the discharge point. Several samples were collected from each production platform. Samples for hydrocarbon analysis were collected into and stored in 1- or 2-L pre-cleaned amber glass bottles that were sealed with Teflon-lined caps. Samples were stored chilled at ≈ 4°C and shipped immediately by air to the analytical laboratory.

Ambient seawater samples were collected at a depth of approximately 1 m below the sea surface approximately 2,000 m down-current from the four production platforms. Additional ambient water samples were collected at the four reference sites. Seawater samples were handled the same as produced water samples.

Bivalve molluscs were collected from legs of the production platforms by divers. Fish and portunid crabs were collected by otter trawl at distances of 200 to 2,000 m from the production platforms. The biological specimens were frozen and shipped by air to the analytical laboratory. Whole soft tissues of bivalves and edible muscle tissues of crabs and fish were analyzed.

Sample Analysis

PAHs in produced water, ambient water, sediment, and tissue samples were analyzed by capillary column gas chromatography with quantification by mass spectrometry in the selected ion monitoring mode (modified EPA Method 8270). Details of the extraction, cleanup, and analytical methods are provided by Sauer and Boehm (1991) and Page et al., 1995). A total of 41 parent PAHs and alkyl-PAH congener groups were quantified, including naphthalene, fluorene, dibenzothiophene, phenanthrene, fluoranthene/pyrene, and chrysene and their C_1 through C_3 or C_4 alkyl homologue groups, acenaphthylene, acenaphthene, biphenyl, anthracene, benz(a)anthracene, benzo(b)fluoranthene, benzo(k)fluoranthene, benzo(e)pyrene, benzo(a)pyrene, perylene, indeno(1,2,3-cd)pyrene, dibenz(a,h)anthracene, and benzo(ghi)perylene.

RESULTS AND DISCUSSION

Monocyclic Aromatic Hydrocarbons

The potential for hydrocarbons and other nonpolar organic chemicals to bioaccumulate in the tissues of marine organisms is inversely correlated to their aqueous solubilities and K_{ow}s (Neff, 1994). The U.S. Environmental Protection Agency (1991) defines bioconcentratable nonpolar organic compounds in industrial effluents as those with a log K_{ow} greater than 3.5. This log K_{ow} criterion was based on the observation that nonpolar organic chemicals with log K_{ow} values lower than 3.5 do not bioconcentrate to any great extent. The BTEX compounds have log K_{ow}s in the range of 2.13 to 3.30 (Eastcott et al., 1988), and so are not expected to bioaccumulate in marine food webs.

Freshwater and marine plants and animals do accumulate BTEX compounds during exposures in the laboratory to relatively high concentrations in solution. There is a strong correlation in the microalga, *Selenastrum capricornutum*, between K_{ow} and BCF of BTEX compounds (Herman et al., 1991). Values for BCF ranged from 42.7 for benzene to 257 for *p*-xylene. BCFs for BTEX compounds in aquatic invertebrates and fish generally are much lower than those for microalgae. Both bioaccumulation and release are rapid, with apparent equilibrium occurring within a few days (Korn et al., 1976; Eldridge et al., 1978; Roubal et al., 1978; Nunes and Benville, 1979; Berry, 1980). Most BCFs for BTEX compounds in tissues of aquatic invertebrates and fish were between 1.1 and 20. Some lipid-rich tissues had BCFs greater than 100. It is probable that under field conditions, tissue residues of BTEX in marine organisms will be much lower than those observed in laboratory exposures, because of rapid active and passive excretion of these compounds and their brief residence times in ocean waters.

Because they are very volatile, BTEX compounds evaporate rapidly from the water following produced water discharge. Under conditions following a pulse input of BTEX into ambient sea water, as occurs during ocean discharge of produced water, where the concentrations in sea water is much greater than the atmospheric concentration divided by the Henry's law constant, the residence time of alkylbenzenes in the upper water column is about 2 days (Gschwend et al., 1982). The half-life of benzene in sea water is about 5 hours (Mackay and Leinonen, 1975).

Brooks et al. (1980) reported that the maximum concentration of benzene measured in seawater immediately below the produced water discharge pipe at a production platform in the Buccaneer Field off Galveston, Texas, was 0.065 μg/L. This represents a nearly 150,000-fold dilution of the concentration of benzene in the produced water effluent (9,500 μg/L), and is well below the U.S. criterion value of 5 μg/L for benzene in drinking water. Concentrations of total gaseous and volatile hydrocarbons, including BTEX aromatics (75 percent of the total) decreased from 22,000 μg/L in the effluent, to 65 μg/L at the air/water interface below the outfall, to less than 2 μg/L in the surface water about 50 m away, indicating very rapid evaporation and dilution of the volatile components of the produced water.

BTEX concentrations in the upper water column near production platforms off Louisiana ranged from 0.008 to 0.332 μg/L (Sauer, 1980), compared to background concentrations of 0.009 to 0.10 μg/l of benzene in surface waters of the outer continental shelf off Texas and Louisiana (Sauer et al., 1978). Concentrations of BTEX compounds in coastal waters of the western Gulf of Mexico at the mouth of the Brazos River, TX, were highly variable (McDonald et al., 1988). Benzene concentrations ranged from 0.004 to 0.233 μg/L; toluene concentrations ranged from 0.005 to 0.11 μg/L; ethylbenzene concentrations ranged from 0.004 to 0.05 μg/L; and concentrations of total xylenes ranged from 0.002 to 0.056 μg/L.

Because of the short residence times of BTEX compounds in sea water and their limited bioaccumulation by marine organisms, it is unlikely that they will be accumulated to potentially toxic concentrations by marine organisms near offshore produced water discharges. Therefore, BTEX compounds were not analyzed in tissues of marine animals in this investigation.

Polycyclic Aromatic Hydrocarbons

PAH Profiles in Produced Water and Marine Animal Tissues. The PAH profile of produced water from the four offshore platforms was dominated by the low molecular weight, more soluble PAHs, particularly naphthalene and alkyl naphthalenes. Trace amounts of fluorenes, dibenzothiophenes, and phenanthrenes also were present. Ambient water samples collected 2,000 m down-current from the discharges contained primarily alkyl naphthalenes and small amounts of phenanthrenes. Traces of higher molecular weigh 4- through 6-ring PAH also were present in some samples, probably derived from aerial deposition of pyrogenic PAH in combustion emissions (soot).

An Atlantic croaker *Micropogonius undulatus* from the vicinity of platform SMI236 in 6 m of water contained high concentrations of PAHs in its muscle tissue. The PAH profile was dominated by naphthalene and alkyl naphthalenes, with lower concentrations of fluorenes, dibenzothiophenes, and phenanthrenes. The PAH profile in the fish closely resembled that in the produced water from platform SMI236, suggesting that the fish had accumulated PAHs from the produced water plume. PAH profiles in other marine animals collected near offshore produced water discharges were highly variable. In most cases, alkyl naphthalenes were the most abundant PAHs in animal tissues, followed by phenanthrene and alkyl phenanthrenes. Occasionally, traces of higher molecular weight 4- through 6-ring PAHs were present in the tissues. For example, most tissues did not contain detectable concentrations of the well-known, carcinogenic PAH, benzo(a)pyrene: however, a few bivalves contained 0.5 to 2.4 μg/kg dry wt. benzo(a)pyrene and a few fish contained up to 1.1 μg/kg dry wt. benzo(a)pyrene.

Bioconcentration Factors for PAHs. All polycyclic aromatic hydrocarbons, with the exception of naphthalene, have log K_{ow}s greater than 3.5 (Mackay et al., 1992) and, therefore, have a strong tendency to bioaccumulate in tissues of marine organisms. The PAHs analyzed in this investigation have log K_{ow}s ranging from 3.33 (naphthalene) to approximately 8.0 (C_4-chrysenes).

Produced water from the four platforms monitored in this study contained mean concentrations of 36.9 μg/L to 578 μg/L total PAHs (Table 1). Mean concentrations of total PAHs in ambient water 2,000 m down-current from the produced water discharges ranged from 0.058 μg/L to 0.118 μg/L (range 0 to 0.337 μg/L) and were directly correlated with concentrations in the produced water. Concentrations of total PAHs in ambient water from the reference stations were similar to those down-current from the produced water dis-

Table 1. PAHs in produced water discharges and range of observed BCFs in fish near the platforms in the western Gulf of Mexico

Platform	PAHs in Produced Water (μg/L)	PAH in Fish Muscle (μg/kg wet)	PAHs in Ambient Water* (μg/L)	BCF Range**
SMI236	158	2.42 - 542	0.071	34.1 - 7,640
VR214	75.4	<0.2 - 3.47	0.060	<3.0 - 57.8
SMI130	578	<0.2 - 7.55	0.118	1.0 - 64.0
HI595	36.9	<0.2 - 40.32	0.058	1.9 - 695

* Ambient water samples collected 2,000 m down-current from the discharge.
** BCF = concentration in fish (μg/kg wet)/concentration in ambient water (μg/L).

charges. Dilution factors (concentration in produced water/concentration in ambient water) for the PAHs in the produced water were in the range of 636 to 4,898 and were highest for the produced water containing the highest concentration of total PAHs. These results indicate that PAHs in produced water were diluted to near background concentrations within 2,000 m down-current of offshore, large-volume produced water discharges.

Pelagic and demersal fish from the vicinity of the four platforms contained <0.2μg/kg to 542 μg/kg wet weight total PAH in edible muscle tissues (Table 1). If it is assumed that the fish were in equilibrium with the ambient water 2,000 m from the platforms, wet weight BCFs were in the range of 1.0 to 7,640. The dominant PAHs in ambient water were alkyl naphthalenes, fluorenes, dibenzothiophenes, and phenanthrenes (average log $K_{ow} \approx 5.0$). Based on the log BCF/log K_{ow} regression of Veith and Kosian (1983) the BCF in fish of a PAH with a log K_{ow} of 5.0 should be approximately 3,590. All but one species of fish had much lower BCFs than the predicted value. An Atlantic croaker *Micropogonias undulatus* collected near platform SMI236 contained 542 μg/kg wet wt. (2,712 μg/kg dry wt.) total PAHs. The PAH assemblage in the croaker tissues was dominated by naphthalene and alkyl naphthalenes, with only traces of higher molecular weight PAHs. Based on the regression of Veith and Kosian (1983), the integrated BCF for the low molecular weight PAHs in the muscle tissue of this fish was 565 and the fish was in approximate equilibrium with 0.96 μg/L total PAHs in the ambient water. Thus, the fish may have been exposed briefly to the concentrated plume of the produced water discharge before capture.

BCFs usually were highest for bivalve molluscs from legs of the platforms, intermediate for portunid crabs, and lowest for fish (Table 2). However, there was considerable within- and between-species variability in BCFs for PAHs in edible tissues and there was no clear relationship between habitat preference of the different species of fish and crustaceans and their BCFs. The highest and lowest BCFs for fish were for demersal species, the flounder *Sciacium* spp. and the Atlantic croaker *Micropogonius undulatus*.

The BCFs for PAHs in the tissues of the bivalves from the legs of the platforms were higher than expected if they were exposed to concentrations of dissolved PAHs similar to those in ambient water 2,000 m from the produced water discharges. Obviously, the bivalves were exposed to higher concentrations of PAHs, probably derived from the produced water plume. BCFs for total PAHs in tissues of crabs and fish were generally much lower than expected, based on equilibrium partitioning theory.

Generally, laboratory-generated BCF/K_{ow} regressions overestimate concentrations of nonpolar organic chemicals in tissues of marine and freshwater organisms in the field, particularly if the chemicals, such as PAHs, are readily metabolized and excreted by the aquatic animals (Hendriks, 1995). Quite frequently, there is a poor correlation between predicted BCFs and actual tissue residues in aquatic animals in the field (e.g., Gossett et al., 1983; Farrington and Westall, 1986). In the field, there usually are large interspecies differences in tissue concentrations of nonpolar organic chemicals that cannot be explained

Table 2. BCFs for PAH in marine animals collected near produced water discharges in the western Gulf of Mexico

Species	Wet Wt BCF	Habitat
Molluscs		
Chama macerophylla	206 - 20,600	Attached Filter Feeder
Oyster *Crassostrea virginica*	5,188	Attached Filter Feeder
Crustaceans		
Blue Crab *Callinectes sapidus*	101 - 2,180	Demersal/Pelagic
Swimming crab *Portunus* spp.	52 - 185	Pelagic
Fish		
Flounder *Syacium* spp.	<3.0 - 3.1	Demersal
Pinfish *Lagodon rhomboides*	1.1 - 33	Demersal/Reef
Seabass *Centropristis philadelphica*	1.0 - 64	Demersal
Porgy *Stenotomus caprinus*	<3.4 - 58	Demersal/Pelagic
Longtail Bass *Hemanthius leptus*	2.4 - 155	Deep-water Demersal
Flatfish (Bothidae)	17 - 110	Demersal
Sea Trout *Cynoscion* spp.	34 - 233	Demersal/Pelagic
Wenchman *Pristipomoides aquilonaris*	1.9 - 695	Reef
Cutlassfish *Trichiurus lepturus*	167 - 284	Demersal/Pelagic
Sea Catfish *Arius felis*	178 - 838	Demersal
Croaker *Micropogonias undulatus*	67 - 7,640	Demersal

simply as differences in tissue lipid concentrations or feeding habits (e.g., Gossett et al., 1983; Farrington and Westall, 1986; Connolly, 1991). Zaroogian et al. (1985) showed that for some species of marine animals, such as oysters *Crassostrea virginica* and sheepshead minnows *Cyprinodon variegatus*, the agreement between predicted and measured BCFs was low ($r^2 = 0.29$ and 0.42, respectively). For all species evaluated, calculated BCFs usually overestimated the measured BCF values. This may be due in part to the fact that only chemicals that are in true solution in the water are modeled by the BCF equations. Much of the nonpolar organic chemicals of environmental concern in natural water bodies are sorbed or complexed with dissolved and particulate organic matter in the water column, decreasing their bioavailability to marine organisms (Servos et al., 1989).

Most BCF/K_{ow} models assume that active metabolic degradation and excretion of accumulated chemicals are negligible (Spacie and Hamelink, 1982; Hawker and Connell, 1985). If active metabolism and excretion do not occur, compounds with log K_{ow}s greater than about 5 may accumulate to high concentrations in aquatic animal tissues. However, most aquatic animals particularly crustaceans and fish, have the ability to metabolize PAHs and convert them to more polar, soluble byproducts that are rapidly excreted by active or passive means (Stegeman, 1981; Kleinow et al., 1987; Stegeman and Kloepper-Sams, 1987; Pritchard and Bend, 1991; Goksøyr and Förlin, 1992). Bivalve molluscs have only a low level of metabolic capability to oxidize PAH (Stegeman, 1985), and so tend to accumulate PAHs to higher concentrations than fish and crustaceans do.

Metabolic degradation and excretion of PAHs may substantially increase the elimination rate constants for the chemicals from tissues, decreasing the equilibrium BCFs. Werner and Kimerle (1982) reported that the empirically determined BCF for a C_{12}-alkylbenzene in bluegills was 35, compared to a predicted value of 6,300. The lower actual than predicted BCF was attributed to metabolic breakdown of the compound. Southworth et al. (1980) showed that bioconcentration of three azaarenes (nitrogen-substituted heterocyclic compounds of the type sometimes found in produced water effluents) by fathead minnows was much less than predicted by K_{ow}/BCF models. Bioconcentration of these compounds

Table 3. Range of concentrations of total PAHs in edible tissues of marine animals collected near four offshore produced water discharges in the western Gulf of Mexico. Concentrations are in µg/kg dry weight

Taxonomic Group	Concentration Range
Bivalve Molluscs	187 - 12,160
Decapod Crustaceans	17 - 1,590
Teleost Fish	<0.1 - 2,712

also was much less in the fish than in *Daphnia pulex*. The authors showed that these differences were due to the rapid metabolic alteration of the azaarenes in the fish tissues.

Concentrations of Total PAH in Tissues of Marine Animals

Bivalve molluscs, crustaceans, and fish from the vicinity of the four offshore produced water discharges contained <0.1 to 12,160 µg/kg dry wt. total PAHs in their edible tissues (Table 3). These concentrations are in the lower part of the range of concentrations of total PAHs in tissues of marine bivalve molluscs, crustaceans, and fish from throughout the world, as reported in the recent scientific literature (Table 4). Much of the data for bivalve molluscs was collected as part of the National Status and Trends, Mussel Watch Program performed in coastal waters of the U.S. by the National Oceanic and Atmospheric Administration (NOAA, 1995) and other coastal mussel watch programs elsewhere in the world. PAH residues in these bivalves reflect nearshore, coastal contamination with PAHs, mainly from runoff from land (National Research Council, 1985). Most of the crustaceans and fish also were collected from coastal waters. Thus, marine animals from the vicinity of offshore production platforms discharging treated produced water to outer continental shelf waters of the western Gulf of Mexico generally have lower concentrations of total PAHs in their tissues than marine animals from coastal waters, world-wide.

Molluscs usually contain lower concentrations of total PAHs than crustaceans and fish, probably due to differences in ability of the different taxa to metabolize and excrete accumulated PAHs. This distribution of tissue residues of PAHs also indicates that PAHs probably do not biomagnify in marine food webs. The limited field studies of PAHs in aquatic food chains support the conclusion that PAHs do not biomagnify. Broman (1990) and Broman et al. (1990) studied the distribution of 19 PAHs (primarily of pyrogenic origin), in a simple food chain from the Baltic Sea consisting of seston (nonliving suspended particles >0.045 µm), blue mussels (*Mytilus edulis*) and common eider ducks (*Somateria mollissima*). The major food of juvenile eider ducks in the Baltic is mussels. Concentrations of the different PAHs were quite different from one another at different levels in the food chain.

Table 4. Range and mean concentrations of total PAHs in whole soft tissues or muscle tissue of marine animals from throughout the world. Data is from the published scientific literature. Concentrations are in µg/kg dry weight

Taxon	Number of samples	Range of concentrations	Arithmatic mean	Geometric mean
Bivalve molluscs	153	3 - 1,730,000	29,370	650
Crustaceans	18	4 - 13,350	1,500	170
Fish	84	2 - 23,450	870	190

This was attributed by the authors to different capabilities of mussels and ducks to metabolize and excrete the different PAHs.

Approximately 80 percent of the PAHs accumulated by the mussels and virtually all the PAHs accumulated by the ducks were accumulated from the food, demonstrating the importance of trophic transfer in accumulation of PAH in this simple natural food chain. However, concentrations of all PAHs decreased from seston to the tissues of the ducks, irrespective of whether PAH concentrations in tissues were expressed on a lipid weight or whole tissue weight basis, indicating that biomagnification was not occurring. PAH concentrations decreased with increasing trophic level in this simple natural food chain.

CONCLUSIONS

Produced water discharges from four platforms on the ourter continental shelf of the western Gulf of Mexico off Louisiana and Texas contained 37 to 578 μg/L total PAHs. The PAH profiles of the produced water samples were dominated by the low molecular weight, more soluble compounds; naphthalene and alkyl naphthalenes were most abundant, followed by alkyl fluorenes, dibenzothiophenes, and phenanthrenes. Only traces of higher molecular weight 4- through 6-ring PAHs were present in some produced water samples.

Ambient water 2,000 m down-current from the produced water discharges contained mean concentrations of total PAHs ranging from 0.058 to 0.118 μg/L, similar to background concentrations at reference locations in the open Gulf of Mexico away from produced water discharges. BCFs in edible tissues of marine animals collected near produced water discharges ranged from about 1 to 20,600. Bivalve molluscs from the legs of the production platforms had the highest BCFs for PAHs, followed by swimming crabs, and then fish. There was no relationship between habitat and BCFs.

Concentrations of total PAHs in tissues of marine animals from the vicinity of offshore produced water discharges ranged from less than 0.1 μg/kg dry wt. to 12,160 μg/kg dry wt. Generally, highest concentrations were in tissues of bivalve molluscs and lowest concentrations were in muscle tissues of fish. The concentrations of PAHs in tissues of marine animals from the vicinity of offshore produced water discharges fell in the lower part of the range of concentrations of PAHs in tissues of similar species from throughout the world, as reported in the scientific literature. The distribution of PAHs in different taxa of marine animals from offshore waters of the Gulf of Mexico indicate that PAHs do not biomagnify in marine food webs.

ACKNOWLEDGMENTS

The work reported in this paper was supported by a contract to Continental Shelf Associates from the U.S. Department of Energy.

REFERENCES

Berry, W.O., 1980. A comparative study of the uptake of toluene by bluegill sunfish *Lepomis macrochirus* and crayfish *Orconectes rusticus*. *Environ. Pollut.* 21:109-119.

Bierman, V.J., Jr., 1990. Equilibrium partitioning and biomagnification of organic chemicals in benthic animals. *Environ. Sci. Technol.* 24:1407-1412.

Black, K.P., Brand, G.W., Grynberg, H., Gwyther, D., Hammond, L.S., Mourtikas, S., Richardson, B.J., and Wardrop, J.A., 1994. Production facilities. Pages 209-407 *In*: J.W. Swan, J.M. Neff, and P.C. Young,

Eds., Environmental Implications of Offshore Oil and Gas Development in Australia. Australian Petroleum Exploration Association Ltd., Canberra, Australia.

Broman, D., 1990. Transport and fate of hydrophobic organic compounds in the Baltic aquatic environment - polycyclic aromatic hydrocarbons, polychlorinated dibenzodioxins and dibenzofurans. Doctoral Thesis, Dept. of Zoology, Universtiy of Stockholm, Stockholm, Sweden. 76 pp.

Broman, D., Näf, C., Lundbergh, I., and Zebühr, Y., 1990. An in situ study on the distribution, biotransformation and flux of polycyclic aromatic hydrocarbons (PAHs) in an aquatic food chain (seston-*Mytilus edulis* L.-*Somateria mollissima*) from the Baltic: an ecotoxicological perspective. *Environ. Toxicol. Chem.* 9:429-442.

Brooks, J.M., Estes, E.L., Wiesenburg, D.A., Schwab, C.R.,and Abdel-Rheim, H.A., 1980. Investigations of surficial sediments, suspended particulates, and volatile hydrocarbons at Buccaneer gas and oil field. 98 pp. *In*: W.B. Jackson and E.P. Wilkens, Eds., Environmental Assessment of the Buccaneer Gas and Oil Field in the Northwestern Gulf of Mexico, 1978-1980. Vol. 1. NOAA Technical Memorandum NMFS-SEFC-47. National Oceanic and Atmospheric Administration, National Marine Fisheries Service, Galveston, TX.

Brown, J.S., Sauer, T.C. Jr., Wade, M.J. and Neff, J.M., 1992. Chemical and toxicological characterization of produced water Freon extracts. Pages 113-132 *In*: J.P. Ray and F.R. Engelhardt, Eds., Produced Water: Technological/Environmental Issues. Plenum Press, New York.

Clayton, J.R., Pavlou, S.P., and Breitner, N.F., 1977. Polychlorinated biphenyls in coastal marine zooplankton: bioaccumulation by equilibrium partitioning. *Environ. Sci. Technol.* 11:676-682.

Collins, A.G., 1975. Geochemistry of Oilfield Waters. Elsevier Scientific Publishers, New York. 496 pp.

Connell, D.W. and Hawker, D.W., 1986. Prediction the distribution of persistent organic chemicals in the environment. *Chem. Austral.*, 53:2428-2430.

Connolly, J.P., 1991. Application of a food chain model to polychlorinated biphenyl contamination of the lobster and winter flounder food chains in New Bedford Harbor. *Environ. Sci. Technol.* 25:760-770.

Davies, R.P. and Dobbs, A.J., 1984. The prediction of bioconcentration in fish. *Wat. Res.* 18:1253-1262.

Eastcott, L., Shiu, W.Y, and Mackay, D., 1988. Environmentally relevant physical-chemical properties of hydrocarbons: a review of data and development of simple correlations. *Oil Chem. Pollut.* 4:191-216.

Eldridge, M.B., Echeverria, T., and Korn, S., 1978. Fate of ^{14}C-benzene in eggs and larvae of Pacific herring (*Clupea harengus pallasi*). *J. Fish. Res. Bd. Canada* 35:861-865.

Farrington, J.W. and Westall, J., 1986. Organic chemical pollutants in the oceans and groundwater: a review of fundamental chemical properties and biogeochemistry. Pages 361-425 *In*: G. Kullenberg, Ed, The Role of the Oceans as a Waste Disposal Option. D. Reidel Publishing Co., New York.

Goksøyr, A. and Förlin, L., 1992. The cytochrome P-450 system in fish, aquatic toxicology and environmental monitoring. *Aquat. Toxicol.* 22:287-312.

Gschwend, P.M., Zafiriou, O.C., Mantoura, R.F.C., Schwarzenbach, R.P., and Gagosian, R.B., 1982. Volatile organic compounds at a coastal site. 1. Seasonal variations. *Environ. Sci. Technol.* 16:31-38.

Gobas, F.A.P.C., McCorquodale, J.R., and Haffner, G.D., 1993. Intestinal absorption and biomagnification of organochlorines. *Environ. Toxicol. Chem.* 12:567-576.

Gossett, R.W., Brown, D.A., and Young, D.R., 1983. Predicting the bioaccumulation of organic compounds in marine organisms using octanol/water partition coefficients. *Mar. Pollut. Bull.* 14:387-392.

Hawker, D.W., and Connell, D.W., 1985. Relationships between partition coefficient, uptake rate constant, clearance rate constant, and time to equilibrium for bioaccumulation. *Chemosphere* 14:1205-1219.

Hendriks, A.J., 1995. Modeling equilibrium concentrations of microcontaminants in organisms of the Rhine delta: can average field residues in the aquatic foodchain be predicted from laboratory accumulation? *Aquat. Toxicol.* 31:1-25.

Herman, D.C., Mayfield, C.I., and Inniss, W.E., 1991. The relationship between toxicity and bioconcentration of volatile aromatic hydrocarbons by the alga *Selenastrum capricornutum*. *Chemosphere* 22:665-676.

Karickhoff, S.W., Brown, D.S., and Scott, T.A., 1979. Sorption of hydrophobic pollutants on natural sediments. *Wat. Res.* 13:241-248.

Kleinow, K.M., Melancon, M.J., and Lech, J.J., 1987. Biotransformation and induction: implications for toxicity, bioaccumulation and monitoring of environmental xenobiotics in fish. *Environ. Hlth. Persp.* 71:105-119.

Knezovich, J.P., Harrison, F.L., and Wilhelm, R.G., 1987. The bioavailability of sediment-sorbed organic chemicals: a review. *Wat. Air Soil Pollut.* 32:233-245.

Korn, S., Hirsch, N., and Struhsaker, J.W., 1976. Uptake, distribution, and depuration of ^{14}C-benzene in northern anchovy, *Engraulis mordax*, and striped bass, *Morone saxatilis*. *Fish. Bull.* 74:545-551.

Mackay, D. and Leinonen, P.J., 1975. Rate of evaporation of low-solubility contaminants from water bodies to atmosphere. *Environ. Sci. Technol.* 9:1178-1180.

Mackay, D., Shiu, W.Y., and Ma, K.C., 1992. Illustrated Handbook of Physical-Chemical Properties and Environmental Fate for Organic Chemicals. Polynuclear Aromatic Hydrocarbons, Polychlorinated Dioxins, and Dibenzofurans. Lewis Publ., Boca Raton, FL. 597 pp.

McAuliffe, C., 1966. Solubility in water of paraffin, cycloparaffin, olefin, acetylene, cycloolefin, and aromatic hydrocarbons. *J. Phys. Chem.* 70:1267-1275.

McCarthy, J.F. and Jiminez, B.D., 1985. Interactions between polycyclic aromatic hydrocarbons and dissolved humic material: binding and dissociation. *Environ. Sci. Technol.* 19:1072-1076.

McDonald, T.J., Kinnicutt, M.C. II, and Brooks, J.M., 1988. Volatile organic compounds at a coastal Gulf of Mexico site. *Chemosphere* 17:123-136.

Middleditch, B.S., 1981. Hydrocarbons and sulfur. Pages 15-54 *In*: B.S. Middleditch, Ed., Environmental Effects of Offshore Oil Production. The Buccaneer Gas and Oil Field Study. Plenum Press, NY.

National Research Council, 1985. Oil in the Sea. Inputs, Fates and Effects. National Academy Press, Washington, DC. 601 pp.

Neff, J.M., 1987. Biological effects of drilling fluids, drill cuttings, and produced waters.Pages 469-538 *In*: D.F. Boesch and N.N. Rabalais, Eds., Long-Term Environmental Effects of Offshore Oil and Gas Development. Elsevier Applied Science Publishers, London.

Neff, J.M., 1994. Critical review of draft EPA guidance on assessment and control of bioconcentratable contaminants in surface waters. API Publication No. 4610. American Petroleum Institute, Washington, DC.

Neff, J.M., Rabalais, N.N., and Boesch, D.F., 1987. Offshore oil and gas development activities potentially causing long-term environmental effects. Pages 149-174 *In*: D.F. Boesch and N.N. Rabalais, Eds., Long-Term Environmental Effects of Offshore Oil and Gas Development. Elsevier Applied Science Publishers, London.

Neff, J.M., Sauer, T.C., and Maciolek, N. 1989. Fate and Effects of Produced Water Discharges in Nearhore Marine Waters. API Publication No. 4472. American Petroleum Institute, Washington, DC. 300 pp.

NOAA (National Oceanic and Atmospheric Administration). 1995. NOAA National Status & Trends Program Mollusk Chemistry Data. Tissue Chemistry Data 86-93. ASCII text file. National Oceanic and Atmospheric Administration, Rockville, MD.

Nunes, P. and Benville, P.E., Jr., 1979. Uptake and depuration of petroleum hydrocarbons in the manila clam, *Tapes semidecussata* Reeve. *Bull. Environ. Contam. Toxicol.* 21:719-726.

Page, D.S., Gilfillan, E.S., Boehm, P.D., and Harner, E.J., 1995. Shoreline ecology program for Prince William Sound, Alaska following the *Exxon Valdez* oil spill: part 1—Study design and methods. Pages — *In*: P.G. Wells, J.N. Butler, and J.S. Hughes, Eds., *Exxon Valdez* Oil Spill: Fate and Effects in Alaskan Waters. ASTM STP 1219. American Society for Testing and Materials, Philadelphia, PA.

Pritchard, J.B. and Bend, J.R., 1991. Relative roles of metabolism and renal excretory mechanisms in xenobiotic elimination in fish. *Environ. Hlth. Persp.* 90:85-92.

Roubal, W.T., Stranahan, R.I., and Malins, D.C., 1978. The accumulation of low molecular weight aromatic hydrocarbons of crude oil by coho salmon (*Oncorhynchus kisutch*) and starry flounder (*Platichthys stellatus*). *Arch. Environ. Contam. Toxicol.* 7:237-244.

Sauer, T.C., Jr., 1980. Volatile liquid hydrocarbons in waters of the Gulf of Mexico and Caribbean Sea. *Limnol. Oceanog.* 25:338-351.

Sauer, T.C. and Boehm, P., 1991. Use of defensible analytical chemical measurements for oil spill natural resource damage assessment. Pages 363-369 *In*: 1991 Oil Spill Conference. Prevention, Behavior, Control, Cleanup. American Petroleum Institute, Washington, DC.

Sauer, T.C., Jr., Sackett, W.M., and Jeffrey, L.M., 1978. Volatile liquid hydrocarbons in the surface coastal waters of the Gulf of Mexico. *Mar. Chem.* 7: 1-16.

Southworth, G.R., Keffer, C.C., and Beauchamp, J.J., 1980. Potential and realized bioconcentration. A compariosn of observed and predicted bioconcentration of azaarenes in fathead minnows (*Pimephales promelas*. *Environ. Sci. Technol.* 14:1529-1531.

Spacie, A. and Hamelink, J.L., 1982. Alternative models for describing the bioconcentration of organics by fish. *Environ. Toxicol. Chem.* 1:309-320.

Stegeman, J.J., 1981. Polynuclear aromatic hydrocarbons and their metabolism in the marine environment. Pages 1-60 *In*: H.V. Gelboin and P.O.P. Ts'o, Eds., Polycyclic Hydrocarbons and Cancer. Volume 3. Academic Press, New York.

Stegeman, J.J., 1985. Benzo(a)pyrene oxidation and microsomal enzyme activity in the mussel (*Mytilus edulis*) and other bivalve mollusc species from the western N. Atlantic. *Mar. Biol.* 89:21-30.

Stegeman, J.J. and Kloepper-Sams, P.J., 1987. Cytochrome P-450 isozymes and monooxygenase activity in aquatic animals. *Environ. Hlth. Persp.* 71:87-95.

Stephenson, M.T., Ayers, R.C., Bickford, L.J., Caudle, D.D., Cline, J.T., Cranmer, G., Duff, A., Garland, E., Herenius, T.A., Jacobs, R.P.W.M., Inglesfield, C., Norris, G., Petersen,J.D., and Read, A.D., 1994. North sea produced water: fate and effects in the marine environment. Report No. 2.62/204. E&P Forum, London, England. 48 pp.

Swan, J.M., Neff, J.M., and Young, P.C. (Eds.)., 1994. Environmental Implications of Offshore Oil and Gas Development in Australia. Australian Petroleum Exploration Association, Ltd., Canberra, Australia. 696 pp.

Terrens, G.W. and Tait, R.D. 1993. Effects on the marine environment of produced formation water discharges from Esso/BHPP's Bass Strait Platforms. Esso Australia Ltd., Melbourne, Australia. 25 pp.

Thomann, R.V., 1989. Bioaccumulation model of organic chemical distribution in aquatic food chains. *Environ. Sci. Technol.* 23:699-707.

U.S. Environmental Protection Agency, 1991. Assessment and Control of Bioconcentratable Contaminants in Surface Waters. Draft, March 1991. U.S. EPA, Office of Water, Washington, DC.

Veith, G.D. and Kosian, P., 1983. Estimating bioconcentration potential from octanol/water partition coefficients. Pages 269-282 *In*: D. Mackay, S. Paterson, and St.J. Eisenreich (eds.), Physical Behavior of PCBs in the great Lakes. Ann Arbor Science Publishers, Ann Arbor, Michigan. pp. 269-282.

Werner, A.F. and Kimerle, R.A., 1982. Uptake and distribution of C_{12} alkylbenzene in bluegill (*Lepomis macrochirus*). *Environ. Toxicol. Chem.* 1:143-146.

Zaroogian, G.E., Heltshe, J.F., and Johnson, M., 1985. Estimation of bioconcentration in marine species using structure-activity models. *Environ. Toxicol. Chem.* 4:3-12.

17

DELINEATION OF BENTHIC IMPACT AND RECOVERY AT TWO PRODUCED WATER DISCHARGE SITES IN INSHORE LOUISIANA

Maureen M. Mulino,[1] Michael F. Rayle,[1] John C. Francis,[2] and Michael A. Poirrier[2]

[1] Steimle & Associates, Inc.
Metairie, Louisiana
[2] Department of Biological Sciences
University of New Orleans
Lakefront, New Orleans, Louisiana

INTRODUCTION

Untreated produced water or oil field brine has for many years been discharged into the estuarine environments of inshore Louisiana. Contaminants contained in these wastewaters include petroleum hydrocarbons, organic acids, heavy metals, radionuclides, treating chemicals and salts (Stevenson, 1992). The relative amount of each of these constituents varies with the formation. Most of the inshore Louisiana discharges are relatively low volume with only 10% of them reported to exceed 5,000 bbl/day (794,850 L/day) (Boesch and Rabalais, 1989).

Produced water discharges in coastal Louisiana are mostly located in the major oil fields. The two fields which are part of this study, Delacroix Island and Bay de Chene, are typical of many fields in coastal Louisiana having been in continual production since the early 1940's. Since that time, produced water has been constantly discharged in varying amounts and in various places in the fields. Current Louisiana Department of Environmental Quality regulations require the termination of most coastal produced water discharges. The termination of discharges began in 1990 in the freshest coastal habitats and is scheduled to be complete by January, 1997 in the coastal open bay habitats.

The estuarine habitats which were examined in this study are characteristically oligohaline to mesohaline and subject to seasonal salinity fluctuations. Benthos in these habitats consist of short-lived, opportunistic species. The dominant organisms in the benthic assemblages include various polychaetes, amphipods and molluscs and are by necessity eurytolerant species which exhibit high reproductive potential.

The goals of this study were to define the area of impact of produced water discharges on sediment chemistry and benthos populations while the discharges were ongoing and to record changes in the produced water related sediment chemistry and benthos populations after discharge termination. Analyses were to be carried out at three selected produced water discharges: an open bay (Bay de Chene Field), a semi-enclosed bay or area of eroded marsh (Delacroix Island Field) and a canal site (Four Isle Dome Field). A significant delay in the termination of the Four Isle Dome discharge prevented the completion of the study of this site.

This study is part of a multi-year, multifaceted study titled ENVIRONMENTAL AND ECONOMIC ASSESSMENT OF DISCHARGES FROM GULF OF MEXICO REGION OIL AND GAS OPERATIONS contracted to Continental Shelf Associates, Inc. by the U.S. Department of Energy. The portion of the study dealing with the recovery of selected estuarine habitats affected by produced water discharges in inshore Louisiana was performed by Steimle and Associates, Inc.

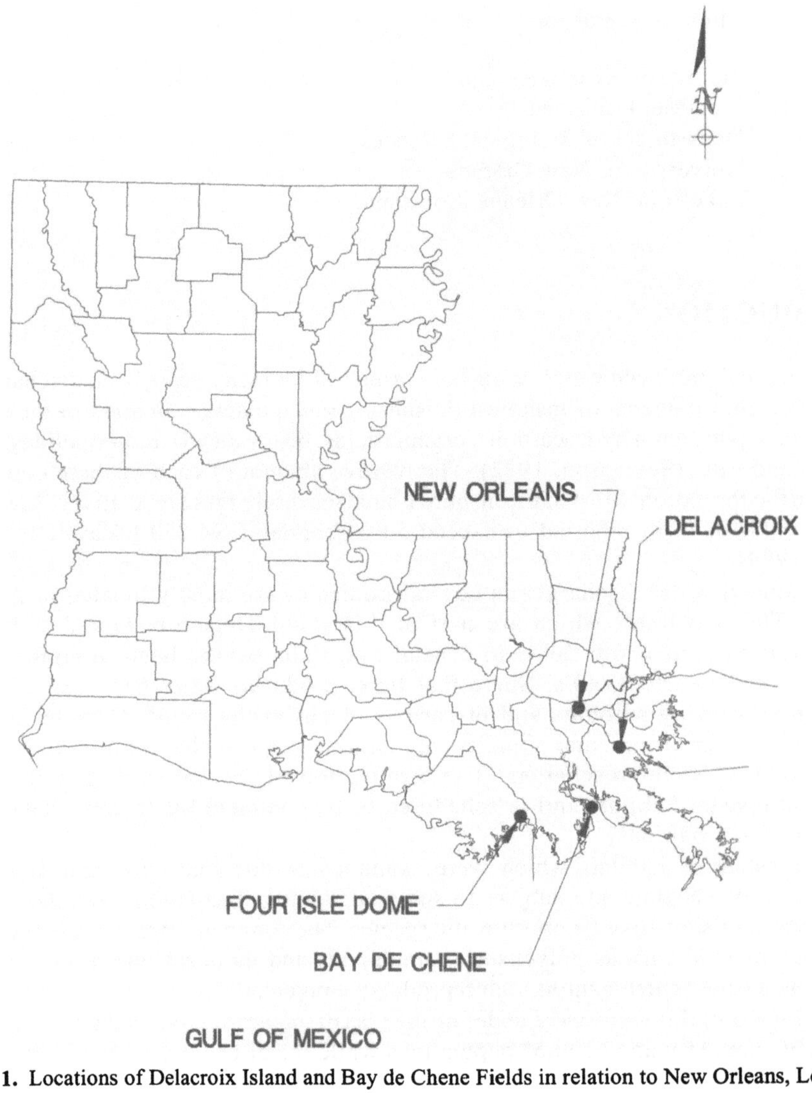

Figure 1. Locations of Delacroix Island and Bay de Chene Fields in relation to New Orleans, Louisiana.

MATERIALS AND METHODS

Description of the Study Areas

Delacroix Island. The Delacroix Island Oil and Gas field is located approximately 30 mi (48 km) southeast of New Orleans, Louisiana (Fig. 1) and has been in constant production since the first well was drilled in 1940. At the time of the pre-termination sampling (April, 1993), there were 11 wells in production in the field and the discharge volume ranged from 1964 bbl/day (312,217 L/day) to 1978 bbl/day (314,442 L/day) for the month prior to discharge termination.

The area, which is subsiding delta, was composed of oil field canals and their remnants and numerous small water bodies with few large open bays. It was highly influenced by the Mississippi River because of its proximity to the Caernarvon diversion located about 15 mi (24 km) northwest of the field. This diversion was intended to bring sediment rich Mississippi River water to the subsiding delta and to reduce salt water intrusion.

Bay de Chene. The Bay de Chene Oil and Gas Field, which is located approximately 42 mi (67 km) south of New Orleans, Louisiana (Fig. 1) is part of the Barataria Basin. This field has been in constant production since the first well was drilled in 1942. At the time of the pre-termination survey (May, 1993), the discharge resulted from four wells in production and the discharge volume was 3,825 bbl/day (608,060 L/day).

The tank battery sampled (Tank Battery #5) is located in Hackberry Bay, a large open bay typical of the Barataria system. The area was influenced seasonally by the Mississippi River discharge although not to the extent that the Delacroix Island area was influenced.

Sediment PAH Sampling and Analysis

Sediment samples were collected at stations shown in Figure 2 at Delacroix Island and Figure 3 at Bay de Chene. Sediment samples were collected in triplicate at the discharge station and the two reference stations. Additionally, during the pre-termination sampling, triplicate sediment cores were taken at all stations. A single replicate was analyzed and all stations on the transect which exhibited the highest PAH concentrations were sampled during post-termination surveys in triplicate.

Sediment samples were collected at each station using a 10 cm (OD) coring device. The cores were iced in the field, frozen within 12 hr and shipped to the laboratory entire where the top five cm were removed for analysis.

The hydrocarbon analyses of the sediments were done by Arthur D. Little, Inc. (ADL). The methods to determine PAH have been established as operating procedures at ADL. ADL's methods were developed or implemented from the literature (Boehm, 1981; NAS, 1985).

Sediment grain size analyses were performed at the University of South Alabama using combined sieve and hydrometer analyses according to recommended American Society for Testing and Materials (ASTM) procedures.

Interstitial chloride was determined from pore water by the Argentometric method (Mohr titration) (APHA, 1992).

Benthos Sampling and Analysis

Benthos sampling was conducted at Delacroix Island at the stations on the three transects designated in Figure 2. Because of the land formations around the discharge, the

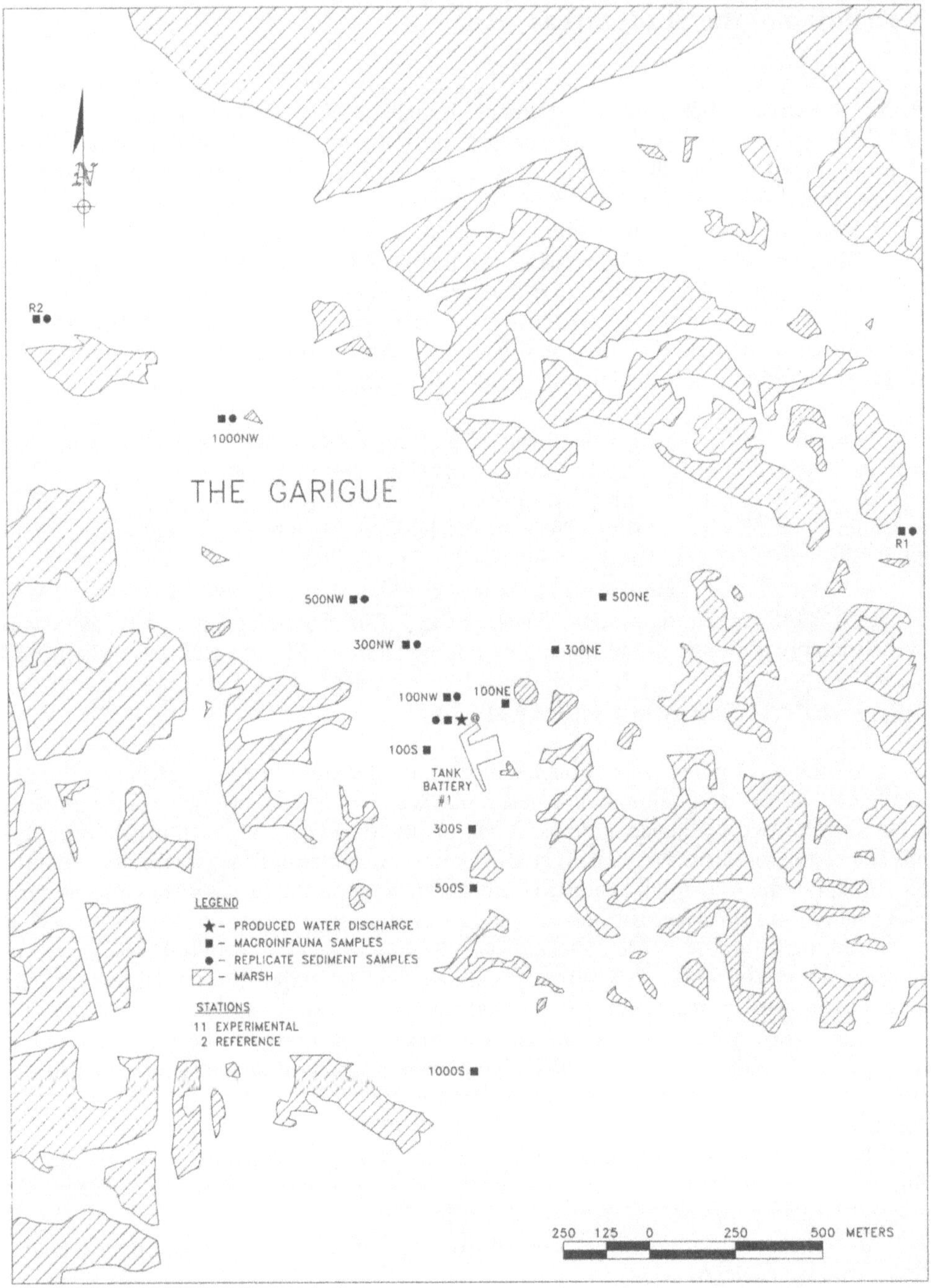

Figure 2. Locations of infauna and sediment hydrocarbon samples at Delacroix Island Field.

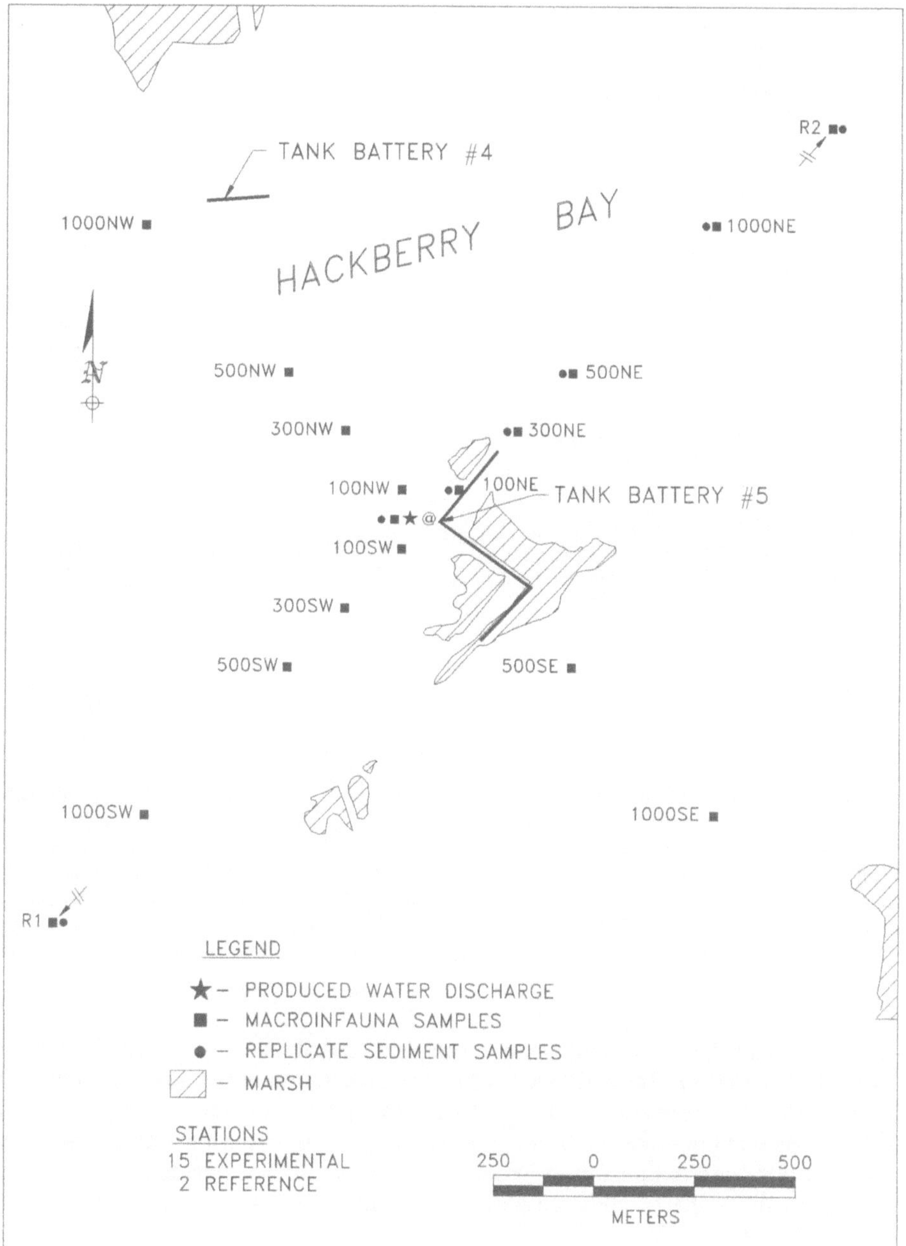

Figure 3. Locations of infauna and sediment hydrocarbon samples at Bay de Chene Field.

northeast transect had no 1000 m station. Benthos were collected during the pre-termination sampling (April, 1993) and the two post-termination samplings (October, 1993 and April, 1994). Termination of the discharge was accomplished approximately one week after the pre-termination survey was completed.

At Bay de Chene, benthos samples were collected at all stations on the four transects shown in Figure 3. Because of the presence of the field facilities on the southeast transect,

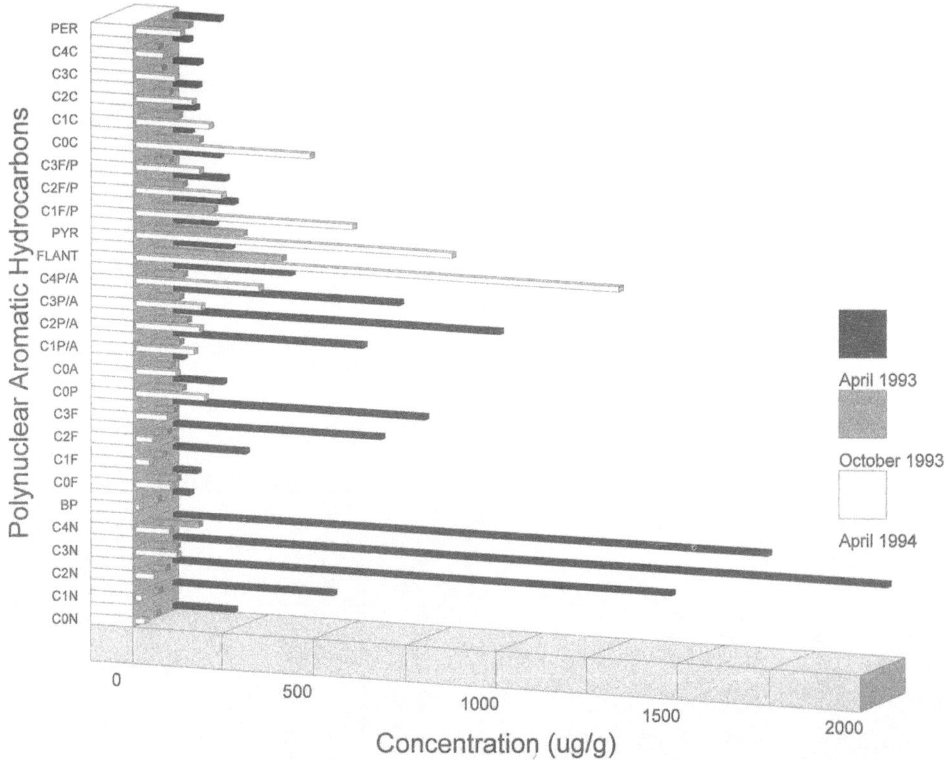

Figure 4. Sediment PAH concentrations at the discharge station at Delacroix Island during all sampling periods. CON=Naphthalene; C1N=C_1-Naphthalenes; C2N=C_2-Naphthalenes; C3N=C_3-Naphthalenes; C4N=C_4-Naphthalenes; BP=Biphenyl; COF=Fluorene; C1F=C_1 -Fluorenes; C2F=C_2-Fluorenes;C3F=C_3-Fluorenes; COD=Dibenzothiophene; C1D=C_1-Dibenzothiophenes; C2D=C_2-Dibenzothiophenes; C3D=C_3-Dibenzothiophenes; COP=Phenanthrene; COA=Anthracene; FLANT=Fluoranthene; PYR=Pyrene; C1FP= C_1Fluoranthenes/Pyrenes; C2FP= C_2Fluoranthenes/Pyrenes; C3FP= C_3Fluoranthenes/Pyrenes; COC=Chrysene; C1C=C_1-Chrysenes; C2C=C_2-Chrysenes; C3C=C_3-Chrysenes; C4C=C_4-Chrysenes; PER=Perylene.

only a 500 m and 1000 m station were sampled. Benthos were collected during the pre-termination sampling (May, 1993) and the two post-termination samplings (May, 1994 and October, 1994). Termination of the discharge was delayed by engineering problems and was accomplished in mid-October, 1993, thus delaying the post-termination samplings. Two reference stations were designated at the Delacroix Island site and the Bay de Chene site to lessen the probability that a reference station had been affected by drilling and/or disposal activities in the past.

All sampling was carried out aboard a 6.5 m MonArk boat. Six replicate benthos samples were collected by Ekman grab (23 cm x 23 cm) and sieved in the field using a WILDCO 0.5 mm screen. Samples were preserved in 10% buffered formalin which had added Rose Bengal stain. Of the six samples collected at each station, three were analyzed from each station at Delacroix Island and four were analyzed from each station at Bay de Chene. The fourth replicate from the Delacroix Island stations is currently being analyzed. Unanalyzed replicates were archived at the University of New Orleans.

Infaunal data were analyzed using analysis of variance procedures described in Sokal and Rohlf (1994). Tukey's post hoc test was used to identify significantly different stations within a group. Dependent variables used in the analysis were number of species and number of individuals. Independent variables for each study site included Sample Period (pre-termination survey, first post-termination survey and second post termination survey) and Zone (discharge site, 100 m, 300 m, 500 m, and 1000 m stations on all transects, and reference stations). The data sets were not completely balanced. At both study locations, there was no transect designation for the discharge site or for the reference sites. At Delacroix Island, there was no 1000 m station on the northeast transect. At Bay de Chene, there were no 500 m and 1000 m stations on the southeast transect.

The assumptions for analysis of variance were verified including normality and homogeneous variances. Both dependent variables (number of species and individuals) were transformed prior to analysis by taking the square root of all sample observations. This procedure was employed to ensure the variables were normally distributed. Analysis of residuals in analysis of variance confirmed that the dependent variables realize essentially the same variance at different levels of the independent variables.

Multiple regression analyses were used to relate the number of species and number of individuals at both study sites to sediment parameters (TOC, grain size, naphthalenes, acenaphthylenes, fluorenes, dibenzothiophenes, anthracenes, fluoranthenes, benzo[a] anthracene and chrysenes). The 100 m station on the northeast at Bay de Chene was excluded from the analysis because it was located at an abandoned fuel dock.

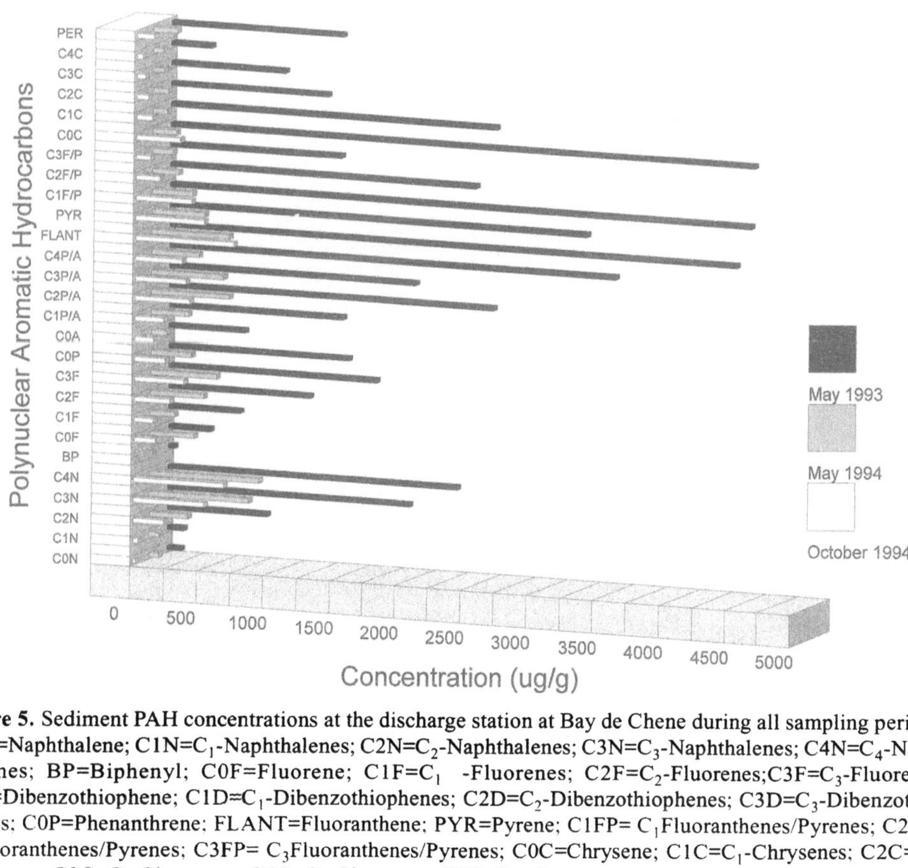

Figure 5. Sediment PAH concentrations at the discharge station at Bay de Chene during all sampling periods. C0N=Naphthalene; C1N=C_1-Naphthalenes; C2N=C_2-Naphthalenes; C3N=C_3-Naphthalenes; C4N=C_4-Naphthalenes; BP=Biphenyl; C0F=Fluorene; C1F=C_1 -Fluorenes; C2F=C_2-Fluorenes;C3F=C_3-Fluorenes; C0D=Dibenzothiophene; C1D=C_1-Dibenzothiophenes; C2D=C_2-Dibenzothiophenes; C3D=C_3-Dibenzothiophenes; C0P=Phenanthrene; FLANT=Fluoranthene; PYR=Pyrene; C1FP= C_1Fluoranthenes/Pyrenes; C2FP= C_2Fluoranthenes/Pyrenes; C3FP= C_3Fluoranthenes/Pyrenes; C0C=Chrysene; C1C=C_1-Chrysenes; C2C=C_2-Chrysenes; C3C=C_3-Chrysenes; C4C=C_4-Chrysenes; PER=Perylene.

RESULTS

Sediment Contamination

Sediment concentrations of polynuclear aromatic hydrocarbons (PAHs) and their homologues are presented in Figure 4 for the Delacroix Island Tank Battery #1 discharge site. The data presented in Figure 4 indicate that the concentrations of most compounds at the discharge site decreased to approximately background levels by six months after discharge termination (October, 1993). Background concentrations in this study were those realized at the reference stations. Other compounds - some chrysenes and some fluoranthenes/pyrenes - realized higher concentrations at both six months and one year post termination (April, 1994). For perspective, coefficients of variation (standard deviation as a percentage of the mean) for the compounds presented in Figure 4 were approximately the same at 40% for all three sampling periods. This result indicates that the means presented in Figure 4 probably are representative of actual concentrations and not the result of aberrant sample points.

Sediment concentrations of PAHs and their homologues are presented in Figure 5 for the Bay de Chene discharge site. The data presented in Figure 5 indicate that the concentrations of all compounds decreased dramatically by one year after discharge termination (May,

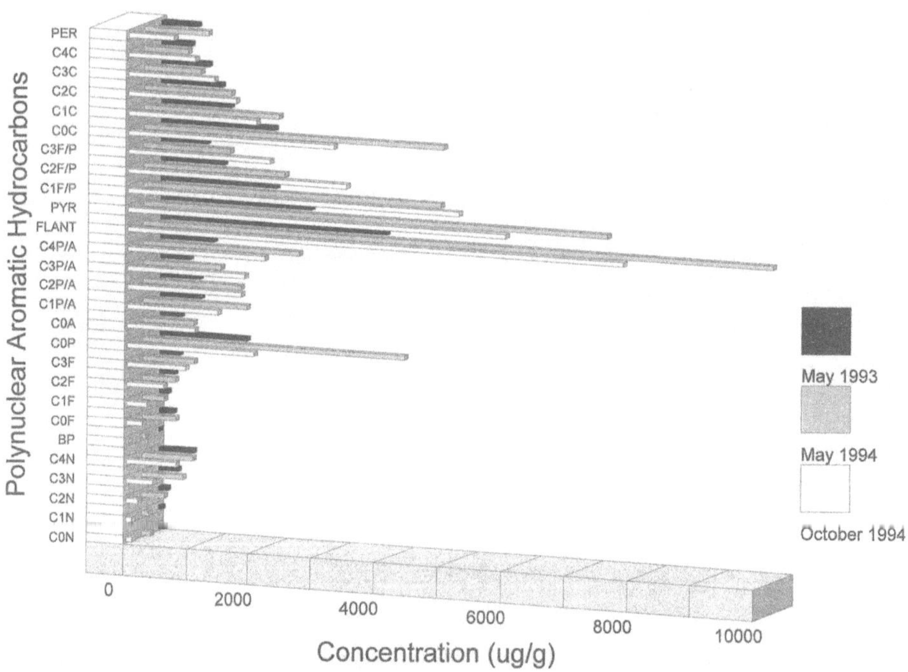

Figure 6. Sediment PAH concentrations at Bay de Chene 100NE station during all sampling periods. CON=Naphthalene; C1N=C_1-Naphthalenes; C2N=C_2-Naphthalenes; C3N=C_3-Naphthalenes; C4N=C_4-Naphthalenes; BP=Biphenyl; COF=Fluorene; C1F=C_1 -Fluorenes; C2F=C_2-Fluorenes;C3F=C_3-Fluorenes; COD=Dibenzothiophene; C1D=C_1-Dibenzothiophenes; C2D=C_2-Dibenzothiophenes; C3D=C_3-Dibenzothiophenes; COP=Phenanthrene; FLANT=Fluoranthene; PYR=Pyrene; C1FP= C_1Fluoranthenes/Pyrenes; C2FP= C_2Fluoranthenes/Pyrenes; C3FP= C_3Fluoranthenes/Pyrenes; COC=Chrysene; C1C=C_1-Chrysenes; C2C=C_2-Chrysenes; C3C=C_3-Chrysenes; C4C=C_4-Chrysenes; PER=Perylene.

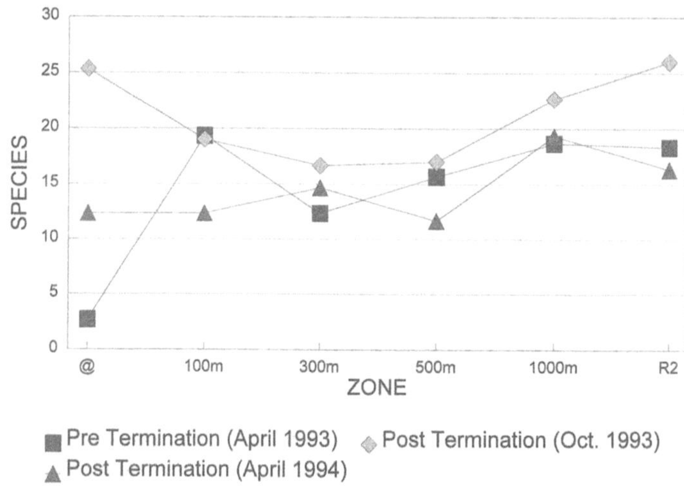

Figure 7. Average number of species collected at stations on the northwest transect at Delacroix Island for all sampling periods.

1994) and that further decreases occurred by the October, 1994 sampling although concentrations were still slightly above the background levels exhibited at the reference stations. For perspective, coefficients of variation (standard deviation as a percentage of the mean) for the compounds presented in Figure 5 were approximately the same within sampling periods but were different for the three sampling periods: 74% for the May, 1993 period, 25% for the May, 1994 period, and 65% for the October, 1994 period. The stable coefficients of variation within sampling periods indicate that the means presented in Figure 5 probably are representative of actual concentrations and not the result of aberrant sample points. We do not have an explanation for the lower coefficients of variation for the May, 1994 period.

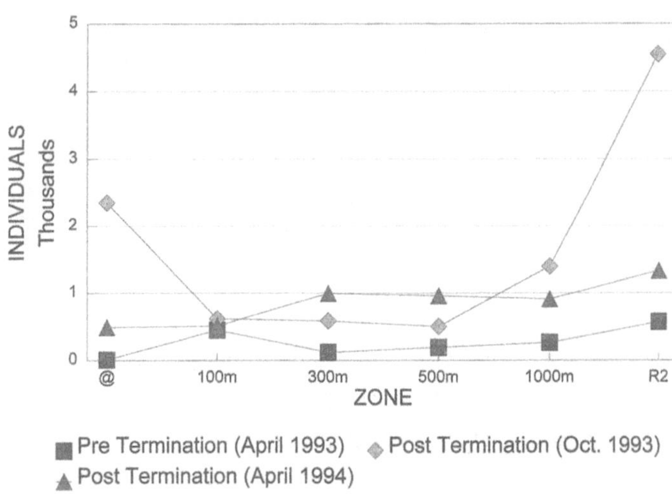

Figure 8. Average number of individuals collected at stations on the northwest transect at Delacroix Island for all sampling periods.

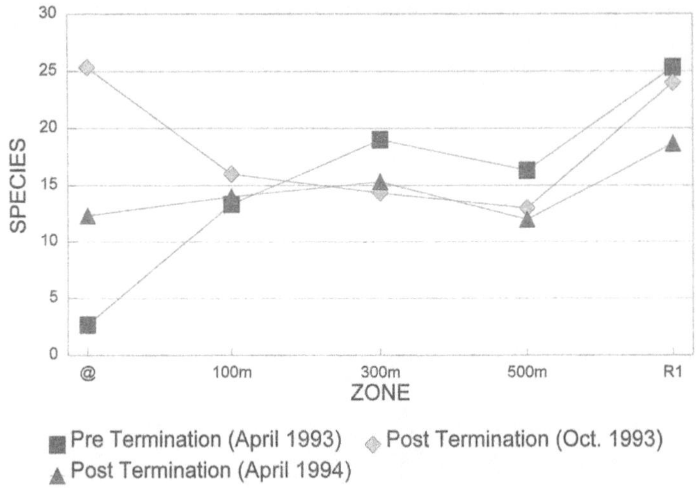

Figure 9. Average number of species collected at stations on the northeast transect at Delacroix Island for all sampling periods.

The most heavily contaminated sediments at Bay de Chene were not at the discharge station, but were found at the Bay de Chene 100 NE station located at an abandoned fuel dock. The data presented in Figure 6 indicate that the 100 NE station had PAH concentrations at least as high as those found at the discharge station, and that concentrations of most compounds did not decrease over the 18 month sampling period from May, 1993 to October, 1994.

Chloride concentrations in the extracted interstitial water from the surface sediments averaged 43.24 g/kg at Delacroix Island and 7.32 g/kg at Bay de Chene during the pre-termination sampling. Chloride concentrations of the produced water averaged 85.27 g/kg at Bay de Chene and 85.97 g/kg at Delacroix Island.

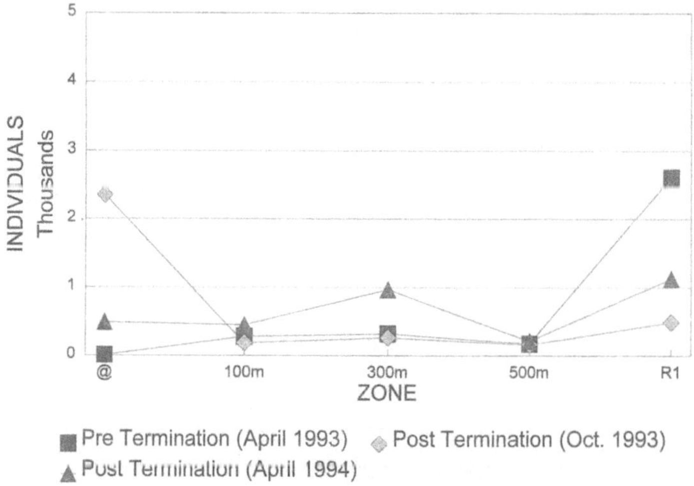

Figure 10. Average number of individuals collected at stations on the northeast transect at Delacroix Island for all sampling periods.

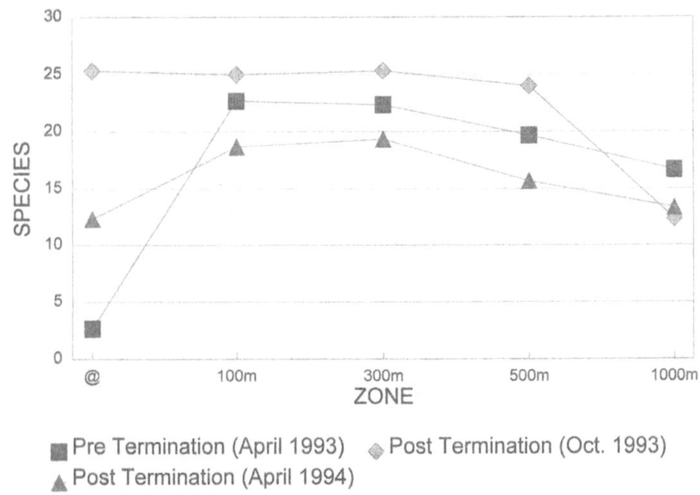

Figure 11. Average number of species collected at stations on the south transect at Delacroix Island for all sampling periods.

Benthos

Delacroix Island. Average number of species at each station in the Delacroix Island study area (by zone) along the northwest transect, the northeast transect, and the south transect are given in Figures 7, 9, and 11. Average number of individuals at each station along the northwest transect, the northeast transect and the south transect are given in Figures 8, 10 and 12. The discharge station was repeated in all transects as the origin of the transect. In these figures, reference station 1 was considered the termination of the northeast transect

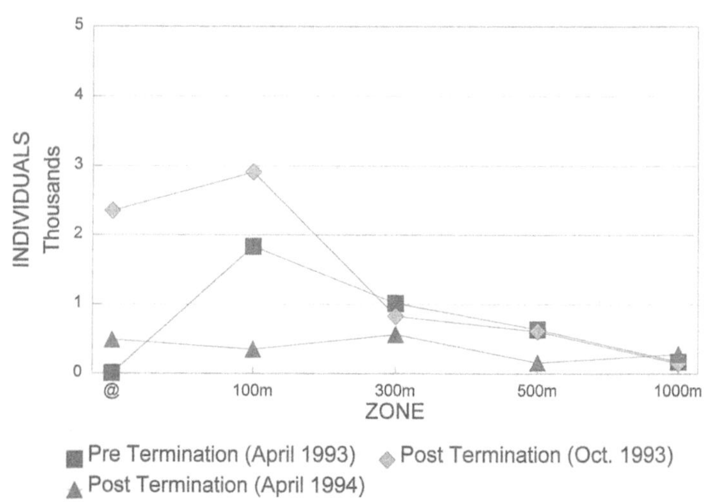

Figure 12. Average number of individuals collected at stations on the south transect at Delacroix Island for all sampling periods.

and reference station 2 was considered the termination of the northwest transect. Zone 5 which does not exist on the northeast transect was not represented in these figures.

The data were analyzed using a two-way factorial analysis of variance for both number of species and number of individuals. Independent variables used in the two-way analysis of variance included zone and sample period. Results of the analysis of variance for number of species showed that both zone and sample period and the interaction of the two were highly significant ($p<.01$). The post hoc test conducted to determine significant differences among zone means showed that the number of species at the discharge site and the number of species at the reference sites were significantly different from other zone means.

A second post hoc test was conducted to determine significant differences among sampling periods. Results of this test showed that the number of species collected in October, 1993 was significantly different from the number of species collected in April, 1993 ($p<.001$) and April, 1994 ($p<.001$).

Results of the two-way analysis of variance for number of individuals show that again the effects of zone and sampling period were highly significant ($p<.01$) and that the interaction of these two variables was significant ($p=.013$). A post hoc test was conducted to determine the significant differences among zone means. Results showed that the number of individuals at the reference stations was significantly different from the number of individuals in all other zones. The number of individuals in Zone 1, the discharge station, was not significantly different from that at Zones 2 (100 m), 3 (500 m) and 4 (1000 m).

A second post hoc test was conducted to determine significant differences among sampling periods. Results of this test showed that the number of individuals collected in October, 1993 was significantly different from that collected in April, 1993 ($p<.01$) and in April, 1994 ($p<.05$).

The multiple regression analysis identified several factors which had a negative influence on the numbers of species and individuals at Delacroix Island. These factors included sample period, dibenzothiophenes and flouranthene. Sediment grain size was not correlated with either numbers of species or individuals.

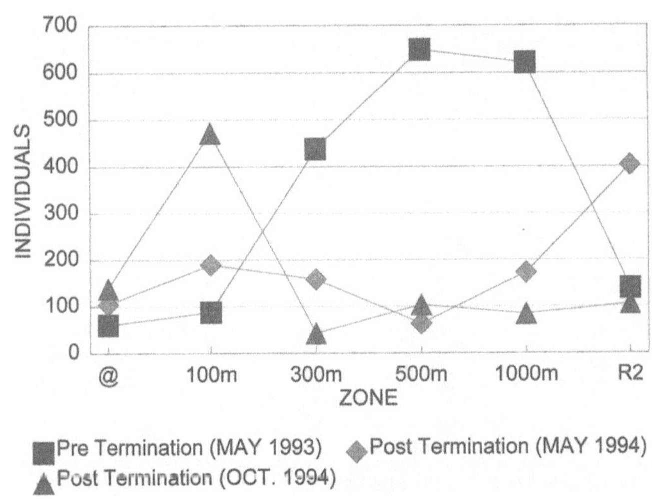

Figure 13. Average number of individuals collected at stations on the northeast transect at Bay de Chene for all sampling periods.

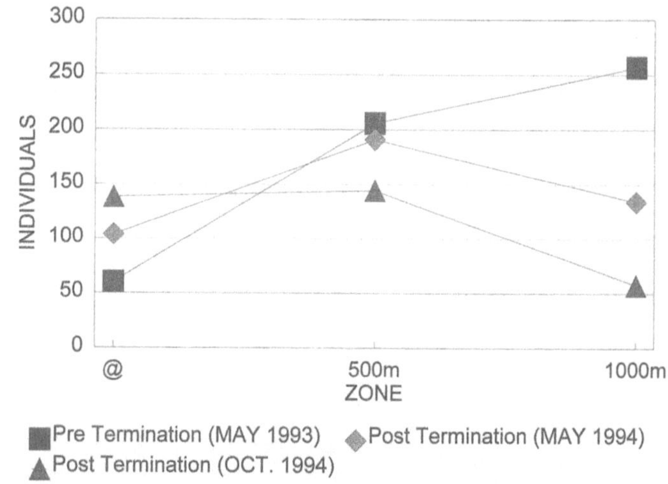

Figure 14. Average number of individuals collected at stations on the southeast transect at Bay de Chene for all sampling periods.

Bay De Chene. Average number of individuals at each station in the Bay de Chene study area (by zone) along the northeast transect, the southeast transect, the northwest transect and the southwest transect are given in Figures 13, 14, 15, and 16, respectively. As in the previous data, the discharge station was considered the origin of all transects. Reference station 1 was considered the termination of the southwest transect and reference station 2 was considered the termination of the northeast transect. The southeast transect was missing the 100 m and 300 m stations. These stations were not represented.

The data were analyzed using a two way factorial analysis of variance for both numbers of species and numbers of individuals. Results of this analysis using numbers of species showed that there were no significant differences among zones and there were no

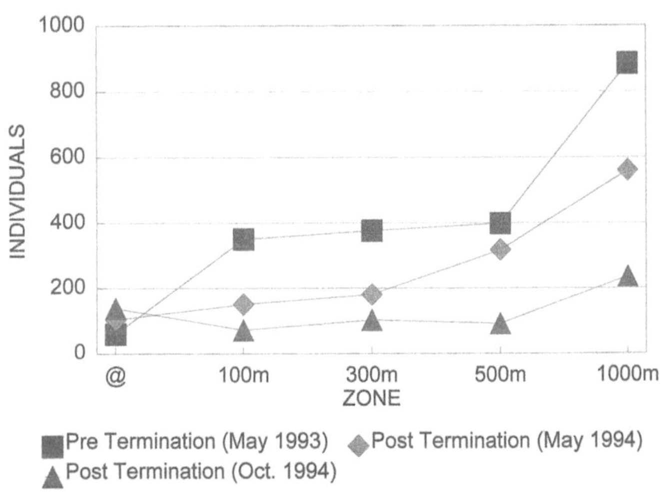

Figure 15. Average number of individuals collected at stations on the northwest transect at Bay de Chene for all sampling periods.

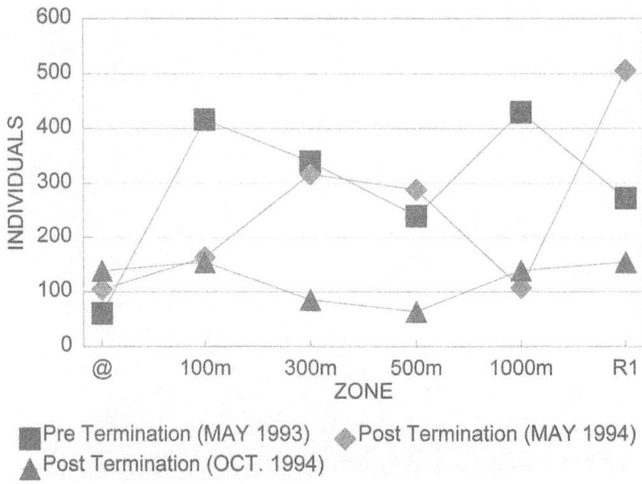

Figure 16. Average number of individuals collected at stations on the southwest transect at Bay de Chene for all sampling periods.

significant differences among sampling periods. Numbers of species are therefore not presented.

With respect to the numbers of individuals, the analysis of variance showed significant effects for zone (p=.001), sampling period (p<.001) and the zone/sampling period interaction (p<.001). The post hoc test for significant differences of numbers of individuals between zone means showed that the discharge station was significantly different from all other stations. No other significant differences were found between zones. The second post hoc test was conducted to determine significant differences in numbers of individuals between sampling periods. Results of this test showed that the May, 1994 sampling was significantly different from the May, 1993 sampling (p<.001) and from the October, 1993 sampling (p<.001) with respect to numbers of individuals. May, 1993 was significantly different from the October, 1994 sampling (p=.057) also with respect to number of individuals.

The multiple regression analysis identified two factors which had negative correlations with the numbers of individuals in the Bay de Chene samples. These factors, sample period and dibenzothiophenes, were two of the factors negatively correlated with the Delacroix Island benthos numbers. Sediment grain size was not correlated with numbers of individuals.

DISCUSSION

Depressed numbers of species and individuals were evident at the discharge site at Delacroix Island Tank Battery #1 during the pre-termination sampling. Statistically, over all three sampling periods, the number of species at the discharge was significantly lower than in all other zones. The benthos data from Delacroix Island appear to exhibit spatial recovery to a stable number of species on all transects by the 100 m station. Figures 7, 9, and 11 show graphically the similarity of the number of species for all sampling periods in Zones 2, 3 and 4 for all transects.

Recovery of the number of individuals from pre-termination levels at the discharge station on the northeast and northwest transects also appears graphically to have been realized by the 100 m stations. The similarity in the numbers of individuals at all sampling times in Zones 2, 3, and 4 are evident in Figures 8, 10 and 12. The south transect, however, appears to exhibit highly variable recovery by the 100 m station. Recovery to a stable number of individuals appears to have been realized by the 300 m station for all discharge times.

In the Bay de Chene study area, numbers of species were relatively constant throughout the zones and no statistically significant differences were found. The discharge station samples exhibited depressed numbers of individuals during the pre-termination sampling although numbers of individuals were much higher than those in the comparable Delacroix Island samples. Statistically, over all sampling periods, the numbers of individuals were different at the discharge station from those in all other zones. The number of individuals at the discharge station was also more stable than at other stations in the area although numbers of individuals increased slightly over time. Numbers of individuals were highly variable at the contaminated 100NE station, the site of an abandoned fuel dock, exhibiting a tolerance by some of the organisms for sediment hydrocarbon accumulations.

Average numbers of individuals in each zone on each transect are shown in Figures 13, 14, 15 and 16. These figures show clearly that beyond the discharge station, numbers of individuals were extremely variable throughout the area. No differences are evident statistically between the reference stations and the other zones. One of the reference stations, R1, was even located in an old oyster seed ground area which has not recently been affected by oil field activities.

The areas affected at Delacroix Island and Bay de Chene were less than 100 m in radius from the discharge. These were similar to the areas reported to be affected in previous studies of coastal environments. Armstrong et al. (1979) identified an area 150 m in radius around the discharge in which the fauna was severely depressed in Galveston Bay. In a summary of the Minerals Management Service studies of the Outer Continental Shelf generated discharges into the coastal areas of Louisiana, Rabalais et al. (1992) reported the area of impacts to benthos varied from nothing in areas of high flow volume such as distributaries of the Mississippi River (Emeline and Romere Passes) and the Empire waterway to 800 m at Pass Fourchon, a dead end canal/dredged bayou environment. In a study of discharges at Eugene Island Block 105 and Lake Pelto Tank Battery #1, both large open coastal bays, Neff et al. (1992) concluded that the benthic fauna was affected up to 20 m from but not as far as 100 m from the discharges.

One of the major factors which was identified by Rabalais et al. (1992) as influencing the area affected by produced water discharge is the hydrology of the receiving environment. This factor may help to explain the differences in magnitude of effect at the two discharge stations in this study. The benthic effects found in the pre-termination samplings were much greater at the Delacroix Island discharge station than at the comparable Bay de Chene station even though the Delacroix discharge was approximately half the volume of the Bay de Chene discharge. Delacroix was a semi-enclosed hydrological environment which did not provide as much opportunity for turbulent mixing or dilution of the discharge as did the open water nature of Bay de Chene. It was therefore more likely that the saline, deoxygenated water of the discharge settled in a non-mixing layer on the bottom.

This was evident in the differences in the chloride concentrations of the interstitial water of the surface sediments at the two discharge stations. Average chloride concentration of the interstitial water at the Delacroix Island discharge station prior to its termination was equivalent to approximately 78 ppt salinity, a level toxic to most estuarine organisms. Chloride concentrations of the interstitial water at the Bay de Chene discharge station prior to discharge termination was equivalent to approximately 13 ppt salinity. Chloride concentrations in the two produced waters were nearly equal.

Hydrocarbons contaminants present in produced water also have the potential to impact benthos. Hydrocarbons adsorb preferentially onto fine grained clays (Boehm, 1987) which predominate in the bottom sediments of Louisiana estuaries. At the Delacroix Island discharge station, sediment PAH concentrations decreased (Fig. 4) over time and between the discharge station and all 100 m stations. The number of benthic species and individuals increased dramatically at the discharge station after termination. Recovery was rapid with a large number of species and individuals inhabiting the former discharge station six months post termination.

Although multiple regression analysis identified two hydrocarbon components, dibenzothiophenes and fluoranthenes, as being negatively correlated with the number of species and individuals, it is not possible to separate these effects from those of the low dissolved oxygen and high salt content of the ongoing discharge. The sediment PAH concentrations at Bay de Chene decreased greatly between the discharge station and the 100 m stations except the 100NE station. Concentrations at this station were similar to those found at the discharge station during the pre-termination survey. Sediment concentrations at the 100NE station did not decrease with time (Fig. 6) although the number of individuals did vary with time (Fig. 13).

At the Bay de Chene discharge station, sediment hydrocarbon concentrations did decrease with time with most of the decrease occurring in the first seven months after termination (Fig. 5). The numbers of individuals and species increased slowly (Figs. 13, 14, 15, 16) through the October, 1994 (one year post termination) sampling. Although the increase in number of individuals was not as dramatic as at Delacroix Island, by October, 1994 the overall decrease in average number of individuals across the area and the slow increase in the number of individuals at the former discharge station make it indistinguishable from other stations. Little prior work is available on the time necessary for recovery of other coastal sites impacted by produced water. Rabalais et al. (1992) reported that an abandoned discharge studied in the Empire Waterway (Pelican Island Terminal) did not show benthic recovery three years after the discharge was abandoned probably because of accumulated contaminants in the sediments. The discharge into this dead-end, facility access canal which emptied into the Empire Waterway was 10,959 bbl/day (1,742,152 L/day), two to three times the volume of either discharge in this study.

Multiple regression analysis identified dibenzothiophene as one of the factors negatively correlated with the number of individuals. As with the Delacroix Island discharge station, however, it is not possible to separate the effects of hydrocarbon accumulation in the sediments from the low dissolved oxygen and high salt content on the initial sampling. The occurrence of benthos at the 100NE station does indicate a tolerance of at least some taxa to sediment hydrocarbon accumulation.

The benthos data from these two study areas show strong seasonal variability as reflected in the statistical differences in sampling periods. Results of the post hoc tests on the two-way analysis of variance for both species and individuals at the Delacroix Island study area clearly showed that the October, 1993 sampling is significantly different from both the April, 1993 and the April, 1994 samplings. Multiple regression analysis identified sampling period as having a negative effect on species and individuals at Delacroix. Although sampling period was entered into the analysis to reflect length of time after discharge termination, seasonal element is present. It is not unusual to find a higher diversity in coastal estuarine habitats in the fall since salinities are higher and more stable than those experienced in the spring.

The taxa which predominated in the Delacroix Island study area also varied seasonally. The October, 1993 sampling (six months post-termination) was characterized by a population explosion at most stations, including the former discharge station, of the amphipod *Grandidierella bonnieroides* and the tube-dwelling amphipods *Corophium louisianum* and *Cerapus benthophilus*. The pre-termination sampling of the discharge station yielded only a few individuals of the gastropod *Texadina sphincstostoma* and a single isopod. The other organisms present during the pre-termination sampling were the epifaunal *Ischadium recurvum*, whose numbers were not included in infaunal counts. Throughout the area the October, 1993 sampling was also characterized by an increase in the numbers of the polychaete *Streblospio benedicti* and the gastropod *Texadina sphincstostoma*.

The April, 1994 sampling was characterized by the dominance of the gastropod *Texadina sphinctostoma*. This common inhabitant of low-salinity estuaries has been observed by one of the authors to undergo population explosions in Lake Pontchartrain following years of introduction of nutrient-enriched Mississippi River water into the lake through the Bonnet Carre' Spillway. The Caernarvon diversion inputs Mississippi River water into the Delacroix area during the spring which is probably related to the increase in numbers of this taxon.

Results of the post hoc tests on the two-way analysis of variance for numbers of individuals at Bay de Chene also show statistically significant differences between sampling periods which may be due to strong seasonal variability. Multiple regression also identified sampling period as having a negative effect on individuals at Bay de Chene. Dominant taxa during the pre-termination sampling included oligochaetes, the polychaete *Streblospio benedicti*, the bivalve *Mulinia lateralis* and the gastropod *Texadina sphinctstoma*. The May, 1994 post-termination sampling was dominated by oligochaetes, and overall numbers of individuals throughout the area were lower than in the previous May.

The taxa in the Delacroix and Bay de Chene study areas are those which are common in coastal Louisiana habitats. Two of these taxa, *Streblospio benedicti* and *Mulinia lateralis*, were also dominant in the sites studied by Neff et al. (1989) and Rabalais et al. (1991). These organisms as all the organisms of inshore Louisiana coastal bays and bayous are eurytolerant, short lived and opportunistic. They have a high reproductive potential which allows them to succeed in the dynamic environment in which they live, the surface of the fine grained sediments which are constantly moved around by storms and currents. This provides a constant input of new colonizers to uncrowded substrate and, therefore, allows rapid recolonization of the areas affected by the discharge.

ACKNOWLEDGMENTS

This work was performed under contract to the Department of Energy (DOE) (DE-AC22-92MT 92001). Thanks are extended to Dr. Brent Smith, the Project Officer. Thanks are also extended to Texaco, Inc. for cooperating fully with this study and allowing us to sample the two facilities.

REFERENCES

American Public Health Association (APHA). 1992. *Standard Methods for the Examination of Water and Wastewater.* 18th Edition. American Public Health Association. Washington, D.C.

Armstrong, H.W., K. Fucik, J. W. Anderson and J.M. Neff. 1979. Effects of oilfield brine on sediments and benthic organisms in Trinity Bay, Texas. Mar. Environ. Res. 2:55-69.

Boehm, P.D. 1981. Petroleum in the marine environment: physical chemical methods. Background paper submitted to Ocean Sciences Board, National Resources Council, Washington, D.C. 152 pp.

Boehm, P.D. 1987. Transport and transformation processes regarding hydrocarbon and metal pollutants in offshore sedimentary environments, *in* "Long-Term Environmental Effects of Offshore Oil and Gas Development", Boesch, D.F. and Rabalais, N.N., eds., Elsevier Applied Science, New York. 708 pp.

Boesch, D.F. and Rabalais, N.N., (eds.), 1989. Environmental Impact of Produced Water Discharges in Coastal Louisiana, Report to La. Div. Mid-continent Oil and Gas Association. LUMCON, Chauvin, La. 287 pp.

National Academy of Sciences. 1985. *Oil in the Sea: Inputs, Fates and Effects*. National Research Council, National Academy of Sciences Press, Washington. D.C. 601 pp.

Neff, J.M., T.C. Sauer and N. Maciolek. 1989. Fate and Effects of Produced Water Discharges in Nearshore Marine Waters. American Petroleum Inst. Publ. No. 4472, American Petroleum Institute, Washington, D.C. 300 pp.

Neff, J.M., T.C. Sauer and N. Maciolek. 1992. Composition, fate and effects of produced water discharges to nearshore marine environments, *in* "Produced Water Technological/ Environmental Issues and Solutions" J.P. Ray and F.R. Englehart (eds.), Plenum Press, New York. 616 pp.

Rabalais, N.N., B.A. McKee, D.J. Reed and J.C. Means. 1991. Fate and Effects of Nearshore Discharges of OCS Produced Waters. Volume II, Technical Report, OCS Study/MMS 91-005, U.S. Dept. Of the Interior, Minerals Management Service, Gulf of Mexico OCS Regional Office, New Orleans, Louisiana. 337 pp.

Rabalais, N.N., B.A.McKee, D.J. Reed and J.C. Means. 1992. Fate and effects of produced water discharges in coastal Louisiana, Gulf of Mexico, USA, *in* "Produced Water Technological/Environmental Issues and Solutions" J.P. Ray and F.R. Englehart (eds.), Plenum Press, New York. 616 pp.

Stevenson, M. 1992. A survey of produced water studies, *in* "Produced Water Technological/Environmental Issues and Solutions" J.P. Ray and F.R. Englehart (eds.), Plenum Press, New York. 616 pp.

Sokal, R. R. And F.J. Rohlf. 1994. *Biometry*. San Francisco; W.H. Freeman.

18

THE EFFECTS OF PRODUCED WATER ON HYDROCARBON LEVELS AND ON P4501A MONOOXYGENASE ACTIVITY IN FISH LARVAE IN THE NORTHERN NORTH SEA

R. M. Stagg, P. A. Gillibrand, A. M. McIntosh, and W. A. Turrell

SOAFD Marine Laboratory
P.O. Box 101, Aberdeen AB9 8DB

INTRODUCTION

Estimates of the input of oil into the North Sea vary between 77 and 211 x 10^3 tonnes per annum (North Sea Task Force, 1993). More than half of this is directly into the offshore marine environment through atmospheric deposition, operational and illegal discharges from shipping and discharges from the offshore oil industry. Estimates (Table 1) are variable and are likely to be subject to high errors but indicate that the offshore industry may contribute between 14 and 25% of the total oil and 27-45% of the oil discharged offshore in the North Sea (North Sea Task Force, 1993). In the UK, oil discharged from the offshore industry occurs from three major sources: in produced water, on cuttings and through accidental spillage. Produced water is the single largest waste stream entering the marine environment from the offshore industry and on the UK continental shelf in 1994 contributed more than 50% to the total amount of oil discharged by the industry (HMSO, 1995).

Production is concentrated where there are geological structures from which oil can be extracted and therefore oil platforms are not distributed randomly in the North Sea (Figure 1). The East Shetland basin, for example, covers approximately 1% of the area (8000 km^2) of the North Sea and yet it contains 29% of all of the oil platforms. As fields age the amount of produced water, and hence oil discharged, increases in response to the need to inject water into the formation to maintain reservoir pressure. Some of the oldest fields in the UK sector of the North Sea may be discharging 98% of the produced fluids as water and again many of these are located in the East Shetland basin. There is therefore a need to consider not just the discharge and impact of individual platforms but also the wider regional impacts of the industry on the North Sea environment. The objectives of this paper are therefore to discuss recent work (Stagg and McIntosh, 1996) on the levels of oil in the water column in the northern North Sea, to comment on the significance of these to fisheries and the environment and to discuss the contribution of the offshore oil industry in the light of some preliminary work (Gillibrand *et al.*, 1995) in which the fate of offshore produced water discharges was modelled.

Produced Water2: Environmental Issues and Mitigation Technologies
Edited by M. Reed and S. Johnsen, Plenum Press, New York, 1996

Table 1. Sources of Oil in the North Sea 10^3 tonnes/year

Coastal	Rivers/Land Run-off	16 - 46
	Industries	5 - 15
	Sewage	3 - 15
	Dredged spoils	2 - 10
	Sewage sludge	1 - 10
	Refineries	4
	Dumped Industrial waste	1 - 2
	Oil-terminals	1
Offshore	Offshore oil and gas	20 - 30
	Illegal discharges from ships	15 - 60
	Atmosphere	7 - 15
	Operational discharges from ships	1 - 2
	Natural seeps	1
Total		86 - 210

Source: North Sea Task Force Quality Status Report 1993

DISTRIBUTION OF OIL LEVELS IN THE NORTHERN NORTH SEA

A variety of analytical methods are available for the determination of hydrocarbons including gravimetry, infra-red spectrometry, ultra-violet fluorescence spectrometry (UVF), high pressure liquid chromatography (HPLC) and gas chromatography (GC) or GC coupled

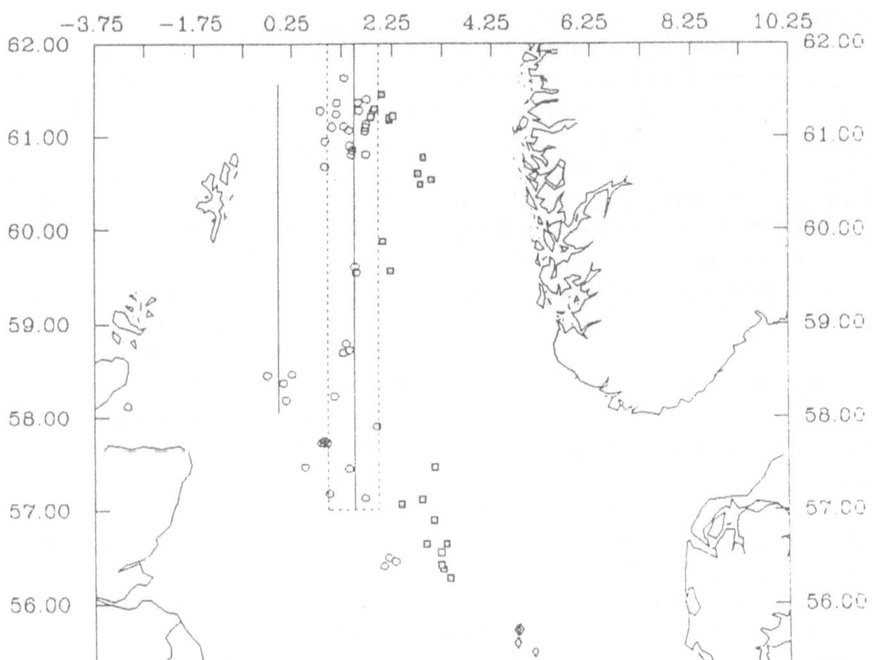

Figure 1. Map showing the location of oil installations in the North Sea, the positions of the transects sampled and the box used to calculate the concentrations used by the particle tracking model. (○ UK, □ Norwegian and ◊ Danish platforms).

to mass spectrometry (GC-MS). In deciding which method to use various factors should be considered including:

1. the ability to obtain representative samples;
2. whether the technique determines hydrocarbons of biological importance such as the polyaromatic hydrocarbons (PAH);
3. the sensitivity of the method; and
4. the quantitative nature of the measurement.

The distribution of hydrocarbons in the water column is patchy and discontinuous (Fondekar and Alagarsamy, 1984) and this either necessitates the collection and analysis of a large number of water samples or the use of an instrument which measures continuously the levels of hydrocarbons *in situ*. PAH are the components of oil which are of most concern because of their ability to cause chronic deleterious effects. In a study of PAH inputs to the North Sea through the Fluxmanche transect across the English Channel (Law and Whinnet, 1993), detection limits for GC-MS techniques and the need for better integrated sampling of the water column were identified as constraints on the methodology. UVF measurements provide a simple, sensitive and rugged tool for screening hydrocarbon concentrations which are selective for the aromatic components of oil. In addition the ability to deploy a fluorimeter in an *in situ* mode (Law, 1984; Dahlmann & Rave, 1988) means that large areas can be monitored, especially in areas with complex hydrography and dynamic water masses. The use of a towed fluorometer to measure hydrocarbon concentrations at the sub-μg l^{-1} levels in the North Sea has been reported previously (Law, 1984; Dahlmann and Rave 1988). However the use of fluorometry does not give the specificity of some other analysis (HPLC, GC or GC-MS) and hydrocarbon determinations by this method will be semi-quantitative because changes in composition may affect the signal measured. Nevertheless the use of the

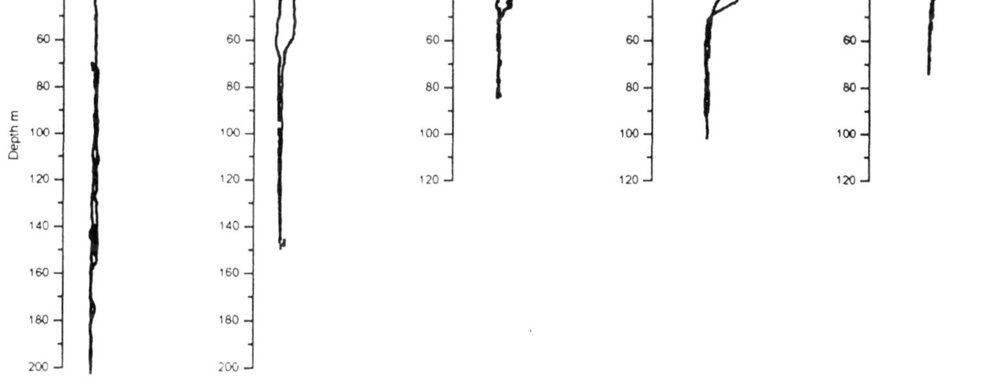

Figure 2. Typical results from single deployments of the fluorimeter at various points on the sections studied to demonstrate the profile of hydrocarbon fluorescence (λEx 310nm, λEm 360nm) with depth. Hydrocarbon data is smooth averaged and expressed as μg l^{-1} Forties equivalents.

technique is justified in that it allows more representative sampling and it does reflect the aromatic hydrocarbon content of the matrix.

With these constraints in mind Stagg and McIntosh (1996) used an *in situ* fluorimeter (calibrated against serial dilutions of a water soluble fraction of Forties crude) to measure hydrocarbon levels on two north-south sections of the North Sea in May 1993 and April 1994. The instrument was deployed on a METHOT net for sampling fish larvae and enabled the simultaneous collection of live fish larvae for determination of exposure to aromatic hydrocarbons and associated effects. The sections were positioned so that they passed

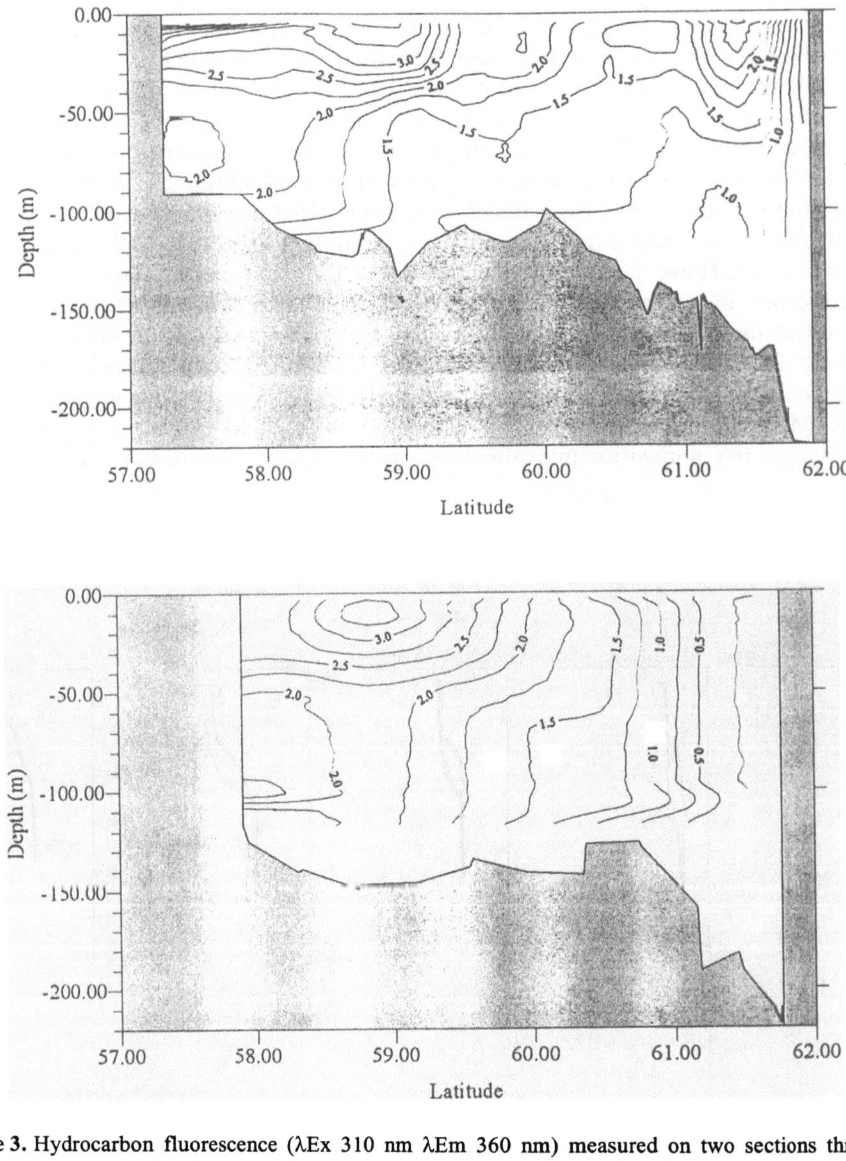

Figure 3. Hydrocarbon fluorescence (λEx 310 nm λEm 360 nm) measured on two sections through the northern North Sea in April 1994 (top: Section 1 30'E; bottom: Section 0.00E). Contours are derived from individual hydrocarbon profiles measured using a Chelsea fluorimeter deployed on an oblique tow from surface to bottom at intervals of approximately 10 miles. The data was used to generate a grid with nodes every 0.15° of latitude and 1 m in depth. Contours were smoothed by averaging every 2 adjacent grid points.

through the major areas of oil production in the UK sector of the North Sea including the East Shetland basin (Figure 1). Stations to the north and west of the East Shetland Basin were also sampled to provide information on the background levels of oil from pristine areas remote from the influence of shipping or oil industry activity. At each station the instrumentation was towed at 4 knots and lowered to the seabed before being hauled back to the ship. This enabled the hydrocarbon fluorescence to be obtained over the full depth range from surface to bottom. Detailed accounts of the methodology have been given elsewhere (Stagg and McIntosh, 1996).

A series of traces from these studies are shown in Figure 2. At the northern and western boundary of the sections worked the hydrocarbon concentration was low (between 0.1 and 0.5 μg l-1). The vertical profile at these clean stations showed no marked changes in concentration with depth (Fig 2 a). In contrast there was a gradient of concentration through the East Shetland Basin (1 - 2 μg l-1) rising to a plateau of between 3 and 4 μg l-1 south of 58° 30'N. At some stations in the close proximity of oil platforms (namely Forties, Brae and those in the east Shetland basin) hydrocarbon concentrations as high as 5 μg l-1 were recorded. At all stations with elevated concentrations of oil there were clear changes in the hydrocarbon concentration with depth. In the upper 50 m levels were invariably higher than in the deeper water (Figure 2b-d). Earlier studies showed a similar fall in concentration with depth (Theobald, 1989; Theobald *et al.*, 1991/2) and elevated concentrations in surface samples (1 - 5 m) were found compared with those taken in deeper water.

Contour diagrams to illustrate the distribution of fluorescence measurements show clearly both the north-south gradients and the depth related changes (Figure 3). Calculation of the averaged hydrocarbon concentration in the top 50 m of the sections (Figure 4) shows a 10 fold increase in concentration between the pristine areas (north and west) and those

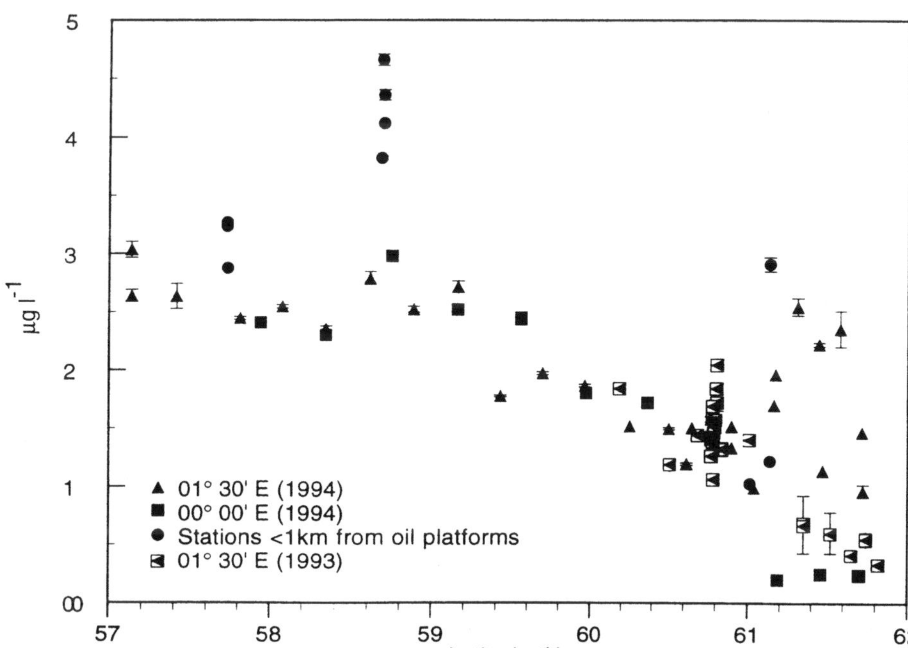

Figure 4. Depth averaged hydrocarbon fluorescence (λEx 310 nm λEm 360 nm) for the top 50 m measured at each station on two sections through the northern North Sea in April 1993 and April 1994. Values are mean ± SE.

areas further south in the North Sea. There are other reports which support these conclusions. For example Theobald (1989) and Theobald *et al.* (1991/2) have also shown hydrocarbon concentrations by UVF of around 0.1 $\mu g\ l^{-1}$ in pristine Atlantic waters compared with levels between 1-2 $\mu g\ l^{-1}$ in the North Sea and 10 $\mu g\ l^{-1}$ in the German Bight. Law (1990) reported background concentrations in the North Sea and the Channel to be 1- 2 $\mu g\ l^{-1}$ but did not specify the precise locations. Likewise Dahlmann and Rave (1988) using a towed *in situ* fluorimeter reported levels of >5 $\mu g\ l^{-1}$ in the German Bight compared to levels of 2 or 3 $\mu g\ l^{-1}$ over much of the northern North Sea. It is likely that the differences in fluorescence observed between the pristine areas and the North Sea are due to contamination with petroleum hydrocarbons. Theobald (1989) and Theobald *et al.* (1991/2) used HPLC techniques to analyse the fluorescence data obtained from offshore seawater samples and showed the presence of a mineral oil UCM on the HPLC trace from North Sea and Elbe samples which was not present in Atlantic water samples. Stagg and McIntosh (1996) presented some evidence which showed that the hydrocarbon fluorescence measurements made were not due to changes in fluorescence yield on the sections studied and were not correlated with chlorophyll measurements.

BIOLOGICAL SIGNIFICANCE

The North Sea is an important basin for fisheries accounting for 3.5% of the worlds catch and yet comprises only 0.18% of the earth's oceans. There is therefore a marked juxtaposition of the oil industry with the fishing industry and this heightens the environmental concern for the impact of discharges from the oil and gas industry on fish and fish populations. In this paper we are considering in particular discharges to the water column and elevated levels of hydrocarbons in the pelagic environment. Most fish species of commercial importance have a pelagic dispersion phase and larvae and eggs are carried in the surface waters. The chronic biological effects of low levels of aromatic hydrocarbon contamination on these stages of fish are poorly understood. Embryos and larvae are thought to be susceptible to contaminants such as hydrocarbons because: they may not have fully developed organ systems and therefore mechanisms to detoxify contaminants (depending on the stage of development); they are in close contact with the environment; have a high surface to volume ratio which facilitates uptake of contaminants; they are not sufficiently mobile to avoid exposure; and they develop in the upper layers of the water column where oil exposure can be expected. Experimental studies of the effects of hydrocarbons have shown: delayed hatching, reductions in length (growth rate); large or deformed yolk-sacs, changes in morphology at the ultra-structural level and gross deformities (reviewed in von Westernhagen, 1988). However, most of these are acute studies and very few have been carried out at environmentally relevant exposure concentrations.

Cytochrome P450-dependant monooxygenases (CYP P450) are membrane-bound enzymes which metabolise a range of exogenous (organic contaminants) and endogenous substrates (steroids, fatty acids). A single form of CYP P450 called CYP1A is induced by environmental exposure to PAH (Addison and Edwards, 1988; Payne *et al.* 1988; Spies *et al.*, 1982; Stagg *et al.*, 1995; Stegeman *et al.*, 1988; Stein *et al.*, 1992; Van Veld *et al.*, 1990) in fish. In the North Sea the induction of CYP 1A has been used to monitor exposure and effects of PAH from the offshore industry either as aryl hydrocarbon hydroxylase (Davies *et al.*, 1984) or as 7-ethoxyresorufin *O*-deethylase (EROD) (Stagg *et al.*, 1995). Induction of CYP1A is also a measure of deleterious biological effects of PAH in that increases in the activity of the enzyme often precedes more severe cellular and organismal effects such as hepatic damage, changes in immune function and reproductive physiology. Activation of PAH by CYP P450 can result in DNA damage (Stegeman, 1987) and there is considerable

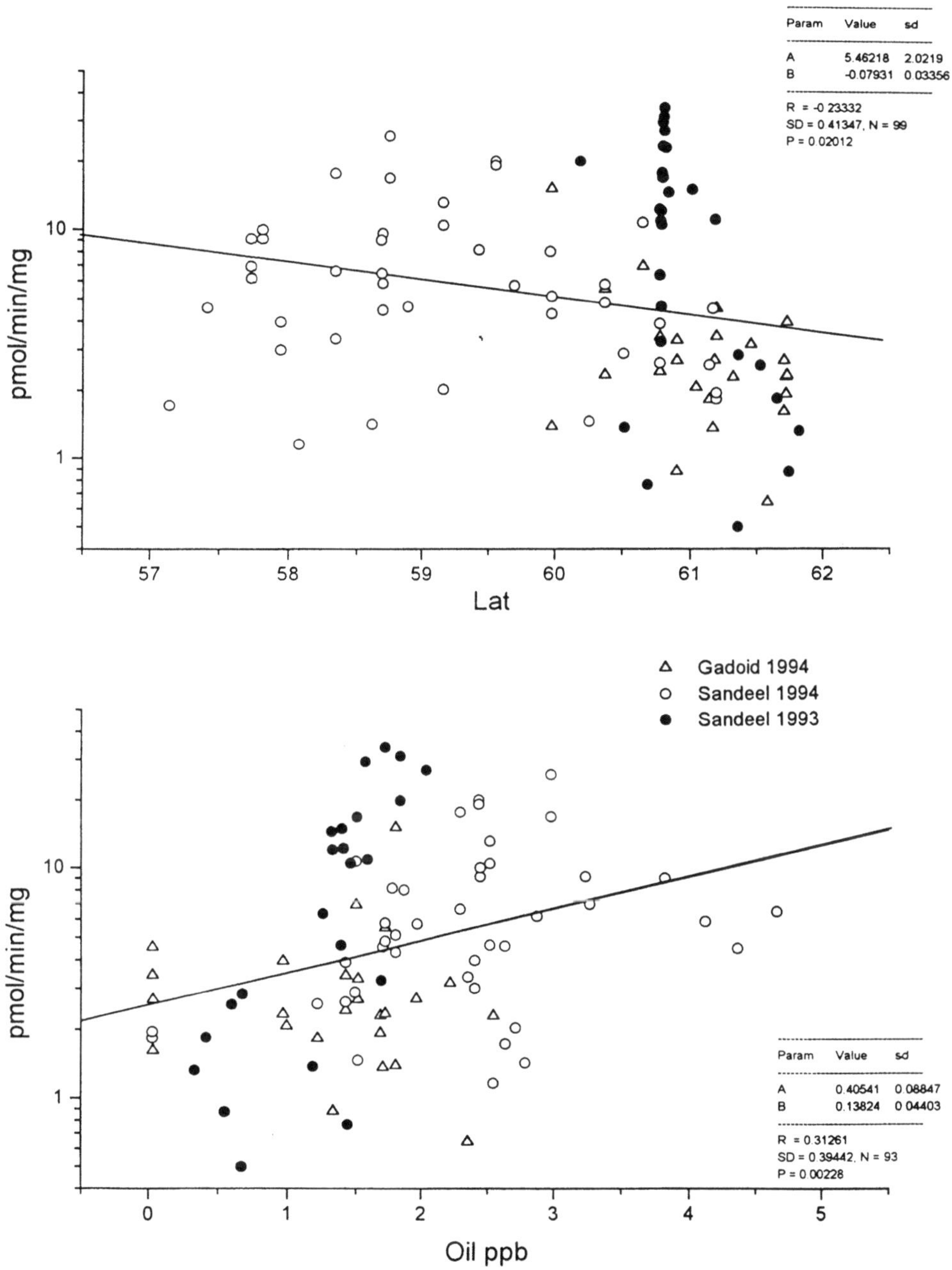

Figure 5. 7-ethoxyresorufin-*O*-deethylase in pooled samples of 3 species of fish larvae caught on oblique surface to bottom tows using a MIKT net: a) relationship to latitude b) relationship to the hydrocarbon fluorescence measured simultaneously and depth averaged between 0 and 50 m. Lines fitted by least squares regression (Gadoid and Sandeel data 1993/4).

support from both field and experimental observations linking CYP1A induction to contaminant levels, DNA damage, hepatic carcinogenesis and other pathological conditions (see reviews by Ionnides and Parke 1993, Livingstone 1993 and Livingstone *et al.*, 1994). Induction of CYP1A activity can therefore be seen as an essential process involved in the aetiology of tumour initiation and carcinogenesis by the activation of compounds such as PAH into pre-carcinogens. Whether or not the reactive intermediates produced will cause damage, in part depends on the capacity of other enzymes (e.g. UDP-glucuronyl transferase (UDP-GT) and glutathione *S*-transferase (GST)) to conjugate and detoxify the compounds formed. Although a role for CYP P450 in chemically-mediated carcinogenesis has been the major focus of recent work it should also be remembered that activation of the Ah receptor (the first step in the induction of P4501A) is also associated with other forms of chemical toxicity such as retinal damage, immune disfunction and atherosclerosis (Bock, 1993).

Late and metamorphosing larvae stages were caught on the same transect used for the hydrocarbon measurements using a Methot-Isaacs-Kidd trawl (MIKT, Methot, 1986) and P4501A activity was measured catalytically by the determination of EROD (7-ethoxyresorufin *O*-deethylase) activity in pooled samples of gadoid and sandeel larvae (Stagg and McIntosh, 1996). EROD activity was extremely variable in the samples analysed despite efforts to ensure that activity was not lost during sample processing. This may be due to net damage to the animals during capture and inevitable delays in sorting larvae from other plankton. Despite this, a significant ($P < 0.02$) gradient of EROD activity with latitude was observed (Fig. 5) along with a positive correlation with hydrocarbon fluorescence ($P < 0.01$). No significant relationship ($P > 0.5$) was observed between EROD activity and temperature (Stagg and McIntosh 1996). These results clearly imply that bioavailable aromatic hydrocarbons are present in the water column at sufficient levels to cause biological effects. Peters *et al.* (1994) also used whole body measurements of EROD activity in the sardine (*Sardinia pilchardus*) as a biomarker of a coastal contamination gradients. Experimental exposure of post-hatch marine fish larvae to fuel oil and the water soluble fraction of crude oil has also been shown to induce benzo-a-pyrene monoxygenase (BPMO) (Binder, 1985) and CYP1A1 protein (Goksøyr *et al.*, 1988; 1991) respectively. Induction of BPMO has also been observed in herring larvae exposed to produced water in experimental mesocosms (Gamble *et al.* 1987). Induction of EROD is an adaptive response in the sense that it will lead to the metabolism and excretion of aromatic compounds but paradoxically the same enzymatic reactions can result in activation of aromatic hydrocarbons to reactive intermediates. In fish larvae few studies have looked for DNA damage or other manifestations of chemical carcinogenesis and mutagenesis in response to PAH. However, deformities have been observed in early embryos of the dab (*Limanda limanda*) in the southern North Sea and the rate of deformities recorded appeared to follow a contamination gradient in the German Bight (Cameron and Berg, 1992). However, although contaminants were implicated as causal agents the experimental and field evidence for a direct link was not compelling.

MODELLING PRODUCED WATER INPUTS

There have been several studies of the nearfield dispersion of produced water, and its more acute toxicological effects (eg Somerville *et al.*, 1987; Murray-Smith *et al.*, 1996) but consideration of the chronic, far field effects has to some extent been neglected. A pilot study using a simple model of advection and diffusion of all the produced water sources in the North Sea has recently been published (Gillibrand *et al.* 1995). This model utilizes a particle tracking model superimposed on a hydrodynamic model of the North Sea.

The hydrodynamic model used was a two-dimensional, depth-integrated numerical model of the North Sea (Prandle, 1984) stretching from 47.5°N63.5°N in latitude and

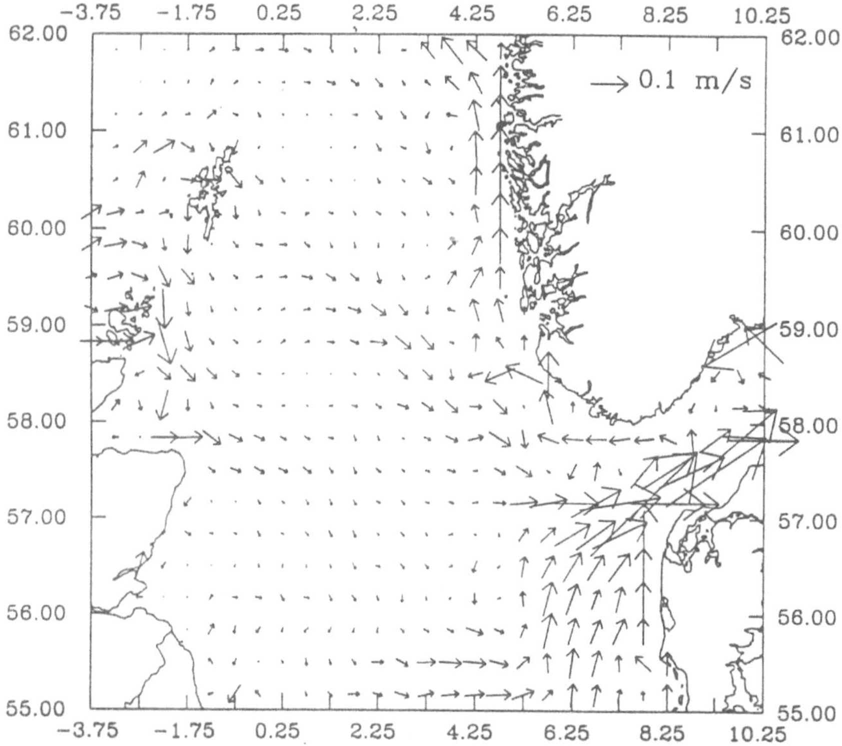

Figure 6. The residual circulation pattern produced by the hydrodynamic model in the northern North Sea. The circulation field was generated by blowing a weak south-westerly wind over the region for 40 tidal cycles.

14.5°W13.0°E in longitude. The model was modified by Needle (1993) to include a radiation condition at the open boundary and to simulate the effects of wind forcing. At the open boundaries, the tidal flow is driven purely by the M2 component. The model was run with a southwesterly wind stress of 0.1414 N m^2 (ie x = y = 0.1 N m^2) until the current regime had settled to steady state (40 tidal cycles). The amplitude and phase of the M2 tide and the east and west components of the residual current, at each grid point, were stored and subsequently used by the particle tracking model. The general pattern of the residual circulation in the northern North Sea generated by this hydrodynamic model (Fig. 6) reproduced the main features of circulation, under a southwesterly wind stress, with inflow occurring between Orkney and Shetland, a cyclonic circulation within the northern North Sea basin, and outflow along the Norwegian coast (e.g. Pingree and Griffiths, 1980; Otto *et al.*, 1993). The cyclonic circulation in the southern North Sea with inflow into the Skaggerak from the German Bight was also reproduced.

The advection and diffusion of the hydrocarbons discharged from each North Sea oil platform (Fig. 1) was simulated using a particle tracking model. The data used was provided by Furnes (1990) and is based on the daily average produced water discharge and hydrocarbon concentration from each of ninety installations. The total oil discharged daily as a component of produced water was approximately 13,000 kg. Particles were driven by the combined effect of tidal and winddriven flows and horizontal and vertical diffusion. At each time step, the tidal and residual currents at each particle position was calculated by interpolating from the hydrodynamic grid extracted from the stored results of the hydrodynamic model. All currents were assumed to be independent of depth. Diffusion was simulated

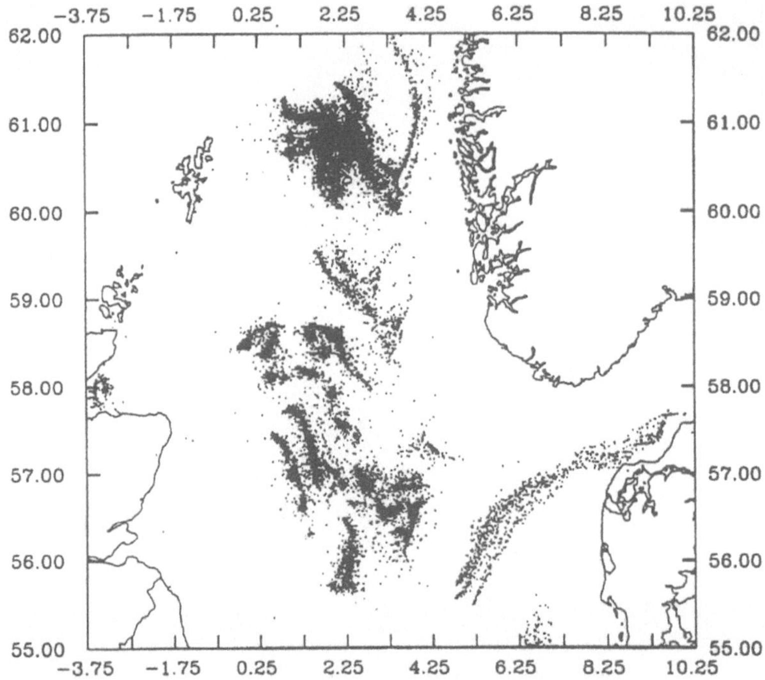

Figure 7. The horizontal distribution of particles after simulation of the discharge of produced water from all platforms for 2 months. Only 15% of the particles are plotted.

by a random walk technique in both the horizontal and vertical directions. The diffusion coefficients used in the present study were based on values given by Talbot and Talbot (1974). Values at the higher end of the given ranges were selected since the observations were taken in early spring, when stratification was still weak and mixing presumably still relatively strong. Horizontal diffusion in both the x and y directions was 10 m^2 s^1 and the vertical diffusivity 0.005 m^2 s^1. Particles were released hourly from each simulated source and the model run for a twomonth period resulting in a final count of particles after two months of 129,600.

The horizontal distribution of particles after a simulation of two months steady discharge is shown in Fig. 7. The dispersion of the particles in the areas surrounding the majority of the sources (Fig. 1) was fairly slow due to the relatively weak residual currents. When particles became trapped in the faster flowing currents of the Norwegian coastal region, and the inflow into the Skaggerak, dispersion was much more rapid. The vertical distribution of particles along 1.5°E and the hydrocarbon concentration that result are shown in Figure 8a. There is a strong vertical stratification with low numbers of particles near the seabed and also a strong north-south pattern of dispersion which reflects the distribution of sources in terms of the mass of oil released from each installation (Fig. 8c). The amount of oil in each particle was proportional to the mass of oil (the product of the produced water discharge rate and the oil concentration) released at each particular installation. Conversion of the particle distribution into oil concentrations was carried out by summing the number of particles in 0.1° boxes along the section and determining the mass of oil in each box. The total mass was then divided by the volume of the box to obtain the concentration. At the northern end of the section, predicted values peak at 1.4 μg/l, whereas at the southern end surface values are of the order of 0.1 μg/l. The observations indicated much higher values

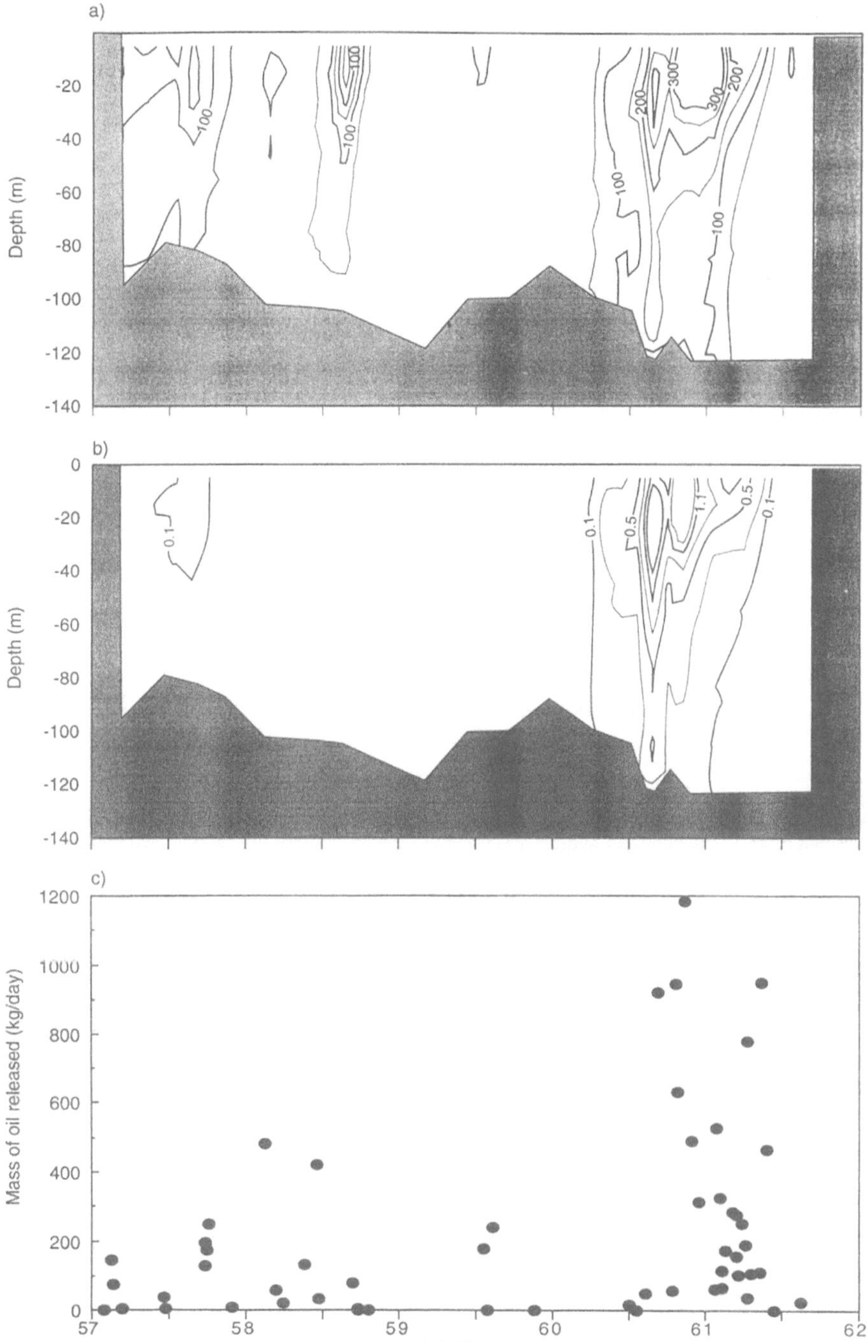

Figure 8. Predicted vertical distribution of particles (a) using a vertical diffusion coefficient of 0.005 m^2 s^{-1} and the calculated resulting hydrocarbon concentration (b) along the sections studied. Also shown (c) is the distribution of sources of hydrocarbons with latitude from the produced water discharges expressed as the mass of oil released per day from each installation between 56 and 62°N.

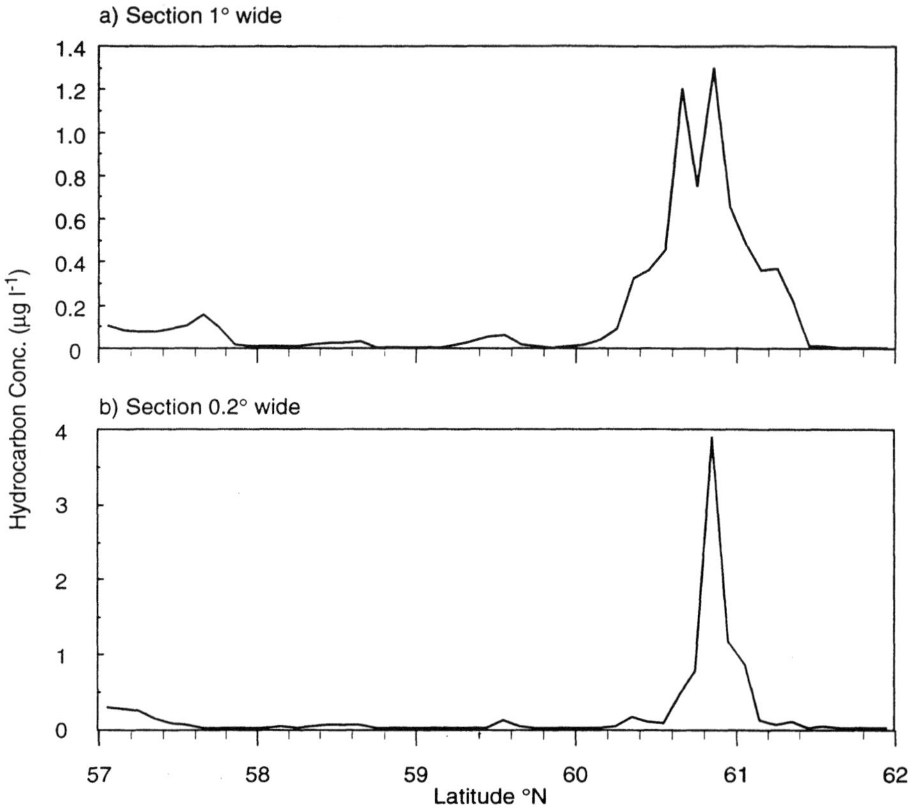

Figure 9. Predicted depth averaged hydrocarbon concentrations between latitude 57 and 62° obtained from the combined particle tracking and hydrodynamic model assuming that the oil is distributed in the upper 50m of the water column (a) in a section between 1° and 2° E and (b) in a section between 1.4° and 1.6°E.

than those predicted by the model. Surface values peaked at about 4.0 µg/l and bottom values were typically 12 µg/l, with the higher levels towards the southern end of the section (Fig. 8b). The absence of oil towards the south in the model results is due to the low levels of discharges (and therefore less oil) in the more southerly locations. Figure 8c shows that some of the northerly platforms are discharging five times as much oil as most of those in the south.

The hydrocarbon concentrations were depth-averaged over the upper 50 m and 100 m of the water column between 57° 62°N along the 1.5°E meridian. This was achieved by summing the number of particles in 0.1° boxes along the section and determining the mass of oil in each box. The total mass was then divided by the volume of the box to obtain the depth-mean concentration. The results are shown in Figure 9. In Figure 9a, the width of the section was 1° of longitude (between 1°2°E, as shown in Fig. 1). The observed depth-mean values (050 m depth) along 1.5°E are shown in Figure 4. There is a peak between 60.5°61.5°N which is captured by the model. However, all modelled values are lower than those observed and, more significantly, a consistent level of relatively high concentrations south of 59°N is not simulated at all. By using a narrower section (1.4°E 1.6°E) in order to reduce the possibility of diluting the oil mass excessively (Fig. 9b), the absolute values of

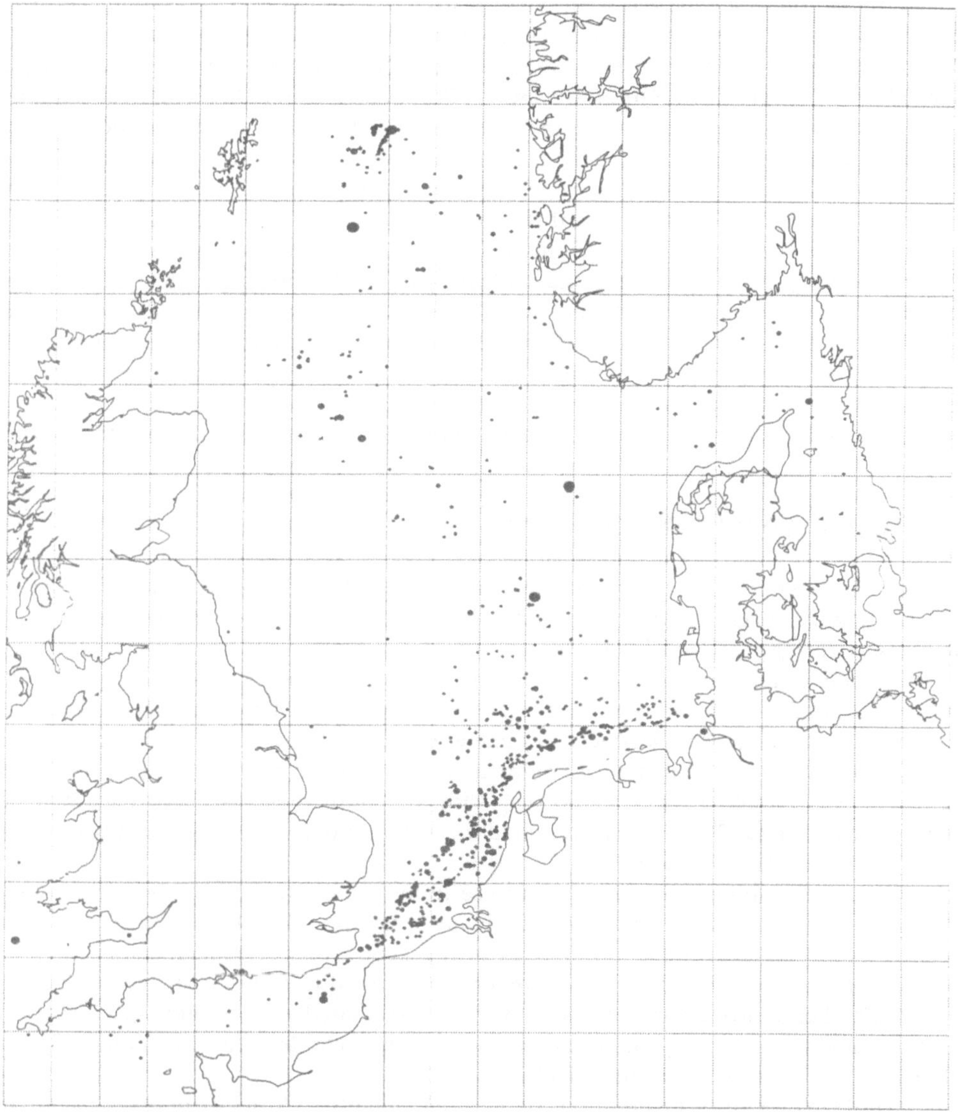

Figure 10. Predicted depth averaged oil concentration ($\mu g\ l^{-1}$) calculated from the output of the particle tracking model and the mass of oil released by each source when using a section either (a) 1° or (b) 0.2° of longitude in width.

the simulation were improved but the southern plateau of high values was still not reproduced.

CONSIDERATION OF OTHER INPUTS

In order to estimate the true effect of the offshore industry on the hydrocarbon concentration gradient of the North Sea, all possible inputs of hydrocarbons to the marine environment must be considered. The contribution and importance of all the sources listed in the introduction should be assessed. With regard to the oil industry, the varying contribution from produced water, cuttings discharges and oil spills must be considered. Leakage of oil from cuttings piles that lie on the seabed must also be quantified. These two sources may have different effects on hydrocarbon levels. Production water discharge, for example, usually occurs above the seasonal thermocline, while sedimentary discharges occur within water that may have longer retention times, permitting higher levels to be accumulated during a season. Oil spills are difficult to quantify but there is some information on their distribution from aerial surveys collected as part of the Bonn Agreement. This is illustrated in Fig. 10 and shows the distribution of oil spills in the North Sea. Whilst many of the spills detected occur in the busy shipping lanes in the southern part of the North Sea there is also quite considerable activity in the north and central North Sea in areas coincident with the oil industry. Atmospheric inputs are also difficult to quantify but may be of particular significance since they are pyrogenic in origin (flaring, smoke stacks and vehicle exhausts) and these may be of particular environmental concern because they are dominated by the larger PAH compounds which are more carcinogenic and less biodegradable in the environment.

SUMMARY AND CONCLUSIONS

Measurements of hydrocarbon concentrations and 7-ethoxyresorufin *O*-deethylase (EROD) activity in selected species of fish larvae were made in the northern North Sea in two surveys during 1993 and 1994. The area selected for study was between 56° 00' N and 62° 00'N and included areas impacted by oil exploration and production together with pristine areas to the north and west. Continuous measurements of hydrocarbons in the water column were collected using a towed fluorimeter deployed on an oblique tow from surface to seabed at stations approximately 10 miles apart on a transect running north to south. The instrument was mounted on a Methot net with a 200μm mesh and live bucket which simultaneously sampled fish larvae present in the water column. The most abundant species caught were sandeel and gadoid larvae.

The results show a progressive increase in hydrocarbon concentration from values around 0.5 $\mu g\ l^{-1}$ in pristine areas to the north and west up to 3 or 4 $\mu g\ l^{-1}$ in areas of most intense exploration and production. In sites with higher levels of contamination there is a distinct depth profile of hydrocarbons with a maximum value between 10 and 50 m depth. Sites with higher hydrocarbon concentrations were associated with elevated EROD activity in larvae caught at the same stations and this would indicate that the hydrocarbon concentrations are sufficient to cause biological effects.

These hydrocarbon distributions imply a regional effect from offshore inputs into the North Sea although at this stage it is unclear which of the sources are of the greatest importance. Discharges from oil installations seem to be a likely candidate although other sources such as atmospheric deposition and spillage from ships should not be neglected. Using 1993 data on the volume of produced water and the concentration of hydrocarbons in the effluent from each installation in the northern North Sea, hydrodynamic modelling work was undertaken to establish whether the increase in hydrocarbon concentrations observed could be explained by the produced water inputs into the North Sea. The results of this work show that in the East Shetland basin some of the observations are simulated by modelling the produced water inputs but that further south the plateau of high oil in water values is not simulated at all. The results are discussed in terms of other inputs likely to contribute to the hydrocarbon levels observed.

In conclusion it would appear that a great deal of research is required if the questions these results pose are to be addressed in particular the following problems should be addressed: identify the sources of the hydrocarbon fluorescence that has been observed;

develop sampling systems and analytical methods to enhance our understanding of the chemistry; develop the modelling approach to see whether we can explain the observations on the basis of other sources from the industry and from terrestrial and atmospheric inputs; and finally, and by no means least, we need to establish whether these low levels of PAHs cause DNA damage and whether this has implications for the survival and well being of fish populations.

ACKNOWLEDGEMENTS

Gunar Furnes for supplying the list of sources of produced water used in the model; the crew of R.V. Scotia and Dana; the Marine Pollution Control Unit in the Coastguard Agency for the data on aerial surveillance under the Bonn Agreement; Mike Heath, John Dunn and the Engineering section for sampling gear and instrument support; Dave Basford and Paul Rankin for larvae identification.

REFERENCES

Addison, R. F., & Edwards, A. J. 1988. Hepatic microsomal monooxygenase activity in flounder *Platichthys flesus* from polluted sites in Langesundfjord and from experimental mesocosms experimentally dosed with diesel oil and copper. Mar. Ecol. Progr. Ser. 46:5154.

Bocke, K.W. 1993. Aryl hydrocarbon or dioxin receptor: biologic and toxic responses. Rev. Physiol. Biochem. Pharmacol. 125: xx-yy

Cameron, P., & Berg, J. 1992. Morphological and chromosomal aberrations during embryonic development in dab (*Limanda limanda*). Mar. Ecol. Prog. Ser. 91: 163169.

Dahlmann, G., & Rave, A. 1988. Continuous insitufluoresence measurements for determination of hydrocarbons in seawater. Dtsch. Hydrog. Z. 41: 7585.

Davies, J., Bell, J., & Houghton, C. 1984. A comparison of levels of hepatic hydrocarbon hydroxylase in fish caught close to and distant from North Sea oil fields. Mar Environ. Res. 14:2345.

Fondekar, S. P., & Alagarsamy, R. 1984. Petroleum hydrocarbon contamination along oil tanker routes in the Arabian Sea. Ind. J. Mar. Sc. 13: 181183.

Furnes, G. 1994. Discharges of produced water from production platforms in the North Sea. Report R064641, E and P Division, Norsk Hydro, Bergen, Norway.

Gamble, J. C., Davies, J. M., Hay, S. J., & Dows, F. K. 1987. Mesocosm experiments on the effects of produced water discharges form oil platforms in the northern North Sea. Sarsia 72:383386.

Gillibrand, P.A., Stagg, R., Turrell, W.R. and Furnes, G. 1995. Observations and modelling of the results of low-level hydrocabon discharges in the North Sea. ICES CM 1995/C:11

Goksøyr, A., Serigstad, B., Solberg, T. S., & Stegeman, J. J. 1988. Response of cod (*Gadus morhua*) larvae and juveniles to oil exposure detected with anticod cytochrome P450c IgG and antiscup cytochrome P450EMAb 1123. Mar. Environ. Res. 24: 3135.

Goksøyr, A., Solberg, T. S., & Serigstad, B. 1991. Immunochemical detection of cytochrome P450 induction in cod larvae and juveniles exposed to a water soluble fraction of North Sea crude oil. Mar. Pollut. Bull. 22: 122127.

HMSO 1995. Development of the Oil and Gas Resources of the United Kingdom. A report to Parliament by the Minister for Energy. Department of Trade And Industry

Ioniddes, C. and Parke, D.V. 1993. Induction of cytochrome P4501 as an indicator of potential chemical carcinogenesis. Drug Metabolism Reviews 25:485-501.

Law, R. J. 1984. The continuous measurement of low concentrations of hydrocarbons in water using a towed fluorimeter. ICES Marine Environmental Quality Committee CM 1984/E:8.

Law, R. J. 1990. Analytical methods used in hydrocarbon oil pollution investigations. Proc. Intl. Symp. Mar. Pollution. 6980.

Law, R. J., & Whinnet, J. A. 1993. The determination of polycylcic aromatic hydrocarbons in seawater from the Fluxmanche trasect (Dover Strait). Oceanologica Acta 16: 593597.

Livingstone, D. R. 1993. Biotechnology and pollution monitoring: use of molecular markers in the aquatic environment. J. Chem. Tech. Biotechnol.57: 195211.

Livingstone, D. R., Forlin, L., & George, S. 1994. Molecular biomarkers and toxic consequences of impact by organic pollution in aquatic organisms. pp 154171. In: Water quality and stress indicators in marine and freshwater systems: linking levels of organisation. (Ed. D.W. Sutcliffe).Ambleside: Freshwater Biological Association.

Methot, R. D. 1986. Frame trawl for sampling pelagic juvenile fish. CalCOFI Rep 27: 267278.

Murray-Smith, R., Gore, D., Flynn, S.A., Vance, I and Stagg, R. 1996. Development of a particle tracking model for the dispersion of produced water discharged form an oil production platform in the North Sea. This volume.

Needle, C.L. 1993. Application of a general purpose hydrodynamic model to the northwest European shelf seas. MSc dissertation, University of Strathclyde/Glasgow Caledonian University, Glasgow, 89pp.

North Sea Task Force 1993 North Sea Quality Status Report. Oslo and Paris Commissions, London.

Otto L., Zimmermann, J.T.F., Furnes, G.K., Mork, M., Saetre, R. and Becker, G. 1990. Review of the physical oceanography of the North Sea. Netherlands Journal of Sea Research, 25(24), 161238.

Payne, J. F., Kiceniuk, J., Fancey, L., Williams, U., Fletcher, G. L., Rahimtula, A. & Fowler, B. 1988. What is a safe level of polycyclic aromatic hydrocarbons for fish: subchronic study on winter flounder (*Pseudopleuronectes americanus*). Can. J. Aquat. Sci. 45:19831993.

Peters, L. D., Porte, C., Albaiges, J., & Livingstone, D. R. 1994. 7 Ethoxyresorufin Odeethylase (EROD) and antioxidant enzyme activities in larvae of sardine (*Sardina pilchardus*) from the north coast of Spain. Mar. Pollut. Bull. 28:299304.

Pingree, R.D. and Griffiths, D.K. 1980. Currents driven by a steady uniform wind stress on the shelf seas around the British Isles. Oceanologica Acta, 3(2), 227236.

Prandle, D. 1984. A modelling study of the mixing of 137Cs in the seas of the European continental shelf. Philosophical Transactions of the Royal Society of London, A 310, 407436.

Somerville, H.J., Bennett, D., Davenport, J.N., Holt, M.S., Lynes, A., Mahieu, A., McCourt, B., Parker, J.G., Stephenson, R.R., Watkinson, R.J. and Wilkinson, T.G. 1987. Environmental effects of produced water from North Sea oil operations. Marine Pollution Bulletin, 18(10), 549558.

Spies, R. B., Felton, J. S., & Dillard, L. 1982. Hepatic mixed function oxidases in California flatfishes are increased in contaminated environments by oil and PCB injestion. Mar. Biol. 70:117127.

Stagg, R. M., McIntosh, A., & Mackie, P. 1995. The induction of hepatic monooxygenase activity in dab (*Limanda limanda* L) in relation to environmental contamination with petroleum hydrocarbons in the northern North Sea. Aquat. Toxicol. 33: 245-264.

Stagg, R.M. and McIntosh, A. 1996. Hydrocarbon concentrations in the northern North Sea and effects on fish larvae. The Science of the Total Environment. *Submitted*

Stegeman, J. J. 1987. P450 systems in fish. Chapter 5 In: Pollutant studies in marine animals. Eds Giam, C.S. and Ray, L.E. CRC Press, Boca Raton, Florida.

Stegeman, J. J., Woodin, B. R., & Goksoyr, A. 1988. Apparent cytochrome P450 induction as an indication of exposure to environmental chemicals in flounder *Platichthys flesus*. Mar. Ecol. Prog. Ser. 46:5560.

Stein, J. E., Collier, T. K., Reichert, W. L., Casillas, E., Hom, T., & Varanasi, U. 1992. Bioindicators of contaminant exposure and sublethal effects: studies with benthic fish in Puget Sound, Washington. Environ. Toxicol. Chem. 11:701714.

Talbot, J.W. and Talbot, G.A. 1974. Diffusion in shallow seas and in English coastal and estuarine waters. ICES Rapports 167: 93-110.

Theobald, N. 1989. Investigations of "Petroleum Hydrocarbons" in seawater using high performance liquid chromatography with fluoresence detection. Mar. Pollut. Bull. 20: 134140.

Theobald, N., Lange, W., & JeryckiBrandes, K. 1991/92. Baseline values of fossile hydrocarbons in the northeast Atlantic Ocean. Dtsch. Hydrogr. Z. 44: 1732.

Van Veld, P. A., Westbrook, D. J., Woodin, B. R., Hale, R. C., Smith, C. L., Huggett, R. J., & Stegman, J. J. 1990. Induced cytochrome P450 in intestine and liver of spot (*Leiostomus xanthrus*) from a polycyclic aromatic hydrocarbon contaminated environment. Aquatic Toxicol. 17:119132.

Westernhagen, H. v. 1988. Sublethal effects of pollutants on fish eggs and larvae. Fish Physiology 11: 253346.

Section IV

DISPERSION MODELS

19

SESSION SUMMARY

Steven A. Flynn

BP Exploration
Farburn Industrial Estate
Dyce, Aberdeen, AB1 9TL, United Kingdom

Mathematical models enhance our understanding and ability to manage possible impacts of our activities in two key ways. Firstly they provide a conceptual framework for interpreting observations. This enables insights into the mechanisms operating and allows the relative significance of various contributing factors to be assessed. Secondly models provide us with a predictive capability. Given a reasonable description of the particular phenomenon of interest, the effects of changes to the operational or environmental parameters can be predicted.

In the case of produced water discharge, models allows us to make the connection between the source conditions and the receiving environment. At the present time a relatively large body of information has been developed on the composition and characteristics of produced water, whereas the understanding of the behaviour of discharges in the environment (and what their significance might ultimately be) is less advanced.

There have been a number of recent successes in describing the dispersion of discharges in the vicinity of offshore platforms. The relatively complex flow behaviour close to the source can now be described using various approaches. Some skill is required to match the most appropriate model type to the conditions. For example, Gaussian type models are most suited to steady state conditions and where longer term average concentrations are required. More complex models incorporate effects such as tidal oscillations and ellipses (eg particle tracking models), and will even model batch discharges well. As the number of studies carried out increases a consensus is beginning to emerge on the likely levels of near field dilution that might be occurring in certain situations. One sign of growing confidence is the fact that dispersion models are beginning to play a part in the regulation of produced water releases (eg USEPA toxicity regulation and the North Sea CHARM project). However, some of the scientific community caution against placing too much faith in some of the more simplified descriptions that ignore complex hydrodynamic and chemical effects. The apparent gaps between model predictions, field measurements and the understanding of the environmental significance of the results still needs to be closed. The key areas for further work on near-field modelling appear to be:

- Field studies to confirm actual levels of dilution and to calibrate model parameters;
- Understanding of the relationship between prevailing environmental concentrations (average and instantaneous) and the effect on ecosystems;

Produced Water2: Environmental Issues and Mitigation Technologies
Edited by M. Reed and S. Johnsen, Plenum Press, New York, 1996

- Description of chemical processes that might occur, particularly for metals coming from a highly reducing environment to an oxidising environment. Although most models have the potential to deal with precipitation, absorption and chemical reactions, the focus until now has been largely on organic components.

Modelling of far-field behaviour presents more of a challenge and there are several areas where further research would be fruitful:

- *Biodegradation processes.* These have been successfully incorporated into dispersion models, but better data on rates of biodegradation are needed. Understanding the effect of temperature and concentration, and factors such as nutrient availability in the receiving environment are research priorities. Under laboratory conditions different biodegradation protocols can give different results. Biodegradation has been discussed in detail during the CHARM project and this may be an area where there is need to incorporate work done in other fora into the models. Modelling studies carried out to date show that, even at the relatively slow biodegradation rates that might be expected in sea water, the reduction in far-field concentration can be substantial.
- *Dispersion coefficients.* Models may rely on dispersion coefficients that are not at all well defined. Worst case values are often selected, and sensitivity or uncertainty analysis is needed. In the far field the problem is compounded. For example, the assumption of constant dispersion coefficients may well be a simplification and it has been proposed that dispersion should be function of length scale to 4/3 power.
- *Time period for exposure.* Models can produce both peak and time weighted averages, and the most appropriate measures for characterising the discharge need to be agreed. It appears that time averaged dose is a promising parameter. However, the latest environmental modelling studies indicate that statistical approaches are likely to be needed to account for the response of populations of exposed organisms.

In summary, mathematical modelling of produced water dispersion has reached an exciting phase. Confidence in near-field predictions is growing, while work on far-field dispersion is beginning to provide tantalising insights into the wider and regional significance of produced water discharges. The science is at the stage where more interaction between those carrying out modelling and those measuring biological effects could provide important breakthroughs.

20

DISPERSION MODELING PERSPECTIVES ON THE ENVIRONMENTAL FATE OF PRODUCED WATER DISCHARGES

Maynard G. Brandsma[1] and Joseph P. Smith[2]

[1] Brandsma Engineering
P.O. Box 378, Durango, Colorado 81302
[2] Exxon Production Research Company
P.O. Box 2189, Houston, Texas 77252-2189

INTRODUCTION

Dispersion modeling provides estimates of the concentration of discharged produced water that take into account the discharge configuration and the hydrographic conditions in the receiving environment. Concentration estimates can be used along with dose-response measurements to assess the potential for environmental effects from discharges. In this paper, we use the Offshore Operators Committee Model to estimate environmental concentrations of discharged produced water based on conditions applicable to the Gulf of Mexico, the North Sea and the Bass Strait (Australia). A summary of published effect concentrations (LC_{50} or EC_{50}) for effluents from these areas was compared with predicted concentrations as an example of the application of dispersion modeling results for environmental assessment. Modeling results showed that rapid dilution reduces the peak effluent concentration to levels below the 50% effect concentration of the great majority of effluents within a short distance from the outfall. More realistic exposure assessments can be developed from modeling based on site-specific current data. Modeling based on time-series current measurements from a field location was used as an example of how constantly changing current conditions can cause the average exposure concentration at a fixed point near the outfall to be approximately an order of magnitude lower than the peak concentration. As a result, comparisons of predicted peak environmental concentrations with long-exposure effect levels are highly conservative and should be considered only as screening-level assessment tools and not be used for final estimates of the potential for environmental effects.

Produced Water 2: Environmental Issues and Mitigation Technologies
Edited by M. Reed and S. Johnsen, Plenum Press, New York, 1996

CHARACTERIZATION OF DISCHARGES AND RECEIVING WATERS

Dispersion modeling uses data on outfall configuration, flow rate, effluent properties and receiving water conditions. Data from published sources were used to develop scenarios for modeling discharges in the North Sea, the Gulf of Mexico, and the Bass Strait.

Gulf of Mexico

Two discharge rates were chosen for simulation: 728 bbl/d (115.8 m^3/d), an industry database value for the median flow rate for Gulf of Mexico facilities (Shannon, 1992); and 25,000 bbl/d (3977.8 m^3/d), the maximum discharge rate from a single port allowable under recent regulations (Environmental Protection Agency, 1993). The outfall pipe diameter was an average value, 0.20 m (8 inches) (Shannon, 1992). The water column distance between outfall and sea floor was set at the median value of 27.4 m (Shannon, 1992).

The effluent salinity and temperature were average values for the Gulf of Mexico: 128 ppt and 29 °C , respectively (Avanti, 1992a).

The current speeds for Gulf of Mexico simulations were 3.3 cm/s, 9.5 cm/s, and 25.3 cm/s. These speeds represent time-weighted 10th, 50th, and 90th percentile values from current meter records from the Central Gulf of Mexico (Smith et al., 1994; and Science Applications International Corp. 1989). The current meter records also provided the raw data for the time series modeling studies. The median speed of 9.5 cm/s is similar to a characteristic speed of 9.9 cm/s cited by Avanti (1993).

Density stratification can vary widely in the Gulf of Mexico, depending on distance from the Mississippi River Delta and season. An intermediate gradient of 0.15 kg/m^3/m with an ambient density of 1017 kg/m^3 was chosen for this study. This choice was based on a review of linear density gradients calculated from published data. Density gradients ranged from relatively weak (0.04 kg/m^3/m , Smith, et al. 1994) to quite strong (0.2 - 0.7 kg/m^3/m; Avanti, 1993; and Ayers et al. 1980).

North Sea

Simulation cases were based on Somerville et al.'s (1987) description of the Brent produced water discharge. The produced water discharge rate of 62,849 bbl/d (10,000 m^3/d) was the peak for each field platform. The effluent, with a bulk density of 1014 kg/m^3 was discharged from a horizontal, near surface outfall 0.76 m in diameter. Discharge depths were in the upper 5 m of the water column so as to promote dispersion in the wave zone.

Water depths and the speed and time dependence of currents characterizing generic site types in the North Sea were identified by Brandsma and McKelvie (1994). The "southern" site was characterized by a 30 m water depth and a rectilinear tidal current with an RMS speed of 37 cm/s. The "central" site had a 100 m depth and an elliptical tidal current with an RMS speed of 26 cm/s. The "northern" site was 150 m deep and had a tidal current with nearly steady speed, rotating at a steady rate around the compass, having an RMS value of 22 cm/s.

Studies at the Brent field (Somerville et al., 1987) show overall residual currents averaging 4 cm/s directed east-northeast. Residual velocities vary from 1 to 4.3 cm/s. Total current speed near the surface in the Brent field (tidal + residual) reaches 30 cm/s approximately 10% of the time. As this field is in the northern North Sea, the RMS speed of 22 cm/s was chosen for modeling.

Table 1. Discharge and receiving environment conditions for dispersion modeling

	Region		
Parameter	Gulf of Mexico	North Sea	Bass Strait †
Effluent Salinity (ppt)	128	24.6	35
Effluent Temperature (°C)	29	30	90
Effluent Density ($kg\text{-}m^{-3}$)	1087.9	1014	988
Ambient Density* ($kg\text{-}m^{-3}$)	1017	1027	1026
Density Gradient ($kg\text{-}m^{-4}$)	0.15	0	0
Pipe Diameter (m)	0.2	0.76	0.2
Pipe Depth (m)	0.3	5	12
Pipe Orientation	Downwards	Horizontal	Downwards
Water Depth (m)	27.4	150	72

* Ambient density refers to water column density at the mouth of the outfall pipe.
† Modeling conditions as used by Terrens and Tait (1994).

Most areas of the North Sea have weak or non-existent density stratification in winter with a thermocline developing in the summer months (Somerville et al., 1987). A constant ambient density was used for modeling.

Bass Strait

Produced water is discharged from platform Halibut in the Bass Strait (Terrens and Tait, 1994), at a rate of 88,057 bbl/d (14000 m^3/d). The effluent salinity approximates that of sea water (35 ppt) and the temperature is approximately 90°C, making the plume positively buoyant. The outfall depth is 12 m and the discharge pipe is oriented vertically downward.

The waters around platform Halibut are characterized by a depth of 72 m and a tidal current sweeping steadily around the compass with a median current speed of 26 cm/s. The water column is frequently well mixed (salinity = 35.45 throughout) with water temperatures in the 16 to 21°C range.

DISPERSION MODELING METHODS

The Offshore Operators' Committee (OOC) mud and produced water discharge model (Brandsma, et al., 1992) was used to simulate the dilution of produced water discharges. Model input data characterizing discharge and receiving environment conditions for Gulf of Mexico, North Sea, and Bass Strait discharges were developed (Table 1) based on the information presented in Section 2. Results reported here are the maximum concentrations of produced water at a given distance from the source for the specified discharge and receiving environment conditons. For the time-series current data case set in the Gulf of Mexico, concentrations were calculated on the surface of a vertical right circular cylinder with a radius of 100 m, having its centerline at the outfall.

COMPARISON OF MAXIMUM PREDICTED EFFLUENT CONCENTRATIONS WITH BIOLOGICAL EFFECT CONCENTRATIONS

A significant body of data has been collected on produced water toxicity. Organism level responses such as mortality (LC_{50}) and growth (EC_{50}), with exposure times ranging

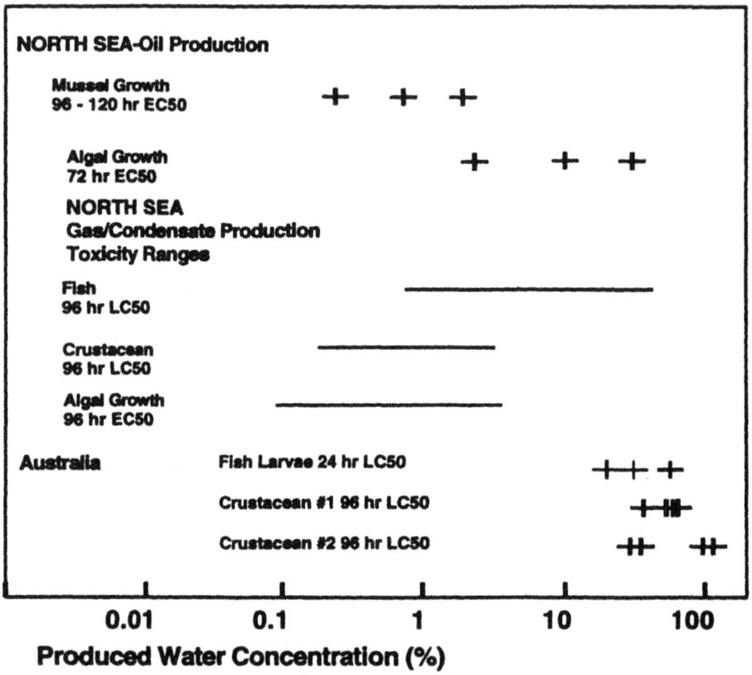

Figure 1. Produced water toxicity effect concentrations for: three North Sea oil production platforms(Sørstrøm et al., 1993); ranges of results for eight North Sea gas/condensate platforms (Jacobs et al., 1991); and four (some data points overlap) Bass Strait oil production platforms (Terrens and Tait, 1994).

from 24 - 96 hours, constitute the bulk of the available data. In Figure 1 we show effect concentrations for produced waters from the North Sea and Australia. North Sea produced water effect concentrations span a broad range. Effect concentrations for produced water from three North Sea oil production platforms (Sørstrøm, et al. 1993) ranged from 0.29 - 30%. Jacobs and Marquenie (1991) reported effect concentrations for produced waters from eight gas/condensate platforms ranging from 0.09% - 42%. Data on four Bass Strait produced waters (Terrens and Tait, 1994) indicate very low toxicity with effect concentrations ranging from 19.6 - 116%.

The larger amount of data available on Gulf of Mexico produced waters reveals a broad distribution of LC_{50}s (Figure 2) for 96 hr exposures of mysid shrimp. The mean LC_{50}

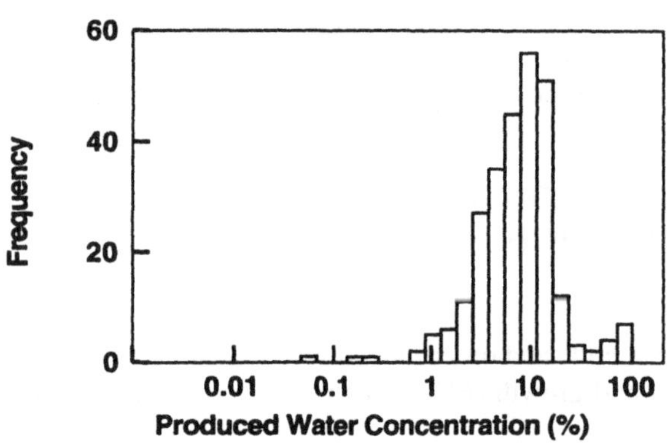

Figure 2. Histogram of 96hr LC_{50}s for exposure of Mysid shrimp to 269 Gulf of Mexico produced waters. (Avanti Corporation, 1992; Moffit et al., 1992; St. Pe, 1990).

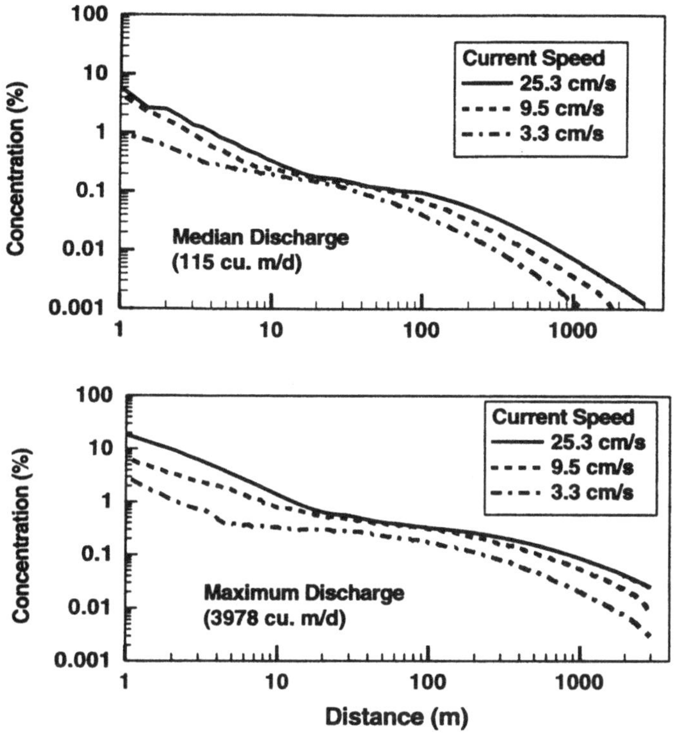

Figure 3. Maximum predicted concentrations for Gulf of Mexico conditions for the median discharge rate (upper) and maximum discharge rate (lower).

is 11.5%, with 98% of the values falling above 1%. There are a few extreme values falling below 0.1%.

Review of the toxicity data indicates that the majority of the effect concentrations for 24 - 120 hour exposures are in excess of 1%. The histogram of LC_{50}s for Gulf of Mexico produced waters reveals a high level of variability in toxicity, suggesting that generalizations about the potential effects of produced water discharges should not be drawn from the results of only a few measurements.

Comparison of the maximum predicted effluent concentration with biological effect concentrations can provide screening-level assessment of the potential for observing biological effects from the discharge. Maximum effluent concentrations were predicted as a function of distance from the outfall for the discharge scenarios presented in Table 1. Maximum effluent concentrations decrease rapidly with distance from the outfall (Figure 3) falling below the great majority of effect levels within a short distance of the outfall. . The maximum concentration at 100 m from the outfall ranges from 0.04% for the 115 m^3/d rate Gulf of Mexico discharge under slow current conditions to 1.3% for the 10000 m^3/d North Sea discharge (Table 2).

Predicted maximum concentrations (Table 2) fall well below LC_{50} or EC_{50} effect concentrations for the great majority of the produced waters, indicating that there is a low potential for observing biological effects corresponding to the laboratory endpoints under field conditions. This assessment is highly conservative for the following reasons. (1) The model predictions represent the highest concentration that an organism is expected to encounter. (2) The exposure to the peak concentrations is likely to be very short in duration

Table 2. Predicted concentrations at 100 m under constant current conditions

Location	Discharge Rate (m^3/d)	Current Speed (cm/s)	Concentration at 100 m (% PW)
Gulf of Mexico	115.8	3.3	0.043
Gulf of Mexico	115.8	9.5	0.073
Gulf of Mexico	115.8	25.3	0.097
Gulf of Mexico	3977.8	3.3	0.18
Gulf of Mexico	3977.8	9.5	0.32
Gulf of Mexico	3977.8	25.3	0.32
North Sea	10000	22	1.3
Bass Strait	14000	26	0.47

due to the constantly shifting current and hydrographic conditions, and in some cases, the motility of the exposed species. (3) The LC_{50} or EC_{50} concentrations used for comparison were taken from laboratory tests involving relatively long (24 - 120 hours) exposures to constant concentrations of produced water. Such exposure regimes are not likely to be encountered under field conditions.

For most discharges, the predicted maximum concentrations are much less than the LC_{50} or EC_{50} levels. These cases present a low risk of incurring the effects corresponding to the laboratory toxicity end points and may be excluded from any further assessment. More detailed modeling of the dilution behavior of the plume, as discussed in the following sections, is helpful in cases where the predicted maximum concentration is similar to LC_{50} or EC_{50} levels.

PREDICTING CONCENTRATIONS WITH FIELD CURRENT DATA

Field measurements of the time dependence of current speed and direction were used to predict the time average effluent concentration under a realistic current regime. This provides an example of the application of dispersion modeling based on actual field conditions.

Current speeds and direction were recorded for 35 days prior to a field verification study of the OOC Model(Smith, et al. 1994). A progressive vector plot of the current meter data (Figure 4) records the displacement of a particle traveling with the current at half hourly intervals. Numerous instances of the vector path looping back and crossing over itself are evident. The current record indicates the presence of some periodic components in the presence of a substantial net displacement, first to the south and then to the east.

The probability of a crossover and the time interval between crossovers must be considered to assess the effect of returned effluent on background concentration. In Figure 4, there are 159 crossing points in 1616 current samples, so approximately 10% of the readings are at a return point. Return paths with shorter circumferences have a greater impact on background concentration since they return effluent to the outfall after less dilution has taken place. A return time of less than three hours can be used as a criterion for loops that will contribute measureably to effluent concentration, as travel for three hours at a median Gulf of Mexico current of 9.5 cm/s would result in dilution to about 0.01% produced water. Analysis of the current meter record indicates that 36% of the loops have durations of three hours or less. Consequently, only about 3.6% of the current record presented in Figure 4 is spent under conditions where significant concentration buildup could occur. Analysis of additional data sets from the central Gulf of Mexico (Science Applications International,

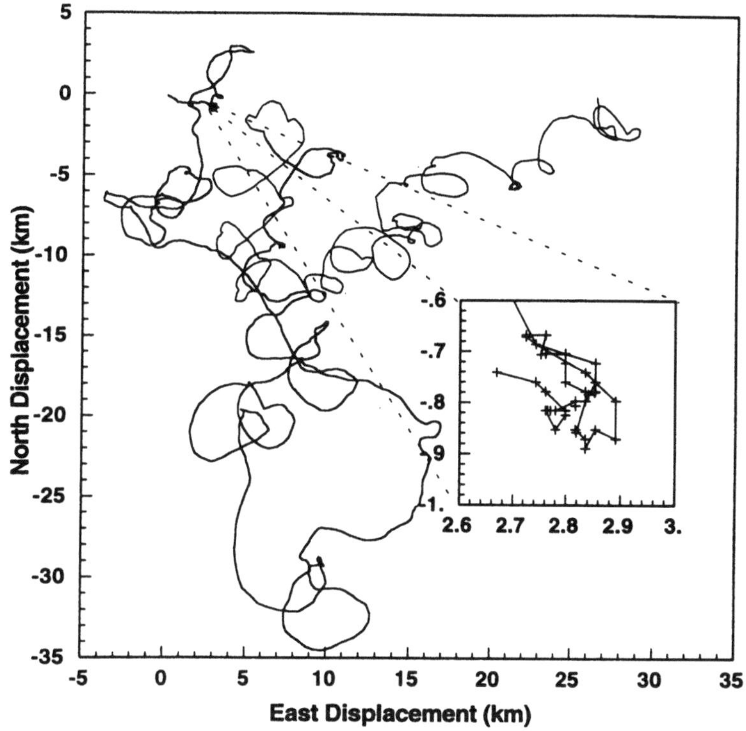

Figure 4. Progressive vector plot of current meter readings for 35 days in the Central Gulf of Mexico. The inset shows the data used for modeling produced water dilution under field current conditions.

1989) yielded similar results: about 6% of the data were on return loops and 14% of the returns occurred in 3 hours or less.

A simulation with real currents was used to predict the produced water concentration at hourly intervals on the surface of vertical cylinder 100 m from the outfall. A region of the current meter data (Figure 4, inset) that displayed multiple returns in 24 hour period was selected and used for an OOC Model simulation of a 3978 m^3/d (25000 bbl/d) discharge under Gulf of Mexico conditions (Table 1). The maximum concentration anywhere on the 100 m cylinder during the 24 hour period was 0.35% (Figure 5). The variations in maximum concentration reflect changes in the current speeds during the test period. Figure 6 shows concentrations as a function of time for four bearings corresponding to the general direction of displacements during the test period. Average concentrations were calculated for these

Figure 5. Maximum predicted concentrations for any bearing at 100 m from the outfall for a 3978 m3/d discharge under Gulf of Mexico conditions. The current time series shown in the inset of Figure 4 was used for the simulation.

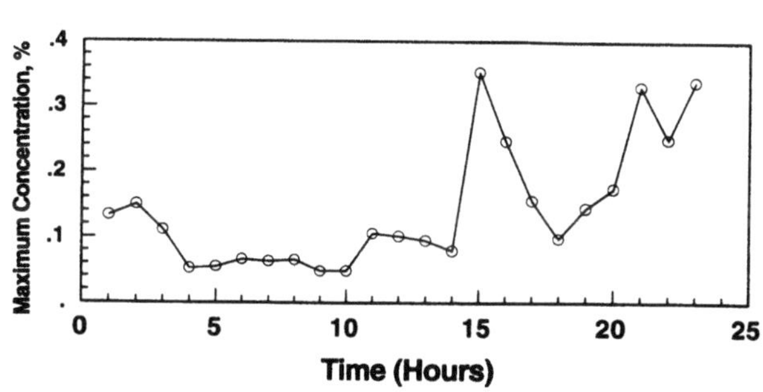

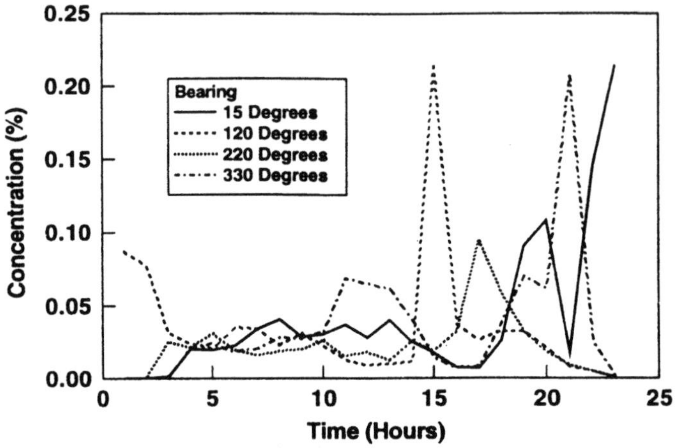

Figure 6. Predicted concentrations at 100 m from the outfall for current time series shown in the inset to Figure 4. Concentrations are calculated for fixed bearings relative to the outfall location.

specific bearings. These values correspond to the average exposure concentration of an organism at a fixed point in the water column. The averages calculated from the data in Figure 6 ranged from 0.02 - 0.04 %. The presence of multiple returns in a short period of time tends to maximize the potential for exposure. This example shows that even under this conservative scenario, the average exposure concentration is approximately an order of magnitude lower than the peak concentration.

DISCUSSION

A combination of modeling results and laboratory measurements of biological effects of produced water exposure as a function of concentration provides a means of assessing the potential for environmental effects from produced water discharges. We have presented modeling results based on published descriptions of discharges in the Gulf of Mexico, the North Sea, and Bass Strait . The predicted concentrations drop to about 1% within a short distance of the outfall. This concentration is lower than the great majority of the reported LC_{50} or EC_{50} levels for exposure durations of 72 - 96 hr. Due to the constantly changing current and hydrographic conditions near ocean outfalls, exposures of this duration to constant, elevated, produced water concentrations are unlikely to occur in the field. As a result, comparison of peak concentrations with reported LC_{50} or EC_{50} levels for long exposure toxicity tests is a highly conservative criterion for assessing the potential for observing the effects corresponding to laboratory toxicity end-points under field conditions. Such comparisons should be regarded only as a screening level assessment tools, i.e. simple methods for determining if a more detailed evaluation is warranted. Based on the results presented here, a screening level evaluation indicates that the majority of produced water discharges can be considered to present a low potential for causing the effects observed in laboratory toxicity tests.

Modeling using more realistic current conditions can be used to assess the potential for effects from discharge of effluents with lower effect concentrations. The results presented here indicate that under realistic current regimes chosen to maximize the potential for accumulation of diluted effluent, the average exposure concentration is approximately an order of magnitude less than the peak concentration. These concentrations are smaller than almost all of the reported LC_{50} or EC_{50} levels. The modeling technique presented here does not consider the possibility that other factors such as biodegradation, volatilization, or precipitation may act to remove produced water components from the water column over the long term. For example, Brendehaug, et al. (1992) report biodegradation rates of North Sea produced waters of approximately 45% in 7 days. Effects such as biodegradation may be sufficiently rapid to offset the contributions of current reversals to to effluent concentration.

REFERENCES

Avanti Corporation 1992a; "Characterization of the data collected from the Louisiana Department of Environmental Quality permit files for development of Texas and Louisiana coastal subcategory NPDES general permits". Report prepared for US EPA Region 6, EPA Contract No. 68-C9-0009.

Avanti Corporation. 1992b. Louisiana DEQ water quality data including water quality analyses for bays and lakes, PW toxicity database, and summary of PW pollutant concentrations. Report prepared for U.S. Environmental Protection Agency, Region VI.

Avanti Corporation. 1993. Ocean discharge criteria evaluation for the NPDES general permit for the western Gulf of Mexico OCS. Report prepared for U.S. Environmental Protection Agency Region VI., EPA Contract No. 68-C9-0009. Dallas, Texas.

Ayers, R.C.,Jr., T.C. Sauer, Jr., D.O Stuebner, R.P. Meek. 1980. An environmental study to assess the effect of drilling fluids on water quality parameters during high rate, high volume discharges to the ocean. Symposium: Research on Environmental Fate and Effects of Drilling Fluids and Cuttings. Lake Buena Vista, Florida. January 21-24.

Brandsma, M.G., Smith, J.P., O'Reilly, J.E., Ayers, R.C., and Holmquist, A.L. 1992. Modeling offshore discharges of produced water. In: *Produced Water*, Ray, J.P., Engelhardt, F.R. Plenum Press. New York. pp. 59-71.

Brandsma, M.G. and McKelvie, S. 1994. Modeling North Sea oil based mud cuttings discharges to assess environmental loading: an E & P Forum study. Paper OTC 7395 presented at the 1994 Offshore Technology Conference. Proc. Vol. 1. Houston, Texas. pp. 215-222.

Brendehaug, J., Johnsen, S., Bryne, K.H., Gjose, A.L., Eide, T.H., Aamot, E. 1992. Toxicity testing and chemical characterization of produced water-a preliminary study. In: *Produced Water*, Ray, J.P., Engelhardt, F.R. Plenum Press. New York. pp. 245-256.

Jacobs, R.P.W.M. Marquenie, J.M. 1991. Produced water discharges from gas/condensate platforms: environmental considerations. Paper SPE 23321 presented at First Conf. on Health, Safety and the Environ ment. The Hague. The Netherlands. 10-14 November 1991.

Environmental Protection Agency (1993);[FRL-4802-1] Final modification of the NPDES general permit for the western portion of the outer continental shelf(OCS) of the Gulf of Mexico(GMG290000) Federal Register Vol 58, No. 231, 12/3/93 , pp 63964-63986.

Jacobs, R.P.W.M., Marquenie, J.M. 1991. Produced water discharges from gas/condensate platforms: environmental considerations, SPE 23321 Presented at First Conf. on Health Safety and the Environment, The Hague, The Netherlands, 11/10-14/1991.

Moffit, C.M. Rhea, M.R., Dorn, P.B., Hall, J.F., Bruney, J.M., Evans, S.H. 1992. Short-term chronic toxicity of produced water and its variability as a function of sample time and discharge rate. In: *Produced Water.* Ray, J.P. and Engelhardt, F.R., Eds. Plenum Press. New York. pp. 235-244.

Science Applications International Corporation (SAIC). 1989. Gulf of Mexico physical oceanography program, final report, year 5, volume II technical report. OCS Report/MMS-89-0068. U.S. Dept. of Interior, Minerals Management Service, Gulf of Mexico, OCS Regional Office. New Orleans, Louisiana. 333 p.

Shannon, B. 1992. OOC produced water discharge configuration database, Submitted to Mr. Jack Ferguson. U.S. Environmental Protection Agency Region 6. Dallas, Texas. 11 November, 1992.

Smith, J.P., Mairs, H.L., Brandsma, M.G., Meek, R.P., and Ayers, R.C.Jr. 1994. Field validation of the offshore operators' committee (OOC) produced water discharge model. SPE paper #28350 presented at SPE Annual Technical Conference and Exhibition. New Orleans, Louisiana. 25-28 Sept. 1994.

Somerville H.J. Bennett, D., Davenport, J.N., Holt, M.S., Lynes, A., Mahieu, A. McCourt, B. Parker, J.G., Stephenson, R.R., Watkinson, R.J., Wilkinson, T.G. 1987. Environmental effect of produced water from North Sea oil operations. *Marine Pollution Bull.* 18:549-558.

Sørstrøm, S., Aunaas, T., Bjerken, S., Brakstad, O.G., Johansen, Ø., Kaarstad, I., Mørk, G.,Strømgren, T., Thendrup. A. (1993); "Produced Water Chemistry and Toxicity Study" IKU Sintef Report for OLF, January 1993.

St. Pe. K.M. 1990. An assessment of produced water impacts to low-energy brackish water systems in southeast Louisiana. Report submitted to Louisiana Dept. of Environmental Quality. July 1990.

Terrens, G.W. and Tait R.D. 1994. Effects on the marine environment of produced formation water discharges from offshore development in Bass Strait, Australia. Paper SPE 27149 presented at the Second Int. Conf. on Health, Safety and Environment in Oil and Gas Exploration and Production. Jakarta, Indonesia. 25-27 January 1994. Proc. Vol 1, pp. 459-470.

21

DEVELOPMENT AND APPRAISAL OF A PARTICLE TRACKING MODEL FOR THE DISPERSION OF PRODUCED WATER DISCHARGED FROM AN OIL PRODUCTION PLATFORM IN THE NORTH SEA

R. Murray-Smith,[1] D. Gore,[2] S. A. Flynn,[3] I. Vance,[4] and R. Stagg[5]

[1] Brixham Environmental Laboratory
Brixham, Devon, TQ5 8BA
[2] Ministry of Agriculture, Fisheries and Food (MAFF)
Burnham-on-Crouch, Essex, CM0 8HA
[3] BP Exploration Europe
Farburn Industrial Estate
Dyce, Aberdeen, AB2 0PB
[4] BP Research and Engineering
Sunbury-on-Thames, Middlesex, TW16 7LN
[5] Scottish Office Agriculture and Fisheries Department (SOAFD)
Marine Laboratory
Aberdeen, AB9 8BD

INTRODUCTION

Produced water is the single largest waste stream from offshore oil and gas production activities which is discharged into the sea. In 1991, for example, some 160 x 10^6 m^3 of produced water was discharged into the North Sea (E&P Forum, 1994). Moreover, since normal production practices result in an increase in the ratio of produced water to hydrocarbon over the life of each reservoir, the volume of produced water discharged into the North Sea is predicted to increase to approximately 330 x 10^6 m^3 by 1998 (E&P Forum, 1994).

Not only does the volume of produced water change over the operational life of a particular reservoir, but the chemical composition of the produced water will vary. Where sea water injection is used as a method of secondary oil recovery, for example, the inorganic constituents of produced water will resemble seawater more closely as more injected seawater breaks through to production wells. The concentration of organic constituents, such as organic acids, associated with the produced water will consequently decrease as a result of the breakthrough of organically depleted sea water. The concentration of production chemicals, however, will not necessarily decrease.

Any environmental impact of produced water discharges will be a combination of effects arising from the dissolved and suspended constituents which are derived from the mineralogy of the reservoir, from the hydrocarbon component and from production chemicals which are added to maximise the efficiency of oil recovery. This environmental impact may be assessed using various different techniques. For example, a matrix of analytical techniques may be used to characterise and quantify the chemical components of the water and then an ecotoxicological response may be assigned based upon published dose-response data for the individual components. Alternatively, ecotoxicity assays could be performed on the whole produced water. Alongside these studies, it is also necessary to estimate what concentrations will be expected in the sea, and the likely timescales over which organisms may be exposed. In this sense, an important factor is the extent of dilution in the sea.

Figure 1. Clyde field location and model area.

This paper describes the development and application of a mathematical model used to estimate the extent of dilution of discharged produced water at various distances from an offshore oil production platform. The model is based on a "random walk" model previously used to estimate the environmental impact of ammonium discharge from methyl methacrylate manufacture on an area of the North Sea 10 km off Tees Bay, UK. (Staples et al., 1993).

CLYDE OIL PRODUCTION PLATFORM

The model was applied to the produced water discharge from the Clyde oil production platform in the UK sector of the North Sea. The BP Exploration Operating Company operates the Clyde platform which first produced oil in March 1987. Current production is around 23.7 mbd (3.8×10^3 m^3 d^{-1}) oil and 50.5 mbd (8.0×10^3 m^3 d^{-1}) water from a single platform located in 80 m of water, some 300 km south east of Aberdeen at 56°27'N, 02°17'E (Figure 1). At this location the tidal range varies from 0.5 m to 0.9 m for a neap and spring tide, respectively. Produced water is discharged at a temperature of 72° from a pipe which terminates 20 m above the sea surface.

Some aspects of the composition of the Clyde produced water are not typical of North Sea produced waters. The total organic carbon (TOC) concentration of Clyde produced water is typically 45 mg l^{-1} which is a low value relative to produced water discharged by some other platforms in the North Sea (Somerville et al., 1987, Tibbetts et al., 1992,). This low value of TOC is explained by the low concentration (25 mg l^{-1}) of volatile fatty acids (VFA). Despite this low concentration, the VFA are actually the major organic component of the Clyde produced water. This is consistent with other case studies where VFA have been shown to contribute 71% to >90% of the organic fraction (Somerville et al., 1987, Tibbetts et al., 1992, Brendehaug et al., 1992). The low value of VFA in Clyde produced water is thought to be a reflection of the relatively high reservoir temperature (148° C) which promotes the decarboxylation of organic acids (Carothers and Kharaka, 1978) to produce gaseous hydrocarbons.

The density of Clyde produced water (1086 kg m^{-3} at 15° C) was similar to that of the Auk reservoir (1085 kg m^{-3} at 20° C) and somewhat greater than the value of approximately 1014 kg m^{-3} given for other North Sea produced waters (Somerville et al., 1987). The effect of the downward velocity of the discharge plume and the high effluent density results in a plume which tends to sink initially and then stabilise at some level in the water column as dilution with sea water increases the buoyancy. This buoyancy effect of the produced water has been incorporated in the model.

MODEL DESCRIPTION

The model used for this study is called TRK (Riddle 1993). It is a plume-type model, developed at Brixham Environmental Laboratory, which runs on a VAX/VMS or Alpha-based computer system. The model is designed to predict surface water concentrations, resulting from the discharge of effluent from one or more sources.

TRK is based on a particle tracking technique with a 'random walk' component to simulate the dispersion. This is a powerful approach which works by releasing several hundred 'particles' each model time step. These particles represent the effluent and are made to move with the advection and dispersion caused by the tidal and wind driven currents, and the mixing processes. This method provides detailed information about the behaviour of an effluent plume, although it can be very computer-intensive since many thousands of particles are tracked simultaneously.

The model can carry out long term (several tidal cycles) simulations and provide 'snapshot' views of the effluent plume at any specified time. It can also predict the concentration variation with time at any specified location (hereafter known as a temporal location).

The model output consists of three dimensional data, which can be used to determine the maximum concentration over the whole water column at each grid cell, or to show the vertical concentration distribution along a given cross-section. The results presented in this paper show some of these features and help to predict the likely behaviour of the effluent, as well as the minimum or 'worst-case' dilutions that might be expected.

The model works on a grid basis and at each grid cell the peak tidal current, residual current and water depth is required. Also required as model input are the mixing coefficients, wind speed and direction, the tidal harmonics and the nature of the discharged material, including its biodegradability. The model was set up for the Clyde field as a 300 x 300 grid covering an area of 30 km x 30 km (i.e. each grid cell measured 100 m x 100 m).

Some of the theoretical considerations are described in more detail below.

Advective Flow

The tidal and wind driven currents are varied on a time-stepping basis. For each particle location, variables are recalculated each time step from the position of the particle at the end of the previous time step. The tidal current, residual current and depth are interpolated from values at the four adjacent grid nodes.

The tidal current variation over depth is simulated by introducing a vertical profile. This variation is expressed by the following formula (Van Veen, 1938):

$$U = Us \cdot (1 - \frac{z}{d})^n$$

where:

Us	=	the instantaneous surface current
U	=	the tidal current at depth z
z	=	distance down from the sea surface (m)
d	=	water depth (m)
n	=	a power coefficient taken as 0.192 (Van Veen, 1938).

The movement of the particle due to the tidal current is then computed as:

$$\Delta X_t = t_s \cdot U \cdot \sin(dir)$$

$$\Delta y_t = t_s \cdot U \cdot \cos(dir)$$

where:

Δx_t	=	particle movement in the x-direction due to the tidal current
Δy_t	=	particle movement in the y-direction due to the tidal current
t_s	=	model time step (sec)
dir	=	direction of the tidal flow (degrees from North).

The direction of tidal flow (dir) is updated each time step according to harmonic coefficients which are specified in the model.

The local wind effects are also incorporated into the model. Wind acting on the water surface induces an additional current component that is greatest at the surface and decreases with depth. This shear effect can result in enhanced spreading of an effluent plume. Also, a counter flow may be generated at depth. The magnitude and form of this wind induced current is poorly understood, but may be taken as 1.4% of the windspeed at the water surface (Dooley and Steele, 1969) and to have a quadratic profile down to a specified depth where the current becomes zero. Below this depth a reverse profile is determined by continuity.

The last advective flow to consider is the tidal residual current, determined by regional scale circulation patterns and longer timescale meteorological conditions. The residual current determines the amount of build-up around the discharge location and the effective residence time of produced water around the oil platform. It is particularly important to consider when assessing far-field concentrations.

The particle movement due to the tidal residual current and wind driven current is superimposed onto the tidal current, so that the overall displacement of the particle in one time step is:

$$\Delta x = \Delta x_t + \Delta x_r + \Delta x_w$$

$$\Delta y = \Delta y_t + \Delta y_r + \Delta y_w$$

where:

Δx_r	=	particle movement in the x-direction due to the residual current
Δy_r	=	particle movement in the y-direction due to the residual current
Δx_w	=	particle movement in the x-direction due to the wind driven current
Δy_w	=	particle movement in the y-direction due to the wind driven current.

Ideally, the time step is chosen such that Δx and Δy are no greater than the grid size, in order that a particle does not jump past more than one grid square at a time.

Dispersion

The random walk displacement on each particle is calculated as follows (Webb, 1982):

$$(6 \cdot Kz \cdot t_s)^{0.5} \cdot R1 \quad \textit{vertical direction}$$

$$(6 \cdot Kh \cdot t_s)^{0.5} \cdot R2 \quad X-\textit{direction}$$

$$(6 \cdot Kh \cdot t_s)^{0.5} \cdot R3 \quad Y-\textit{direction}$$

where:

Kz	=	vertical mixing coefficient (m^2/s)
Kh	=	horizontal mixing coefficient (m^2/s)

R1, R2 and R3 are random numbers selected from a uniform distribution between -1 and 1.

For the majority of the model runs the horizontal and vertical mixing coefficients (Kh and Kz respectively) are assumed to be constant over the whole model area and constant in time. However, a formulation of the horizontal mixing based on the 'diffusion velocity' concept was also tested and the results are discussed in a later section.

If the vertical diffusion step takes the particle above the water surface or below the seabed, then reflection at the surface or bottom is applied to keep the particle within the water column.

Calculation of Concentration

The concentration of effluent over the modelled area is determined using a counting grid. The number of particles in a grid square and over a specified depth interval, d_c (m), are counted and converted to a concentration by:

$$C = \frac{N_p \cdot P_{mass}}{dx \cdot dx \cdot d_c}$$

where:

C	=	concentration (mg l^{-1})
N_p	=	number of particles counted
P_{mass}	=	mass of effluent that each particle represents (g)
dx	=	horizontal grid size of counting grid (m)
dc	=	vertical grid size of counting grid (m).

Degradation

The model allows a first order degradation rate to simulate biodegradation or other removal processes. The probability that a particle will be lost by degradation in a specific time step is given by (Webb, 1982):

$$p = 1 - 10^{-\left(\frac{t_s}{R}\right)}$$

where:

p	=	probability of loss
t_s	=	model time step (hrs)
R	=	degradation rate (hrs).

The model can simulate the effect of different degradation rates at different times if these are known (e.g. if a time lag is required before bacteria acclimatise to the effluent once discharged).

Sinking and Dilution of a Dense Plume

An effluent discharged into the sea that is denser than the sea water will sink due to its negative buoyancy and its initial vertical momentum. The momentum effect will be dissipated very rapidly (usually within 20 m of the outfall), but the buoyancy effect can persist for 1-2 km from the source, but with a diminishing rate of sinking.

As the plume sinks, seawater will be entrained into the plume, and mix with the effluent, causing dilution and hence a reduction in the excess density. As the plume dilutes the ambient turbulence in the sea will start to mix the plume, enhancing the dilution. The plume density will gradually approach the seawater density and the boundaries of the plume will disappear.

These processes have been described by Luhar (1991). Luhar's theory was formulated for the rise of a buoyant gas plume in the atmosphere. However, application of the theory to the dense plume case follows when the densities are interchanged.

MODEL INPUT DATA

Tidal Currents

A continuous recording current meter record over a 20 day period in 1971 has been used to provide baseline information for the model. The major axis of the tidal ellipse is typically NNE/SSW (flooding towards the SSW), depending on meteorological conditions and residual water movements. The peak spring tide current is typically $0.43 m.s^{-1}$, corresponding to a tidal range of 0.9m. The same values are specified everywhere on the model grid and are varied according to the harmonic coefficients which describe the longitudinal and transverse components of the tidal ellipse. It is not known precisely what these are, and they may vary according to tidal and meteorological conditions. However, available tidal current records show a ratio of 5:1 for the longitudinal to transverse components with a clockwise rotation throughout the tide. A clockwise rotation is consistent with what might be expected for large scale eddies in the Northern Hemisphere.

Residual Current

The tidal residual current determines the long term current drift and hence the gradual removal of effluent from the discharge area over several tidal cycles. A value of $0.03 m.s^{-1}$ (calculated from the same current record described above) has been taken everywhere on the model grid and this is superimposed onto the main tidal current to compute the overall water movement.

Depths

The depth of water is taken to be 80m everywhere on the model grid.

Mixing Coefficients

The mixing coefficients determine how much the effluent will be dispersed, and how wide and deep the plume will become. They are measures of the turbulence in the water column and are given as K_h and K_z, horizontal and vertical mixing coefficients respectively. The theory of how they are used is given in the model description.

Local mixing coefficients are not known for the Clyde area. Some measurements taken about 100 m away have been reported by Talbot and Talbot (1974). However, these data were not used since they were recorded in the midst of severe gales and the mixing was considerably enhanced.

The 'worst-case' situation in terms of minimum dilution is when the mixing coefficients are low, and this occurs during calm sea conditions. Some consistent data, taken in calm conditions, were obtained for the Southern North Sea (Talbot and Talbot, 1974) and these have been used to represent low mixing conditions in the model:

$$K_h = 0.3 \ m^2 \ s^{-1}$$

$$K_z = 0.002 \ m^2 \ s^{-1}$$

Effluent Parameters

The following effluent parameters were used in the model:

flow rate:	0.1 $m^3 s^{-1}$ (50,500 barrels day^{-1})
salinity:	132‰
temperature:	72°C
density:	1086 kg m^{-3} (at 15°C)
Predicted biodegradation rates:	27% day $^{-1}$ (at summer temperature of 14.7°C) 7% day $^{-1}$ (at winter temperature of 6°C)

The predicted biodegradation rates have been extrapolated from experimental measurements of biodegradation of the produced water at 10°C and 20°C. The experimental measurements (obtained from manometric respirometry and closed bottle tests) show that most of the biochemical oxygen demand (BOD) occured within about 10 days at 20°C and about 30 days at 10°C (see Figure 2). The biodegradation rate calculated from these curves is 35% per day and 15% per day respectively for 20°C and 10°C. Interpolating from these

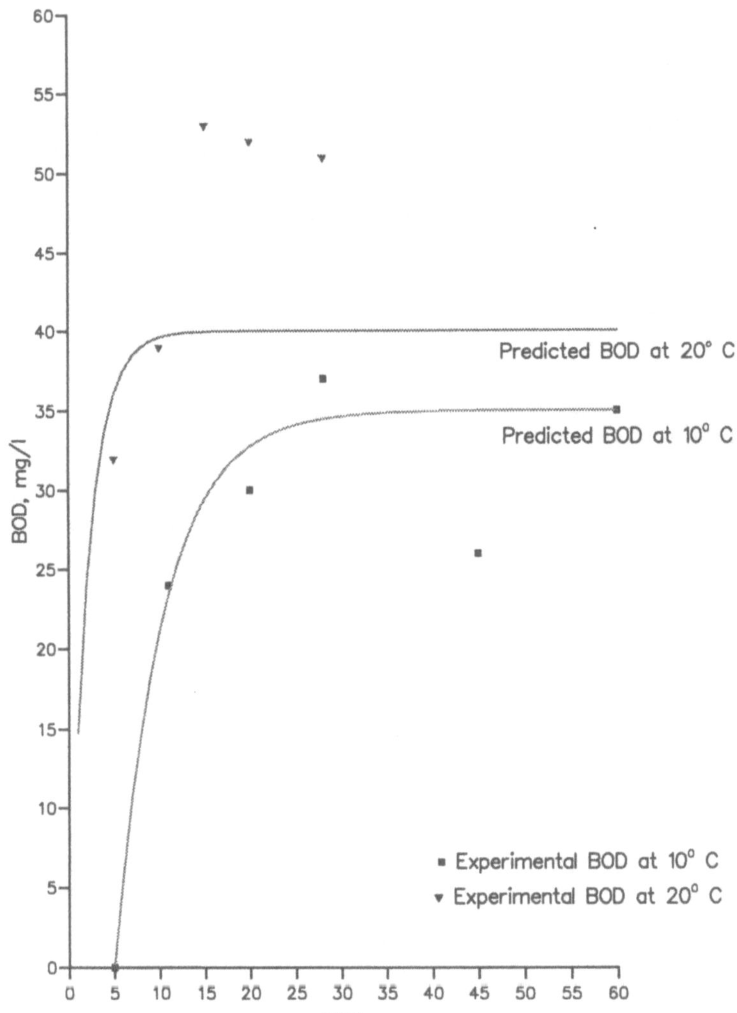

Figure 2. BOD curves assumed by the model.

data, using Arrhenius' relationship (k proportional to 1/T(K)) gives removal rates of 27%/day at 14.7°C (summer) and 7%/day at 6°C (winter). These interpolated rates have been used in the model.

ASSUMPTIONS AND LIMITATIONS OF THE MODEL

The model assumes that the theory outlined in the model description is a valid approach for simulating the dilution of a discharged plume. In general, the process descriptions are founded on basic principles of mass conservation and continuity, and the random walk concept is now widely accepted. However, some aspects need special consideration.

Density Effects

The formulae for calculating the initial dilution, and the sinking of a dense effluent plume, take no account of the effect of impact on the sea surface (given that the Clyde outfall is 20m above sea level), nor of any possible spreading or loss of volatile components as the effluent falls through the air. Instead, the model assumes that the discharge is just below the sea surface, although its nature in terms of pipe diameter, exit velocity and orientation can be specified.

Also, the model does not compute the density directly from the temperature and salinity values. Instead, the initial density (at 15°C) is specified, and this is then diluted directly as the plume entrains clean seawater. Thus the model cannot predict the fine detail in the near field to describe the individual contributions of temperature and salinity changes to the effluent density. This would not be expected to significantly effect the predicted dilutions, but it could change the position and trajectory of the plume within the water column.

Tidal and Residual Currents

The tidal currents in the model have not been hydrodynamically generated. Instead, data recorded from a point source has been extrapolated across the entire model area and the tidal and residual currents are assumed to be the same everywhere. The harmonic coefficients are used to give the current speed and direction at each model time step, but again, this is assumed to apply simultaneously across the whole model area. This approach is probably not too unreasonable, given the small size and location of the model area in relation to the surrounding sea. However, it should be stressed that local and regional circlulation patterns will vary according to different tidal and meteorological conditions. This variation may have relatively little effect in the near field, but further away (>5 km), larger scale circulation patterns may become important, effectively providing an additional dilution. Such large scale patterns are not considered in the model, although the contribution of larger eddies can be considered implicitly through the use of a mixing velocity as discussed below.

Mixing Coefficients

The model generally assumes that the mixing is the same everywhere. It takes no account of spatial variation, horizontally or with depth. This is probably conservative for the reason that as the plume moves further away from the discharge location and spreads laterally, larger scale eddies, as well as the smaller scale eddies, will contribute to the effluent dilution. One approach to overcome this problem is to use the concept of a diffusion velocity which effectively allows the horizontal mixing coefficient to increase with distance from the discharge location. This is considered further in the sensitivity analysis.

In the very near field, it is also possible that wake effects associated with the legs of the oil platform will enhance the initial dilution. However, this is not taken into account in the model.

In the model simulations, no account of a thermocline is taken for summer conditions. A thermocline would inhibit the vertical movement of effluent and may confine the plume to the layer either above or below the thermocline where concentrations could become elevated.

Wind

The wind is incorporated into the model to take into account the additional shear that can enhance spreading of the effluent plume (see model description). Different sensitivity runs assume the quadratic profile to apply over 10m or 30m, although the actual form of this profile is not known. The work by Dooley and Steele (1969) suggest that for coastal waters the wind effect is very shallow, (1-2m depth), but no data are given for deeper oceans. The selected values of 10m and 30m are rather arbitrary, based on judgement. Similarly, the surface wind-induced current is taken to be 1.4% of the wind speed (Dooley and Steele, 1969), although the actual value may depend on other factors such as sea state and wave height, which in turn tend to be dependant on wind speed.

Far field meteorological effects that effect the residual current patterns are not considered in the model.

Biodegradation

The biodegradation rates computed from the experimental work are based on the premise that most of the BOD occured within 10 days and 30 days at 20°C and 10°C respectively. Although the fitted curves are not unreasonable (Figure 2) they are derived from only a small amount of data. The biodegradation tests were conducted using water where the temperature and salinity were close to the seawater value. Both of these factors help to make the predictions more realistic and it is assumed that the interpolated biodegradation rates can be applied to the field situation. This may not always be valid (especially at very low concentrations, or if the nutrients are a limiting factor) and is an area for further research.

Particle Resolution

One limitation of the particle tracking approach is the number of particles that can be tracked at any one time. This becomes relevant for continuous discharge situations as the number of particles in the water steadily increases, particularly if few are lost due to biodegradation. For short model runs, more particles can be released at each time step whereas for longer runs the number has to be reduced so that the total number does not exceed the maximum allowed in the model. The advantage of having more particles is that the resolution of the plume is better and that more precise concentrations are predicted at any given location.

Associated with this is the selection of the 'counting grid' used to calculate the effluent concentration (see model description). This needs to be selected carefully in order to represent the volume of water that the plume is occupying. If the counting

Table 1a. Sensitivity analysis runs: description and minimum dilutions predicted at different distances from the discharge location. Sensitivity to environmental and discharge parameters

Run	Description	100m	500m	1000m	5000m	10000m
1	Baseline run	584	1754	2380	4761	–
2	No harmonics	292	775	1041	3333	–
3	Summer biodegradation (27%/day)	606	1754	2380	5441	–
4	Winter biodegradation (7%/day)	591	1754	2221	4988	–
5	Neap tide	483	1666	3033	–	–
6	Kh x 5	1075	3030	4761	8333	–
7	Kh x 10	1754	3703	5555	11111	–
8	Kz x 5	925	2380	3333	8333	–
9	Kz x 10	1010	2564	4166	8333	–
10	Kh x 10 and Kz x 10	2083	4166	5555	11100	–
11	80% wastewater flow rate	595	1984	2314	10416	–
12	120% wastewater flow rate	487	1543	1984	3968	–
13	Kh x 5 and Kz x 5, 80% eff. flow rate	1736	4166	5952	13900	–
14	Exit velocity x 4 & port diameter/2	606	1851	2564	4761	–
15	Exit velocity x 100 & port diameter/10	606	1960	2777	8333	–
16	wind 5m/s, 292.5, 10m profile	512	1754	3333	8333	–
17	wind 5m/s, 292.5, 30m profile	529	1754	3333	11111	–
18	wind 5m/s. 22.5, 30m profile	680	2083	3030	6666	–
19	wind 5m/s 202.5, 30m profile	512	1960	2777	8333	–
20	Horizontal Diffusion velocity V=0.01	370	1111	2564	11111	–

Table 1b. Sensitivity to counting grid size, number of particles and run length

Run	Description	100m	500m	1000m	5000m	10000m
21	Kh x 10 and Kz x 10 & counting grid 50m x 50m x 1.5m	2705	8333	16025	34700	–
22	Counting grid 50m x 50m x 1.5m	1010	2564	4166	8333	–
23	Counting grid 20m x 20m x 5m	918	2777	3584	12345	–
24	No harmonics & counting grid 50m x 50m x 1.5m	432	1111	1587	3703	–
25	No harmonics & counting grid 20m x 20 x 3m	294	913	1257	3921	–
26	80% wastewater flow rate, 4 tides, 250 parts/time step	496	1488	2083	2604	5208
27	80% wastewater flow rate, 1 tide, 250 parts/time step	578	1736	2083	5208	–
28	Baseline run, 4 tides, 250 parts/time step	378	833	1388	2777	8333
29	Baseline run, 1 tide, 250 parts/time step	462	1190	1388	4166	–
30	Kh x 5, 4 tides, 250 parts/time step	694	1666	1666	2777	4166
31	Kh x 5, 1 tide, 250 parts/time step	833	2083	2777	2777	–
32	Diffusion velocity V=0.01, 4 tides, 250 parts/time step	277	757	1041	2777	8333
33	Diffusion velocity V=0.01, 1 tide, 250 parts/time step	308	925	1666	2777	–

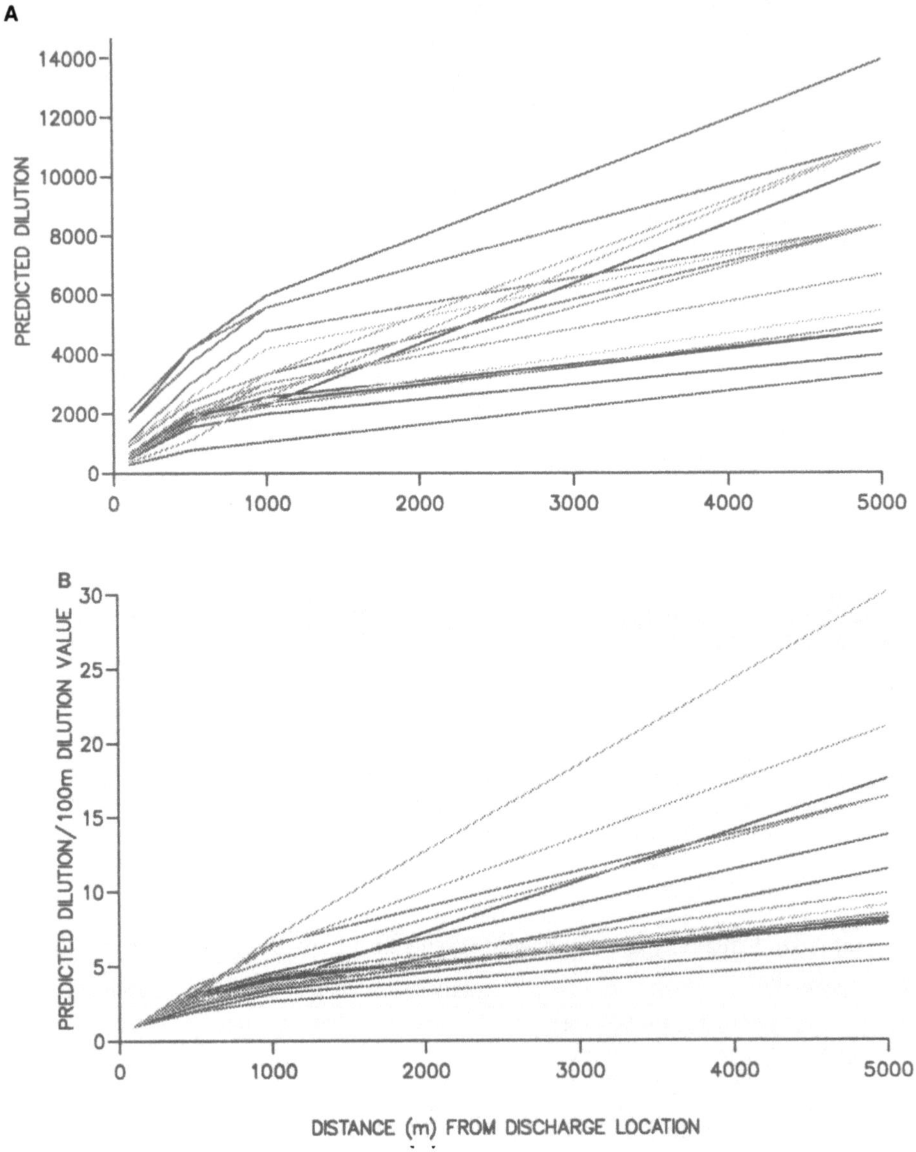

Figure 3. Graphs showing the predicted effluent dilution against distance for a range of different environmental and discharge conditions. A. Total effluent dilution; B. relative dilution after 100 m.

grid is too large, it may contain a large proportion of clean water and hence give an artificially high dilution. Conversely, if the counting grid is too small, it will not contain a sufficient number of particles to give a good resolution of the predicted concentration. Ideally, the counting grid should be small in the near field where the plume is narrow and the particle density is high, but larger in the far field where the plume is more spread and the particle density is lower. Some of these aspects are discussed further in the sensitivity analysis.

SENSITIVITY ANALYSIS

A sensitivity analysis was carried out to investigate the importance of different model parameters on the predicted dilution of the produced water after discharge into the sea. A selection of 33 (out of a total of 78) model runs is described in Tables 1a and 1b. Table 1a represents a reasonable range of real environmental and discharge conditions that may occur in the field whereas Table 1b represents the range of predicted dilutions that depend largely on the choice of model control parameters (counting grid size, no. particles and no. tides simulated). Both types of variation are discussed below and should be considered when assessing the range of dilutions predicted by the model. For each model run, an outline description of the run parameters is given, together with the predicted dilution at 100 m, 500 m, 1000 m, 5000 m and 10000 m from the discharge location.

Figure 3a also shows graphically the range of predicted dilutions at each distance and Table 2 gives the range of values and the mean and standard deviation associated with the different types of runs. It should be noted that these statistics only reflect the runs that were undertaken. These used mostly worst-case parameters which explains why the distribution appears rather skew and why the mean is more biased towards the minimum value. Figure 3b shows the relative dilution obtained after the first 100m which, for all of the runs, is clearly small compared to the initial dilution obtained within the first 100m of the discharge location. It should be stressed that Figures 3a and 3b are designed to give an impression of the range of expected dilutions and hence no key is given to identify specific runs.

The following sections describe the runs and main observations in more detail.

Baseline Run

The baseline run (Run 1) represents ‘typical’ conditions, assuming the basic data described in the Model Input Data section above. It has a high resolution, with 1000 particles released per model time step (equivalent to every three minutes in ‘real time’) and simulates a continuous release of produced water for one tidal cycle, starting with a zero background concentration. The relevant baseline run parameters are:

Table 2. Sensitivity analysis runs: predicted dilution statistics at different distances from the discharge location

Runs (1-20)	100m	500m	1000m	5000m
Minimum	292	775	1041	3333
Maximum	2083	4166	5952	13900
Mean	802	2186	3253	8035
Standard Deviation	498	920	1308	2947
Runs (21-33)				
Minimum	277	757	1041	2604
Maximum	2705	8333	16025	34700
Mean	722	2029	3132	6836
Standard Deviation	643	2003	3984	8838
Runs (1-33)				
Minimum	277	757	1041	2604
Maximum	2705	8333	16025	34700
Mean	770	2124	3205	7548
Standard Deviation	551	1419	2640	5970

Tidal range	0.9 m (spring)
Horizontal mixing (kh)	0.3 m^2s^{-1}
Vertical mixing (kz)	0.001 m^2s^{-1}
Harmonics	included
effluent flow rate	0.1m s^{-1}
wind	none
biodegradation	none
counting grid	20m x 20m x 1.5m
no. particles per time step	1000
no. tides simulated	1

These are the main parameters that were varied for the sensitivity analysis. For all of the other runs in Table 1, conditions were exactly the same as the baseline conditions apart from the factors outlined in the description.

Effects of Varying Environmental and Discharge Parameters

Tidal Range. Comparing Run 1 with Run 5 shows that the predicted dilution is reduced by about 18% at 100m, principally because the effluent is being discharged into a smaller volume of water. Further away from the source the dilution appears to increase, although this is partly due to equilibrium not being reached, particularly at 1000 m from the outfall. It is worth noting also that reduced dilution due to the slower advective currents on a neap tide is partly compensated by increased mixing due to the longer plume travel time.

Note that the dilution at 5000m cannot be calculated since the plume does not travel far enough on a neap tide.

Mixing. Runs 6 and 7 show the horizontal mixing increased by a factor of 5 and 10 respectively , while runs 8 and 9 show the vertical mixing increased by the same factors. An order of magnitude variation is a realistic range for the mixing coefficients and reflects the uncertainty in characterising the mixing process. The model is generally sensitive to the magnitude of the horizontal mixing coefficient, adding an additional dilution factor of up to 3 (with Kh x 10) at 100 m from the source. In comparison, using Kz x 10 only gives a maximum additional dilution of 1.8, possibly because the vertical mixing is inhibited by the relative density difference of the plume and seawater.

With the way the mixing is formulated in the model, the effect of the mixing tends to be reduced as the plume moves further away from the source and spreads out, effectively causing the 'middle' part of the plume to mix with itself (although in fact this is hardly noticeable in Runs 6-10). However, using the concept of a horizontal diffusion velocity (see model description), the plume width will continue to increase as it moves away from the source, and a greater dilution should be observed in the mid to far field. This is demonstrated in Run 20 which should be compared against Run 2 since it contains no harmonics. Even at 100 m a greater dilution is observed, but the relative dilution increases by over a factor of 3 at 5000 m.

Wind. Runs 16-19 show the effect of different wind conditions. Run 16 simulates a wind acting at 90° across the plume and demonstrates the effect of lateral shear. Run 17 simulates the same, but with a profile over the top 30 m of water. Run 18 simulates a wind along the main axis of the plume, opposing the tidal current. Run 19 simulates a wind acting along the main axis of the plume but in the same direction as the tidal current (i.e. opposite to run 18). Both runs 18 and 19 simulate a 30 m profile.

In the near field, none of the wind simulations vary greatly from the baseline run, although further away from the discharge location the effect of the wind becomes more significant, adding a factor of two or more to the predicted dilution. However, the dynamics are clearly complex (for example the cross-wind simulations actually predict higher concentrations at 100 m than the baseline run which has no wind. Interestingly, an opposing wind (Run 18) will slow the effective current and thus give more time for mixing to take place before the plume reaches the monitoring points (analogous to the neap tide effect mentioned above). There seems to be little difference between a profile depth of 10 m and 30 m, at least when the wind is acting across the width of the plume. Other profile depths were not investigated.

Harmonics. Inclusion of the harmonic coefficients that describe the tidal ellipse is one of the most important factors for determining the predicted dilution, particularly within the first 100m or so of the outfall. Comparing Run 1 with Run 2 clearly demonstrates this and indicates that adequate data describing the tidal ellipse are important for obtaining realistic predictions of near field dilution.

Effluent Flow Rate. The effect of varying the effluent flow rate is shown in runs 11 and 12. Run 11 shows the flow rate reduced by 20% and Run 12 shows the flow rate increased by the same amount. The model is sensitive to small variations in this parameter, with a lower flow rate allowing a significantly greater dilution. This is consistent with predictions from other similar models (Rye et al, 1995).

Effects of Varying Model Control Parameters

Number of Tides Simulated. One of the advantages of the particle tracking approach is that build-up of effluent over a period of time can be simulated. This factor will be particularly important if the residual currents are slow and it may be necessary to simulate the discharge for several tidal cycles in order to obtain predicted dilutions which are representative of equilibrium conditions.

Runs 26, 28, 30 and 32 in Table 1b are comparable to Runs 27, 29, 31 and 33 respectively and show the differences between running for one or four tidal cycles under different conditions.

Figure 4 shows the temporal output for Run 28. Each graph shows the predicted dilution with time at each distance from the discharge location. In order to help interpret this figure, consider the graph for 100 m.

Starting from time zero, no effluent reaches this temporal location for about 6.5 hours. This is because the tidal ellipse causes the plume to move around this point and effectively miss it altogether. After 6.5 hours, however, the plume swings around, passing over the temporal monitoring point, when a high concentration (low dilution) is detected. The plume then continues to move around and eventually change direction as the tide turns. During this time the temporal location becomes less exposed to the plume and much of the time (when the tide is flowing in the opposite direction) lies in clean seawater. Only when the tide has turned a full cycle does the dilution minimum occur again at this point. Indeed, the time period for exposure of a temporal location to high concentrations is typically only about 20-30 minutes during any one tide, a feature which reflects these harmonic variations.

After four tidal cycles, it is possible to see the effects of build-up, when effluent residues from previous tides are detected at each temporal location. In order to obtain a true picture of equilibrium conditions it is necessary to simulate a long enough period so that the dilution pattern does not vary from one tide to the next. However, for the purposes of this

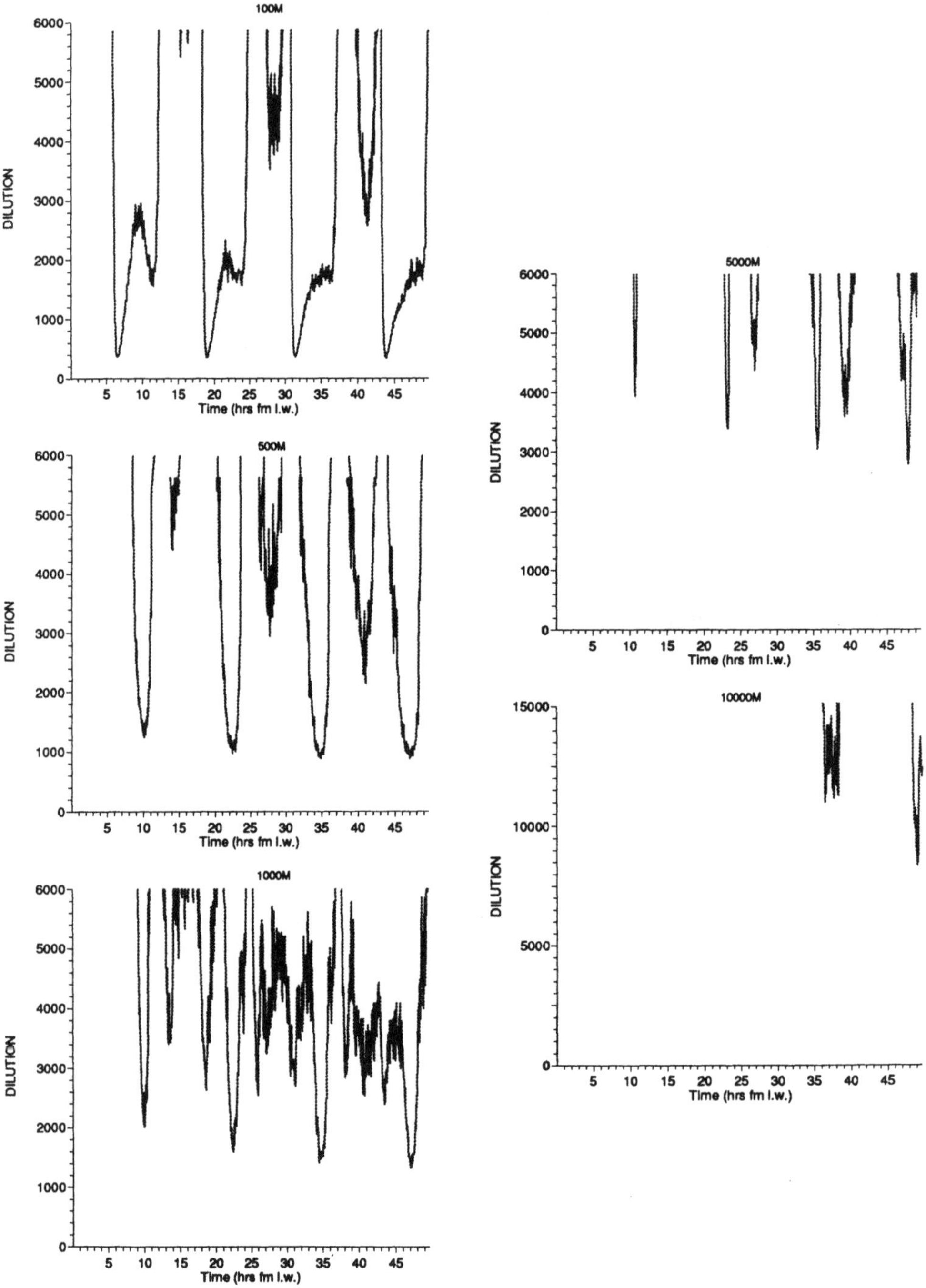

Figure 4. Run 28 predictions. Temporal output at different distances from discharge location. Each graph shows the predicted dilution against time.

study, only the minimum dilution value has been taken, and it can be seen that for distances up to 1000 m or so the minimum predicted dilution does not vary greatly from one tide to the next. For 5000 m and 10000 m the build up is more significant and the minimum dilutions are still decreasing after four tidal cycles. The 10000 m result is a good example; nothing is detected for nearly three tidal cycles, and clearly if a true dilution estimate representative of a continuous release is required at this distance it would be necessary to run a longer simulation.

Number of Particles. Runs 1, 8, 11 and 20 all represent a release of 1000 particles per model time step. Runs 29, 31, 27 and 33 respectively are run under exactly the same conditions but only with 250 particles per time step. Using fewer particles tended to under-predict the dilution in these runs by between 0.3% and 21% at 100 m, depending on how the particles were spread. At further distances, the number of particles occupying a given volume is reduced, causing the resolution to become more erratic. Hence, similar numbers keep appearing for different runs beyond about 1km from the discharge point. At 5000 m the predicted dilution using 250 particles is between about 10% and 75% less than predicted using 1000 particles. This is a significant variation and sound judgement is needed to optimise the number of particles and the counting grid size, depending on how far away from the discharge location the predicted dilutions are required.

It does appear that using too few particles will tend to underestimate the dilution. However, further investigation is required since it is not clear why this should be the case and very occasionally the converse is true.

Counting Grid Size. Runs 21 to 25 show the effects of varying the counting grid size. In all cases the predicted dilution is shown to increase if the counting grid size is increased. Thus it is important to ensure that the grid size is small enough so as to not artificially inflate the predicted dilutions. Moreover, as outlined in the assumptions and limitations, it may be necessary to select different grid sizes for different distances, so as to avoid poor resolution. The fact that the same numbers appear several times in the predicted dilutions, particularly in the far-field, is indicative that a relatively small number of particles are present in a relatively small volume of water.

FIELD RESULTS AND MODEL COMPARISON

In 1994 a joint research cruise was conducted between SOAFD, MAFF and industry representatives and the opportunity was taken to assess the actual dilutions and dispersion around specific North Sea production installations, including the Clyde platform. Details of the sampling exercise are given by Stagg et al, 1996.

Of specific interest were the measured concentrations of hydrocarbons and various other organic components within the effluent plume in relation to the concentrations in the discharged effluent. Both fixed station sampling and continuously towed fluorimeter measuring techniques were utilised to collect these data. The model was also used to predict the likely location and depth of the plume at different times in order to optimise the sampling locations.

Figure 5 shows the 'worst-case' (i.e. minimum) measured dilutions with distance for alkanes and methylnaphthalenes. All of the dilution values recorded were extremely low and comparing these with the predicted range of dilutions given by the model in Figure 3 shows that the measured values were much greater than predicted. The concentrations of other organics such as phenols and methylphenols in the water column were below detection limits (0.01 - 0.05 $\mu g\ l^{-1}$) which supports the evidence for high dilution factors at all locations

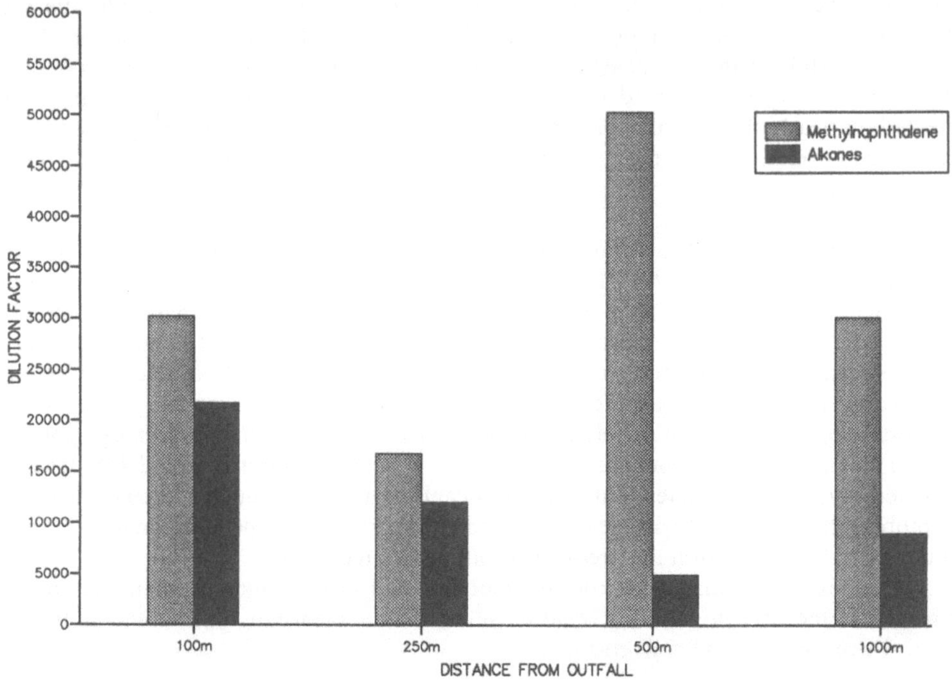

Figure 5. Clyde produced water - measured dilution values.

sampled. The results from the towed fluorimeter measurements also indicated approximately a 5000 fold dilution at 100 m from the platform (Stagg et al, 1996) which is much higher than any of the predicted dilutions given by the model.

The concentration gradient away from the outfall is also poorly defined in the field data whereas an increasing dilution with distance is predicted by the model.

It should also be noted that the dilutions for each component are different for the same distances. This is not readily explained unless there are different removal processes which are active, such as volatilisation or some chemical reaction, which could remove some of the components faster than others.

There could be two main reasons for the discrepancy between the measured and predicted values. Either the actual dilutions are much greater than predicted by the model or the measured data is not representative of the true minimum-plume-centre-line dilution which is what the model predicts. Generally, the model would not be expected to produce such high dilutions unless there is some other mechanism present that could account for this concentration decrease.

Effluent plumes are typically long and narrow, particularly for the first few hundred metres of a discharge point. Depending on the mixing conditions, the plume cross sectional areas predicted by the model for Clyde vary from about 20 m x 2 m to 50 m x 4 m at 100 m from the discharge point, taking into account the tidal harmonics. Moreover, the bulk of the effluent will lie along the plume centre-line, with relatively low concentrations at the plume edges.

Also, depending on the precise run conditions, the plume typically continues sinking, due to its high density, for about 300 m, after which it tends to stabilise under neutral buoyancy. The depth of the plume centre-line at 100 m was predicted to be about 10 m deep.

At 500 m this increased to about 15 m. However, these values are only very approximate and depend on the tidal currents and discharge conditions.

Given that the bulk of the plume is predicted to occupy a relatively small volume of water and a small depth range, this leaves little room for error in locating the position of the plume. The samples in Figure 5 were taken at about 15m, as suggested by the model. However, given the uncertainties involved this may not have been the optimum depth.

In summary, it is not clear why there are such differences between the observed and predicted values. They may be due to aspects not taken into account in the model or associated with difficulties in sampling in the open sea. Determining the cause of the discrepancies will require further study.

SUMMARY AND CONCLUSIONS

A quantitative understanding of the extent of dispersion processes is a prerequisite for the assessment of the environmental impact of components of produced water discharged into the marine environment. To this end, a computer model was developed to predict the dispersion of produced water discharged from the Clyde oil production platform, situated in 80m of water , approximately 300 km south east of Aberdeen.

The model is based on a particle tracking technique with a 'random walk' component to simulate dispersion. Within the model, several hundred 'particles' are released at each time step. These 'particles' represent components of the produced water and are made to move in relation to the advection and dispersion, caused by the tidal currents, wind and mixing processes. Longer term biodegradation effects were incorporated by using rate constants derived from laboratory experiments. Based on a sensitivity analysis, minimum dilution factors were predicted to be between 300 and 3000 at 100 m from the discharge point, and between 1000 and 16000 at 1km.

A field survey was carried out during which samples of undiluted produced water were obtained. Discrete samples of sea water were also taken, within the predicted position of the effluent plume. Hydrocarbon concentrations in the discrete samples were measured. Additional, continuous, measurements were made using a towed fluorimeter which was undulated through the predicted depth range of the predicted plume.

Measured dilution factors greatly exceeded the minimum values predicted by the model. Reasons for this discrepancy are discussed in relation to the predicted narrow plume dimensions, the uncertainties in the model predictions and the practical difficulties of sampling such a plume in the open sea.

An important consideration in assessing the environmental impact of produced water discharges, is the reduction in concentration due to dilution in the sea. This paper describes the theory and application of a particle tracking model for predicting this dilution under a range of different conditions for the Clyde oil production platform in the North Sea.

The initial dilution can be very rapid, typically reducing the initial concentration by a minimum factor of between 300 and 3000 within 100 m from the outfall. Further away the physical dilution is less rapid but other removal processes such as biodegradation may become important. At 1km from the discharge point, the predicted range of dilutions is between 1000 and 16000. Note that these values represent the minimum dilutions predicted throughout the water column and throughout a tidal cycle. The plume is relatively narrow and most of the water column remains unaffected by it. Even along the plume centre-line, however, the temporal predictions show that the peaks of high concentration are predicted to persist for only a short period of time (typically 20-30 minutes), owing to the elliptical motion of the tidal currents.

These dilution factors have been determined from the sensitivity analysis presented in this paper which provides predicted dilutions for a range of possible environmental and discharge conditions. Also considered are the range of other model parameters which need to be specified by the user of such a model, and whose values may be critical to the analysis.

The model has been compared with data from a survey conducted in 1994 which obtained measurements of hydrocarbon concentration and concentrations of specific organic components.

Measured dilutions were generally much higher than predicted although the sampling was undertaken extensively in the vicinity of the plume. However, there are various anomalies surrounding the measured data which are difficult to explain in the context of what is predicted, including the absence of a longitudinal gradient and the different dilution values obtained for different organic components.

Differences between the observed and predicted values may be due to aspects not taken into account in the model or associated with the difficulties in sampling and analysis during the study. Further research would be required to explore these issues further.

ACKNOWLEDGMENTS

The authors would like to thank the BP Exploration Operating Company Ltd. for permission to publish this work.

REFERENCES

An environmental risk evaluation model for offshore E&P chemicals. Draft document. MoU Project prepared by TNO and Aquateam. August 1993.

Brendehaug, J., Johnsen, S., Byrne, K.H., Gjose, A.L., Eide, T.H. and Aamot, E., 1992 toxicity testing and chemical characterization of produced water - a preliminary study. In, J.P. Ray and F.R. Engelhardt (eds.), produced water technological /environmental issues and solutions. Environmental Science Research Volume 46. Proceedings of the 1992 International Produced Water Symposium, Feb. 4-7, 1992 San Diego, California. Plenum Press, pp 245 - 266.

Carothers W.W. and Kharaka Y.F. (1978). Aliphatic acid anions in oil-field waters - Implications for the origin of natural gas. The American Association of Petroleum Geologists Bulletin, 62, 12, pp 2441-2453.

DOE, 1983. Methods for the examination of waters and associated materials, assessment of biodegradability, 1981. HMSO

Dooley H D and Steele J H, 1969. Wind driven currents near a coast. Deutsche Hydrographische Zeitschrift 22 (5), pp 213-223

Luhar K L, 1991. Random walk modelling of air pollution dispersion. A dissertation submitted to the University of Cambridge for the degree of Doctor of Philosophy, July 1991.

OECD Guideline 301/D, 1993, OECD guideline for testing of chemicals adopted 17/7/92, Ready Biodegradability.

Riddle A M, 1993. Program setup and user manual; TRK water quality model. Brixham Environmental Laboratory, Manual SW111/U.

Rye H., Reed M., Melbye A., and Johnsen, S. 1995. Dilution factors for produced water Releases. 6th International Symposium: Oil Field Chemicals.

Somerville, H.J., Bennett, D., Davenport, J.N., Holt, M.S., Lynes, A., Mahieau, A., McCourt, B., Parker, J.G., Stephenson, R.R., Watkinson, R.J. and Wilkinson, T.G., 1987. Environmental effect of produced water from North Sea oil operations. Marine Poll.Bull.18.10:549-558.

Stagg R., Gore D. J., Whale G. F., Kirby M. K., Blackburn M., Bifield S, McIntosh A.D., Vance I., Flynn S.A. and Foster A. (1996'). Field evaluation of acute toxic effects of produced water discharges from North Sea oil production platforms: Implications for monitoring acute impact in the environment.

Staples, T.B., Lewis, R.E., Owens, N.J.P., Woodward, E.M.S., Rees, A.P. and Plummer, D.H. ,1993. A modelling study of the contribution of riverine and anthropogenic inputs of ammonium to a North Sea coastal area. Estuarine, Coastal and Shelf Science 37, 497 - 507.

Talbot J W and Talbot G A, 1974. Diffusion in shallow seas and english coastal and estuarine waters. In: physical processes responsible for dispersal of pollutants in the sea. Kullenberg G and Talbot J W, eds. 259pp

Tibbetts, P.J.C., Buchanan, I.T.. Gawel, L.J. and Large, R. ,1992. A comprehensive determination of produced water composition. In, J.P. Ray and F.R. Engelhardt (eds.), Produced Water. Technological/Environmental Issues and Solutions. Environmental Science Research Volume 46. Proceedings of the 1992 International Produced Water Symposium, Feb. 4-7, 1992 San Diego, California. Plenum Press, pp 97 - 112.

The Oil Industry International Exploration and Production Forum (E&P Forum), 1994. North Sea produced water: Fate and effects in the marine environment. Report No. 2.62/204, May 1994. E&P Forum, 25-28 Old Burlington Street, London W1X 1LB.

Van Veen J, 1938. Water movements in the Straights of Dover. J. Cons., Cons. Perm. Int. Explor. Mer. v.13 pp 738.

Webb A J, 1982. A random walk model of the dispersion of caesium137 in the Irish Sea. Thesis for MSc Univ. Wales Feb. 1982.

22

A BUOYANT JET/PLUME MODEL FOR SUBSEA RELEASES

Helge Mohn Skåtun

Det Norske Veritas Industry AS
Veritasvn. 1
N-1322 Høvik, Norway

INTRODUCTION

Numerical models for dispersion of effluents from e.g. offshore installations is a valuable tool to use for predicting the concentrations of e.g. produced water in the vicinity of the discharges, and hence the effect of produced water to marine life.

Some of the models have previously been used nearshore, e.g. for predicting dispersion of sub-surface discharges from industry or municipalities into fjords and lakes. Comparison with measurements have shown good correspondence with modelled results. One such model is the BJET model, a jet/plume model which will be described here.

The porting of such models offshore has raised questions on how well the models perform offshore. Unfortunately, offshore measurements are rare, due to the difficulties and high costs associated. Fortunately, a few such measurement programs or field experiments have been reported in the open literature, and a comparison with this

In order to elucidate which parameters are the most important, a sensitivity analysis for the model has been performed. The major results of this sensitivity analysis will be presented.

THE MODEL

The following describes the model BJET, originally developed at the Norwegian Institute of Technology (Nilsson, 1991) and the modifications applied to this model. The model consists of three sub-models; the model for the Zone of Flow Establishment (ZFE), the Gaussian model zone, and the Zone of Limited Entrainment (ZLE) which extends out to the far-field zone. These three sub-models will be described in this paper.

A sketch of the flow is shown in Fig.1. The figure also shows the transition points between the three zones in the model.

Produced Water2: Environmental Issues and Mitigation Technologies
Edited by M. Reed and S. Johnsen, Plenum Press, New York, 1996

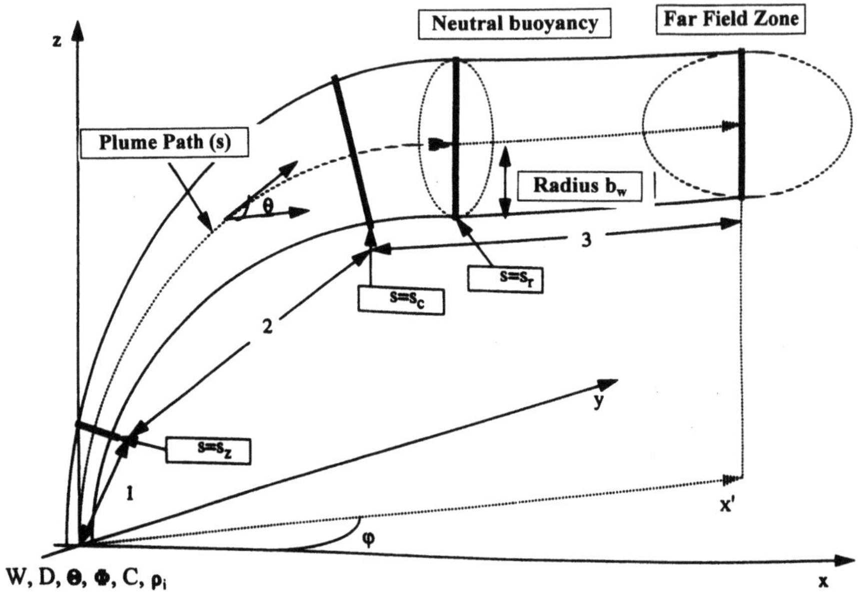

Figure 1. Schematic view of the jet/plume model. The numbers show the three zones; 1) the zone of flow establishment (ZFE), 2) the Gaussian zone and 3) the zone of limited entrainment (ZLE). φ is the horizontal angle between the released flow and the ambient flow, θ is the vertical angle between the released flow and the horizontal.

Zone of Flow Establishment (ZFE)

This sub-model covers the initial stage of the release, from the discharge orifice to the position where the jet/plume obtains a Gaussian profile. At this stage the flow is governed by the release geometry, and is characterised by increasing turbulence as the flow leaves the discharge.

Gaussian Model Zone

At this stage of the flow the flow variables such as concentrations and momentum are described by Gaussian profiles as function of radial distances from the plume centre. The input variables to the Gaussian flow are obtained from the ZFE relations.

The conservation equations for the volume flux q, the momentum fluxes m_x, m_y and m_z in the x, y and z-directions respectively, and the buoyancy flux b' are given by the equations:

$$\frac{d}{ds}q = \frac{d}{ds}\left[\pi b_w^2\left(2\overline{U_a i_s} + u_s\right)\right] = 2\pi\alpha b_w\left|\overline{U_j} - \overline{U_a}\right|$$

$$\frac{d}{ds}(m_x) = \frac{d}{ds}\left[\pi b_w^2\left(2\overline{U_a i_s} + u_s\right)\cos\theta \cos\varphi\right] = 2\pi\alpha b_w U_a \cos\gamma\left|\overline{U_j} - \overline{U_a}\right| + F_x$$

$$\frac{d}{ds}(m_y) = \frac{d}{ds}\left[\pi b_w^2\left(2\overline{U_a i_s} + u_s\right)\cos\theta \sin\varphi\right] = 2\pi\alpha b_w U_a \sin\gamma\left|\overline{U_j} - \overline{U_a}\right| + F_y$$

$$\frac{d}{ds}(m_z) = \frac{d}{ds}\left[\pi b_w^2\left(2\overline{U_a i_s} + u_s\right)\sin\theta\right] = g\pi\lambda^2 b_w\left(\frac{\rho_a - \rho}{\rho_0}\right) + F_x$$

$$\frac{d}{ds}b' = \frac{d}{ds}\left[\pi b_w^2(\rho_a - \rho)\left(2\overline{U_a i_s} + u_s\right)\right] = \pi\frac{1+\lambda^2}{\lambda}b_w\left(2\overline{U_a i_s} + u_s\right)\frac{d\rho_a}{ds}$$

These conservation equations for the Gaussian flow may be solved to yield the following expressions:

$$q = \pi\, b_w^2\, (2\overline{U_a} \cdot \overline{i_s} + u_s)$$

$$m_x = \frac{1}{2}\,(2\overline{U_a} \cdot \overline{i_s} + u_s)\, q \cos\theta \cos\phi$$

$$m_y = \frac{1}{2}\,(2\overline{U_a} \cdot \overline{i_s} + u_s)\, q \cos\theta\, s\, in\, \phi$$

$$m_z = \frac{1}{2}\,(2\overline{U_a} \cdot \overline{i_s} + u_s)\, q \sin\theta$$

$$b' = \pi\, b_w^2\, (\rho_a - \rho)\, (2\overline{U_a} \cdot \overline{i_s} + u_s)$$

The equations are solved by means of a 4th order Runge-Kutta scheme.

Zone of Limited Entrainment (ZLE)

In this zone the horizontal stratification becomes important. The vertical mixing is reduced due to the effect that the vertical kinetic eddy energy is used to generate internal waves in the fluid rather than mixing the surrounding water into the plume. The lateral mixing is not affected by this effect, and hence the plume becomes elliptical rather than circular as in the initial stages.

The criterion for the transition from the Gaussian Zone to the ZLE is that the frequency of the energetic large scale eddies in the plume reaches the Brunt-Väisäla frequency (N) of the ambient flow.

A turbulent Froude number F_t is defined by Thorpe (1982) as

$$F_t = \frac{w'}{L_t N}.$$

When F_t approaches 1 the mixing is significantly reduced. In terms of the flow variables this criterion can be expressed as

$$u_s \leq b_w \cos(\theta)\, N$$

In this zone the vertical radius (b_v) is decreasing and the horizontal radius (b_h) is increasing as the distance from the release orifice increases. The radius concept in the Gaussian zone may still be used by introducing the relation:

$$b_w{}^2 = (b_h b_v)$$

Entrainment Formulation

The entrainment coefficient used in the original BJET model as presented by Nilsson (1991) is formulated as

$$\alpha = 0.082\left(1 + 0.4\frac{U_a}{u_s}\right)$$

Sørgård (1992b) presents a discussion of various pararameterizations of the entrainment function. For the ambient fluid at rest, Fisher (1979) proposed the relation

$$\alpha = \alpha_j \exp\left[\ln\left(\frac{\alpha_p}{\alpha_j}\right)\left(\frac{R}{R_p}\right)^2\right]$$

Lee and Cheung (1991) investigated the mixing characteristics of buoyant plumes in crossflows in laboratory experiments over a wide range of ambient flow velocities. Based on these experiments, Sørgård (1992b) introduced the following expression for the entrainment coefficient.

$$\alpha = \alpha_j \exp\left[\ln\left(\frac{\alpha_p}{\alpha_j}\right)\left(\frac{R}{R_p}\right)^2\right] + 0.45\left[1 - \exp\left(-1.25\frac{U_a}{u_s}\right)\right]\exp\left[\ln(1.2)\left(\frac{R}{R_p}\right)^2\right]$$

The first term is identical to the expression used by Fisher et al. (1979) for the ambient fluid at rest, while the last term expresses the effect of the ambient flow.

The drag coefficient C_D is suggested by Sørgård (1992b) to be formulated as

$$C_D = 3\exp\left[-0.4\frac{u_s}{U_a}\right].$$

SENSITIVITY ANALYSIS

Release Scenarios

The release scenarios were set up by modifying three aspects of the experiments. Below are listed the three aspects and their alternative formulations:

- Jet / plume options:

  Jet: High release velocity (0.6 m/s), low density difference
  Plume: Low release velocity (0.1 m/s), high density difference

- Ambient horizontal stratification

  Simple linear profile, constant vertical density gradient
  Variable density gradient, upper layer has increased stratification.

- Ambient flow velocity

  Ambient fluid at rest ($U_s = 0$)
  Moving ambient fluid ($U_s = 0.2$ m/s)

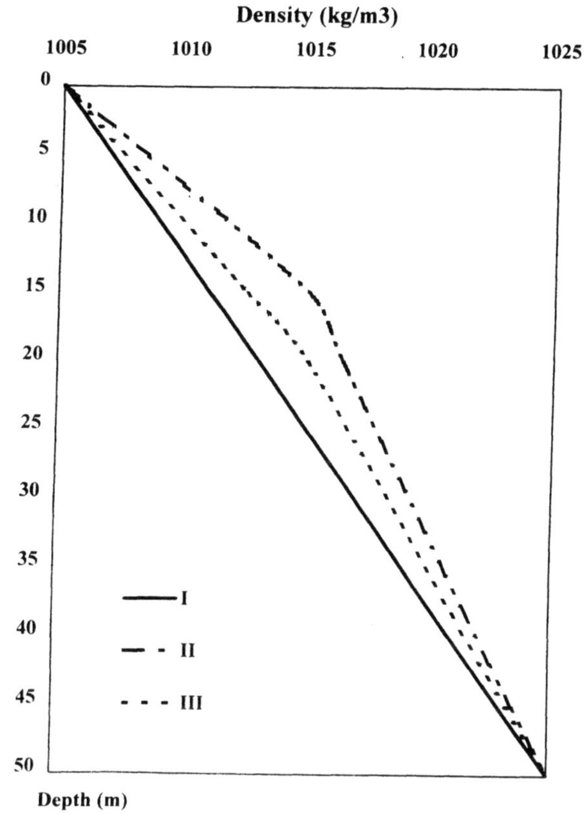

Figure 2. Vertical density profiles used in the simulations; I: linear profile (solid); II: profile with increased stratification in upper layer (dashed); III: modified profile with increased stratification in upper layer (dash-dotted).

By combining the two options for each aspect, a set of 8 different release scenarios may be defined.

Sensitivity Analysis

For these release scenarios the 7 parameters shown in table 1 were perturbated to find the sensitivity of the results to each parameter.

Table 1. Parameters used for perturbation in the sensitivity analysis

Parameter	Symbol	Initial value	Perturbation
Release velocity	W	0.1 (0.6) m/s	+/- 10%
Release density	ρ_i	1000 (1017) kg/m^3	+/- 3 kg/m^3
Entrainment coefficient	α	Fisher Sørgård	+/- 10%
Drag coefficient	C_D	Sørgård	+/- 10%
Tilting angle	θ	90 (vertical)	+10/-5°
Ambient flow velocity	U_a	0 (0.2) m/s	+/- 0.03 m/s
Stratification of ambient fluid	$\rho_a(Z)$	Cf. Fig. 2	+/- 5 kg/m^3

Table 2. Range of plume centre concentrations observed at various distances from discharge

	Distance from release (m)						
	4	6	13	28	58	78	103
Average	1.9	1.54	0.91	4.47	0.36	0.23	0.10
Upper	30	10.7	2.2	92	3.2	2.0	0.37
Lower	0.13	0.18	0.36	0.22	0.03	0.03	0.02

COMPARISON WITH FIELD EXPERIMENTS

Field experiments have been conducted by Offshore Operators Committee (OOC) in the Gulf of Mexico (Smith et al., 1993). Produced water containing a tracer (Radium) and a fluorizing tracer was released from an offshore platform.

The released produced water was very warm and saline; T = 32°C and S = 84 ppt. No values were given for the density of the release, which is computed to be 1058 kg/m^3.

In the experiment, water samples were collected by means of divers and a fixed grid arrangement of sampling hoses. This grid was positioned downstream from the release point, and the sampling depth was chosen to correspond to the depth of neutral buoyancy. The determination of this depth was based on fluorescence measurements. The samples were taken at 4, 6, 13, 28, 58, 78 and 103 m downstream from the platform. At each distance there were three sampling sites. Effort was made to locate the sampling sites to the centre of the plume. Approx. 30 samples were taken at each distance from the release, and these were fitted to a log-normal distribution. The measured concentrations, given as a percentage of the release concentration, can be summarised in table 2.

The variation of the measuring depths are shown in table 3. "Shallow" and "Deep" denote the upper and lower limit for the vertical level of the measurements.

Model Setup

The model was set up in four experiments where some model parameters were varied from one experiment to the other. Table 4 shows the parameters to be varied in the experiments.

Experiment C is the experiment that is supposed to correspond best to the field experiments. Due to some uncertainty regarding the ambient flow direction which changes with depth, experiment A is considered to take care of the somewhat reduced flow component in the downstream direction.

As the density of the released water was not given directly, and the temperature and salinity are outside the range of the common formulas for computing density of sea water, experiment B is considered to point out how sensitive the results are to changes in this density.

Table 3. Range of plume centre depths observed at various distances from discharge

	Distance from release (m)						
	4	6	13	28	58	78	103
Shallow	7.6	7.6	7.7	8.0	9.6	8.3	7.9
Deep	9.0	9.0	9.2	10.2	10.2	13.0	9.3

Table 4. Parameters used in release simulations compared with field experiments

Experiment	Release density, ρ (kg/m³)	Ambient velocity, U_a (m/s)	Entrainment parameterization
A	1058	0.15	Sørgård
B	1050	0.20	Sørgård
C	1058	0.20	Sørgård
D	1058	0.20	Fisher

Experiment D uses the same basic parameters as experiment C, but in this experiment the entrainment coefficient does not include the effect of the ambient flow (Sørgård, 1992b), but uses the formulation proposed by Fisher (1979).

Results

The results of the experiments are given as concentration values at various distances from the release, table 5, and as computed depth of neutral buoyancy and plume diameter at 103 m downstream.

Table 5. Computed concentrations (% relative to initial concentration) in the plume centre at various distances for the experiments A-D

	Distance from release (m)						
	4	6	13	28	58	78	103
A	2.02	1.25	0.57	0.42	0.31	0.28	0.25
B	3.00	1.90	0.83	0.58	0.44	0.39	0.36
C	2.57	1.61	0.69	0.48	0.36	0.32	0.29
D	9.17	6.15	2.81	1.35	0.77	0.62	0.53

DISCUSSION AND CONCLUSION

Sensitivity of Model Parameters

The results of the sensitivity analysis show that the results are most sensitive to $\rho_a(z)$, the vertical density profile of the ambient fluid. The second most important factors are the entrainment parameter α and the density of the discharge, ρ_i. This parameter is more important for plume driven flows than for jet driven flows.

Table 6. Computed depth of neutral buoyancy, plume diameter, dilution factor and concentration at 103 m downstream

Experiment	Depth of neutral buoyancy	Plume Diameter	Dilution factor	Relative concentration %
A	8.6	7.85	394.4	0.25
B	7.6	5.75	280.2	0.36
C	8.1	6.36	343.9	0.29
D	16.3	4.51	187.7	0.53

Orientation of the release tube, i.e. the vertical angle θ, is only important for stagnant flows.

The release velocity and the ambient velocity are of equal importance for the results.

Comparisons with Field Experiments

The results from the simulations of the field experiments (Model case C) show good agreement with the observed results. The depth of neutral buoyancy correspond well to the observations. The concentrations compare very well with the observed ones at mid-range, but at the most distant location, 103 m, the concentration is 3 times higher than the observed. This suggests that the BJET model does not take fully into account the turbulence in the ambient flow.

The two simulations with modified discharge density (Exp. A) and ambient flow (Exp. B) show that these parameters do not change the results dramatically. The concentrations are higher, and the levels of neutral buoyancy are closer to the surface.

The simulation in which the entrainment coefficient does not take into account the ambient flow (Exp. D) show a significant difference from the three other simulations. The concentrations are now 2-5 times the observed ones at all observation points, and the depth of the plume is now 16.3 m below the surface at 103 m. This clearly shows that the plume does not mix sufficiently with the ambient flow.

The conclusion of these comparisons is that the model compares well with the observations. The importance of a correct formulation of the entrainment coefficient is clearly demonstrated. The comparisons also show that the concentrations are too high when the flow approaches the far field zone, i.e. the plume velocity approaches the ambient velocity.

Limitations of the Model

The present version of the model considers only the density in the mixing process, not the salinity and temperature. This does not take into account that the density is a non-linear function of temperature and salinity. This is important when dealing with produced water, which very often has high temperatures and large salinity. High temperature leads to decreased density, and large salinity lead to increased density.

The model does not take into account horizontal variations of the ambient flow. Behind Gravity Based Structures (GBS) a local flow field may develop, consisting of eddies generated by the GBS. These eddies are usually of the same scale as the structure itself. In such cases it is difficult to determine the centre for the plume, and whether these eddies should be considered as a part of the "mean" flow or as a part of the turbulent flow. The releases of large scale eddies is not encountered to the same degree with jacket platforms, and the scale of the eddies are much smaller since the flow can go "through" the structure.

Another aspect to consider with this type of model is the influence of surface and bottom. The model terminates when the plume centre reaches either the surface or the bottom.

Recommendations

The results show that the BJET model is applicable for computing concentrations and vertical levels of produced water plumes provided the ambient flow is sufficiently linear. It is also of importance to find the exact plume centre when conducting measurements. The concentrations predicted by Gaussian models will decrease significantly at some distance away from the plume centre.

ACKNOWLEDGEMENTS

The model experiments described herein were computed by Mr. Dag E. Hagen while preparing his M.Sc. thesis during the summer 1994. The data from the field experiments described were kindly provided by Mr. Geir Indrebø of Exxon (Norway). The author has also benefited from discussion with Drs. Eirik Sørgård and Ketil Skogen.

REFERENCES

Fan, L., Turbulent buoyant jet into stratified or flowing ambient fluids, Report no. KH-R-15, California Institute of Technology, Pasadena, California, 196 pp.,1967.

Fisher, H.B., List, E.J., Koh, R.C.Y, Imberger, J., Brooks, N.H., Mixing in inland and coastal waters, Academic Press, 483 pp., 1979.

Hagen, D. E., 1994: Sensitivity analysis of the BJET program. M.Sc. thesis. The Norwegian Institute of Technology, Trondheim (in Norwegian).

Lee, J. H. W. & Cheung, V. , 1991. Mixing of buoyancy-dominated jets in a weak current. *Proc. Instn Civ. Engrs*, Part 2, 91, Mar., 113-129.

Nilsson, J. R. , 1991. Innlagring av oppdriftsstråle, M.Sc. thesis. The Norwegian Institute of Technology, Trondheim (in Norwegian). 157 pp..

Smith, J.P. , Mairs, H. L. , Brandsma, M.G, Meek, R.P.& Ayers Jr., R.C. , Field Observations of Produced Water Dilution: Comparison With Dispersion Model Predictions, Report prepared for the Offshore Operators Committee, 1993.

Sørgård, E., 1992, BJET: The subsea buoyant jet model, Det Norske Veritas Research AS, Tech. rep. no. 92-2055.

Sørgård, E. , 1992., Validations and suggested improvements for the BJET jet/plume model for subsea releases, Det Norske Veritas Research AS, Tech. rep. no. 92-2050.

Thorpe, S. A., 1982. On the layers produced by rapidly oscillating a vertical grid in a uniformly stratified fluid. *J. Fluid Mech., 124, 391-409.*

23

AN INTERCOMPARISON BETWEEN FIELD MEASUREMENTS AND THREE DIFFERENT CALCULATION MODELS FOR ESTIMATES OF DILUTION FACTORS

Henrik Rye,[1] Mark Reed,[1] Alf Melbye,[2] and Ståle Johnsen[2]

[1] IKU Petroleum Research
SINTEF, Trondheim, Norway
[2] STATOIL
Trondheim, Norway

INTRODUCTION

Produced water releases have recently received increased attention due to its potential toxicity and its expected increase in the amounts on the Norwegian continental shelf for the years to come. Factors that govern the toxicity of the release when mixed into the recipient water becomes therefore essential. This paper deals with one such factor, namely the mixing rate of the release. When sufficiently mixed with the ambient water, the concentrations of the toxic components will fall below the "No Effect Level" (NEC level), and harmful effects from the release are no longer expected. It is therefore of interest to determine how far off the release site (and how fast) this NEC level is surpassed for produced water releases.

The mixing of the release in the recipient may be expressed through dilution factors. In the CHARM model, the dilution factor is assumed to be 1:1000 at a distance of 500 m from the release site in the hazard assessment mode (for description of the CHARM system, see papers elsewhere in this proceedings). In the CHARM risk analysis mode, a dilution formula is included which is identical to the analytical model given in this paper. The resulting differences in the dilution using the various formulas may be several orders of magnitude. It is therefore important to clarify what factors or formulas that should be used in order to assess a proper dilution factor.

Various numerical models have been proposed to compute the dilution at different distances from the source. Sometimes models are used without proper verification against field data. This paper considers results from measurements of dilution factors on the Oseberg field in September 1993 and their implications for the use of numerical models to calculate the dilution of produced water releases. The paper compares results from three different models with the field experiences in order to arrive at what factors should be included in the models to reproduce measured concentrations fairly well.

Produced Water2: Environmental Issues and Mitigation Technologies
Edited by M. Reed and S. Johnsen, Plenum Press, New York, 1996

The results from the field trials are summarized in the OCEANOR report written by Thendrup et. al., 1994. The comparisons are mainly based on the results as presented in the OCEANOR report.

COLLECTION OF FIELD DATA

A proper data collection program is important for an intercomparison with field data. The plume generated from the release is generally narrow and also sometimes confined to only some depth interval only. Field trials based on sampling of water masses only may therefore give results with a limited value because the plume may be located outside the recording spots.

The field trials were conducted by adding a tracer to the produced water release. For the Oseberg Field Center case, 150 kg tracer (Basonyl Rot 555 Flussig, 40 % solution) was injected continuously in the discharge pipeline together with 75 l Red spirit during a time span of 18 hours. The release is located at 20 m depth. The injection started at 2. Sept. 1993 at 2200 and stopped at 3. Sept. 1993 at 1600.

The concentrations (or dilutions) in the recipient may then be checked upon by recording the diluted tracer along transects or cross sections downstream or in the vicinity of the release site. The concentrations of the tracer was measured with an in situ fluormeter, manufactured by Q-instrument I/S, Denmark. Concentration levels down to order 10^{-12} g/cm^3 (or order 0.001 ppb) are assumed to be detectable, provided that the temperature variations are not too large.

At the same time, the release rate of the produced water, the winds, waves and currents at various depths as well as hydrography were recorded in order to the control the environmental conditions during the tracer release.

The models applied for simulating the concentrations in the water is mainly based upon the assumption of a continuous release. As a result of this assumption, the flux of release through the oulet opening will be equal to the flux of release through a vertical section encircling the release point not too far away from the platform. The distance away from the source where this condition will be fulfilled will be somewhat dependent on the time variability of the currents on the site, but within the order of some hundred meters (or lower than one kilometer from the platform) will in most cases be a reasonable distance for the fulfillment of this condition. This means that it is possible to check the fulfillment of continuity (or to what extent the tracer is conserved) by inspection of the concentrations recorded along a path encircling the platform.

In general, the condition of continuity for the tracer release may in principle be formulated as

$$q = Q_0 c_0 = r \int_0^{2\pi} U(\theta) H(\theta) c(\theta) d\theta \qquad (1)$$

where

$q =$ flux of tracer through the outlet opening (g/s)
$U =$ current velocity (m/s)
$H =$ height of the plume diluted in the sea water (m)
$c =$ concentration of tracer measured in the recipient (g/m^3)
$Q_0 =$ the release rate through the outlet opening (m^3/s)
$c_0 =$ concentration of tracer in the outlet opening (g/m^3)

and the integration is performed along a path (θ) encircling the platform. r is the distance from the platform where the integration is performed. Both U and c may be considered as averaged over the depth interval H where concentrations are larger than zero.

THE OSEBERG CONCENTRATIONS

Two different experiments were carried out, one at the Oseberg field and one at the Statfjord field. It turns out that the results from the Oseberg trial are reasonable consistent with the continuity requirement outlined in the precious chapter. Therefore, the Oseberg results are used to compare results from calculations with the measured concentrations.

The conservation of continuity may be checked upon by noting that the width of the plume (L) times the vertical extension of the plume (H) times the current velocity away from the release site (U) times the concentration in the plume (c) must be equal to the flux of the tracer (q) through the outlet opening. Or, to put it in a more formal manner:

$$q = ULHc = Qc = Q_0c_0 \tag{2}$$

where

q =	flux of tracer through the outlet opening (g/s)
U =	average current velocity (m/s)
L =	width of the plume diluted in the sea water (m)
H =	height of the plume diluted in the sea water (m)
c =	concentration of tracer measured in the recipient (g/m^3)
Q =	the amount of sea water made available for mixing (m^3/s)
Q_0 =	the release rate through the outlet opening (m^3/s)
c_0 =	concentration of tracer in the outlet opening (g/m^3).

Equation (2) may be considered as an approximation to Equation (1). The dilution is then defined as:

$$\text{Dilution} = \frac{Q}{Q_0} = \frac{c_0}{c} = \frac{ULH}{Q_0} \tag{3}$$

The relation

$$\frac{c_0}{c} = \frac{ULH}{Q_0} \tag{4}$$

has the property that all parameters are known (more or less) from the field trial. U, L, H and c are all measured in the field, while C_0 and Q_0 are both known from the dosage of the release. The Equation (4) can then be used to check whether the principle of continuity (or mass conservation of tracer) of the release is fulfilled or not (assuming a conservative tracer).

For the Oseberg case, the discharge rate in total is estimated to be 0.007 m^3/s (Q_0). The amount of tracer was close to 2.31 g/s (q), which means that the concentration at the outlet opening is close to 334.5 g/m^3 or ppm (c_0) for the tracer. All numbers referring to dilution are related to the concentration at the outlet.

The release velocity on the Oseberg field was rather low, order 0.007 m/s. The outlet diameter is rather large (1.168 m) and the release flux rather low (0.007 m^3/sec) so that considerable mixing must be anticipated to take place inside the outlet pipe as well. The near-field mixing outside the outlet opening is therefore relatively moderate and may be

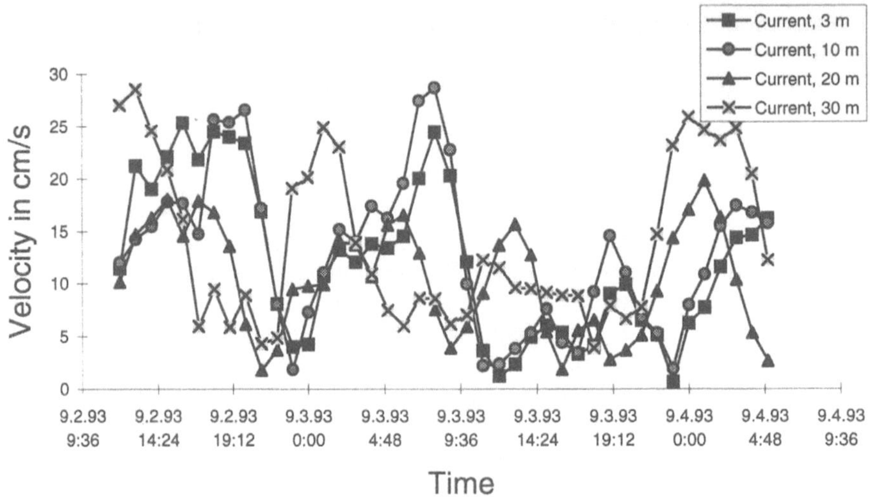

Figure 1. The current velocities (in cm/s) at the Oseberg site recorded at 3, 10, 20 and 30 m depth during the field trial. Time of track is located between 93 09 03, 14:24 and 93 09 03, 19:12.

assumed not present at all. The release merely behaves like a passive tracer added to the recipient in the release depth of 20 m at the Oseberg field. In the following, all influences from near-field mixing are therefore neglected.

The Oseberg measurements were made at 3. September 1993 at 1600 - 2000. Winds were from NNE with velocities of order 10 - 12 m/s. Currents at this date are shown in Figures 1 - 2 (velocity and direction, respectively).

Different sections were recorded, located between 0.6 and 3.9 km from the source. Maximum concentrations in the plume were found within the range 2.0 - 0.1 ppb. The time period of the recording appears to coincide with a time period where the currents were

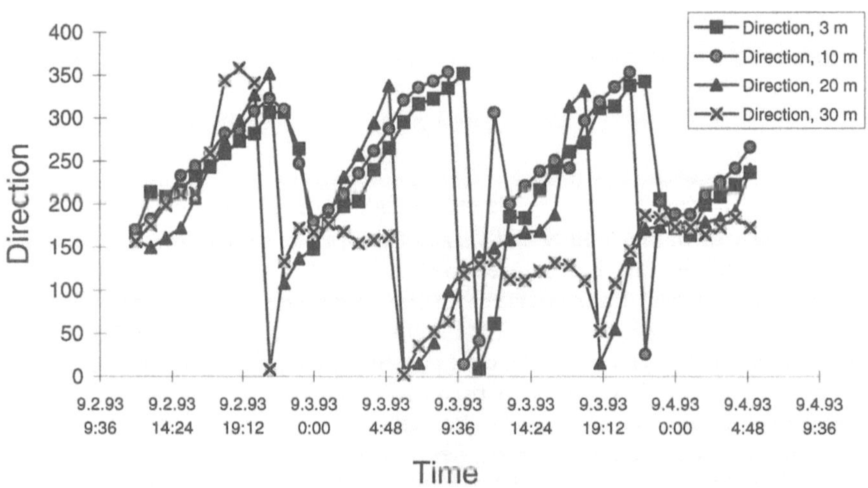

Figure 2. The current directions (0 - 360 deg) at the Oseberg site recorded at 3, 10, 20 and 30 m depth during the field trial. Time of track is located between 93 09 03, 14:24 and 93 09 03, 19:12.

Table 1. Dimensions of the plume and maximum concentrations recorded at 3. September 1993 at the Oseberg field at 1600 - 2000 p.m. Currents (U) = 0.05 m/s. Initial flux of release (Q_0) = 0.007 m^3/s. Flux of tracer (q) = 2.31 g/s. Initial concentration (c0) = 334.5 ppm or g/m^3. No near-field mixing

Section No.	Distance from the source x (m)	Max. concentrations c (ppb)	Height of plume H (m)	Width of plume L (m)
1	600	1 - 2	30 - 35	600 - 700
2	800	1 - 2	35 (?)	600 - 700
3	2000	0.5 - 1.0	25 - 30	500
4	2400	0.1 - 0.5	> 30	600
5	3300	0.1 - 0.2	30	> 600
6	3900	0.1 - 0.2	15 - 20 (?)	> 600

reasonably uni-directional (towards WSW) and of uniform strength (typically 0.05 m/s) above 20 m depth or so, see Figures 1 and 2. The following table gives an indication of the dimensions of the plume recorded.

The numbers indicate a width of plume (L) of order 600 m and a vertical extent of the plume (H) of order 30 m. The current velocities (U) in the recording period were typically 0.05 m/s. Flux of recipient water is therefore of order (600 m × 30 m × 0.05 m/s) = 900 m^3/s (estimate of Q). Assuming conservation of tracer, the concentration (c) in a plume with a flux of order 900 m^3/s becomes, on the average, 2 - 3 ppb (using Equation 4). This number seems also the be of the same order-of-magnitude, but somewhat larger than what was recorded in the field within the first two kilometers from the release site. For larger distances, the deviation is larger.

Judged from the field data, the flux of tracer away from the platform seems therefore to be somewhat lower than the flux of release of tracer through the outlet opening. It may therefore be that the tracer is not conserved, but reduces somewhat in the field. Probably the tracer is not conservative, but is absorbed by, or adsorbed to, other constituents.

The actual dilution can be found by taking the ratio between the measured concentrations (c) and the concentration in the outlet opening (c_0). For c values in the area 2.0 -0.1 ppb, the dilution becomes larger than 1:100 000. For a 0.1 ppb concentration, the dilution is of order 1:3 mill.

COMPARISON WITH THE ANALYTICAL APPROACH

In this chapter a comparison with an analytical approach is given.
Equation (3) or (4) can be rewritten

$$c = c_0 \frac{Q_0}{ULH} \tag{5}$$

which determines the concentration in the recipient as a function of the initial conditions (c_0, Q_0) and the dimensions of the plume (U, L and H). In Rye et. al. (1995), an analytical model is proposed which makes it possible to evaluate the numbers for the dimensions of the plume. Assuming that the current velocity (U) is known, the width (L) and the height (H) can be estimated from

$$2\sqrt{3}\,\frac{V}{U}X \tag{6}$$

$$H = \sqrt{8K_Z\frac{x}{U}} \tag{7}$$

where

x = distance from the source (m)
V = horizontal diffusion velocity (m/s)
K_z = vertical diffusion coefficient (m^2/s).

The parameter V expresses the lateral spread of the release (the sidespread). The parameter K_z expresses the ability for the ambient water of vertical mixing. Further details of these parameters and the choice of numbers are given in Rye et. al. (1995).

Combining the Equations (5), (6) and (7), we obtain the dilution formula as presently used in the CHARM risk assessment mode:

$$\text{Dilution} = \frac{c_0}{c} = \frac{ULH}{Q_0} = \frac{V}{Q_0}\sqrt{96K_z\frac{x^3}{U}} \tag{8}$$

which then expresses the dilution as a function of the 5 parameters V, Q_0, K_z, x and U.

Assuming representative numbers during the Oseberg trial for the current velocity (U = 0.05 m/s), the vertical diffusion coefficient ($K_z = 0.01\ m^2/s$ for a 10 m/s wind velocity) and a reasonable horizontal (lateral) diffusion velocity (V = 0.013 m/s), we obtain the following numbers for the concentration (c), the height (H) and the width (L) of the plume as a function of distance (x) from the source ($Q_0 = 0.007\ m^3/s$).

We note that the average concentrations appear to be within the same range of magnitude as what were measured in the field (order 2.0 - 0.1 ppb). Also note that the dilution of order 500 m from the source is two orders of magnitude larger than 1:1000.

On the other hand, the depth of the plume is very much larger than what was measured. The reason for this is probably that the analytical model does not take into account the effect of stratification in the water masses at 25 - 30 m depth. This stratification will tend to damp out the vertical turbulence, and then reduce the vertical diffusion coefficient down to considerably below the 0.01 m^2/s value that was used in the calculations. This would, in turn, reduce the vertical extent of the plume in the calculations.

The measurements indicated that the currents in 30 m depth were both stronger and also had another direction than the currents above. It may therefore be that the part of the release that diffused downwards was catched up with this underwater current and then simply

Table 2. Dimensions of the plume and average concentrations calculated by means of an analytical model. Currents (U) = 0.05 m/s. Vertical diffusion coefficient (K_z) = 0.01 m^2/s. Horizontal (lateral) diffusion velocity (V) = 0.013 m/s. Initial flux of release (Q_0) = 0.007 m^3/s. Initial concentration (c_0) = 334.5 ppm or g/m^3. No near-field mixing

Distance from the source (x) m	Dilution factor (c_0/c) -	Average concentrations (c) ppb	Height of plume (H) m	Width of plume (L) m
600	120,000	2.8	31	540
800	185,000	1.8	36	720
2000	733,000	0.46	57	1,800
2400	955,000	0.35	62	2,160
3300	1,550,000	0.22	73	2,970
3900	2,000,000	0.17	79	3,510

advected away in another direction (mainly towards E). This would in fact explain the very low concentrations measured below 30 m depth, and also in part explain why the amount of tracer was apparently lower in the water masses, as compared to what was released through the outlet opening (continuity). In fact, later simulations (with the PROVANN model), which are able to include effects from vertical shear in the currents, indicate that this was indeed the case for the Oseberg event. The results from these simulations are considered further below.

COMPARISON WITH THE EULERIAN NUMERICAL MODEL

The Eulerian numerical model is characterized by a continuous, stationary approach with no particles involved. It merely models the concentrations from a passive release, given the vertical velocity profile, the horizontal and the vertical diffusion coefficients and the amount of release. The Eulerian model may be considered as an extension of the analytical model in the sense that the parameters U and K_z are both allowed to vary in the vertical dimension. Thus, it is possible to include effects from stratification on both velocity profile and diffusion coefficients as well.

Outline of the model approach can be found in Rye (1995). A more comprensive report on the model approach applied to surface oil spills is given in Rye (1993). A brief outline is given below.

The spreading process in the ocean is modelled by means of a numerical solution of the differential equation

$$U \frac{\partial c}{\partial x} = \frac{\partial}{\partial z}\left(K_z \frac{\partial c}{\partial z} \right) + \frac{\partial}{\partial y}\left(K_h \frac{\partial c}{\partial y} \right) \tag{9}$$

where

x = horizontal coordinate directed downstream
y = lateral horizontal coordinate normal to x
z = vertical coordinate
U = U(z) = water current velocity in the x direction
c = c(x,y,z) = concentration of release or tracer
K_z = $K_z(z)$ = vertical diffusion coefficient
K_h = $K_h(x)$ = horizontal diffusion coefficient.

In addition, an initial distribution of c(0,y,z) has to be given. This distribution is given by the release conditions. The influx concentration is assumed distributed laterally and verticaly across the outlet opening. Once the influx is given, the tracer will be redistributed in the water masses due to horizontal turbulence (expressed through the horizontal diffusion coefficient K_h), vertical turbulence (expressed through the vertical diffusion coefficient K_z) and effects from vertical variation in the horizontal velocity profile (expressed through the vertical variation in the current velocity U). The equation gives the concentrations c calculated in space (x,y,z) downstream from the source. Stationary conditions are assumed, the flux of release as well as wind and current conditions are all assumed not to vary in time during the simulation period.

Although the Equation (9) is three-dimensional, it may be reduced to a set of 2 two-dimensional equations which are then solved simultaneously. This is done by integrating analytically the Equation (9) in the y (lateral) direction, assuming that the tracer (or hydrocarbon) concentration in the y direction follows a normal distribution function (a

Gaussian plume model assumption in the lateral direction). The two equations to be solved represent then one equation for the laterally integrated content of tracer in the water masses and one equation for the lateral variance of the tracer content. The actual three- dimensional distribution is then obtained from the normal distribution function. This function describes then the concentration in the y direction, while the parameters in the normal distribution function (the average concentration and the variance) will vary with x and z only. This technique thus reduces the problem from a three-dimensional one into an integration in the x- and z- space only. The two equations derived from (9) is solved numerically by means of a differential scheme (Crank-Nicolson).

For the Oseberg case, the near-field mixing can be neglected. In fact, for the case to be considered, the release velocity (0.007 m/s) is considerably lower than the current velocity (0.05 m/s or larger). The release is therefore assumed to be located at the release depth, even though the density of the release (1022 kg/m^3) is somewhat lower than the ambient (1026 kg/m^3). The observations indicate, however, that the boyancy effects of this plume are negligible.

In order to verify the model calculations, the field case on the Oseberg trial conducted at 3. September 1993 was compared to model simulation results. The results from the simulations are shown in the Figures 3 and 4. Figure 3 shows a vertical section of the plume, while Figure 4 shows the horizontal extent. Currents (U) are assumed equal to 0.05 m/s at all depths, while the vertical diffusion coefficient (K_z) is assumed equal to 0.01 m^2/s close to the surface and decreasing down to 0.0001 m^2/s below 30 m depth. Horizontal (lateral) diffusion velocity (V) was assumed equal to 0.013 m/s. The relation between the horizontal diffusion velocity (V) and the horizontal diffusion coefficient (K_h) is given by Rye et. al. (1995), which states that

$$K_h = \frac{V^2}{U} x \tag{10}$$

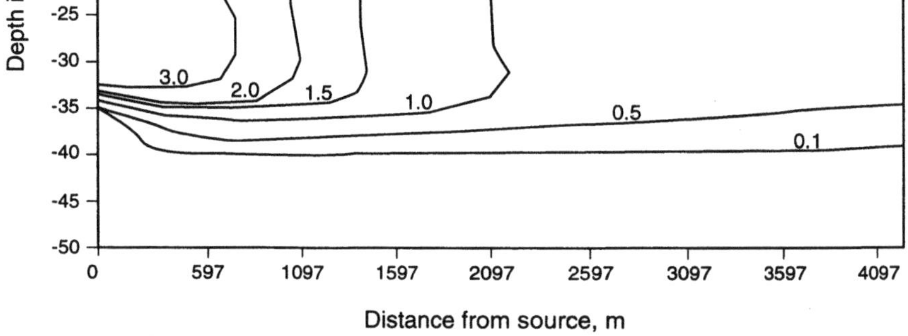

Figure 3. Numerical simulation of the release of tracer at the Oseberg field trial at 3. September 1993. Simulations are based on a numerical solution of Equation (9). Concentrations in ppb. Vertical cross section across the center of the plume. Release of tracer (q) = 2.31 g/s at 20 m depth. Initial flux of release (Q_0) = 0.007 m^3/s. Initial average concentration (c_0) = 334.5 ppm or g/m^3. No near-field mixing. Current velocity (U) = 0.05 m/s. Diffusion coefficients (K_z and K_h) are variable.

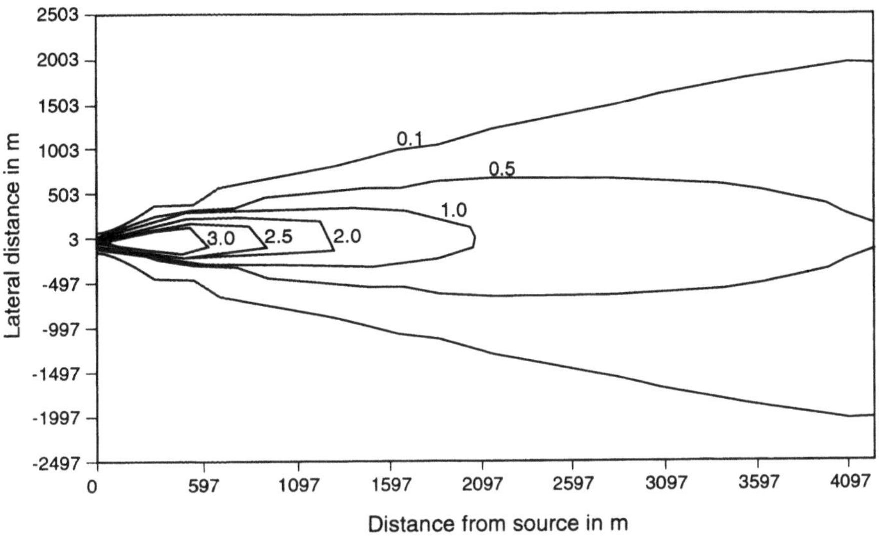

Figure 4. Numerical simulation of the release of tracer at the Oseberg field trial at 3. September 1993. Simulations are based on a numerical solution of Equation (9). Concentrations in ppb. Horizontal cross section across the center of the plume at 20 m depth. Release of tracer (q) = 2.31 g/s at 20 m depth. Initial flux of release (Q_0) = 0.007 m^3/s. Initial average concentration (c_0) = 334.5 ppm or g/m^3. No near-field mixing. Current velocity (U) = 0.05 m/s. Diffusion coefficients (K_z and K_h) are variable.

The horizontal diffusion coefficient is therefore not a constant, but is varying with the distance (x) downstream from the source (Bowden, 1983).

We note that the concentrations simulated are somewhat larger than what the measurements indicate. In the distance interval 600 - 4000 m from the source, the maximum concentrations are calculated to be decreasing from 3 to 0.5 ppb, while those observed were within the interval 2 - 0.2 ppb. However, the deviations are within a factor of 2 - 3. Some of the explanation may be that in the simulations, the vertical diffusion is reduced in the 25 - 35 m depth interval so that the tracer does not diffuse across this interval. The tracer is therefore "trapped" above this depth interval. We know, however, that the currents were changing both strength and direction in this depth interval. The tracer may therefore have been advected away in another direction. This would have reduced the concentrations in the upper layer. The Eulerian model does not account for this effect. This may explain why concentrations are somewhat larger in the model calculations than in the field experiment.

However, the deviations are within the factor of 2 - 3, which is considered to be acceptable. The deviations may also originate from the fact that the tracer is not conservative, but is subjected to some (unknown) decay.

The width of the plume (Figure 4) is of the same order-of-magnitude as the distance from the source. This result is consistent with the width of the plume calculated by means of the analytical model, see Table 2.

COMPARISON WITH THE PROVANN PARTICLE BASED MODEL

The PROVANN model is a three-dimensional model which is able to simulate the spreading of particles in three dimensions for a time-varying wind input. The particles are spread due to vertical and horizontal diffusion and advected away due to residual currents, tidal currents and the wind induced Ekman current. The concentrations of the release are estimated from the calculated particle density of the release at various distances from the source.

The model is based on solving the generalized transport equation:

$$\frac{\delta c}{\delta t} + \vec{V} \cdot \vec{\nabla} c = \vec{\nabla} \cdot K \vec{\nabla} c + \sum_{j=1}^{n} r_j c \qquad (11)$$

where

c	=	concentration in the water masses,
t	=	time,
$\vec{V}$	=	advective transport vector,
$\vec{\nabla}$	=	gradient operator, and
K	=	turbulent dispersion coefficient.

The objective in solving Equation 11 is to produce for a biological effects model a spatial time history of the distribution of a contaminant in the environment. The first term on the left hand side is the temporal rate of change of the concentration at a particular location in space. This rate of change is computed by solving the other terms in Equation 11: The second term on the left (the advective transport term); the first term on the right (the diffusive transport term); the series of process specific terms at the end of the equation.

The terms r_j in Equation (11) are process rates, including:

- addition of mass from continuous release,
- evaporation from surface,
- dissolution from oil droplets,
- deposition from water column to bottom sediments,
- degradation (surface, water column, shoreline, sediments).

Equation 11 is central to the model in that it summarizes all the processes simulated by the model. The algorithms used to simulate the spreading and dilution are described in some more detail below.

The model applied here is a version of the more general PROVANN model which has been modified in order to simulate dilution and dissolution of produced water in particular. In the actual simulations, fenol is used as a "tracer" because of its large solubility in water. Further details of the more general model approach can be found in Reed et. al. 1995 (short version) and in Reed et. al. (1992) for a more extensive outline of the PROVANN model.

The location of each particle released is determined as a sum of 6 different contributions which are as follows.

For each time step, the next position of the particle is determined from:

- Advection due to the residual current (not time varying, constant with the depth)
- Advection due to the tidal currents (time varying, constant with the depth)
- Advection due to the wind-induced Ekman transport (dependent on the time-varying wind, varying with depth in both strength and direction)
- Statistical spread in the vertical direction due to vertical turbulent diffusion.
- Statistical spread in the horizontal direction (two directions) due to horizontal turbulent diffusion.

Transport currents in the model include mean flow and tides, and wind-driven currents. A wind-driven "slab flow" model (Reed, 1980) has been incorporated to supply wind-driven transport currents. The hydrodynamic equations of motion are averaged over d, the depth of the wind-driven layer. The non-linear acceleration terms are neglected, and the equations of motion become

$$\frac{du}{dt} = fv + \frac{\tau_x^S}{\rho_w d} - \frac{\tau_x^B}{\rho_w d} \tag{12}$$

$$\frac{dv}{dt} = fu + \frac{\tau_y^S}{\rho_w d} - \frac{\tau_y^B}{\rho_w d} \tag{13}$$

where

u, v	=	horizontal current velocity components
$\tau_{x,y}^{S,B}$	=	horizontal stress in x or y direction at surface (S) of bottom(B)
f	=	coriolis parameter
ρ_w	=	density of water
d	=	depth of the wind-driven layer
d/dt	=	time derivative

Horizontal shears in the current field contribute to dispersion of contaminants in the modeled water column. Additional dispersion is parameterized through horizontal diffusivity parameters, K_x and K_y. For the particle based approach, the horizontal diffusivity parameters are both related to the time that has elapsed since the release of the particle according to the relation (Bowden, 1983) :

$$K_x = K_y = 1.17\ 10^{-6}\ t^{1.34} \tag{14}$$

which is valid for $K_{x,y}$ in m^2/s and t in seconds.

The vertical diffusion coefficient above the pycnocline layer is related to the wind velocity Wind (Thorpe, 1984):

$$K_z = 0.0015\ \text{Wind} \tag{15}$$

where Wind is given in m/s and K_z in m^2/s.

A particle based random walk algorithm (Reed, 1980) is used to simulate both horizontal and vertical dispersion in the water column. Particles diffuse with velocities v_i,

$$V_i = R^* \sqrt{6K_i/\Delta t} \tag{16}$$

where the subscript i = 1,2,3 corresponds to the horizontal and vertical directions x,y,z, K_i is associated diffusivity, Δt is the model timestep, and R^* is a random variate uniformly distributed over the interval $-1.0 \leq R^* \leq 1.0$. The value of K_z is computed from Equation (15), and the values of K_x and K_y are computed from Equation (14). The concentrations of contaminant averaged over subintervals of the total mixing depth are then used to compute an exposure field in the water column.

For the application on the Oseberg trial, the currents measured on the site were used as a basis to specify the most realistic input. Near-field mixing is neglected. The particles are released at 20 m depth on the Oseberg location.

This model has the ability to include vertical variations in the currents as well as time variations. The current conditions measured at the site as shown in the Figures 1 and 2 were used as a basis for specifying the current input. A residual current at 30 m depth equal to 0.1 m/s directed towards E were superposed on an Ekman-driven current generated from the

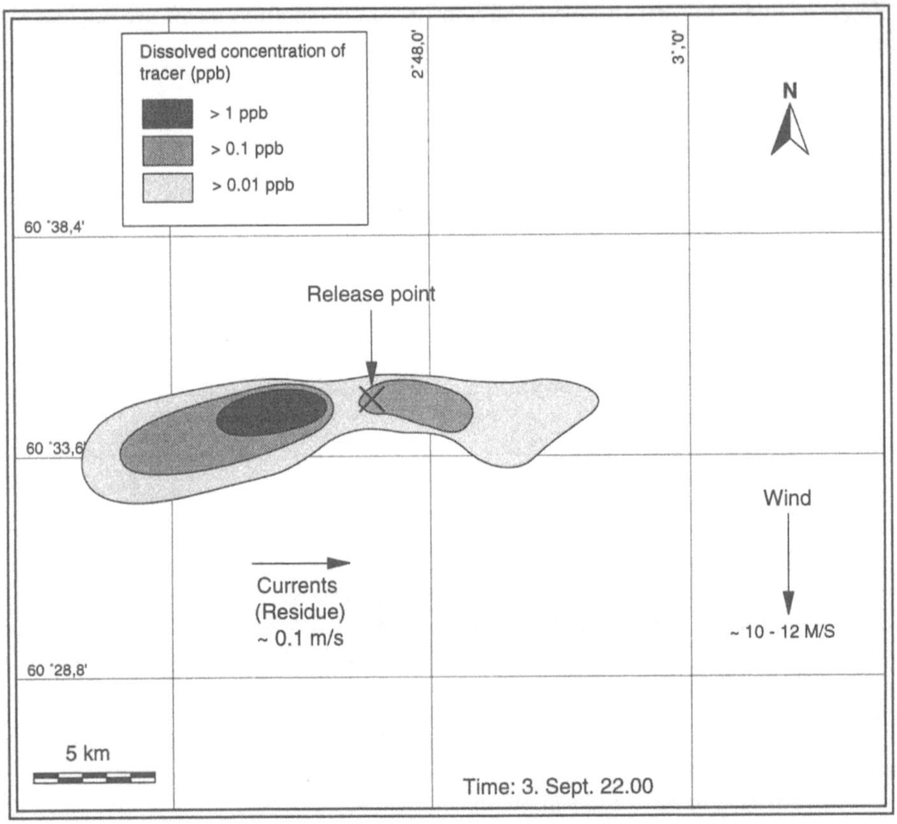

Figure 5. Release of tracer at 20 m depth at the Oseberg field in the North Sea. Release conditions as for the Oseberg trial. Winds from N of order 10 m/s. Residual currents towards E, 0.1 m/s. Pycnocline depth 25 m. Snapshot of the release at about 24 hours after the initiation of the release.

prevailing winds from N. The resulting currents show contributions close to the surface towards W and S, while under the pycnocline (assumed located in 25 m depth) the prevailing currents are directed towards E.

The actual release of the tracer was then simulated. An example result is shown in Figure 5. The concentration levels are shown on the figure. The concentrations are generally below 1 ppb but occasionally above 1 ppb, as measured. Also, the figure shows some of the release advected towards the E, which is interpretated as some of the droplets have penetrated into the lower layer (below 25 m depth), where the currents are towards E. The droplets are then advected away with this current. The results obtained with this model turn out to give the best fit, both in reproducing concentration levels, but also to represent the geographic spread of the release around the release point.

CONCLUSIONS

Field data on the dilution of produced water released from the Statfjord and Oseberg oil fields are evaluated and compared to different analytical and numerical models.

The Statfjord field data were found not to be suitable for comparison with model calculations. The Oseberg data were however found to be reasonably consistent with continuity requirements.

Three different models were used for the comparison with the field data. An analytical model approach, one Eulerian model for a continuous and stationary description of the release and one particle based approach (Lagrangian model) for the calculation of the time-varying spread of the release in the recipient.

Coefficients for the spreading values of the effluent in the models were all selected within the intervals of expected values found in the literature. All the three approaches (the analytical, the Eulerian and the particle-based approach) gave results that indicated concentrations levels which are of the the same order of magnitude as what was measured. Deviations from the field data in the dilution calculations were within order 2 - 3. However, it may also be that the tracer used in the field was not conservative, and that some of the deviations were caused by adsorbsion of tracer to other constituents in the water mass or inside the outlet arrangement.

Best fit was found with the use of the particle based PROVANN model, which also has the largest ability to reproduce actual conditions.

The actual dilution of the release on the Oseberg field trial was both measured and calculated to be of the order of 1:100 000 - 1:3 mill. It should be noted that this number is dependent on release rate and also the mixing and transport conditions in the recipient.

REFERENCES

Bowden, 1983: "Physical Oceanography of Coastal Waters". Ellis Horwood Ltd. John Wiley & Sons.

Reed. M., 1980. "An oil spill fishery interaction model: development and applications". Ph.D. Dissertation, Department of Ocean Engineering, U. Rhode Island, Kingston, R.I., 235 p.

Reed, M., et.al., 1992: "A Natural Resource Damage Assessment Model for Coastal and Marine Environments". Technical Documentation. Report prepared for Office of Environmental Affairs, U.S. Department of the Interior, Washington, DC 20240, USA. Applied Science Associates, USA.

Reed, M. et. al., 1995: "A Three-dimensional Oil and Chemical Spill Model for Environmental Impact Assessment". Paper presented at the 1995 Oil Spill Conference, Long Beach, California, 27 February - 2. March 1995.

Rye, H. (1993): "Concentrations of Oil in Water generated from Surface Spills". Technical Report, Det Norske Veritas, Norway. DNV Report No. 93-3674 dated 7. April 1993.

Rye, H. (1995): "A Multi-component Oil Spill Model for Dissolved Aromatic Concentrations". Paper presented at the 1995 Oil Spill Conference, Long Beach, California, 27 February - 2. March 1995.

Rye et. al. (1995): "Dilution Factors for Produced Water Releases". 6th Int. Symp. on Oil Field Chemicals, 19-22 March 1995, Geilo, Norway. Edited by NIF, Norway.

Thendrup et. al., 1994: "Field Study of Dispersion of Produced Water. Data Report". OCEANOR Report OCN R-94011 dated 11. March 1994.

Thorpe, 1984. On the determination of K_v in the near-surface ocean from acoustic measurements of bubbles. J. Phys. Ocean. 14:855-863.

Section V

ENVIRONMENTAL MODELLING

24

SESSION SUMMARY FOR SECTION V

Environmental Modelling

Mark Reed

Environmental modelling provides a logical and internally consistent framework for evaluating and comparing alternative management strategies. The papers presented in this session reflect distinctly different approaches to the development and use of models. The first paper, by Edgar Furuholt of Statoil, describes a method for risk analysis of marine discharges using the concept of dose. The subsequent three papers, respectively by Albert Bos of Servo Delden, Gro Øfjord et al. of Aquateam, and Henk Schobben et al of TNO, all deal with the CHARM (Chemical Hazard Assessment and Risk Management) model. CHARM was developed to provide a simple screening tool for the use and discharge of chemicals offshore, and relies solely on the comparison of potential environmental and no-effect concentrations, rather than actual exposure or dose. The final paper, by Reed et al. of IKU Petroleum Research represents an effort to provide realistic exposures and resulting internal body burdens for both passive and actively swimming marine organisms. Each of these modelling approaches has a clear place in the environmental manager's toolbox, providing risk assessment capabilities over a range of complexity and specificity.

The environmental effects of produced water can be divided into acute and chronic categories. Comparison of toxicity, field, and modelling study results discussed elsewhere in this volume support the contention that concentration levels necessary to induce acute effects may occur only within a few hundred meters of discharges from offshore platforms. This is true primarily because of very rapid vertical and horizontal turbulent mixing and dilution of discharge plumes. However, organisms will not generally remain for more than a few tens of minutes or at most hours within this near field zone. The exposure- and dose-based models are therefore likely to give more accurate estimates of environmental risk than models based solely on concentrations, whereas the latter fulfill the need for rapid and simplified screening.

The question which remains to be answered is whether or not the low levels of produced water constituents found over large areas of some of the world's most biologically productive seas, are likely to have any adverse ecological effects. Although a significant amount of research remains to be carried out in the area of low level chronic effects, this does appear to be the correct focus for future work. A research program which coordinates laboratory and field work with numerical modelling studies appears to provide the most promising avenue for success.

Produced Water2: Environmental Issues and Mitigation Technologies
Edited by M. Reed and S. Johnsen, Plenum Press, New York, 1996

ENVIRONMENTAL EFFECTS OF DISCHARGE AND REINJECTION OF PRODUCED WATER

Edgar Furuholt

STATOIL
Trondheim, Norway

INTRODUCTION

Since new, strict regulations for oil content on drill cuttings were introduced in 1993, produced water has become the major contributor of oil discharge to the sea from the Norwegian oil industry. As a concequence, much focus has been placed on possible acute toxicity of produced water and the identification of natural components and chemicals which contribute most to that toxicity.

The amount of produced water from the fields in the Norwegian part of the North Sea is steadily increasing as new fields come on stream and the water cut increases on older fields. In 1994, the discharge was approx. 34 Mm^3, and is expected to increase to around 90 Mm^3 in the year 2000 (OLF, 1995). Typical discharge volumes are from 500 m^3/day to 25,000 m^3/day from one platform. A typical produced water profile for a large field is shown in fig 1.

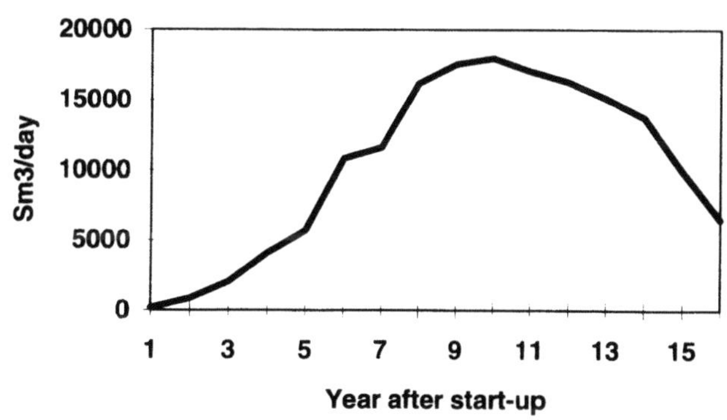

Figure 1. Produced water profile for a large field.

Produced Water2: Environmental Issues and Mitigation Technologies
Edited by M. Reed and S. Johnsen, Plenum Press, New York, 1996

DILUTION AND DEGRADATION OF PRODUCED WATER

Field tests have shown that the produced water is diluted more rapidly than earlier anticipated (Riksheim et al, 1994). These field tests have been used to calibrate computer models. A simplified formula for calculating the dilution is (Rye et al, 1995):

$$\frac{c(x)}{c_0} = \frac{Q_0}{V}\sqrt{\frac{U}{96x^3K_z^{\,2}}} \qquad (1)$$

where

c_0	concentration of a component in the discharge
$c(x)$	concentration at distance x
x	distance from discharge point
Q_0	discharge volume pr time unit (flux)
U	current velocity
K_z	vertical diffusion coefficient.(typically 0,01 m^2/s)
V	vertical dispersion (typically 0,013 m/s).

The dilution for two typical discharge rates is shown in fig. 2. The dilution is fairly rapid in both cases, but it is worth while to note that the concentration drops significantly slower when the discharge rate is high. According to formula (1), a 1000 time dilution is obtained approx 50 m from the discharge point for the low discharge rate and approx 250 m for the high rate. On some platforms, the produced water is discharged together with cooling water. The discharge rate of cooling water may be 1 - 20 times higher than the produced water, depending a.o. on the age of the field, cf. fig 1. Formula (1) then predicts a slower dilution for this situation. For technical reasons, however, air is mixed into the water flow before it is discharged. This leads to heavy turbulence around the discharge and a virtually instantaneous dilution of in the order of 10 times. In addition, the water plume will quickly rise to the surface, resulting in increased evaporation of volatile components. In these cases, formula (1) will therefore significantly underestimate the dilution.

Components of the produced water are removed from the water coloumn by a number of processes including:

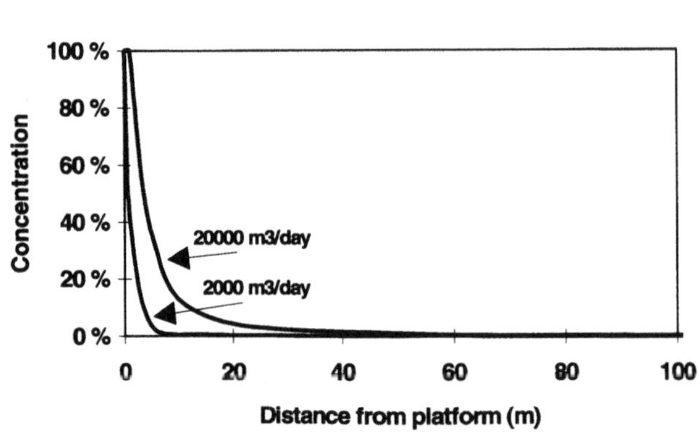

Figure 2. Dilution of produced water for a high and a low discharge rate.

- evaporation of volatile components like benzene if it comes in contact with the air
- biodegradation
- adsorption to particles and sedimentation
- chemical transformation like hydrolysis, oxidation and formation of chemical complexes

These processes, exept evaporation, are fairly slow compared to the initial dilution. They will therefore hardly affect the acute toxicity of the produced water, but be decisive for possible chronic effects.

POSSIBLE EFFECTS OF PRODUCED WATER

Produced water shows a fairly low acute toxicity to marine organisms. The toxicity of a chemical is normally quantified as an LC50 or EC50 value, defined as the concentration causing 50 % of the organisms in a test population to die or show a significant negative effect when they are exposed to a fixed concentration of the chemical for a defined time period. EC50/LC50 values for produced waters from North Sea oil fields range from 2 % for 24 hour exposure to over 60 % for 48 hour exposure depending on test organism and field (E&P Forum, 1994). This corresponds to 50 to 1.7 times dilution. Produced water from gas fields will generally be somewhat more toxic than from oil fields. The amount of produced water from gas fields will, on the other hand, be much smaller than from oil fields, with a following faster dilution, c.f. formula (1).

Many petroleum hydrocarbons and chemicals applied by the oil industry have potential to accumulate in marine organisms. The organisms will excrete or metabolise these components to a varying degree. Possible effects will also depend on the availability of the components to the organisms, again depending on whether they are removed from the water coloumn, as described above.

In shallow water fields, the produced water plume may reach the sea floor, leading to a steady exposure to benthic organisms in the vicinity of the discharge point. The production platforms in the Norwegian sector of the North Sea are situated in water depths from 70 to 350 m. The produced water is discharged at depths from 0 to 20 m below sea level and will drift away from the platforms at fairly constant depth or rising to the surface. Benthic organisms will therefore not be affected by the produced water or any of its components, at least not in the vicinity of the platform where the concentration of toxic components is significant. Further away from the platform, some of the dispersed oil and heavier components that adheres to particles in the water, will sink to the bottom. As this will be distributed over a very large area, the concentration in the sediments will be extremely low and probably of no significance.

Little knowledge exists on the fate in the marine environment and possible chronic effects of those (natural) components in the produced water that show the lowest degradation rates and have potential for accumulation, like PAH and alcylated phenols.

PRODUCED WATER COMPOSITION

Typical values for the most important components in the produced water are shown below (E&P Forum, 1994):

Total organic matter	50 - 800 ppm	
Oil (dispersed)		20 - 40 ppm
Fatty acids		20 - 700 ppm
Phenols		1 - 10 ppm
Aromatics		1 - 5 ppm
Production chemicals	0 - 20 ppm	
Biocides		
Corrosion inhibitors		
Scale inhibitors		
Emulsion breakers		
Anti foam		
Oxygen scavengers		
Flocculants		
Wax inhibitors		
Gas treatment chemicals		
Metals	Concentrations similar to sea water	

The concentration of the production chemicals in the discharge water depends both on the amount used and how they follow the gas, oil and water phases.

The average oil concentration in the produced water discharge in 1994 was 23.5 mg/l, resulting in a total discharge of 790 tonnes of oil (OLF, 1995).

CONVENTIONAL METHOD FOR RISK ASSESSMENT

Risk or hazard assessment of chemicals and products is rapidly gaining interest and is in some cases mandated by the authorities. The most common method of quantifying the health or environmental risk of a chemical is to calculate a PEC/NEC ratio. PEC is the Predicted Environmental Concentration of the chemical, found eg. from dilution models. NEC is the No Effect Concentration, defined as the concentration under which there will be no adverse effect to organisms. The NEC value is derived from LC50 or EC 50 values by application of a suitable safety factor. This safety factor shall account for differences between the test species and local biota, differences between laboratory and real environment, indirect effects due to species interactions and take some account for chronic effects. There are various methods for establishing a safety factor. In a method for risk analyses developed by TNO in The Netherlands (Karman et al, 1994), the NEC value is found by dividing the geometric mean of all available LC50 and EC50 values by the safety factor $1000/\sqrt{n}$, where n is the number of species for which LC50 or EC50 values exist.

A PEC/NEC ratio of 1 will generally correspond to a probability of 5 % that biota will be affected (Karman et al, 1994). The combined risk of simultaneous exposure to a mixture of chemicals, like produced water, is estimated by statistical summation of the risks from individual chemicals. Synergistic and antagonistic effects will normally cancel each other for mixtures with more than 5 toxicants (Karman et al, 1994).

This method is applied in the CHARM model (Øfjord et al, 1995) for risk assessment of offshore production chemicals.

EXPOSURE SITUATION FOR FOR MARINE ORGANISMS TO PRODUCED WATER DISCHARGE

As described above, benthic organisms will not experience produced water concentrations that may lead to acute effects. Passive organisms drifting past the discharge point

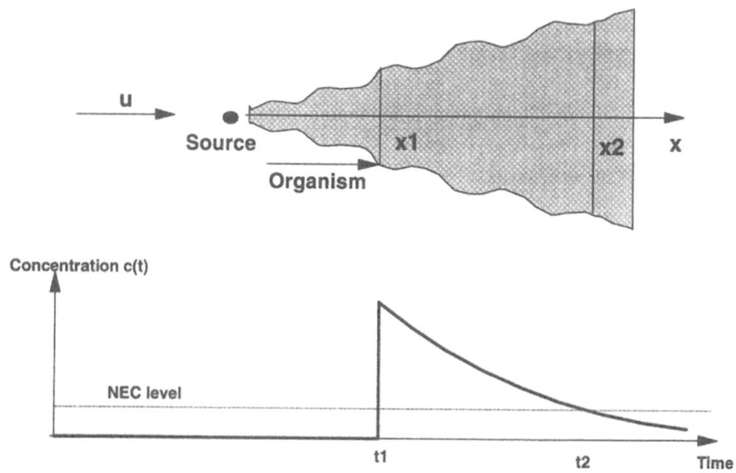

Figure 3. Schematic representation of the concentration a passive organism is exposed to when drifting past a produced water discharge.

will experience a time varying concentration as indicated in fig 3. The distance to the point where the concentration in the center of the plume drops below the NEC level is typically 200 - 600 m, depending on discharge flux, current velocity, temperature and salinity gradients etc. With a typical current velocity of 0.05 - 0.1 m/s, the corresponding drifting time will be 0.5 - 3 hours.

Standard tests for toxicity are performed for durations of 24 to 120 hours (1 - 5 days), with 96 hours as the most common period. It is well known that the LC50 and EC50 values depend on the exposure time. Few data exist, however, that describe this connection. In particular, little data exist for toxicity of short exposure times, in the range of 0.5 - 5 hours. The few data that are available indicate that the exposure dose is decisive, the dose being defined as the time integral of the concentration over the exposure period. This means that a certain dose will give the same effect irrespective of the combination of time and concentration that has led to this dose. At least this seems to be valid for reasonable variations

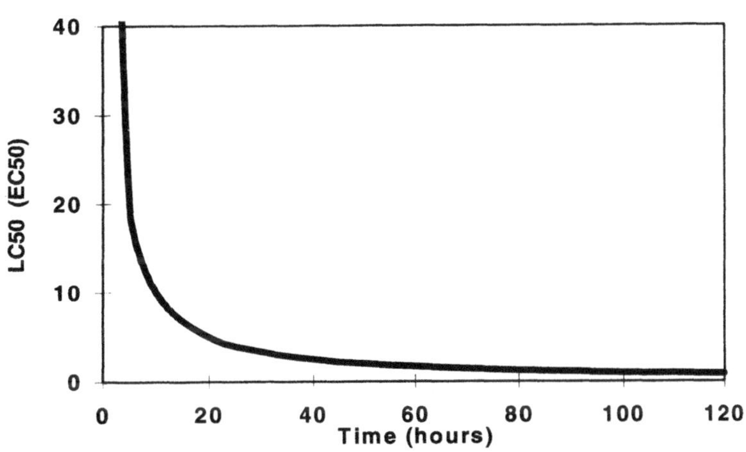

Figure 4. LC50 (or EC50) value as function of exposure period.

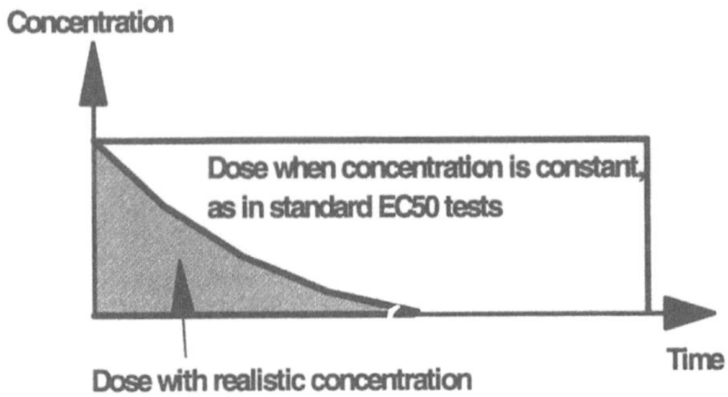

Figure 5. Exposure dose for constant exposure concentration, asin standard LC50 tests, and for a time varying concentration as experienced by a passive organism drifting past a discharge of produced water.

in the exposure time. It follows from this, that for exposure to a constant concentration, the relation between LC50 or EC50 and exposure time can be described by a hyperbola, as indicated in fig 4. This gives a possiblity to convert existing data from standard (48 - 120 hour) LC50 and EC50 tests to short exposure times through a simple calculation procedure as discussed below.

In the produced water discharge case, the exposure dose is the integral under the concentration curve indicated in fig. 3 and 5. If we assume that a passive organism is exposed to a concentration exceeding the NEC level for t hours and that the concentration drops linearly with time, then the exposure dose can be calculated as the area of a triangle:

$$\text{Dose}_{\text{discharge}} = 0.5 * t * c_0 \text{ ppm-hours} \tag{2}$$

where c_0 = concentration at the discharge point, i.e. in undiluted produced water.

As indicated above, typical values of t will be 0.5 - 3 hours. By using a value of 1 hour, the dose will be:

$$\text{Dose}_{\text{discharge}} = 0.5 * c_0 \text{ ppm-hours} \tag{3}$$

This is a fairly conservative estimate of the exposure dose because:

- the concentration drops much faster than linearly, see fig 2
- most exposed organisms will enter the discharge plume some distance downstream from the discharge point and will therefore experience elevated concentrations for a shorter time than those drifting directly past the discharge point
- most exposed organisms will not enter the center of the plume where the concentration is highest.

In a standard 96 hour test with constant concentration c_0, the exposure dose will be

$$\text{Dose}_{\text{test}} \approx 96 * c_0 \text{ ppm-hours} \tag{4}$$

The ratio between these two doses is

$$\frac{\text{Dose}_{\text{test}}}{\text{Dose}_{\text{discharge}}} \approx 200 \tag{5}$$

If, therefore, an organism can tolerate the same dose irrespective of the combination of exposure time and concentration that lead to this dose, the organism can in our case tolerate a 200 times higher concentration in the discharge (i.e. the undiluted produced water) than what would be concluded from a standard test. Naturally, this has a significant effect on the

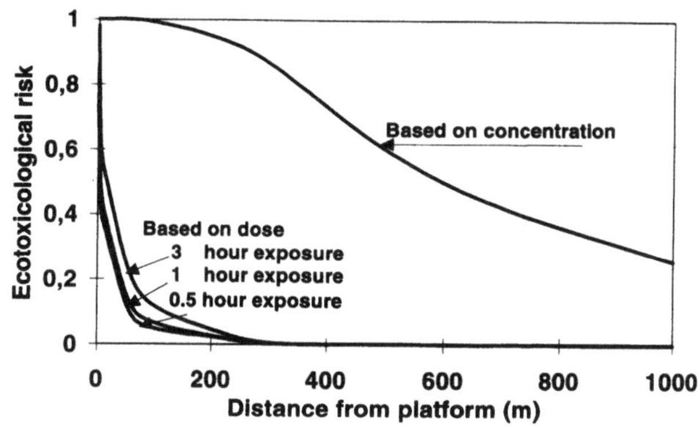

Figure 6. Ecotoxicological risk from discharge of produced water as calculated based on constant exposure concentration and on realistic exposure dose (time varying concentration).

calculated ecotoxicolgical risk from discharge of produced water, as illustrated in fig 6 for a typical large volume discharge.

Fig 6 indicates that if stationary organisms had been present, i.e. if they had something to attach to, there would have been a risk for acute effects up to 1000 - 2000 m downstream of the platform. It should be noted that this high risk level would be confined to a few meters horizontally and vertically around the center of the discharge plume and that the direction of the discharge plume will change with the direction of the current as this change with the tide. Fig 6 also indicates that for passive organisms drifting past the platform, the risk for acute effects would be restricted to those organisms entering the center of the plume within 50 - 100 m downstream of the discharge point. Fig 6 also shows that the risk level is only marginally affected by the assumed exposure time used in formula (2) above (or, equivalently, the assumed exposure time in the LC50 test in formula (4)).

Fig 7 shows the ecotoxicological risk vs. distance from the platform for three different oil fields. Although there are significant differences between the fields, the risk is small in all cases and confined to 50 - 100 m from the discharge point.

MAIN CONTRIBUTORS TO ACUTE TOXICITY

Fig 8 shows how the most important groups of components in the produced water from four fields in the Norwegian part of the North Sea contribute to the acute toxicity. The NEC values are deduced from TNO's database MARITOX, which includes virtually all

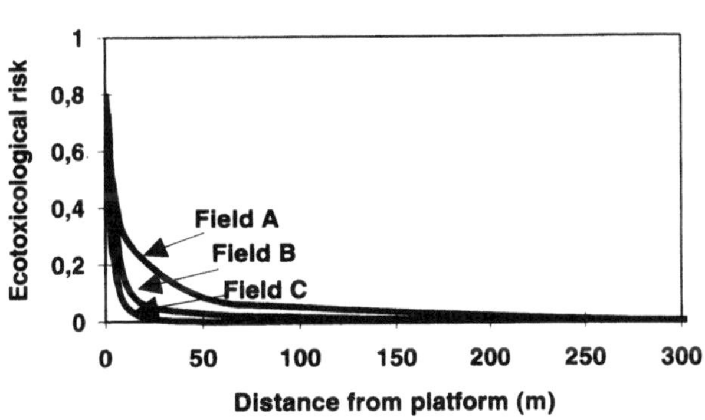

Figure 7. Ecotoxicological risk from discharge of produced water from three different oil fields.

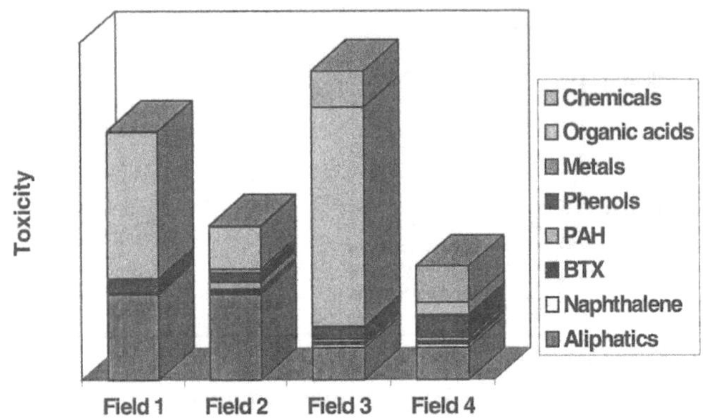

Figure 8. Relative contributions from the most important groups of components to the total toxicity of produced water.

published data on toxicity for marine organisms. The toxicity of a component is in this case calculated as the PEC/NEC ratio of that component in undiluted produced water, and the toxicity of a group is simply calculated as the sum of PEC/NEC ratios for all components in that group.

Two major comments arise from fig 8:

- Aliphatics, i.e. dispersed oil, seem to be one of the main contributors. This is an effect of the standard way of testing toxicity of crude oil, where the oil is mixed with water, then left to settle, and the toxicity of the water phase is tested. The resulting toxicity is ascribed to the aliphatics, although what is remaining in the water phase is the water soluble components like aromatics and phenols. The available data are therefore clearly of little relevance in risk analysis of produced water and may lead to wrong conclusions (e.g. massive use of chemicals to reduce the amount of dispersed oil). It is difficult to find the true toxicity of dispersed oil (aliphatics) for the following reasons:
- the oil droplets tend to adhere to the walls of the test vessels instead of staying uniformly distributed in the water
- given sufficient time, there will be an equilibrium between the concentrations of the variuos components in the oil and the water phase. Little is known about the chemical composition of the oil droplets at the time of discharge and how it may change in the first minutes or hours as the discharge spreads in the sea
- the biological effect mechanisms of oil droplets on organims may be very different from one species to another, depending on several factors, including whether they filter water for food or how large they are compared to oil droplets that may adhere to them.

In spite of these problems, there is a need for relevant toxocity data for dispersed oil.

For Field 1 and Field 3, the organic acids seems to be the dominating contributors to toxicity. The reason for this is the high concentration of acetic acid in the water from these fields, 600 - 700 ppm, which is 10 - 20 times higher than for any other component. Acetic acid biodegrades very rapidly and is not thought to be of any concern in produced water discharges. In areas of low water circulation, the rapid biodegradation of large amounts of organic matter might lead to concern over the oxygen content in the water. There is no indication that this should cause any concern in the open waters of the North Sea where the major discharges take place. The large discharges of acetic acid will lead to increased concentrations of micro organisms feeding on it. Again, there is no indication that this causes any problems in the areas in question.

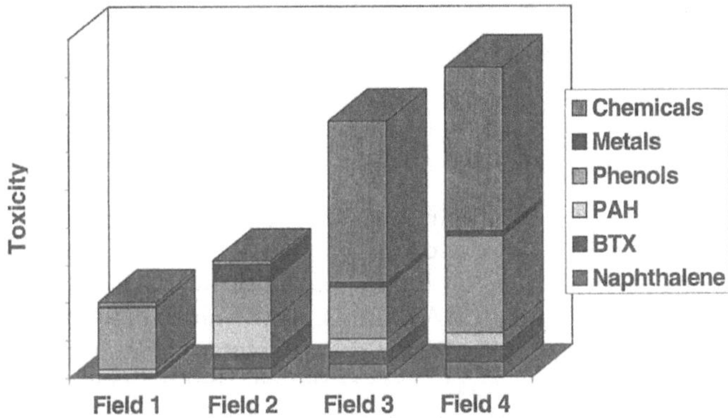

Figure 9. Relative contributions from the most important groups of components to the total toxicity of produced water when aliphates (dispersed oil) and organic acids are excluded.

These two comments serve to illustrate that the results of risk analyses need to be interpreted with care at the current state of the art.

Fig 9 shows the same results as fig 8 when the contributions from aliphatics and organic acids are disregarded. In Field 3 and Field 4, the production chemicals are the dominant contributors to the toxicity, which in this case is mainly caused by an extensive use of corrosion inhibitors. Alternative corrosion inhibitors with much lower toxicity exist and will normally be used. It illustrates, however, that chemicals should be chosen with care and their use should be minimized. Of the natural components, the phenols are seen to contribute most to acute toxicity, followed by PAH, although there are marked differences between the four fields. This conforms well with tests performed to determine the sources of toxicity in produced water (Johnsen et al., 1993). The contributions from BTX (benzene, toluene, xylene) and metals are low. For platforms with large discharges, the plume normally rises to the surface close to the discharge point. This causes volatile compounds like the BTX's to evaporate, reducing their concentration and possible harmful effects in the sea. Field 2 shows some contribution from metals, caused by high concentrations of barium. As a matter of curiosity and illustration of the necessity for careful interpretation of the results, the toxicity of metals naturally present in sea water would also have been visible on the figure.

Fig 10 shows how the most important groups of components contribute to the total toxicity of produced water from one field when five different corrosion inhibitors are used.

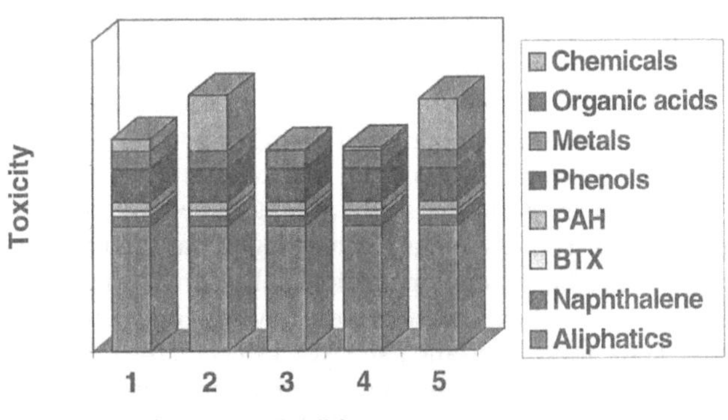

Figure 10. Relative contributions from the most important groups of components to the total toxicity of produced water from one field, using different corrosion inhibitors.

Corrosion inhibitors are among the most toxic production chemicals in use. The figure clearly shows that it is important to choose corrosion inhibitors with care.

CHRONIC EFFECTS

The method for risk analyses applied in this paper will only account for possible chronic effects through the safety factor used for arriving at the NEC values. This is clearly inadequate for components that have a low biodegradability and have potential for bioaccumulation and biomagnification. In the CHARM model, chemicals with potential for bioaccumulation are sorted out on the basis of their octanol water partitioning coefficient (log P_{ow}). This will exclude some of the cumbersome chemicals, but the method is irrelevant for surface active chemicals for example. Further, the use of log P_{ow} will not lead to a quantification of the risk, or effects, of bioaccumulation.

A quantification of the risk for chronic effects would require knowledge of the "body burden" of an organism (dose or concentration in the body or organs of the organism), which would be a function of direct uptake from the water, uptake through the food chain, metabolism, excretion and exposure time and concentration. We are developing computer models for this purpose (Reed, 1995). The models seem to give interesting and valuable results, but currently they suffer from lack of data to describe the many chemical and biological processes.

REINJECTION OF PRODUCED WATER

As an alternative to discharge, the produced water may be reinjected either into the producing reservoir or into a separate geological formation. Sea water is injected into all Norwegian oil fields to increase the oil production. Although little experience exists with reinjection from the North Sea oil fields, there is a concern that massive reinjection of produced water may lead to reduced injectivity, i.e. that more energy will be required to inject produced water than sea water. As the energy is generated in gas turbines, this will lead to increased emissions to air of CO_2, NO_x, CH_4 and VOC, which all lead to environmental impacts. If the produced water is injected into a separate reservoir, all of the energy consumption and the resulting emissions to air will an increase compared to the situation when the water is discharged. The impacts of the discharged water on the marine life will therefore be substituted with the environmental impacts of emissions to air.

As an example, we have performed calculations for two fields (field I and field II) where we have assumed that the injectivity will be reduced by 50 % when produced water is reinjected into the producing reservoir, replacing an equal amount of sea water that would otherwise have been injected. The two reservoirs have different properties, including different injectivity for sea water. The energy required for injection of 1 m^3 of sea water is therefore different, and is assumed to be 3.7 kWh/m^3 and 6.1 kWh/m^3 respectively. The emissions to air are found from data on emissions factors for the turbines and efficiency of turbines, generators and pumps. The above pumping energy of 3.7 kWh/m^3 is, for instance, found to result in a CO_2 emission of 2.6 kg/m^3.

There are no universally accepted methods for comparing different environmental impacts. This is, however, an area of significant international effort and interest, and a few methods have been developed, especially for the purpose of impact analyses in life cycle analyses. Two of these methods were developed by CML in The Netherlands (Wit et al, 1993) and BUWAL in Switzerland (SETAC, 1992). In the CML system, the emissions are classified and quantified according to their contribution to potential environmental impacts, like global

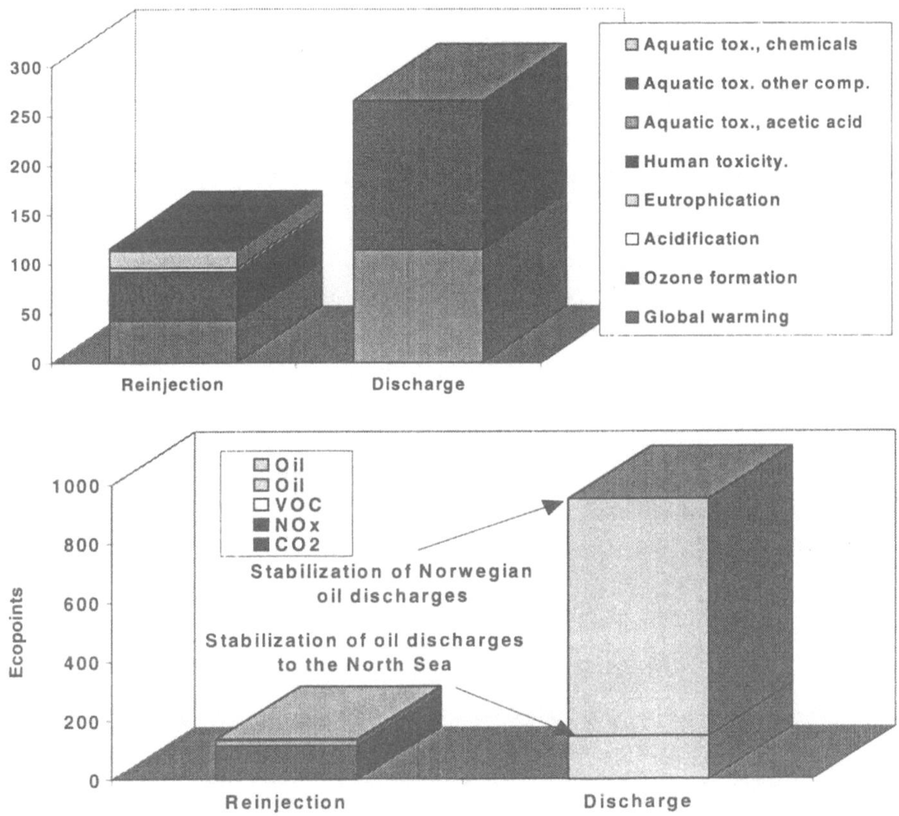

Figure 11. Total environmental impacts from reinjection and discharge of produced water from Oil Field I as evaluated by two valuation systems (top figure is the CML system, the bottom figure is the BUWAL system).

warming, acidification, photo oxidant formation, aquatic toxicity etc. The contribution to global warming of an emission is quantified as CO_2 equivalents, the contribution to aquatic toxicity is quantified as the inverse of the LC50 value etc. The system also contains a method to weigh the impact categories against each other according to the total anthropogenic contribution to those impacts.

In the BUWAL method, the various emissions are weighted against set (politically) goals for allowable emissions in an area. This will mean that the emission of NO_x for example may be given a different weight in different countries according the goals for total NO_x emissions in those countries. We have used Norwegian goals in our calculations, except for emissions to sea, where there is no clearly stated goal. We have therefore used two "goals", one is stabilizing the total oil discharges to the North Sea from Norway, the other is stabilizing the total oil discharges to the North Sea.

The results from applying these two methods to compare discharge and reinjection on produced water on field I and field II are shown in fig 11 and 12. Even if the results from such calculations must be treated with great care at the current state of the art, some interesting observations can be made:

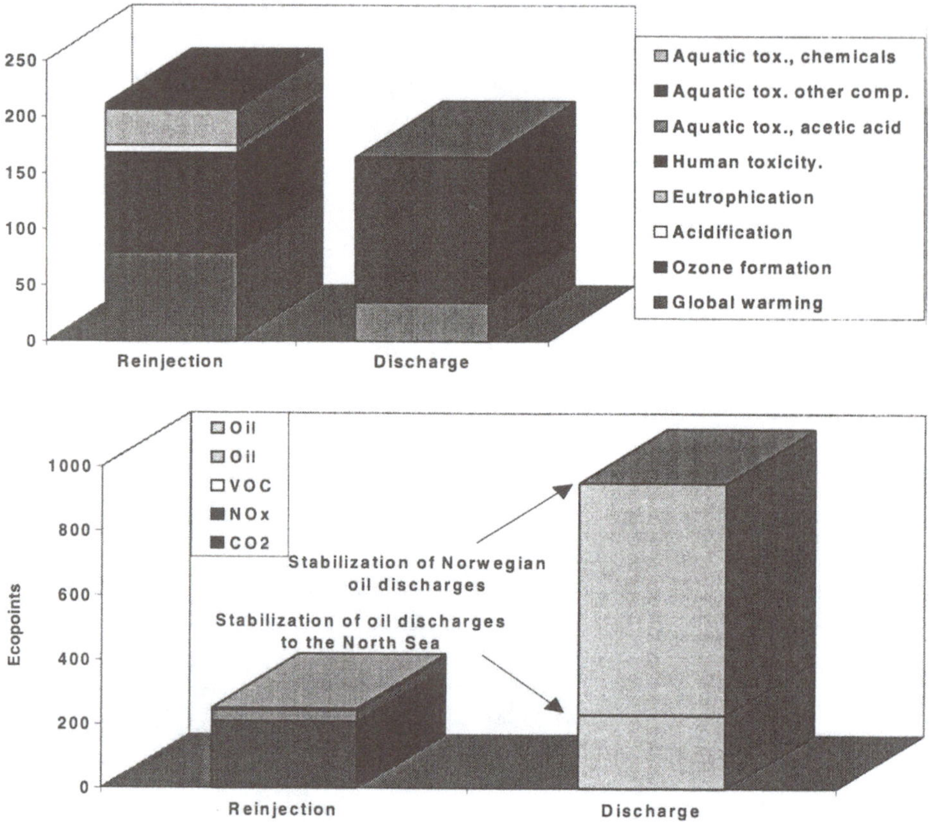

Figure 12. Total environmental impacts from reinjection and discharge of produced water from Oil Field II as evaluated by two valuation systems (top figure is the CML system, the bottom figure is the BUWAL system).

- With our assumptions for reduction of injectivity, the potential environmental impacts of discharge and reinjection of sea water seem to be of the same order of magnitude. This is maybe the most important observation, as it clearly indicates that reinjection of produced water may not be "the answer" in all situations.
- According to the CML system, reinjection gives the lowest environmental impacts for Field I while discharge is the most favourable solution for Field II. These conclusions are of course very sensitive to the assumptions on reduced injectivity, but indicates that differences between fields may be important when selecting the best solution.
- For emissions to the air, the CML system puts an almost equal emphasis on global warming and photo oxidant formation (a result of NO_X and VOC emissions). The BUWAL system ranks global warming as the most important impact, which is to be expected with the strict Norwegian goal of stabilizing its total CO_2 emissions.
- With the BUWAL method, the two different goals for oil emissions to sea give significantly different results, indicating that the method is sensitive to the assumptions made. If stabilizing the Norwegian oil discharges is paramount, then reinjection is "best" for both fields, where as if stabilizing the total oil discharges to the

North Sea is adequate, then disharge of the produced water would be the "best" solution.

- Different methods illustrate different aspects of the environmental issue, e.g. "true" environmental impacts (as far as can be described with the current knowledge), political preferences, economic valuations etc. If one chooses to use methods for quantification and comparison of environmental impacts, one should therefore use several methods to illustrate the different aspects.

CONCLUSIONS

The calculation of ecotoxicological risk from discharges of produced water based on the common method of PEC/NEC ratios, does not give a good representation of the true risk for deep water fields. The use of exposure dose is shown to give a much better description of the real situation.

Relatively little data exist that describe the effect of exposure to a dose of a chemical and the effect of different combinations of exposure times and concentrations that results in the same dose. Indications are, however, that dose can be used to predict the effects irrespective of exposure time. It is then simple to apply existing LC50 or EC50 data from standard (48 or 96 hour) tests.

The risk for acute effects of produced water on marine life is low and confined to a few tens of meters downstream of the discharge point and in the center of the discharge plume. Acute toxicity is therefore of no real concern, meaning that the focus should be on possible chronic effects.

The results from risk analyses should be interpreted with care. Two examples that illustrate this are the apparent high toxicities of aliphatics (dispersed oil) caused by the standard test method for crude oil, which has little relevance for the produced water situation, and acetic acid, which is of little environmental concern.

Conventional risk analyses, as applied in this paper, do not adequately model chronic effects. Computer models can, and will be, developed, but much test work remains to be done to supply such models with all necessary data.

Two different methods were applied to compare the environmental impacts of reinjection and discharge of produced water. Although such models are controversial and at an early stage of development, they indicate that reinjection of produced water may not be "the final solution" in all cases. As these models have widely different basis, several should be applied to illustrate different environmental aspects.

REFERENCES

E&P Forum: "North Sea Produced Water. Fate and Effects in the Marine Environment". Draft Report, 1992.

E&P Forum: North Sea Produced Water: Fate and Effects in the Marine Environment. Report No. 2.62/204, 1994.

Johnsen, S., Smith, A. Th., Brendehaug, J., Riksheim, H., Gjøse, A. L.: "Identification of Sources of Acute Toxicity in Produced Water." paper SPE 27138 presented at the 1993 SPE Conference in Djakarta, Jan. 1994.

Karman, C.. C., Schobben, H. P. M., Scholten, M. C. Th: "Ecotoxicological risk of produced water from oil production platforms in the Statfjord and Gullfaks area." TNO 1994, (Restricted).

OLF: Environmental report 1993 - 1995. OLF, Stavanger, 1995.

Reed, Mark. "Bioaccumulation modelling. " 1995 International Seminar on Produced Water, Trondheim, Sept. 1995.

Riksheim, H., Johnsen, S. "Determination of Produced Water Constituents in the Vicinity of Production Fields in the North Sea", paper SPE 27151 presented at the 1993 SPE Conference in Djakarta, Jan. 1994.

Rye, H., Reed, M., Melbye, A., Johnsen, S.: "Dilution Factors for Produced Water Releases. " 6th International Symposium Oil Field Chemicals, Geilo, Norway, March 1995.

SETAC: "Life Cycle Assessment". Report from workshop in Leiden Dec. 1991. SETAC, 1992.

Wit, R., Taselaar, H., Heijungs, R., Huppes, G.: "REIM: LCA-based Ranking of Environmental Investment Models." CML Report 103, Leiden 1993.

Øfjord, G. D., Vik, E. A., Bakke, S., Johnson, D. R., Edvardson, T.: "CHARM - Chemical Hazard Assessment and Risk Management of Offshore E&P Chemicals." 6th International Symposium Oil Field Chemicals, Geilo, Norway, March 1995.

26

THE ROLE OF THE SUPPLIER OF PRODUCTION CHEMICALS IN MANAGING THE ENVIRONMENTAL IMPACT OF PRODUCED WATER IN OFFSHORE OIL AND GAS PRODUCTION

Albert Bos

Servo Delden B.V.
P.O. Box 1
7490 AA Delden, The Netherlands

INTRODUCTION

The intent of this paper is to put a perspective on the issue of the regulation of offshore chemicals and to illuminate the role of the specialty chemicals industry in this respect. It is the concern of the regulators that the use of offshore chemicals may have an impact on the environment. The nature of the specialty chemicals industry which produces and supplies these chemicals (and are thus also referred to in this paper as the suppliers) will be described as well as its role in the management of the environmental impact of the offshore industry through development of better products. A way of implementing this development process and the implications will be discussed as well as the possible role of the CHARM model. Finally the need for cooperation between regulators and industry will be stressed.

THE CONCERN WITH THE ENVIRONMENTAL IMPACT OF THE OFFSHORE OIL AND GAS PRODUCTION INDUSTRY

In the offshore oil and gas production industry production chemicals, also referred to as chemical additives, are used on a continuous basis for various purposes in the production process usually resulting in a continuous emission of relatively small quantities into the environment. In this they are different from drilling chemicals which are used on a temporary basis and may be discharged batchwise at much higher concentrations. The most important and most widely used production chemicals are the corrosion inhibitors, emulsion breakers, scale inhibitors and to a lesser extent biocides. These product groups are complemented with a multitude of products for more specific applications such as drag reducers, pour-point depressors, hydrate inhibitors, de-foamers, reverse emulsion breakers, H_2S and oxygen scavengers etc..

Although there is a relatively long history of several decades of usage of these chemicals in the offshore industry there is in recent years a growing concern whether the marine environment of the North East Atlantic can sustain this usage. This concern is growing as the total amount of produced water emitted into the North Sea is increasing due to the exploitation of more wells and to an increasing water production of the existing wells which are nearing the end of their lifetime. In the third North Sea Ministers Conference (1993) the ministers of the participating countries of the Oslo and Paris commissions requested the development and adoption of a mandatory control system for the use and discharge of chemicals offshore. In the fourth North Sea Ministers Conference (1995) this request was extended to take into account the CHARM model (Chemical Hazard Assessment and Risk Management) and the HOCNF (Harmonized Offshore Chemicals Notification Format). Once implemented, this harmonized system will be used to reduce and ultimately phase out those products which pose a possible threat for the marine environment. On basis of the precautionary principle, products or chemicals of which not enough is known with respect to their environmental effects will be assumed to pose a threat. This system is planned to be in effect in 1996 and is now being worked upon by the participating national authorities.

Production chemicals play a very definite role in the operation of oil and gas platforms and are vital to the production processes involved. In a sense one could compare the importance of production chemicals for offshore operations to the importance of vitamins for living organisms: without them proper functioning is simply not possible. The regulatory drive for the reduction or even phasing out of these chemicals to reduce the environmental risks will thus prove to be a very difficult exercise. In the same way enforcing the substitution of products with others which are preferable from the environmental point of view but lack a high technological performance may ultimately lead to operational breakdown and to increased environmental impact or even environmental disasters. It is evident that the future move towards lower and lower environmental impact of the offshore E&P industry will require large scale cooperation between the three parties involved, i.e. the authorities of the participating countries, the producing companies and their operators, and the suppliers. Simultaneously large scale research must be performed to gain insight into the environmental risks of production chemicals and to provide the basis of the development of better products.

THE SPECIALTY CHEMICALS INDUSTRY

It is a misconception to think of the specialty chemicals industry in the same way as of the bulk chemical industry and to apply the same regulations to both. The specialty chemicals industry is characterized by a very large diversity in highly specialized chemicals combined with low production quantities thereof. As an example, in our company about 1500 different chemicals are produced or bought as components for our product package. Of these, about 400 are used for a great number of products developed for oil and gas production. Typical yearly production for specific chemicals may range from an exceptional several thousand tons to as low as several hundred kilos. Most chemicals, however, are in the middle range of one to several hundreds of tons per year.

Together with the emerging environmental awareness there is also an urge of the regulators and of the oil and gas industry to know the exact composition of the products used. The refusal of the specialty chemicals industry to give this information could erroneously be interpreted as that there is something to hide. This is however not the reason. The difficulty for the supplying industry is that it in most cases it is not simple chemicals which are sold but formulated products aimed at solving specific operational problems. The technical performance of these products is partly due to the chemistry and physical behavior of the components but may also be largely influenced by combined or the so-called synergistic effects (in physical behavior) of the formulation or blend. The added value of the product is thus for a large part determined by the effectiveness (performance/price ratio) and secondary properties of the formulation and to a lesser extent by the chemistry of the components in the formulation.

In order to achieve a product range with very high effectiveness for the multitude of operational problems encountered in the production process, a relatively large R&D overhead is required as well as a strong service support to the customer thus giving a large overhead. Disclosing the exact composition of a product and risking that this information would fall into the wrong hands would impair the competitiveness of supplying companies and jeopardize the expected revenues of the costly and time-intensive R&D and service. A successful product could in that way be copied and produced by companies who don't invest as heavily in R&D and service and can thus sell with much lower profit margins. This might be welcome for the oil and gas industry from an economical point of view but would also mean the death sentence for continued R&D and support. It is therefore a firm point of view of the supplying industry to only disclose exact compositions of products to regulators only, given the guarantee that this information will be treated as strictly confidential.

DEVELOPING BETTER PRODUCTS – TOWARDS A 'GREENER' FUTURE

Confronted with the complexity of this issue and a dire need to gain insight into the environmental aspects of its products, Servo Delden BV initiated the "Green Chemistry" project in 1993. This project is aimed to generate knowledge on the impact and fate of production chemicals, to integrate this knowledge in the R&D process and in this way to enable development of a new product package with reduced environmental impact for marketing on the North Sea and eventually the rest of the world. This project has been reviewed by the Dutch ministry of economical affairs and has been awarded a subsidy in the advanced environmental technology research stimulation program.

The development process of a new product is depicted schematically and in simplified form in figure 1. On basis of the desired performance and behavior existing components

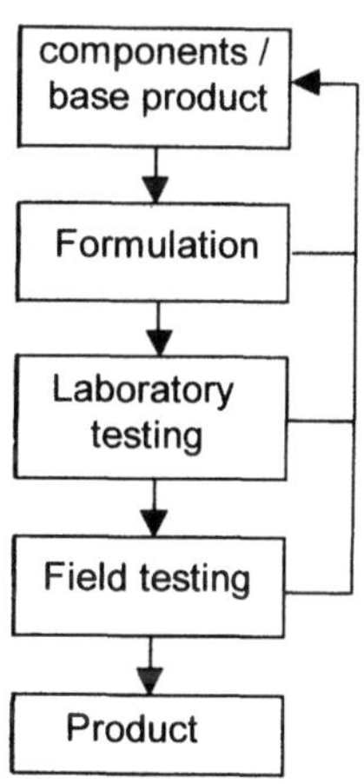

Figure 1. The steps involved in the development of a new product.

are selected and formulated into a development product which is then tested in the laboratory. Together with the technical performance also a multitude of other parameters are taken into account such as stability, ease of handling, flash points, pour points etc.. Once the product has been developed to satisfaction field tests must be performed to evaluate the performance under real field conditions. The final result of this work is an optimized product which has taken from months to years of research to develop. Of course the development of a total product package consisting of hundreds of products will take much longer, even up to decades, as the capacity of the R&D department is limited.

Until recently, environmental considerations were not taken into account in this development process or only in a very limited form such as for example avoiding carcinogenic components. The aim of the regulators to adopt a regulating framework which will induce the industry to *continuously develop* and use chemicals with lower environmental impact has a profound impact on this process.

The approach to integrating environmental considerations in the development process can be divided into two distinct parts. First of all a framework is needed for evaluating products, preferably in a quantitative and non-discontinuous way. Such an evaluation enables the comparison of possible alternatives. Secondly, on basis of this evaluation framework, selection of existing components and development of new components can be performed resulting in a base of preferred components for product development. Large scale research is then needed to develop new or improved components which combine the desired physical behavior with better intrinsic ecological properties. This continuous development and selection of better components when making formulations for new products will then also lead to a continuous improvement of offshore chemicals from the environmental point of view.

Although stated simply, practical aspects make this process very difficult. It implies that the intrinsic characteristics such as toxicity, biodegradation and partitioning behavior of the basic components must be determined as well as their behavior and fate in the environment. Starting from scratch this must be done for the existing components as well as for the newly developed components. The sheer number of components, possible derivatives thereof and as yet unresearched alternatives to them makes it impossible to acquire the necessary data in a direct way by ecological testing. The costs and amount of time involved are absolutely inhibitory, even if there would be enough capacity available at the testing laboratoria, therefore other methods are required. A significant part of Servo's Green Chemistry project is devoted to the acquisition and development of rapid screening techniques and QSAR models for screening and predicting specific characteristics of chemicals beforehand. This acquisition and development is itself costly and time consuming, but is in the long run the only effective route available. One of the screening techniques which was introduced with good success is the photoluminescent bacteria test, commonly referred to as the Microtox or Lumistox test. The advantages of this test are that it is relatively cheap and rapid, that it can be performed under normal laboratory conditions by regular laboratory assistants, and that it is very robust and gives highly reproducible results. As an example figure 2 is presented which shows the correlation of this test with the acute toxicity for Acartia Tonsa of a group of alternative surfactants with the same intended technical function when used in offshore products. Similar correlation's are also found for algae and sediment reworkers and have been reported in literature for fish.

THE ROLE OF THE CHARM MODEL

It has been stated above that a framework is needed for evaluating products in a quantitative and discontinuous way. This is where the CHARM model can be a very valuable

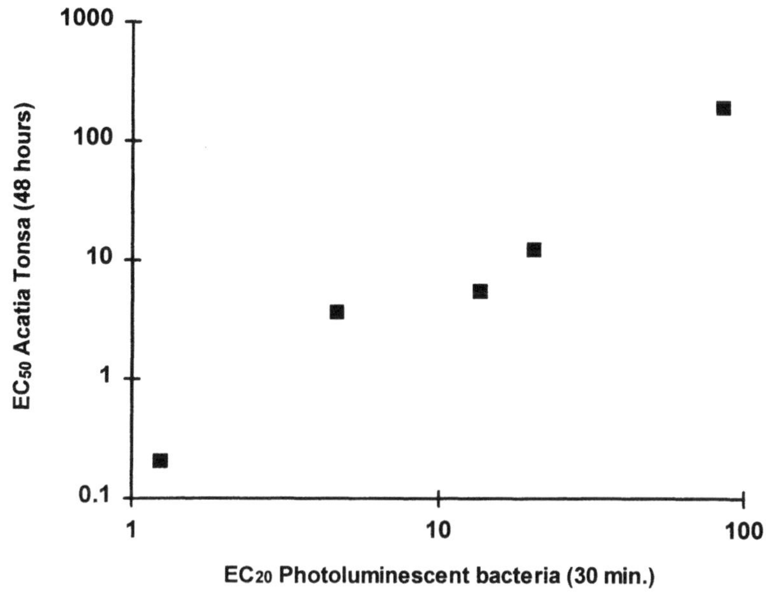

Figure 2. The observed correlation between acute toxicity for photoluminescent bacteria and the marine crustancean Acartia tonsa for a group of surfactants with the same technical application.

tool for research. The CHARM model is intended to be a transparent and simple model which estimates the environmental hazard or risk of a chemical used offshore and expresses this as the ratio of the PEC (Predicted Environmental Concentration) and the PNEC (Predicted No-Effect Concentration). In the model the intrinsic characteristics of a chemical are taken into account together with the quantity used, an estimated or measured release from the platform and the fate of the chemical once it has found its way into the environment. Although originating as a private project of the Dutch and Norwegian authorities, the development of the CHARM model is now a combined effort of the authorities of the North Sea countries and of industry, the latter being represented by E&P Forum (the association of operators) and EOSCA (the association of suppliers). At this moment two and a half years have been invested resulting in version 2.1 of the model. This version has been evaluated by all parties involved and the limitations and weaknesses have been identified. Most notably among these are the problems involving the octanol-water partitioning coefficient, the rules applied for the estimation of accumulation, and the evaluation of surfactants. The third and final phase which started in July of 1995 is aimed to significantly improve on the model and come to a final workable version which can then be used in regulating the use and discharge of offshore chemicals. One must however be aware of the fact that the CHARM model only takes a limited number of aspects into account and can therefore not be the sole means of evaluation. Aspects as blacklisted chemicals, undesired biodegradation metabolites, reactions occurring during use and secondary effects as increased carry-over to the water phase of chemicals which are intrinsic to the production process such as aromatics and PAH's must be evaluated separately.

FINAL DISCUSSION

Although it is the preference of the regulators to reduce the use of chemical additives in the oil and gas production industry to ultimately zero in the future, this is not very likely to happen as the use of these chemicals is simply unavoidable unless revolutionary and as yet unforeseen new developments in technology emerge. This is also recognized by the regulators which may explain their wish to develop a framework which will enforce a continuous effort from industry to develop better and better products and technologies. A central aspect in such a system is a framework for evaluating and comparing alternative products. Such a framework has to be workable and agreed upon by all parties involved and can make good use of the CHARM model currently being developed by combining it with a means to evaluate the aspects which are not covered by the model itself.

It is in principle possible to implement a system which will enforce a continuous improvement but one must realize that the main effort of the actual process of this development will have to be done by the suppliers of the chemical additives. Cooperation between the suppliers, the operators and regulators is a prerequisite as the objectives must be clearly defined for all three parties involved. Due to the structure and functioning of the specialty chemicals industry with the large diversity and low quantities involved this will give an increased burden which may even be inhibitory to further research and development. This would be the case if after the generation of an initial limited number of products intended for use on the North Sea all further research is abandoned by the suppliers and attention is directed to the less difficult and more profitable markets elsewhere, thus resulting in exactly the opposite effect as was intended by the regulators. It is therefore necessary that an agreement between regulators, operators and suppliers can be found as to the objectives and goal of the regulation of offshore chemicals and the route to follow to achieve them. Without such a combined agreement a lot of confusion will arise which may do more damage than good and an improvement of the current situation is not likely to occur.

27

AN INTRODUCTION TO THE CHARM MODEL

Henk P. M. Schobben,[1] Eilen A. Vik,[2] Guus G. Hutjes,[3] Chris C. Karman,[1] and Gro D. Øfjord[2]

[1] TNO Institute of Environmental Sciences, Energy Research and Process Innovation, Environmental Quality and Safety Division
Department for Ecological Risk Studies
P.O. Box 57
NL-1780 AB Den Helder
The Netherlands
[2] Aquateam Norwegian Water Technology Centre a/s
P.O. Box 6326 Etterstad
0604 Oslo
Norway
[3] IRO, Association of Dutch Suppliers in the Oil and Gas Industry
P.O. Box 7261
NL-2701 AG Zoetermeer
The Netherlands

BACKGROUND AND OBJECTIVES

At the Third International Conference on the Protection of the North Sea (March 1990), the ministers of the participating countries agreed (paragraph 28.3) 'to develop and adopt a harmonised mandatory control system for the use and discharge of offshore E&P chemicals'. The Paris Commission (PARCOM) was requested to develop such a system.

In this context, the Dutch and Norwegian authorities, together with the branch associations of the international offshore industry, E&P Forum (The oil industry international Exploration & Production Forum) and EOSCA (European Oilfield Speciality Chemicals Association), initiated the development of a decision supporting model called CHARM (Chemical Hazard Assessment and Risk Management). IRO, Aquateam and TNO where contracted to develop the model.

The objective of the CHARM project is to develop an environmental risk evaluation model for the use and discharge of offshore E&P chemicals (including Hazard Assessment, Risk Analysis and Risk Management procedures). This model can serve as a tool in the implementation of the legislation regarding E&P chemicals.

The basic requirement for the CHARM model was to provide a transparent and simple calculation of hazard and risk levels for the marine environment on the basis of relevant properties of chemicals, with logical (i.e. programmable) rules for data handling, default definitions and calculation principles.

PROJECT ORGANISATION

The project was initiated in 1992 by the Dutch and Norwegian authorities, in co-operation with the representatives of the E&P industry in the Netherlands, Norway, Denmark and the United Kingdom (NOGEPA, OLF, NSOC-D and UKOOA respectively) and the branch organisations E&P Forum and EOSCA. Since 1994 the authorities of the United Kingdom and Denmark actively participate in the project. The authorities of other member countries of OSPAR are invited as observers to the project.

The following participants (in alphabetical order) financially supported the second phase of the project:

- Danish EPA (Environmental Protection Agency)
- EOSCA (European Oilfield Speciality Chemicals Association)
- Netherlands Ministry of Economic Affairs
- Netherlands Ministry of Housing, Physical Planning and Environment
- Netherlands Ministry of Transport, Public Works and Water Management, Rijkswaterstaat
- NOGEPA (Netherlands Oil & Gas Exploration and Production Association)
- NFR (The Norwegian Research Council)
- NSOC-D (North Sea Operators Committee - Denmark)
- OLF (Norwegian Oil Operators Association)
- Norwegian SFT (State Pollution Control Authority)
- UK DTI (Department of Trade and Industry)
- UKOOA (United Kingdom Oil Operators Association)

IRO serves as the secretary of the project while Aquateam a/s and TNO were responsible for the development of the model.

PROJECT HISTORY

The project was initiated in 1992. A first phase was conducted in 1993. In 1994, a second phase was executed. During the spring of 1995 an interim phase was organised in which participants and their affiliates got the opportunity to test and evaluate the model. Based on the results of this evaluation a scope of work for the third phase, which started in the summer of 1995 and will end in September 1996, was defined.

During the first phase of the CHARM project (1993), a framework for Hazard Assessment, Risk Analysis and Risk Management was developed. In this phase most attention was paid to the development of the overall concept and the Hazard Assessment module in particular. Less energy was put into the development of Risk Analysis and Risk Management. At the end of the first phase, the model was restricted to the environmental evaluation of the use and discharge of production chemicals (Schobben et al., 1994a). A provisional version of the software entitled 'CHARM-Wizard' (V1.0) was released (Karman & Schobben, 1994).

The conceptual model was evaluated by means of a sensitivity analysis in the second phase of the project (1994), and verified by participants in the project and external experts. Necessary adaptations were defined and executed (Schobben et al., 1994b). The model was further extended to encompass categories such as drilling, cementing, completion and workover chemicals. Modules for Risk Analysis and Risk Management were developed. All these subjects were discussed during four workshops which were organised during this phase. Finally, a much more sophisticated version of the CHARM-Wizard (V2.1) software was released, together with an accompanying manual (Karman, 1994).

The present CHARM model (Schobben et al., 1994c) is a more or less complete tool for the evaluation of offshore E&P chemicals. There are, however, important differences in the status of the different modules. The structure of the Hazard Assessment module is agreed upon by all the participants. The Risk Analysis module has only recently been developed and has not yet been verified by all the participants. While the model is fully capable of handling production chemicals, it's ability to evaluate drilling chemicals (and especially cementing, completion and workover chemicals) is not yet fully established. The model will be elaborated in full detail in the current third phase of the project.

During a specialist workshop of GOP (Paris Commission) in Aberdeen (September 1993), the first discussion on CHARM was initiated. The results of the first phase were presented and discussed at the GOP meeting held in February 1994 (GOP 18/4/9). In June 1994, the OSPAR Commission concluded that a combination of CHARM and the PARCOM Harmonised Offshore Chemical Notification Format was a good basis for establishing the harmonised mandatory control system requested by the North Sea ministers. The Commission also concluded that the dialogue and co-operation between national authorities and industry carried out within this CHARM project had been very constructive. During the 4th International Conference on the protection of the North Sea the ministers agreed to invite OSPAR "to adopt a Harmonised Mandatory Control System for the use and discharge of chemicals offshore, if possible at OSPAR 1996, taking account of the CHARM model and its related developments" (Anonymous, 1995).

SCOPE OF THE SYSTEM

CHARM is a decision supporting tool and not a decision imposing method. CHARM is based on the best available scientific knowledge. However as a tool to support the regulation of presently used E&P chemicals decisions were taken as to reach a result. This means that it is based on the present situation concerning availability of information. Due to the use of default values and safety factors the model can run on a minimum of information. CHARM is capable of performing all standard calculations on the basis of the PARCOM Harmonised Offshore Chemical Notification Format. The more information available however, the more accurate the result. The user of the model is able to define criteria for the use of CHARM which of course has an influence on the result. The model is focused on the environmental evaluation of the use and discharge of E&P chemicals, on the basis of the data available on these chemicals and platform related conditions. It has to be noted that CHARM does not cover potential harm occurring during the production and transportation of chemicals or during the handling of unused remainders. It only produces information on potential harm occurring in the marine environment (in casu the North and Norwegian seas). Potential air pollution problems and human health problems are not within the scope of CHARM.

OUTLINE OF THE SYSTEM

CHARM enables a stepwise environmental evaluation of E&P chemicals by means of a successive Pre-screening - Hazard Assessment - Risk Analysis - Risk Management.

Product	HQ Pelagic	HQ Benthic	HQ Foodchain	HQ Ecosystem
B	1.3	0.7	2.1	2.1
D	2.5	3.1	0.8	3.1
F	3.8	1.0	2.9	3.8
H	6.2	4.8	2.4	4.8
G	1.7	2.3	5.2	5.2
E	0.5	2.1	5.3	5.3
C	8.4	2.8	1.2	8.4
A	0.8	1.4	9.5	9.5

Figure 1. Typical result of the Hazard Assessment module.

Pre-screening. Based on hazardous properties that fall beyond the scope of a 'PEC:PNEC' comparison (i.e. long term persistency and/or accumulation potential), will be executed prior to the Hazard Assessment and Risk Analysis being carried out. The user will be notified (by means of a 'flagging-system') as to when the data will not allow a valid and accurate hazard or risk assessment to be arrived at .

Hazard Assessment. The assessment of the potential of a substance to cause harm to the target groups exposed to it (according to EC-definitions for hazard identification: the identification of the adverse effects which a substance has an inherent capacity to cause: EC Directive 93/67/EC). Within CHARM, Hazard Assessment covers a general environmental evaluation of a chemical on the basis of the intrinsic properties of that chemical. The hazard is expressed in one general unity, with a magnitude related to (non exceptional or accidental) 'realistic worst case' conditions for use at a reference oil or gas platform. Hazard Assessment is primarily intended for the relative comparison of single chemicals, i.e. for ranking or classifying chemicals.

Risk Analysis. The estimation of the probability of the occurrence of harmful effects (according to EC definitions for risk characterisation: the estimation of the incidence and

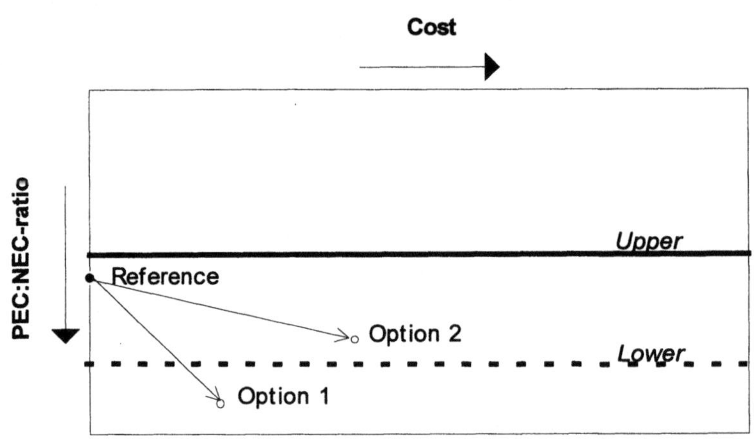

Figure 2. Example of a plot of the cost versus risk benefit of 2 options for risk reduction (o). For reference, an Upper and Lower Risk Level are shown. The angle of each line with the y-axis is a measure of the cost effectiveness of each option.

severity of the adverse effects likely to occur in a human population or environmental compartment due to actual or predicted exposure to a substance, EC Directive 93/67/EC). Within CHARM, Risk Analysis covers a more specific evaluation of the environmental impact from the discharge of a chemical, or a combination of chemicals, under actual conditions. Such a specific analysis enables Risk Management to be implemented on the basis of various scenarios for environmental care options and the input of various cost options.

Risk Management means to take measures that are appropriate for at least diminishing significantly the presence of and/or exposure to a substance in situations of concern. Within CHARM risk management is based on the interpretation of a two-dimensional graph in which the risk reduction of a measure is plotted against its cost (see Figure 2).

THE PEC/PNEC APPROACH

The environmental Hazard Assessment and Risk Analysis are based on a comparison of the intensity of the environmental exposure to released chemicals and the concurrent sensitivity of marine biota. This is worked out within CHARM by the internationally adapted method of proportional comparison of a PEC (= Predicted Environmental Concentration) with a PNEC (= Predicted No Effect 'threshold' Concentration). The PEC is estimated from the use, release, dilution and fate of the chemical, on the basis of a logical combination of properties such as P_{ow} and biodegradation. The PNEC is estimated from the toxicity data available on the chemicals, on the basis of agreed OECD principles. Both are combined in the PEC/PNEC ratio which is the central quantity in the model.

The approach is extended to three environmental compartments: water, sediment and biota. Release of offshore E&P chemicals to the air is not taken into consideration. In the water compartment, the occurrence of a chemical can have adverse effects on the pelagic biota (i.e. organisms that live in the water column of the sea, such as plankton and most fish species). In the sediment, a chemical can affect the benthic biota (i.e. the organisms that live in and on the sediment layer of the sea, such as worms, echinoderms, crabs and bivalve molluscs). The third compartment, 'biota', is different from water and sediment, in that it has no physical boundaries. It is important because a certain concentration of a chemical in an organism (due to accumulation) can have adverse effects on its predators, and on the predators' predators subsequently. The highest of the three PEC/PNEC ratios is used to

Table 1. Synoptic list of the required data, related to chemical characteristics

Parameter	Unity
Trade name	–
Application group	–
Type of platform	–
Dosage	$mg.l^{-1}$, $mg.kg^{-1}$, m^3
Specific gravity	$kg.m^{-3}$
N-octanol/water partition coefficient (P_{ow})	–
Molecular weight	–
Aerobe biodegradation	fraction
Fraction non-solvent	fraction
Toxicity tests	
Pelagic	$mg.l^{-1}$
Benthic	$mg.kg^{-1}$ sediment (DW)
Food chain	$mg.kg^{-1}$ food (FW)

Table 2. Synoptic list of required data, related to platform characteristics

	Parameter	Unity
Production		
	Water production	$m^3.d^{-1}$
	Oil or gas production	$m^3.d^{-1}$
	Condensate production	$m^3.d^{-1}$
Drilling site		
	Discharge of water miscible muds	$m^3.well^{-1}$
	Discharge of water immiscible muds	$m^3.well^{-1}$
	Discharge of cement	$m^3.well^{-1}$
	No. of wells per year	–
	Drilling time per well per site	d
	Specific gravity of sediment	$kg.m^{-3}$
Environment		
	Dilution for PEC_{water}	–
	Refreshment rate	$fraction.d^{-1}$
	Platform density	km^{-2}
	Water depth	m
	Fraction organic carbon in sediment	fraction
	lipid content biota	fraction

characterise the hazard or risk of the discharge of a chemical, as this approach avoids arbitrary weighting amongst compartments and still ensures protection of the other compartments by means of measures to minimise or reduce hazard and risk.

DATA REQUIREMENTS

The data requirements of the CHARM model are related to chemical products or to platforms. With regard to chemical characteristics, the data required by CHARM are all included in the PARCOM Harmonised Offshore Chemical Notification Format. The most important are data on dosage, N-octanol/water partition coefficient (P_{ow}) and toxicity of the product.

In case of Hazard Assessment the platform characteristics are defined as a realistic worst case situation (based on the 95 percentile values for the different parameters). In Risk Analysis or Risk Management it is possible to define a specific platform instead of the standard platforms defined for Hazard Assessment. The necessary parameters are presented in the following table (if no data is available, values from the standard platforms can be used as default values).

FURTHER DEVELOPMENTS

The CHARM project is planned to finish in September 1996. By than a new report with a full description of the model and its technical backgrounds will become available. There will also be a new CHARM Wizard (version 3.0). The model will be completed and the value of its results can be seen in the light of the results of the uncertainty analysis and the validation studies which were part of the third phase of the project.

REFERENCES

Anonymous (1995): Esbjerg declaration. 4th International Conference on the protection of the North Sea.

Karman C.C. (1994): CHARM Wizard V2.1. TNO-MW R94/315.

Karman C.C. & H.P.M. Schobben (1994): CHARM Wizard: Tool for Chemical Hazard Assessment and Risk Management. Manual. TNO report.

Schobben H.P.M., M.C.Th. Scholten, E.A. Vik & S. Bakke (1994a): CHARM: An environmental risk evaluation model for offshore E&P chemicals. Report TNO, Den Helder (NL) & Aquateam A/S, Oslo (N).

Schobben H.P.M., C.C. Karman, M.C.Th. Scholten & H. van het Groenewoud (1994b): CHARM 2.0. TNO-report R94/189.

Schobben H.P.M., C.C. Karman & M.C.Th. Scholten (1994c): CHARM 2.1: Chemical Hazard Assessment and Risk Management of offshore Exploration & Production chemicals. TNO-MW R94/315.

28

THE CHARM MODEL USED IN ENVIRONMENTAL RISK MANAGEMENT OF PRODUCED WATER ON ULA

Gro D. Øfjord, Siri Bakke, and Eilen A. Vik

Aquateam - Norwegian Water Technology Centre A/S
P.O. Box 6326 Etterstad, 0604 Oslo, Norway

INTRODUCTION

The offshore exploration and production (E&P) industry is facing increasing costs from meeting its environmental responsibilities. Environmental risk management is therefore becoming an increasingly important issue, especially in view of the discharge of a large number of pollutants potentially harmful to the environment. Environmental risk management is a step-wise process that generally starts with risk analysis of alternative remediation strategies. Through risk analysis the probability of the occurrence of adverse biological effects as a result of exposure to a chemical can be estimated. Fate/risk models are generally applied to predict environmental concentrations. The next steps in the risk management process can cover political decision-making for determination of acceptability criteria and taking risk reducing measures (van Leeuwen et al., 1992). The third step is cost determination of alternative options, and the last step will be a cost/environmental risk benefit comparison.

Chemicals play an important role in offshore oil production. Unfortunately, some of these chemicals - amounting to several hundred tons a year - will end up in the marine environment with the produced water. Produced water is a mixture of different compounds that may have a negative effect on the marine environment. In addition to the chemicals, the produced water also contains various amounts of other environmental pollutants such as oil, aromatic and aliphatic hydrocarbons and heavy metals (Johnsen et la., 1993; Schobben et al., 1994 and Slager et al., 1992). As the oilfields mature, the produced water volumes will increase, more chemicals are used, and the environmental risks increase. The risk can be reduced by choosing less harmful chemicals, improving the water treatment, reinjecting produced water or selecting alternative (corrosion resistant) materials. These choices are, however, not always clear cut. Which option is the better for the environment and to which cost? The Oslo and Paris Commission (OSPAR) was in 1990 requested to develop and adopt a harmonized mandatory control system for the use and discharge of offshore E&P chemicals. The CHARM (Chemical Hazard Assessment and Risk Management) model was developed on initiative from the authorities and the offshore industry in Norway, the Netherlands, the UK and Denmark. The model aims to give the oil companies, chemical suppliers and legislators a scientific framework for analyzing the environmental risk of chemicals used and discharged during drilling and production activities.

This paper describes environmental risk analysis and risk management, and gives an example on how the CHARM model can be used. The information on water qualities, volumes and types of chemicals used, options evaluated and the costs of various management options are based on data from the Ula platform (operated by BP Norway Ltd.). The data have been provided by BP. The example presented is based on various options for produced water handling.

DESCRIPTION OF THE CHARM MODEL

The CHARM model is developed by Aquateam - Norwegian Water Technology Centre A/S (N) and TNO - Marine Laboratory (NL), (Schobben et al. 1994 and Øfjord et al. 1995).

The model enables a stepwise environmental evaluation by means of a successive Pre-screening, Hazard Assessment, Risk Analysis, and Risk Management on the basis of available data on chemicals and platform related conditions.

The environmental Hazard Assessment and Risk Analysis are based on both a comparison of the intensity of the environmental exposure to released chemicals and the concurrent sensitivity of the marine biota. This is based on the internationally adapted method of proportional comparison of an estimated PEC (Predicted Environmental Concentration) with an estimated NEC (No Effect Concentration). The PEC is estimated from the use, release, dilution and fate of the chemicals. The NEC is estimated from the available toxicity data on the basis of agreed OECD principles (OECD, 1993). Both are combined in the PEC:NEC ratio (risk quotient) which is the central unit in the model. The PEC:NEC approach includes three environmental compartments: water (pelagic), sediment (benthic) and food chain (biota). The highest of the three PEC:NEC ratios is used to characterize the risk (PEC/NEC for the ecosystem). This approach avoids arbitrary weighting amongst compartments and still ensures protection of the other compartments by means of measures to minimize or reduce risk.

In the Hazard Assessment module, chemicals can be ranked or classified based on specific information of the chemicals combined with general information on a standard reference platform. The Risk Analysis is based on a combination of information on a (or a number of) chemical(s) and a site specific platform. Risk Analysis is possible on the basis of a single chemical, but it is also possible to evaluate a total package of all the chemicals used on a platform, together with other toxicants in produced water.

The Risk Management module implies the definition and evaluation of measures to be taken in order to reduce the environmental risk in a cost-effective way. Within CHARM, Risk Management is restricted to evaluating measures defined by the user (scenarios), by using the Risk Analysis module.

ENVIRONMENTAL RISK

Risk Analysis

The Risk Analysis can be performed on the basis of the concentrations and ecotoxicological information of individual substances in produced water together with the same type of information for the chemicals. The risk of these combined toxicants, however, cannot simply be estimated from the sum of the PEC:NEC ratios of the separate substances. If an organism dies due to the exposure to one toxicant, it cannot die again when exposed to another. The PEC:NEC ratio is no real measure of risk, but is only an (non-linear) indication of the probability that the ecosystem is affected (Schobben et al., 1994). A simple calculation rule is used to translate the PEC:NEC ratio of individual chemicals/toxicants into an equivalent risk estimate:

$$P(A+B) = P(A) + P(B) - P(A) \bullet P(B) \quad (1)$$

where P(X) is the probability for the ecosystem of being affected by exposure to toxicant (X).

The relationship between a PEC:NEC ratio (i.e. the risk quotient) and the risk is a function of the sensitivity of all species. In CHARM, the risk of a package of chemicals and other substances in produced water is calculated by the application of the following steps:

1. calculation of the PEC:NEC ratio for all individual chemicals and all substances in produced water,
2. translation of the PEC:NEC ratios into risk estimates for individual chemicals using a calibrated relation on the basis of a lognormal distribution of species sensitivity,
3. combination of the risk estimates of all substances into one risk estimate for the package of chemicals and the composition of produced water using equation 1, and finally,
4. translation of the overall risk estimate into one integrated PEC:NEC ratio using a fixed relation.

Risk Management

Once Risk Analysis is performed, Risk Management is enabled on the basis of various scenarios for environmental care options and input of various cost options. Risk Management means to take measures which are appropriate at least to diminish the presence of, and/or exposure to, a substance in the situation of concern. The environmental Risk Management,

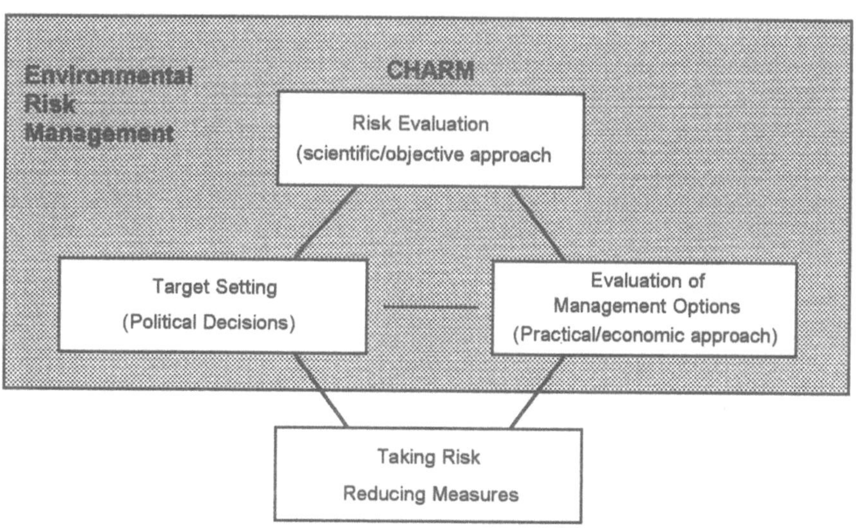

Figure 1. Environmental Risk Management process in CHARM.

as it is presently used in CHARM, covers potential management options for continuous discharges of chemicals and other produced water toxicants from offshore E&P activities to the marine environment. The basic principle is to imply measures to be taken to reduce environmental risk in a cost-effective way. The environmental risk management process includes the steps shown in Figure 1. The process is described below.

Risk Evaluation. The risk evaluation is based on the Risk Analyses performed for all defined management options. This has previously been described in Chapter 3.1.

Target Setting. The target setting includes defining the maximum tolerable risk (MTR) and a negligible risk (NR). Such targets can be defined by the environmental authorities, by the operator or by the chemical supplier. The target setting is usually combined with time frames and is defined based on observed environmental effects and acceptable costs to combat environmental problems.

Environmental Management Options. Environmental management options are defined by the operator on the basis of available practical solutions and the economic implications.

Practical solutions. Available measures to comply with potential negative environmental effects of produced water may include changes in the application of a chemical (e.g. dosage), the selection of alternative chemicals and alternative technical solutions for produced water handling. These technical solutions may include selection of chemicals, alternative materials, improved water treatment or reinjection of produced water.

Economical approach. The cost/benefit philosophy is based on the identification of those measures which give the highest risk reduction per unit cost. The measures can then be ranked according to their cost/benefit ratio. A two-dimensional plot is used within CHARM to enable a simple evaluation of alternative measures (Figure 2). This implies that no weighting is done between economical and ecological effects. By changing the dosage or the fraction released of a chemical, or by selecting an alternative chemical, the user of the

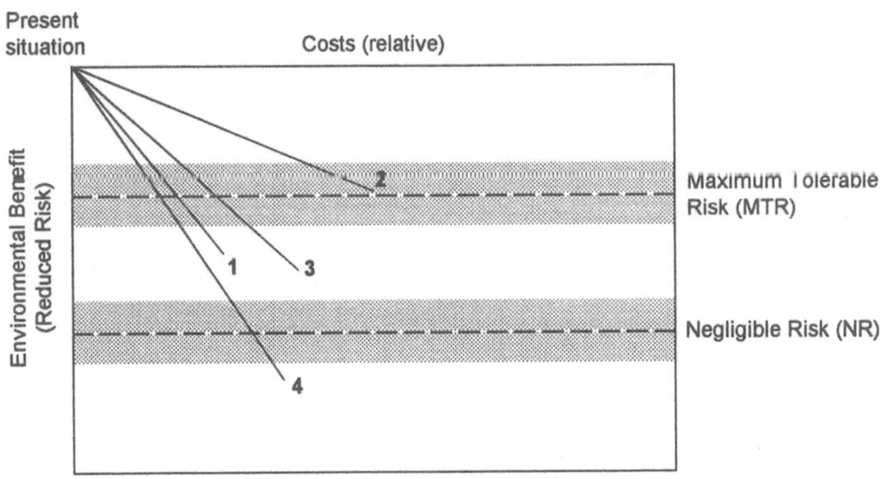

Figure 2. Illustration of environmental cost-effectiveness of four different management options using CHARM.

model can define measures for which the risk reduction is calculated by the model. The different measures are plotted as points in the "ecology-economy" graph. These points are linked to the "present risk" point, so that the angle between each line and the y-axis is a measure of the cost-effectiveness of each scenario. The interpretation of the points in this graph can be improved by the introduction of target settings related to the necessity for intervention (maximum tolerable risk, MTR) and the security of environmental sustainability (negligible risk, NR). These target settings are usually based on political decisions.

Cost-effectiveness analyses are generally used to determine the optimum environmental solution to obtain the maximum environmental protection at the lowest possible cost. Holgate et al. (1993) proposes to use such a cost-benefit model to define the best use (environmental cost-effectiveness) to derive a cost/environmental benefit ratio based on the cost of meeting the risk reduction targets.

The cost/environmental benefit ratio for an environmental solution (of known cost in net-present value) can be calculated when the reduction in environmental risk (of known environmental benefit/effectiveness) is known.

$$\text{Cost/environmental benefit ratio} = \frac{\text{Cost of environmental solution}}{\Delta \frac{\text{PEC}}{\text{NEC}}\ \text{reduction}} \left(\frac{\text{ECU}}{\text{PEC/NEC units}} \right) \quad (2)$$

The cost/environmental benefit ratios are useful to operators as they can be used for direct ranking of environmental projects alongside commercial projects. These values can also be used to help the governments assess the costs of legislation and provide the basis for target settings.

ENVIRONMENTAL MANAGEMENT OPTIONS FOR HANDLING OF PRODUCED WATER AT ULA

General

The produced water quality/composition varies from oilfield to oilfield. Also, there are large variations in the volumes produced. On a typical oil production platform in the North Sea, seawater is injected to maintain the production pressure in the reservoir. Production and injection chemicals are used to handle operational problems (e.g. scaling, corrosion, bacteriological growth), and to improve the water treatment process (e.g. oxygen removal, oil/water separation). The type of chemicals needed in production and injection will, to a large extent, depend on the water and oil composition, the technical facilities selected and vary from field to field. From an environmental point of view, the implementation of the 40 mg/l hydrocarbon discharge limit for produced water, has been the driving regulation for produced water handling. The focus which have been pointed towards the chemicals in use, opens up for other management considerations. Produced water also contains other constituents than oil, many of which might have an environmental impact. Typical produced water contains hydrocarbons, dissolved organics (residual chemicals, aromatic compounds such as PAH, BETX (benzene, ethylbenzene, toulene, xylene), phenols and organic acids), as well as heavy metals.

The degree of liberty regarding environmental management is much larger on the planning stage for a new field than for a field which has already been in operation for some time. Handling of produced water to reduce the environmental impact of discharges may include the following alternatives:

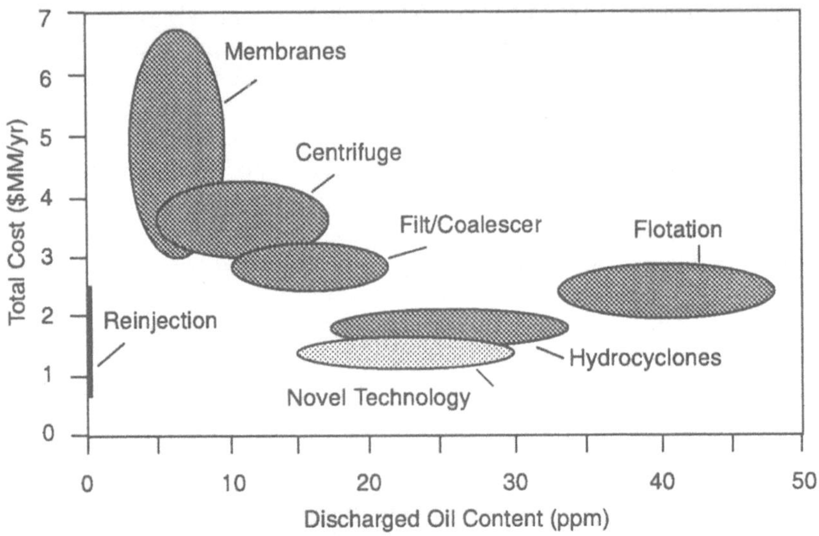

Figure 3. Various options and costs for improved water treatment (according to Paige and Sweeney, 1993).

Selection of Alternative Chemicals. The environmental risk of various chemicals varies with the amount used, the chemicals' intrinsic properties, and the ecotoxicological characteristics. Selection of an alternative to a very toxic chemical or a persistent chemical would be the first choice. Optimization of the dosage is also important.

Improved Water Treatment. Improved water treatment will reduce the content of hydrocarbons in the produced water. Since the focus has been on oil reduction, the techniques considered have so far mainly been concentrated on hydrocyclones and gas flotation units. When the focus is changed to include dissolved organics in produced water, there are several other processes which should be considered.

To the extent where these compounds are adsorbed to particulates or to oil, the centrifuge and membrane processes are relevant alternatives. Finer membranes, ion exchange processes, and biological treatment are typical processes which also could be considered.

The costs for the various water treatment options vary greatly. Figure 3 summarizes the relative costs for installing and operating various water treatment options in an offshore system (Paige and Sweeney, 1993).

Produced Water Reinjection (PWRI). The environmental risks associated with produced water discharges will increase with increasing water production unless improved treatment (see above) or reinjection of produced water (PWRI) is implemented. Reinjection can be part of the existing water injection system (a mixed sea/produced water system) or it can be two separate systems (one for seawater and one for produced water). However, limited experiences exist with respect to costs and technical practicality of PWRI. Several technical problems need to be considered. These are especially connected to injectivity, scaling, reservoir souring and corrosion. Each field needs to be considered separately relative to reservoir conditions and quality of produced water and seawater.

Options Evaluated on Ula

The Ula field is operated by BP Norway Ltd. and is located in the Norwegian sector of the North Sea. The oil production started in 1986, the seawater injection in 1988, and the water production in 1992. When the water production started, the produced water was treated in a gas flotation unit (GFU). This water treatment plant was designed and built long before the water production started. The oil reserves in the reservoirs connected to Ula turned out to be approximately four times larger than originally anticipated. The result was a higher water production. As the produced water volume increased, the original water treatment plant had too low capacity to handle all the water needed on Ula throughout the lifecycle of production. Hydrocyclones were therefore installed in 1994. The hydrocyclones were designed to treat all the produced water down to 20-30 mg/l hydrocarbons (Bakke et al., 1995). Some of the production chemicals were replaced.

BP has at several occasions looked into the possibilities of reducing the volumes of water discharged by produced water reinjection (PWRI) at the Forties, at the Prudoe Bay and at the Wytch Farm fields. There are different reasons why PWRI has been introduced, and the practical experiences have been variable. BP decided to look closely into this environmental management option for Ula. The positive results obtained from a single well trial during the spring of 1994 on Ula, was the push to initiate a full scale PWRI trial using the existing water injection system. This full scale reinjection trial was initiated in February 1995, where a mixed produced water/seawater stream is being injected.

The situation on the Ula platform before replacement of the gas flotation unit with the hydrocyclones, and therefore the change in chemical usage, has been defined as the reference situation. Two options have been elaborated for comparison to the reference situation using the Risk Management module in CHARM.

The Reference Situation. This was the situation on the Ula platform in 1994 before the hydrocyclones were installed. The water treatment plant consisted of GFU. The treatment system had problems meeting the environmental legislations of 40 mg/l hydrocarbons in produced water. Six process chemicals were used; a scale inhibitor, a demulsifier, and a surfactant in the production system and a biocide, an oxygen scavenger, and an antifoam in the injection system.

Option 1 - Improved Water Treatment. This involved replacement of the GFU with hydrocyclones. The surfactant, the scale inhibitor, and the demulsifier were replaced. The new chemicals included a combined scale/corrosion inhibitor and a new demulsifier. The introduction of a the water treatment system implied large investments, but an anticipated reduction in operational costs.

Option 2 - Reinjection of Produced Water. The same water treatment system and the same chemicals as in option 1 are used for the PWRI. In theory, no discharge of chemicals and other produced water constituents will occur. It is, however, assumed that due to technical problems that may occur during injection, the actual produced water discharge will be approximately 10% of the discharge in option 1, i.e. a 90% reinjection of all water produced. This entailed investments for PWRI equipment, but also some reduction in the operational costs. The use of oxygen scavenger and antifoam has also been reduced due to treatment of less seawater.

Table 1. Environmental data for the injection and production chemicals used on Ula

Chemical	Dosage (mg/l)	Biodegradation (%)	Log P_{ow}	Toxicity $E(L)C_{50}$ Algae (mg/l)	Crustaceans (mg/l)	Sediment reworker (mg/µg)	Rat (mg/µg)
Injection							
Biocide (1,2)	6	40	0	0.5	1	0.1	2,700
Oxygen Scavenger (1,2)	5	98	0	280	550	13,000	2,000
Antifoam (1,2)	0.2	0	3.8	50	200	4,200	200
Production							
Scale inhibitor (1)	30	2	0	3,000	350	820	100,000
Demulsifier (1)	8	35	4.4	20	70	290	15,000
Surfactant (1)	40	70	2.9	4	8	2,700	25
Corrosion/Scale inhibitor (2)	25	5	3.1	2	3	500	100,000
Demulsifier (2)	10	45	4.2	70	80	5,000	100,000

(1) Reference situation
(2) Options 1 and 2

RESULTS AND DISCUSSION

Environmental Risk Benefit

Ecotoxicological data and information on the use of the Ula chemicals are presented in Table 2. The produced water from Ula has been analyzed for several organic substances and heavy metals. Data on the concentrations of these substances, as well as literature data both for concentrations in produced water and ecotoxicological information, are presented in Table 2.

The data in Tables 1 and 2 are used in the Risk Analysis of the produced water discharges on Ula for the scenarios described in Chapter 4.2. As seen from Table 2, the literature toxicity data on different substances in produced water varies depending on the exact substances and on the organisms studied. PAH includes a whole range of aromatic compounds with different toxicity, biodegradation and bioaccumulation potential, and so does BETX. Limited information is available on the toxicity to marine ecosystems, including the OSPAR organisms. Sometimes the variation in toxicity is up to two orders of magnitude. The values used in the risk analysis may have a significant impact on the result. The toxicity values used in this example are chosen from the lower range.

The individual risk quotients (PEC/NEC ratios) have been calculated for three environmental compartments, for the production chemicals, as well as for the other produced water constituents, as described in Chapter 2. The highest risk quotient of the three compartments equals the risk quotient to the ecosystem. The individual risks of each substance have been combined and translated into a risk quotient for the total produced water. The combined risk is calculated for the reference situation and the two options described in Chapter 4 (see Table 3).

From the table it can be seen that the combined risk in option 1 is reduced by the improved water treatment and subsequently, significantly reduced in option 2 when produced water reinjection is implemented. The risk in the reference situation is 0.57, indicating that there is a 57 % probability that the produced water discharge will pose a risk to the marine

Table 2. Toxicity data and concentrations of some selected constituents in produced water*

	Concentrations in produced water (mg/l)					Toxicity (mg/l) $E(L)C_{50}$ Literature data		
	Ula data							
Parameter	Reference situation	Options 1 and 2	Literature data	Biodegradation (%)	Log P_{ow}	Algae	Crustaceans	Fish
BETX	12.0	14.3	0.4-5	50	2.7	20-80	11.5-1200	
PAH	3.0	0.66	0.1-3	64	3.6	0.633	0.8-110	
Phenol	0.09	1.7	1.4-23	80	1.5	—	0.4-30	
Organic Acids	13.0	8.2-11.2	0.5-800	90	1.0	90	50-250	
Cadmium	0.02	<0.05	0.01-0.1	0	0	681	0.02-11.6	0.8-50
Copper	0.02	0.12	0.06-0.8	0	0	0.05-0.5	0.5-100	0.03-0.5
Lead	0.27	<0.01	0.05-0.2	0	0	—	0.7-4.4	
Zinc	0.09	0.15	0.49-4.59	0	0	0.1-0.25	0.07-4	60-70
Oil	50	20	20-200	70	3	16	0.74-36	

*Data from Ula and the literature (Jacobs, 1992; Schiff, 1992; Slager, 1992; U.S. Geological Survey, 1989)

Table 3. The combined risk and PEC/NEC ratio of produced water discharges for the reference situation, for option 1 and for option 2

	Reference situation	Option 1: Improved water treatment	Option 2: Produced water reinjection
Combined risk	0.57	0.49	0.03
PEC/NEC ratio	23	17	0.6

environment. The risk of the produced water discharges is reduced from 0.57 to 0.49 by the improved water treatment. The reduction is partly caused by the reduction in oil concentration (from 50 mg/l to 20 mg/l), partly by the general improvement in water quality, and also by the replacement of the demulsifier. When the produced water is reinjected the risk is reduced to 3% and the corresponding PEC/NEC ratio is reduced from 23, for the reference situation, to 0.6. It is generally considered that when PEC < NEC, the ecosystem is likely not to be severely affected by the discharge. Data for the dilution rates around the Ula platform was not available for the risk evaluation. The default value in the CHARM model of 1:1000 at 500 m has therefore been used. Dilution factors up to 1:100,000 have, however, been observed at 500 m around North Sea platforms. The dilution factor of 1:1000 is very conservative and the PEC/NEC ratios are therefore expected to be lower than shown in this example.

The aromatics (BETX, PAH), organic acids and heavy metals do not contribute much to the total risk of the produced water discharge. The concentrations of phenols, organic acids, and zinc (Zn) in the water produced on Ula are, however, low compared to other produced waters in the North Sea (Tibbets et al., 1992). Phenol is expected to have the highest contribution to the environmental risk.

Each constituent in produced water has a different contribution to the total risk of the produced water discharged, as presented in Figure 4.

As seen from Figure 4, the biocide has the far highest PEC/NEC ratio of produced water. It should, however, be questionned whether this is real or if the existing CHARM model needs to be modified. Biocides are added to the seawater injection system and is

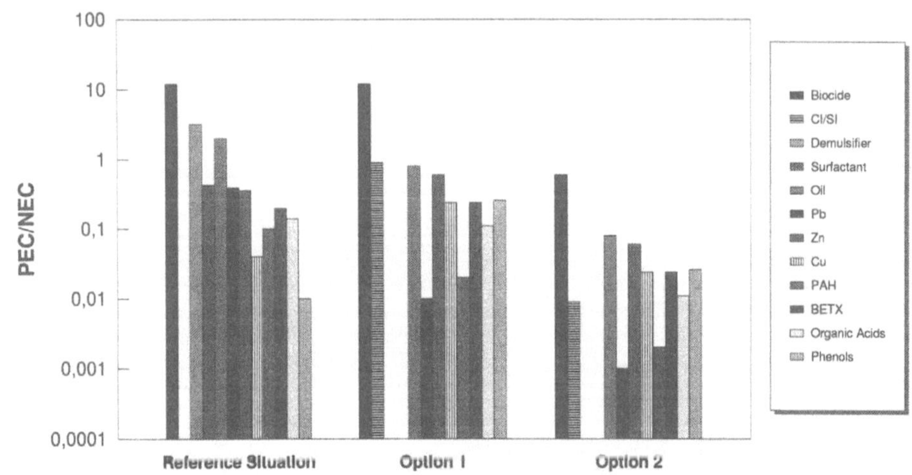

Figure 4. PEC/NEC ratio of the different constituents in produced water for the reference situation and for the two options.

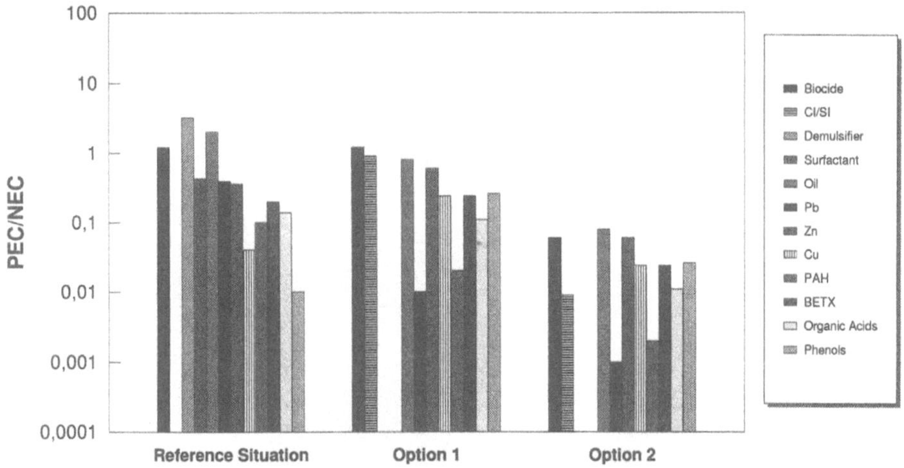

Figure 5. PEC/NEC ratio for the different constituents in produced water when the fraction of biocide released is reduced to 1/10.

injected with the water to the reservoir batchwise. Biocides are consumed in the reservoir. At the present stage in the development of the CHARM model, this has not been accounted for. The shortest retention time registered in the Norwegian sector of the North Sea is 3 months. However, most reservoirs in the North Sea have much higher retention times. The half-life of biocides is estimated to 1 month (Graversen et al., 1993). Assuming that the retention time is in the range of 3 months, the fraction of biocide released with the produced water will be reduced to approximately 1/10 of the original concentration. This is significantly reducing the contribution of the biocide to the total risk. Figure 5 shows the PEC/NEC ratios when the biocide concentration is reduced to 1/10.

The corrosion/scale inhibitor and the demulsifier (used in the reference situation) also contribute significantly compared to the other chemicals. When looking at the risk of the toxicants in the produced water, the oil content gives the highest risk quotient. The oil content of produced water was reduced when the flotation units were replaced with hydrocyclones. Taking these facts into consideration, the possibility for reducing the risk of produced water discharges can be improved by improved water treatment.

The relative contribution of the chemicals and the other toxicants to the risk can be presented using a pie chart, see Figure 6. The biocide gives a significantly higher contribution to the risk than any other chemical or toxicant in the produced water. When the relative contribution of the biocide to risk is calculated for the two options, it can be seen that even though the discharge is reduced by 90%, the relative risk of the biocide is increased. The explanation to this is that the relationship between the PEC/NEC ratio and the risk is non-linear.

Different options exist with respect to reducing the environmental impact of production chemicals (Holgate, 1993). The need for biocides can be reduced by the use of ultrafiltration or disinfection to eliminate microorganisms and nutrients. Use of corrosion resistant materials reduces the need for corrosion inhibitors. Replacement of the flotation cells with centrifuges or hydrocyclones, or reduction of the turbulence in the production flow may reduce the need for the demulsifier. In all cases, for all production chemicals, it is important to optimize the dosage.

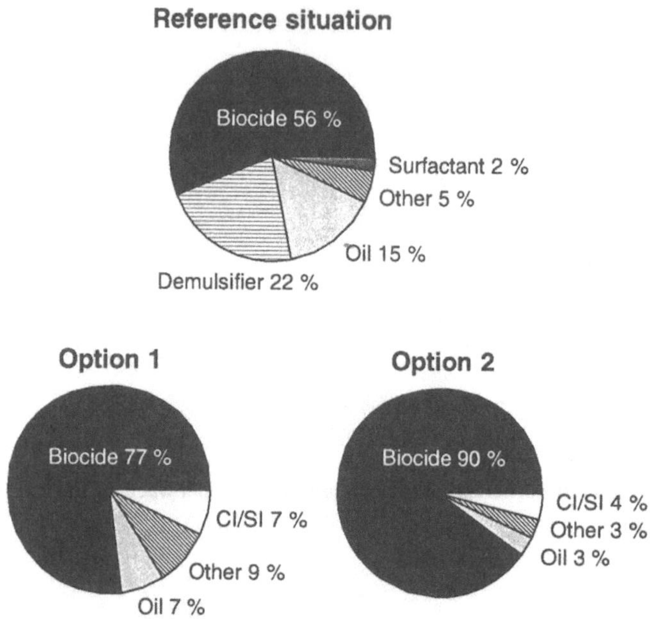

Figure 6. The relative contribution to the total risk of constituents in produced water. Reference situation (GFU and 6 chemicals), Option 1 (improved water treatment, hydrocyclones and 5 chemicals) and Option 2 (PWRI).

Cost Effectiveness

Most options for reduction of environmental impact of discharges of produced water imply additional costs.

For the reference situation, the net-present value is defined as 0. The different options imply additional costs relative to the reference situation. The additional costs for the improved water treatment and the PWRI include investment costs and expected increases/reductions in operational costs. However, not all of the needed information was available for the calculation of the net-present values in this example. Table 4 summarizes the information and gives the resulting estimated cost/environmental benefit ratio for the three different options.

Table 4. Estimated cost/environmental benefit ratio for the two options

Option evaluated	Additional costs compared with reference situation (ΔNOK)	Environmental benefit compared with reference situation (ΔPEC/NEC units)	Cost/environmental benefit ratio (ΔECU/(ΔPEC/NEC units)
Present situation	0	0	0
1. Improved water treatment	1.4	6	0.23
2A. Produced water reinjection	1.5	22	0.07
2B. Produced water reinjection excluding reduced operational costs	1.8	22	0.08

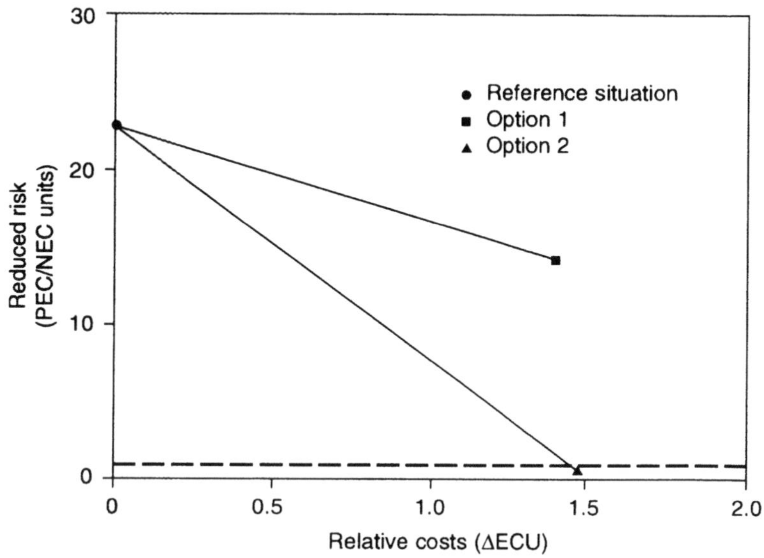

Figure 7. Cost/environmental benefit of the two different options used in the example.

The net-present value for the PWRI - option 2 is only slightly higher than the value for option 1. This is due to the expected savings in operational costs. If these savings are not achieved, the net-present value is increased, as described in Table 4.

The costs associated with options 1 and 2 are estimated costs per year for the Ula platform. The environmental/cost effectiveness relations of the two options can be compared to a reference situation in a graphical presentation, see Figure 7. The Y-axis corresponds to environmental benefit/effectiveness (= reduced risk) and the X-axis corresponds to the net-present value associated with this effectiveness. The slope of each option equals the cost/environmental benefit ratio. The lower the cost/environmental benefit ratio, the higher the cost/ environmental benefit/effectiveness.

The cost/environmental benefit ratio for option 2 - including improved water treatment and 90% produced water reinjection - is the lowest, and this option gives the best environmental/ cost-effectiveness. It must, however, be emphasized that the costs used in the calculations are estimated values. Net-present values for investment costs and expected savings per year due to reduced chemical need and energy consumption are not yet available, and more experience from the full scale PWRI is needed to come up with these figures.

ACKNOWLEDGMENTS

The authors would like to express their gratitude to BP for letting us use the data from the Ula platform and to the oil operators (OLF, NOGEPA, UKOOA), chemical suppliers (EOSCA), and environmental authorities in Norway, the Netherlands, UK and Denmark for their financial support in the CHARM project. Also, the financial support of the Norwegian Research Council (NFR) is gratefully acknowledged.

REFERENCES

Bakke, S.; Vik, E.A.; Grüner, H. and Hjelmås, T.A., 1995: Produced water reinjection - experiences from the Ula field. Paper to be presented at the 1995 International Seminar on Produced Water, Trondheim, Norway, 25-28 September.

E&P Forum, 1994: North Sea produced water. Fate and effects in the marine environment. Report No. 2.62/204, London, May.

Evans, R.C., 1994: Development of environmental protection related to produced water treatments and disposal (produced water reinjection). Second International Conference on Health, Safety & Environment, Jakarta, Indonesia, Volume 1, SPE 27179, 25-27 January.

Graversen, P., Hansen, N. and Christensen, F.B., 1993: Mass balance of production chemicals. Norwegian Oil Industry Association (OLF) Final Report.

Holgate, M.; Mentzer, E.; Davies, G.; Verburgh, J.; Whitehead, A. and Shepherd, J., 1993: Realizing the benefits of environmental risk management. SPE Paper 27610.

Hughes, S.W.; Sehsuvaroglu, S.A. and Slater, J.M., 1994: Produced water treatment technologies: A case study. Second International Conference on Health, Safety & Environment, Jakarta, Indonesia, Volume 1, SPE 27131, 25-27 January.

Jacobs, R.P.W.M.; Grant, R.O.H.; Kwant, J.; Marquenie, J.M. and Mentzer, E., 1992: The composition of produced water from Shell operated oil and gas production in the North Sea. In: Produced water (eds. J.P. Ray and F.R. Engelhart), Plenum Press, New York

OECD, 1993: Guidance document for aquatic effects assessment, Paris.

OLF, 1993: Summary report - Environmental programme - Phase II.

OSPAR, 1995: Harmonised offshore chemical notification format. Oslo and Paris Commissions - Summary record of the meeting of the programme and measures committee (PRAM), Annex 10, Oviedo, Italy, 20-24 February.

Paige, R.W. and Sweeney, F.M., 1993: Produced water reinjection - Understanding the problems. Paper presented at the Water Management Offshore Conference.

Schiff, K.C.; Reish, D.J.; Anderson, J.W. and Bay, S.M., 1992: A comparative evaluation of produced water toxicity. In: Produced water (eds. J.P. Ray and F.R. Engelhart), Plenum Press, New York

Schobben, H.P.M.; Karman, C.C. and Scholten , M.C.Th., 1994: CHARM 2.1: Chemical hazard assessment and risk management of offshore exploration & production chemicals. Final Report, TNO, den Helder.

Slager, L.K.; Van Hattum, B.; Tromp Meesters, M.; Scholten, M.C.Th., Kaag, N.H.B.M.; Cofino, W.P. and Feenstra, J.F., 1992: Environmental aspects of produced water discharges from oil and gas production on the Dutch continental shelf. Part III. Environmental effects. E-92119, Institute of Environmental Studies.

Tibbetts, P.J.C.; Buchanan, I.T.; Gawel, L.J. and Large, R., 1992: A comprehensive determination of produced water composition. In: Produced water (eds. J.P. Ray and F.R. Engelhart), Plenum Press, New York

U.S. Geological Survey, 1989: Applicability of ambient toxicity testing to national or regional water quality assessment. Open file report 89-55.

van der Zandt, P. and van Leeuwen, 1992: A proposal for priority setting of existing chemical substances. EU DG Environment, Nuclear Safety and Civil Protection.

Verschueren, K., 1983: Handbook of environmental data on organic chemicals. 2nd Edition, Van Nostrand Reinhold, New York.

Øfjord, G.D.; Bakke, S.; Vik, E.A.; Edvardsson, T. and Johnson, D., 1995. CHARM-Chemical Hazard Assessment and Risk Management of offshore E&P chemicals. 6th International Symposium on Oil Field Chemicals, 19-22 March, Geilo, Norway.

29

PROVANN

A Model System for Assessing Potential Chronic Effects of Produced Water

Mark Reed,[1] Ståle Johnsen,[2] Alf Melbye,[2] and Henrik Rye[1]

[1] IKU Petroleum Research
[2] STATOIL Research Center
Trondheim, Norway

INTRODUCTION

Produced water from offshore oil and gas production platforms contains a variety of hydrocarbons, heavy metals, and production chemicals. Vertical and horizontal mixing generally brings concentrations in discharge plumes below levels associated with acute effects within 500 or 1000 m of the source (Rye et al., 1996). Chronic effects outside this region, on the other hand, remain a potential problem.

The purpose of PROVANN, the system of models described here, is to assess the potential for chronic effects from produced water. The preliminary focus is on potential bioaccumulation and biomagnification of produced water constituents in the marine food web. Other possible types of chronic effects, such as reduced fecundity, or pheromone response interference, can also be assessed with the model described here to the extent that such effects may be correlated with exposure.

PROVANN simulates 3-dimensional transport, dilution, and degradation of chemicals released into the water, from one or more simultaneous sources. The resulting time series of concentration fields is saved in a file, which then serves as input to a simulation of fish movement (swimming) and fish egg and larval transport.

The PROVANN model system described here consists of four components:

- a near-field release model,
- a far-field transport model,
- a biological exposure model, and
- a bioaccumulation and biomagnification model.

The near-field and far-field models are integrated into one system component and the biological processes into a second, as depicted schematically in Figure 1. Potential effects at the population level could be incorporated through linkage to population dynamics models for selected species of interest.

Produced Water2: Environmental Issues and Mitigation Technologies
Edited by M. Reed and S. Johnsen, Plenum Press, New York, 1996

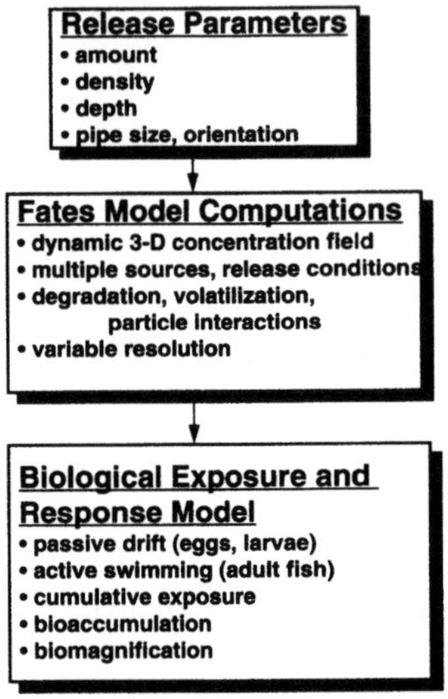

Figure 1. Schematic overview of the PROVANN model system.

The near-field release model is based on the buoyant plume equations developed by Koh and Chang (1973) and Brandsma et al., (1980, 1992). As the plume forms, water is entrained, the plume increases in cross section, and its density approaches that of the surrounding water. In the presence of a density gradient or thermocline, the plume may reach an equilibrium level in the water column. Otherwise, the plume will eventually arrive at the sea floor or the sea surface, depending on its initial buoyancy, depth, strength and orientation.

The far-field model uses individual particles released from the near-field plume to simulate the further transport and fate of contaminants in the water column. A fraction of the contaminant mass is associated with each particle. These particles are Lagrangian in the sense that they move with the surrounding water, unless buoyant oil droplets are being simulated. The use of Lagrangian particles in the contaminant transport calculations effectively eliminates numerical diffusion problems associated with finite difference calculations in the presence of steep gradients (e.g. at the edge of contaminant plumes), and also provides a versatile modeling framework. Both the near-field and far-field models are described in detail in Reed et al. (1995).

The far-field model solves the generalized transport equation in three physical dimensions and time:

$$\frac{\delta C}{\delta t} + \vec{V} \cdot \vec{\nabla} C = \vec{\nabla} \cdot D \vec{\nabla} C + \sum_{i=1}^{n} r_i C$$

Relevant processes (r_i) included in the physical fates component are:

- adsorption/dissolution kinetics,
- entrainment and dissolution of oil droplets into the water column,
- volatilization from water column,
- degradation,
- addition of mass from continuous releases, multiple sources,
- dissolution from sediments to water column,
- deposition from water column to sediments.

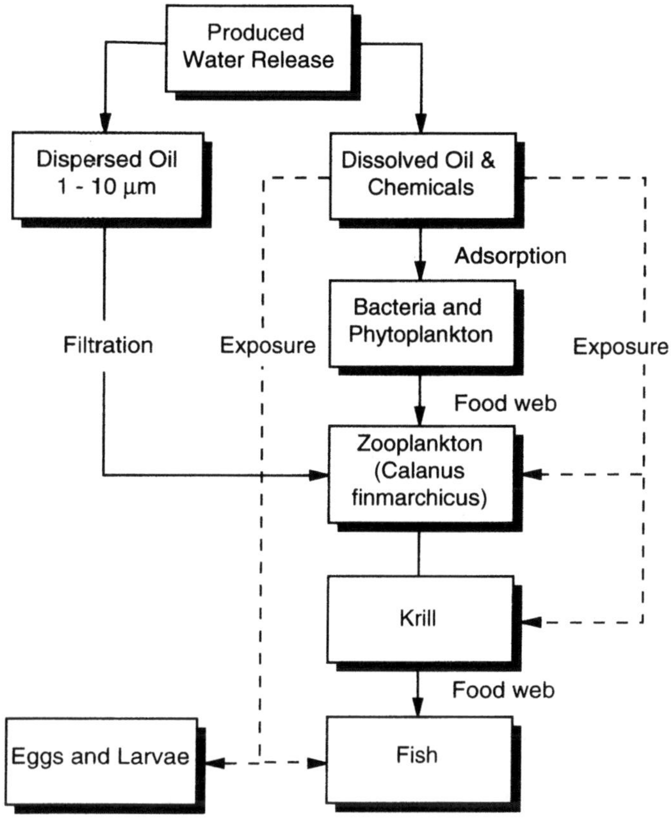

Figure 2. Exposure pathways in the bioaccumulation/biomagnification model. (The uptake of oil droplets by Calanus is not yet included.)

These process rates (r_i) result in mass transfers from one model state to another.

The biological exposure model simulates exposure pathways of planktonic organisms and fish (Figure 2). Phytoplankton, fish eggs and larvae, as well as other zooplankton upon which fish may feed, are simulated as passive drifters. Swimming by pelagic and demersal fish is modeled as a random walk process, with the speed being a function of fish size and type (Figure 3). Some fish within the geographical region being modeled will swim through the plume, which in the present version is assumed not to affect their behavior. Contaminated food organisms which they consume while within the plume, as well as direct uptake from dissolved contaminants in the water will contribute to their body burden.

For screening purposes a relatively simple food chain model has been incorporated into PROVANN. Bacteria and phytoplankton are assumed to adsorb pollutants in accordance with simple equilibrium partitioning between adsorption and dissolution. Calanus copepods, krill, and pelagic fish, which form the higher trophic levels of this simple food chain, are then potentially exposed to pollutants through absorption across gill membranes, as well as through ingestion of lower trophic organisms.

If it is assumed that equilibrium partitioning kinetics are sufficient to estimate concentrations in the phytoplankton, then the ratio of the adsorbed to the dissolved concentration is given by:

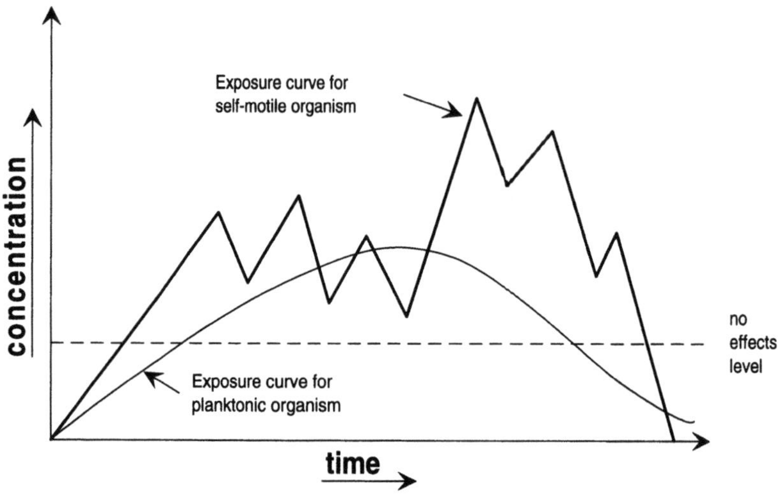

Figure 3. Conceptional comparison of exposure patterns for freely swiming and passively drifting organisms in the vicinity of a contaminated release plume.

$$C_p/C_d = K_p\, C_{phyto} \tag{1}$$

where

C_p	=	adsorbed concentration on phytoplankton
C_d	=	dissolved concentration in the water column
K_p	=	equilibrium partition coefficient for phytoplankton
C_{phyto}	=	concentration of phytoplankton in water column.

The partition coefficient K_p has units "per unit concentration (e.g. mg/l) adsorbing material." If phytoplankton represent the major adsorptive particles present, then the total concentration is the sum of $C_p + C_d$, and the concentration on the phytoplankton can be calculated from

$$C_p = K_p\, C_{phyto}\, C_{total} / (1 + K_p\, C_{phyto}). \tag{2}$$

In the case of a system with phytoplankton, herbivorous zooplankton, krill, and fish present, the total concentration of a substance averaged over a given volume of water is distributed according to

$$C_{total} = C_d + C_p + C_z + C_k + C_f\,, \tag{3}$$

where the meaning of the subscripts are:

d	=	dissolved
p	=	phytoplankton
z	=	herbivorous zooplankton (e.g. *Calanus finmarchicus*)
k	=	krill
f	=	planktivorous fish (e.g. herring).

The concentrations in (3) are values averaged over a given volume of water, and need to be divided by the mass density per unit mass water of each organism class to compute typical concentrations in individual organisms.

Within a given volume of water, the following equations, in conjunction with (3), govern the distribution of the chemical substance being modeled:

concentration adsorbed to phytoplankton, present in concentration C_{phyto}:

$$C_p = K_p\, C_{phyto}\, C_d \tag{4}$$

concentration adsorbed to herbivorous zooplankton, present in concentration C_{zoo}:

$$C_z = K_z\, C_{zoo}\, C_d \tag{5}$$

concentration internal to herbivorous zooplankton:

$$dC_{zi}/dt = k_z\, C_d - k_{z\text{-}dep}\, C_{zi} + k_{grazing\ z\text{-}p}\, C_p \tag{6}$$

concentration internal to krill, assumed to feed primarily on herbivorous zooplankton:

$$dC_k/dt = k_k\, C_d - k_{k\text{-}dep}\, C_k + k_{grazing\ k\text{-}z}\, C_z \tag{7}$$

internal concentration in fish, assumed to feed primarily on krill:

$$dC_f / dt = k_f\, C_d - k_{f\text{-}dep}\, C_f + k_{grazing\ f\text{-}k}\, C_k \tag{8}$$

Note that equations (6) - (8) give body burden concentrations for individuals, and need therefore to be corrected for the mean mass density of individuals per unit mass of water.

For a known value of C_{total}, and given values of the partition coefficients, uptake, and depuration and grazing rate parameters, the set of equations (3) - (8) can be solved. For the simple model postulated here, it is assumed that, since the metabolic processes in zooplankton occur much more rapidly than in fish, it is reasonable to use the equilibrium solutions to (6) and (7):

$$C_z = (k_z C_d + k_{grazing\ z\text{-}p} C_p) / k_{z\text{-}dep}, \tag{9}$$

and

$$C_k = (k_k C_d + k_{grazing\ k\text{-}z} C_z) / k_{k\text{-}dep}, \tag{10}$$

where the effect of external adsorption is neglected for krill and larger organisms.

The dynamic equation for bioaccumulation and depuration for fish can then be solved analytically as a function of time t as the simulation proceeds:

$$C_f(t) = (-(k_f C_d + k_{grazing\ f\text{-}k}\, C_k) / k_{f\text{-}dep} + C_{f0})\, \exp(-k_{f\text{-}dep}\, t) + (k_f C_d + k_{grazing\ f\text{-}k}\, C_k) / k_{f\text{-}dep}. \tag{11}$$

where C_{f0} is the internal concentration in the fish at the beginning of the exposure interval.

EXAMPLE SIMULATIONS

Two pairs of simulations are used to demonstrate model performance and capabilities. Both sets involve releases of produced water from two neighboring platforms offshore Trondheim, Norway.

The first pair of simulations traces the evolution of the concentration fields of C_7-phenol and naphthalene in produced water, and covers the 60-day time period April - May. The total assumed release rate for produced water was 4610 m^3 per day, divided 30% - 70% between two platforms. Assuming a concentration of 0.3 ppm naphthalene in the produced water gave a total release of 83 kg naphthalene over 60 days. For phenol, the release was 332 kg over 60 days. Hydrodynamic input for these first simulations is supplied by a 3D hydrodynamic model covering the region of interest.

The second pair of simulations again traces phenols from two simultaneous sources on Haltenbanken, but compares results with and without degradation, and with a simpler hydrodynamic input. Release rates for total produced water are 22,500 and 13,700 m^3/day, containing concentrations of 16 and 5 mg/l phenols, corresponding to 400 and 70 kg/day respectively. Release depths are 10 and 40 meters for the larger and smaller releases. A pycnocline is assumed to limit vertical mixing below a depth of 20 m. This is a typical stratification depth in early spring, which reduces to some extent the interactions between the two releases. The releases are simulated over a 90 day period, to allow full formation of the combined release field. The background current field used is based on Sætre and Bjørke (1988), as are spawning locations in 1986 for herring off the coast of mid-Norway (Figure 4). A wind record for 1986 supplied by the Norwegian Meteorological Institute was used to drive a time-varying wind-driven current field, averaged over the upper mixed layer.

Biological and chemical input parameters used in the simulations are given in Tables 1-3.

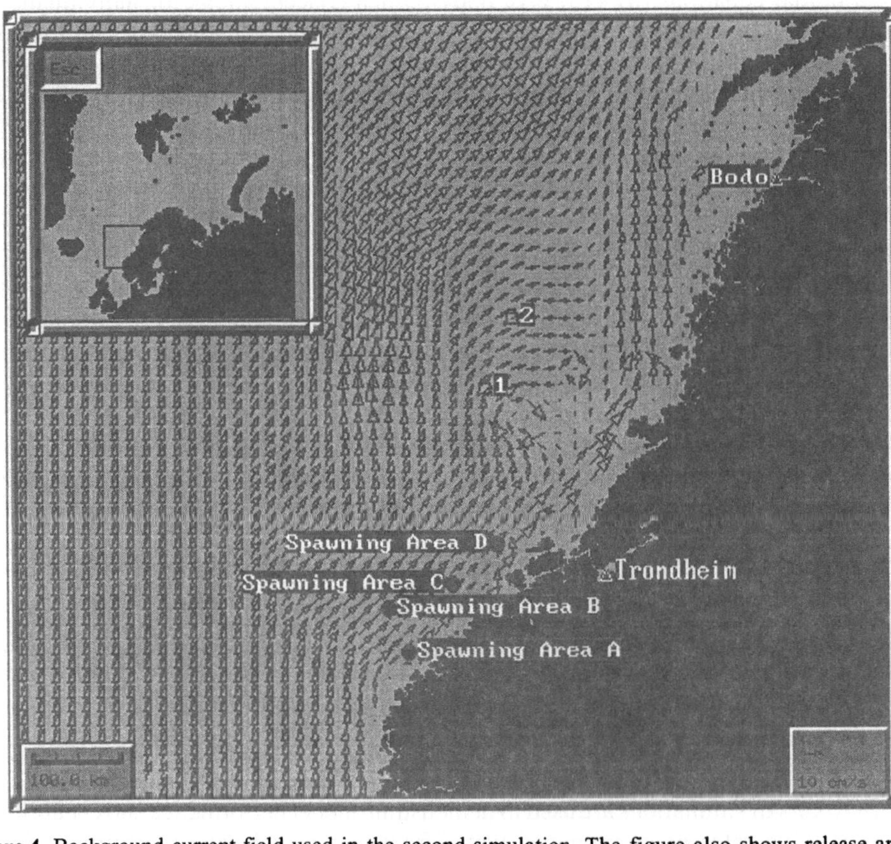

Figure 4. Background current field used in the second simulation. The figure also shows release areas for spring spawning herring eggs and larvae, and the locations of the produced water releases.

Table 1. Biological uptake, depuration, and predation (grazing) parameters used in the calculations. It is assumed that rates for krill and herbivorous zooplankton are similar

	k_z (day^{-1})	$k_{z\text{-dep}}$ (day^{-1})	$k_{gr\ z\text{-}p}$ (day^{-1})	k_k (day^{-1})	$k_{k\text{-dep}}$ (day^{-1})	$k_{gr\ k\text{-}z}$ (day^{-1})	k_f (day^{-1})	$k_{f\text{-dep}}$ (day^{-1})	$k_{gr\ f\text{-}k}$ (day^{-1})
Naphthalene	7200	1.2	1.0	7200	12	0.5	720	0.4	0.1
C_7 Phenol	7200	12	1.0	7200	12	0.5	720	1.2	0.1

Table 2. Mean biological parameters assumed to characterize phytoplankton, herbivorous zooplankton, krill, and small pelagic fish

	Mean Concentration (mg/l)	Mass/Individual (gm)	Number Individuals per m^3
Phytoplankton	5	0.000005	1,000,000
Zooplankton	3.5	0.001	3500
Krill	0.25	0.5	0.5
Fish	0.1	500	0.0002

In the simple biological model used here it is assumed that typical spring concentrations of zooplankton and phytoplankton in the Haltenbanken area are on the order of 3.5 and 5 mg/l, respectively. To the extent that these values are on the high side, model output will be weighted towards increased uptake of pollutants in the food chain. The adsorption-dissolution coefficients for phytoplankton and zooplankton are both set equal to the partition coefficient K_p. Faunal organism weights have been estimated by assuming that dry weight is about 1/3 of wet weight, and that carbon content is about 40% of dry weight (Sakshaug et al., 1991).

MODEL RESULTS

PROVANN simulates 3-dimensional transport, dilution, and degradation of chemicals released into the water, from one or more simultaneous sources. The resulting time series of concentration fields is saved in a file, which then serves as input to a simulation of fish movement (swimming) and advective transport of fish eggs and larvae.

First Pair of Simulations

Figures 5a - 5d are snapshots of the evolution of the naphthalene concentrations in the produced water plume from the two platforms. Five days after the start of the simulation (Figure 5a), the two concentration fields remain separated and distinct from each other. After 12 days (Figure 5b), the plumes have joined together, such that the smaller release to the north is revealed simply as an area of higher concentration in the overall concentration field.

Table 3. Chemical parameters used in the model calculations

	Solubility (ppb@25C)	Vapor Pressure (atm@25C)	Adsorbed/Dissolved Partition Coeff. K_p	Degradation Rate (day^{-1})	Density (ton/m^3)
Naphthalene	22000	0.0001	1380	0.02	1.16
C_7 Phenol	8.2×10^7	0.0006	140	0.001	1.11

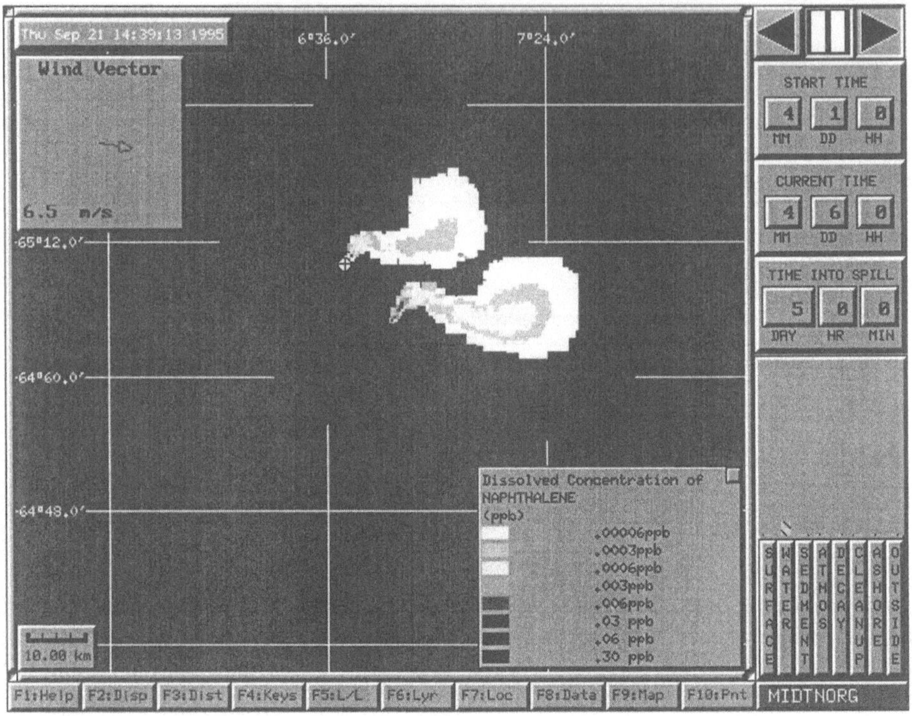

Figure 5a. Concentration field of naphthalene 5 days after beginning of the simulation.

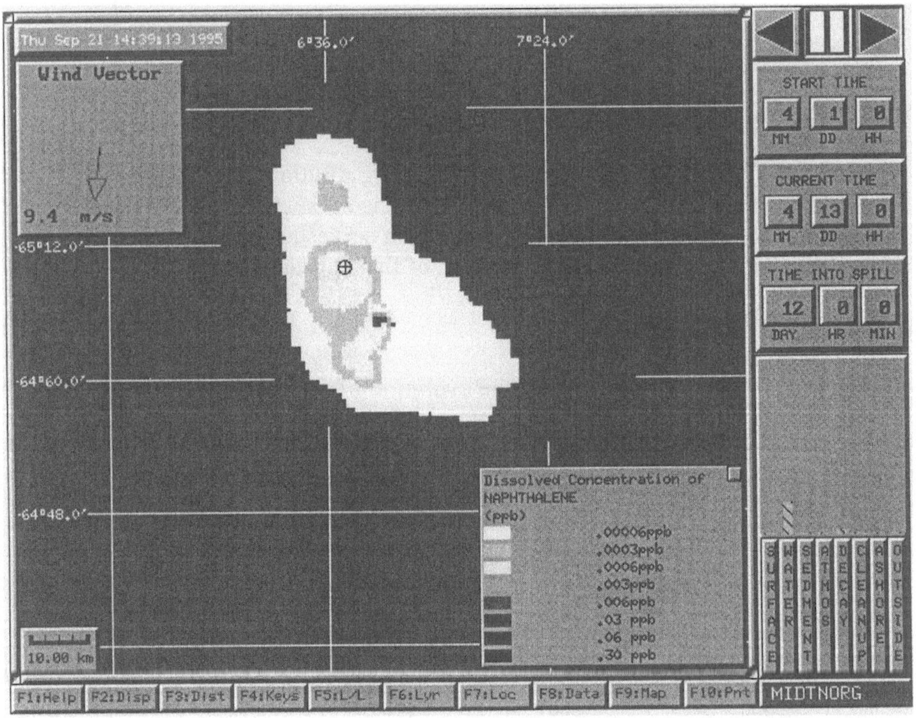

Figure 5b. Concentration field of naphthalene 12 days after beginning of the simulation.

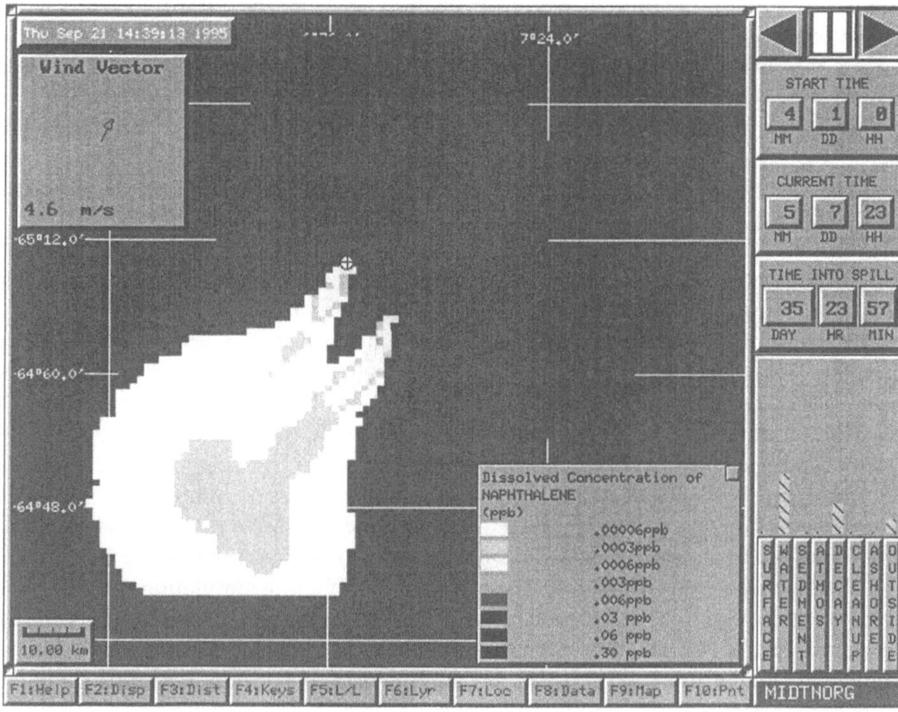

Figure 5c. Concentration field of naphthalene 36 days after beginning of the simulation.

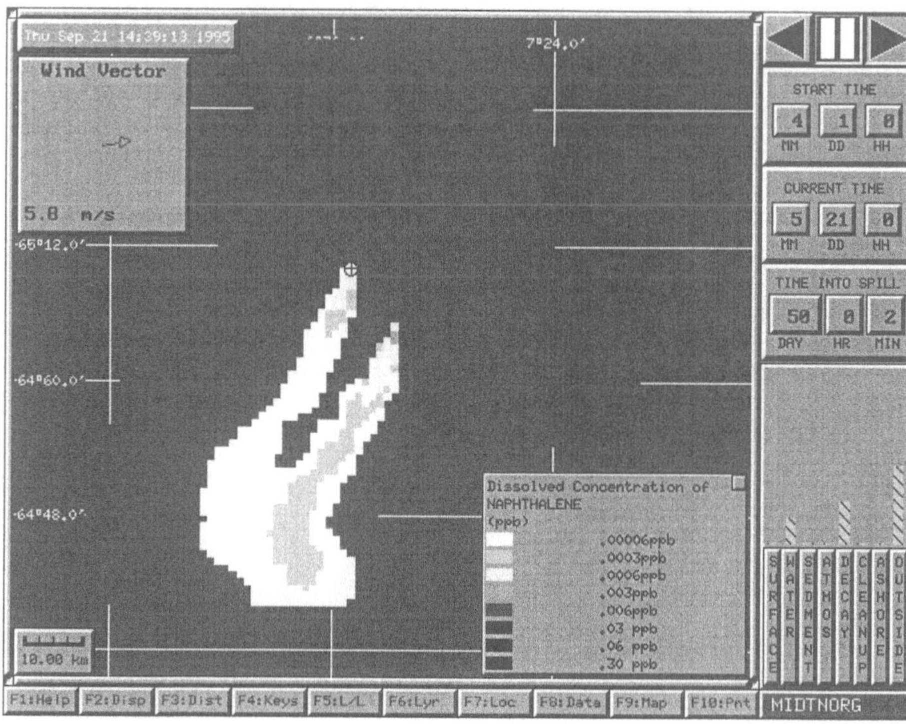

Figure 5d. Concentration field of naphthalene 50 days after beginning of the simulation.

At 36 days (Figure 5c), the plumes begin to separate again under the influence of a stable weather pattern, and remain nearly separate for the next 2 weeks (Figure 5d).

Note that the concentrations recorded are extremely low, the white areas in Figures 5a - 5d representing a concentration of 6×10^{-14}.

The simulation of fish exposures is carried out within a three-dimensional Eulerian (spatially fixed) grid. Fish are represented by particles which move stochastically within defined speed ranges, depending on fish type. Demersal and pelagic fish swim in the lower and upper water column, respectively. The model checks at each time step whether each fish particle is within the release plume, which is itself resolved on a second expanding, translating three-dimensional grid, independent of the Eulerian grid. Thus each particle accumulates and depurates as the simulation proceeds, depending on its physical location relative to the plume.

The size of the Eulerian grid affects the absolute results of a simulation, in both the fraction of the simulated fish particles which encounter the plume, and the magnitude of exposures. If the fixed grid is very large relative to the size of the plume, relatively fewer fish particles will encounter the plume than if the fixed grid is small. In the extreme, one could create a very small fixed grid very near the source, such that all of the fish were exposed virtually all the time, simulating perhaps fish in a moored cage. Thus the fraction of the fish represented in the model which are exposed to the produced water plume is relative to the selected grid size; what fraction of the total population is exposed depends on the fraction enclosed by the grid.

The small magnitude of the released quantities in this first simulation resulted in very limited areas within which environmental effects occurred. Figure 6 shows time series of whole body concentration for fish encountering naphthalene in the produced water plume. Maximum body burdens are on the order of 6 ppb. The distribution of body burdens in the local population during the simulation is shown in Figure 7. The majority of fish remain unaffected, with a maximum of about 1% reflecting an increase in body burden over background at any one time. During the first 25 days of the simulation, when the plume is becoming established, the fraction of fish with non-zero (above background) body burdens increases. This fraction then begins to decrease as stronger winds result in more rapid dilution of the plumes.

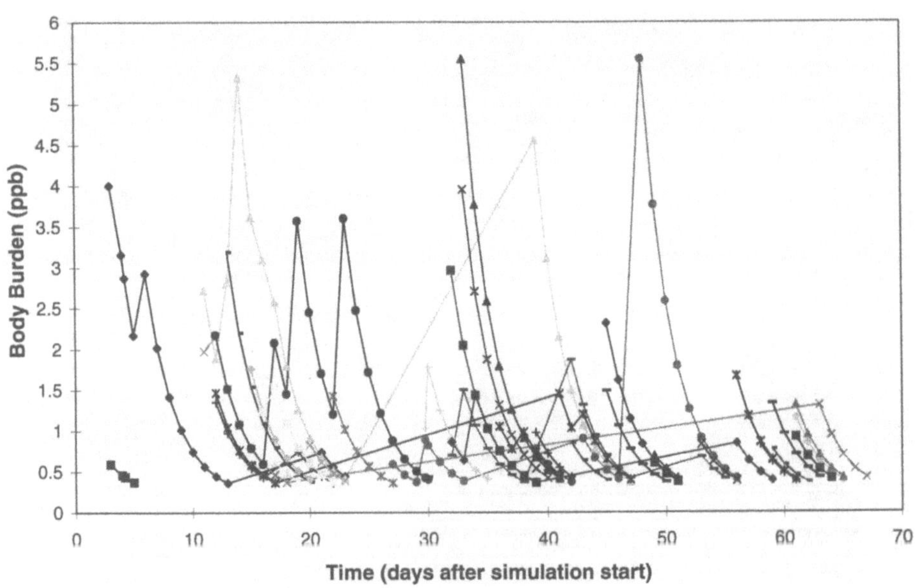

Figure 6. Time history of body burdens (ppb) in selected fish during simulated release of naphthalene.

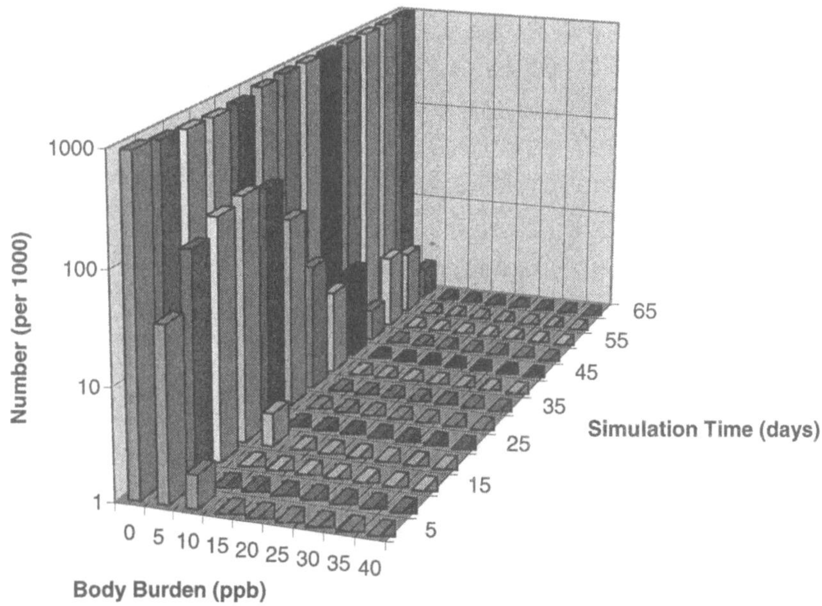

Figure 7. Distribution of body burdens of naphthalene in small pelagic fish (assumed herring). Swimming speed 1 km/day.

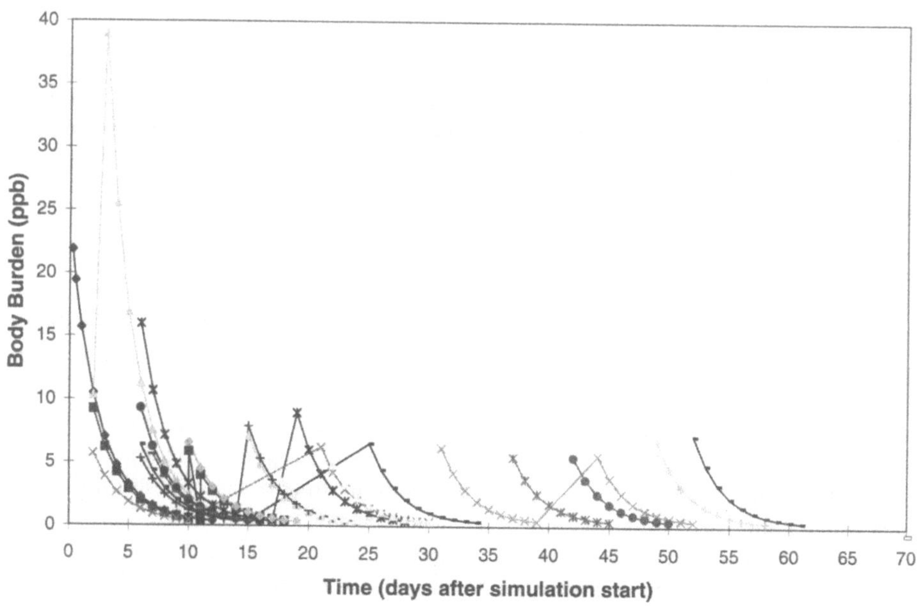

Figure 8. Time history of body burdens (ppb) in selected fish during simulated release of C_7-phenol.

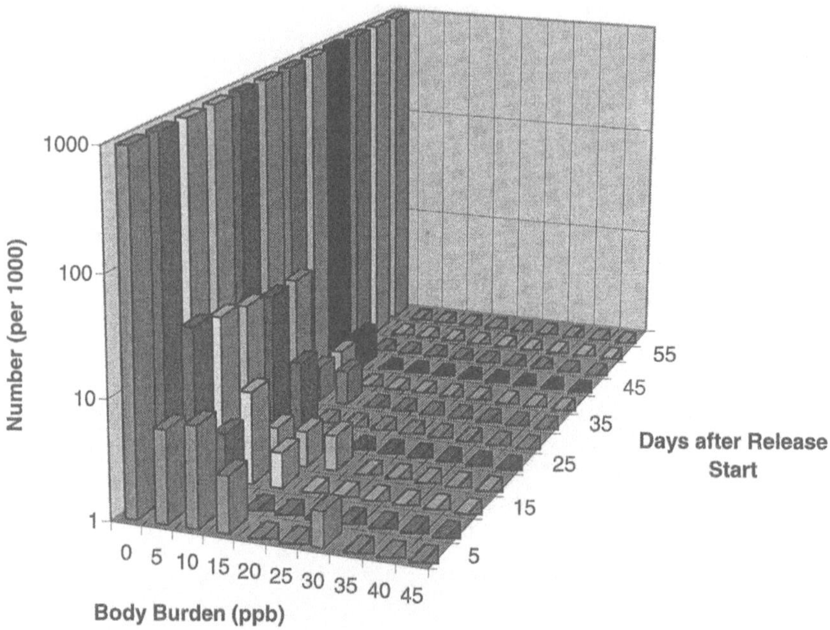

Figure 9. Distribution of body burdens of C_7-phenol in small pelagic fish (assumed herring). Swimming speed 1 km/day.

Figure 8 depicts time series of C_7-phenol body burdens for selected pelagic fish. Although the release rate for phenol is about 4 times greater than that for naphthalene, maximum body burdens are similar as a result of a higher depuration rate (Table 1). However, Figure 9 shows that individual fish may achieve higher short term body burdens of phenol than naphthalene, since the high concentration core of the plume is larger.

Both C_7-phenol and naphthalene have relatively high solubilities and low adsorbed-dissolved partition coefficients (Table 3), as reflected in the uptake and depuration rates in the food chain (Table 1). As a result, the modelled body burdens discussed above for fish are primarily the result of direct absorbtion of chemical across gill tissues from the water, with only a small percentage coming through feeding.

Second Pair of Simulations

To observe the relative importance of degradation in PROVANN, we follow the advection of spring-spawning coastal herring eggs and larvae into the produced water plumes. Particles representing the ichthyoplankton are released at four spawning areas identified by Sætre and Bjørke (1988), as shown in Figure 4. Advection due to wind and background current carries the larvae northward, and some of them encounter the plume. The model then computes the time-integrated exposure in units of concentration-time, of all particles which enter the plume.

Figure 10a is a histogram of the distribution of exposures for the released particles during the simulation without degradation. About 33% of the released particles never enter the plume. Many particles which do enter the plume are captured within the gyre over Haltenbanken (Figure 4), and experience relatively high exposures.

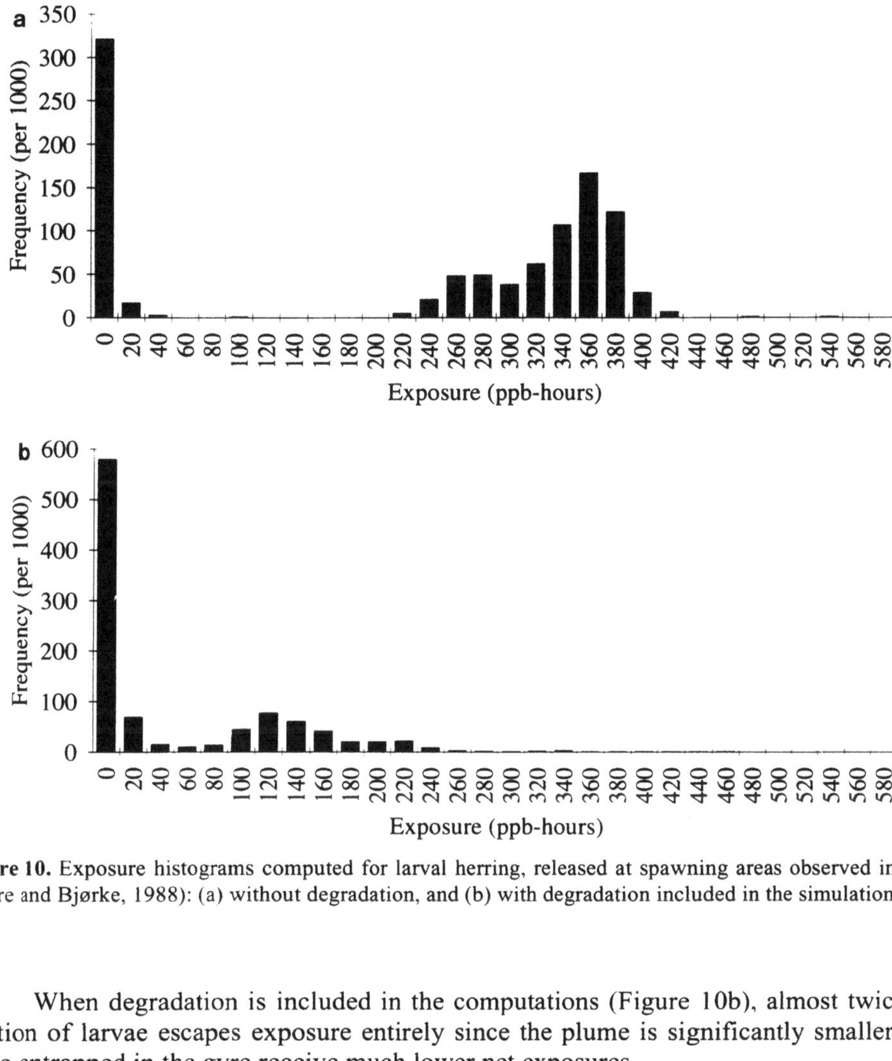

Figure 10. Exposure histograms computed for larval herring, released at spawning areas observed in 1986 (Sætre and Bjørke, 1988): (a) without degradation, and (b) with degradation included in the simulation.

When degradation is included in the computations (Figure 10b), almost twice the fraction of larvae escapes exposure entirely since the plume is significantly smaller, and those entrapped in the gyre receive much lower net exposures.

SUMMARY AND CONCLUSIONS

The PROVANN system of models introduced here is designed tu assess the potential for chronic effects from produced water. The preliminary focus is on potential bioaccumulation and biomagnification of produced water constituents in the marine food web. Other possible types of chronic effects, such as reduced fecundity, or pheromone response interference, can also be assessed with the model described here to the extent that such effects may be correlated with exposure. PROVANN simulates 3-dimensional transport, dilution, and degradation of chemicals released into the water, from one or more simultaneous sources. The resulting time series of concentration fields is saved in a file, which then serves as input to a simulation of fish movement (swimming) and fish egg and larval transport.

Two sets of simulations have been discussed, one using a time-dependent 3-dimensional hydrodynamic field computed by a numerical model, the second using a mean background current with a superimposed vertically averaged time-dependent wind-driven current. Although the two sets of simulations are not strictly comparable due to different release rates, it is clear that the full 3-dimensional field results in significantly more mixing than the simpler approximation. This is not surprising, in that vertical and horizontal shears are much weaker or lacking in the simpler case. Compensation through increased dispersion coefficients is a potential and usual remedy, but the correct scaling of these coefficients remains a problem (Rye et al, 1996).

The inclusion of degradation is clearly an important factor in modelling long term exposures. The same is probably true for other slow processes, such as evaporation from the water to the atmosphere, which are often neglected in assessment of acute impacts. Thus it is important to implement relatively comprehensive models for chronic assessment problems.

Simulations of adult fish, eggs, and larvae exposed to two individual components of produced water (C_7 phenol and naphthalene) from multiple sources on Haltenbanken suggest that bioaccumulation and biomagnification of these two substances will be small. There may be components which will have a greater tendency to biomagnify than those evaluated here. Furthermore, the biota are actually exposed to the mixture of chemicals which compose the produced water stream. The possible effects of multiple components therefore remains to be addressed.

Finally, it should be noted that only the physical fates component of PROVANN has in any sense been verified (Rye et al, 1996). No testing, calibration, or verification work has yet been performed on the preliminary biological components of the system.

REFERENCES

Brandsma, M.G ., L. Davis, R. Ayers, T. Sauer, 1980. A computer model to predict the short term fate of drilling discharges in the marine environment. Proceedings of Symposium on Environmental Fate and Effects of Drilling Fluids and Cuttings. January 21-24, 1980. Lake Buewna vista, Florida. 0.588-610.

Brandsma, M.G ., L. Davis, R. Ayers, T. Sauer, 1992. Modeling Offshore Discharges of Produced Water. In *Produced Water*, edited by J.P. Ray and F.R. Engelhart, Plenum Press, New York, USA.

Koh, R. C. and Y. C. Chang, 1973. Mathematical model for barged ocean disposal of waste. Report to Office of Research and Development, U. S. Environmental Protection Agency, Wash. DC. circa 400 p.

Reed, M., D. French, H. Rines, and H. Rye, 1995. A three dimensional oil and chemical spill model for environmental impact assessment. 1995 International Oil Spill Conference Proceedings, p. 61-66.

Rye, H., M. Reed, A. Melbye, S. Johnson, 1996. An intercomparison between field measurements and three different calculation models for estimates of dilution factors. In Reed, M. and S. Johnson (eds), 1996. Produced Water: Environmental and Mitigation Issues. Plenum Press, NY.

Sakshaug, E., A. Bjørge, B. Gulliksen, H. Loeng, F. Mehlum, 1992: Økosystem Barentshavet. ISBN 82-90565-17-8, Mesna-Trykk a.s., Lillehammer, Norway, 304 p.

Staples, C., K. Dickson, J. Rodgers, F. Saleh, 1985: A model for predicting the influence of suspended sediments on the bioavailability of neutral organic chemicals in the water compartment. In R. Cardwell, R. Purdy and R. Bahner (eds*), Aquatic Toxicology and Hazard Assessment: Seventh Symposium.* ASTM Pub. 04-854000-16. p. 417-428.

Sætre, R. og Bjørke, H., 1988: Oljevirksomhet på Norne. Konsekvenser for fiskeressursene. HELP Report no. 19, 1988. Marine Research Institute, Bergen. 87-7461-005-9.

Section VI

U.S. DEPARTMENT OF ENERGY: GULF OF MEXICO STUDIES

30

SESSION SUMMARY

David S. Gettleson

Continental Shelf Association, Inc.
Jupiter, Florida

This session includes papers that present results from the U.S. Department of Energy (DOE) Gulf of Mexico Region Oil and Gas Operations. This multi-year study was initiated in 1992 with completion in 1996.

The first paper presented in this session provides an overview of the DOE oil exploration and production environmental research program and the general technical approach of this multi-year study. The following three papers present data on radionuclides, metals, and polycyclic aromatic hydrocarbons (PAH's) in produced water, amblent water, sediments (metals and PAH's), and tissues of marine invertebrates and fish offshore in the Gulf of Mexico. The fifth paper discusses the catch, distribution, and consumption patterns of seafood from coastal and offshore waters of Louisiana and Texas. The final paper examines the health and ecological risks associated with the discharge of produced water to Louisiana open bays.

This DOE study is the largest, most comprehensive study of the environmental and human health risks associated with the discharge of produced water in coastal and offshore Gulf of Mexico waters. This study indicates that naturally occuring radioactive materials (NORM) in produced water discharges are rapidly diluted in the receiving waters, and do not accumulate in edible tissues of organisms living in the vicinity of offshore produced water discharges.The study also indicates that while several metals are enriched in produced water relative to receiving waters, dilution in receiving waters is rapid, resulting in limited elevated levels of some metals in sediments and tissues in the vicinity of offshore produced water discharges. The study also shows that while PAH's are present in produced water discharges, they are rapidly diluted in receiving waters. The study indicates that sediment concentrations of total PAH's are elevated within 20 m of discharges, but reach background concentrations at 2000m. Tissue concentrations of total PAH's range from 0,1 to 12,160 μg/kg dry wt in the vicinity of offshore produced water discharges. Ecological risk assessment suggests that PAH's in the receiving water are well below US EPA chronic water quality criteria and in general, sediments PAH's levels are also well below threshold concentrations, indicating that they should be considered safe for benthic marine animals.Concentrations of PAH's in tissues are below critical body residues associated with possible adverse effects based on an estimated average concentration causing harmful responses in five persent of the organisms in an aquatic community during chronic exposure.

Personal interviews with Gulf of Mexico Louisiana and Texas fishermen detailed the species and quantities oaught, location of catch, mode of fishing, distribution of catch, family consumption pattern, and demographics of the fisherman. Surveys with wholesaler/processor, retail outlets, and restaurants further characterized distribution and consumption pattern of seafood. It was found that a large percentage of seafood was harvested commercially and recreationally near oilfield structures. These data will be used in conjunction with the levels of radionuclide, metals, and PAH's in tissues of fishes to predict human health risk from seafood consumption.

A conservative screening-level human health risk assessment for contaminants discharged in produced water to open bays in Louisiana is a cost-effective way to focus on those contaminants that may be of concern. Two contaminants of possible concern, radium and lead, are further analyzed through a quantitative risk assessment. This assessment suggests radium discharged to open bays in produced water is not expected to affect aquatic fish or shellfish nor does it present an important risk to human health.

Additional ecological and human health quantitative risk assessments are being performed for radionuclides, metals, and organics as part of this study. Once these assessments are completed, any needs for additional information relative to the ecological and human health risk associated with the dicharge of produced water in coastal and offshore waters of the Gulf of Mexico can be determined.

U.S. DEPARTMENT OF ENERGY SPONSORED STUDY - ENVIRONMENTAL AND ECONOMIC ASSESSMENT OF DISCHARGES FROM GULF OF MEXICO REGION OIL AND GAS OPERATIONS

An Overview

Brent W. Smith[1] and David A. Gettleson[2]

[1] U.S. Department of Energy
900 Commerce Road East
New Orleans, Louisiana 70123
[2] Continental Shelf Associates, Inc.
759 Parkway Street
Jupiter, Florida 33477

INTRODUCTION

U.S. Department of Energy Oil and Gas Environmental Research Program

The U.S. Department of Energy initiated a new Oil and Gas Environmental Research Program in 1991. The mission of the program is to maximize the production and ultimate recovery of U.S. oil and gas resources by reducing the cost of effective environmental protection (Ham and Hochheiser, 1994). The mission is accomplished by setting priorities among relevant environmental issues and, for the highest priorities, conducting research to better understand environmental risks, developing new, affordable environmental compliance technologies, promoting sound environmental practices through outreach and assistance to industry, and working with regulators to ensure a proper balance between environmental protection and economic and energy impacts. The Department of Energy also seeks to enhance the competitive position of United States companies in exporting environmental exploration and production technologies (Smith, 1995a).

Produced Water2: Environmental Issues and Mitigation Technologies
Edited by M. Reed and S. Johnsen, Plenum Press, New York, 1996

Oil Exploration and Production Environmental Research Program

A component of the overall program is the Oil Exploration and Production Environmental Research Program. There are five principal research areas that form the basis of this program: Area of Review Variance Methodology/Risk Assessment, Streamlining Regulations, Environmental Technologies and Practices, Environmental Outreach and Program Planning, and Gulf of Mexico Discharges (Smith, 1995b).

The Gulf of Mexico Discharges research area is designed to assess the environmental and human health risks associated with produced water and sand discharges from offshore and coastal oil and gas operations in the Gulf of Mexico Region (Smith, 1995a). This research area includes four related projects. The first project is the one discussed in this paper. The second is a produced water risk assessment by Brookhaven National Laboratory that uses information provided by the first project. The third and fourth projects are a quality assurance study by Lawrence Livermore National Laboratory of laboratory analyses of naturally occurring radioactive materials and participation in a produced water toxicity testing study by Lumitox, Inc. The results of these projects will support *Domestic Natural Gas and Oil Initiative* (U.S. Department of Energy, 1993) goals to assist regulatory agencies in developing technically sound and justifiable regulations.

GENERAL TECHNICAL APPROACH OF STUDY

In 1992, the U.S. Department of Energy (DOE) contracted with Continental Shelf Associates, Inc. (CSA) and its team of subcontractors and independent consultants to conduct the multiyear study entitled "*Environmental and Economic Assessment of Discharges from Gulf of Mexico Region Oil and Gas Operations.*"

The study is designed to increase the present state of scientific knowledge concerning the following topics:

- The fate and environmental effects of pollutants found in produced water;
- The catch, consumption, and human use patterns of seafood species collected from coastal and offshore waters of the Gulf of Mexico; and
- The economic impacts of current and potential future regulations on offshore oil and gas producers of the Gulf of Mexico region.

Separate study components were implemented to address these topics. The components generated data in the following areas:

- The characteristics of offshore produced water and produced sand discharges as they pertain to naturally occurring radioactive materials (NORM), trace metals, and organics variably found in association with these discharges;
- The fate of NORM, trace metals, and organics in water, sediment, and biota near several offshore oil and gas facilities;
- The catch, consumption, and human use patterns of seafood species collected from coastal and offshore waters;
- The recovery of two terminated produced water discharge sites located in open bay and semi-enclosed bay (broken marsh area) environments of coastal Louisiana; and
- The economic and energy supply impacts of existing and anticipated Federal and State offshore and coastal discharge regulations.

Risk assessment for the produced water discharges is being evaluated within this study by Brookhaven National Laboratory.

The results from the first two study components are presented in the four papers following in this session. Produced sand data are provided in a report being prepared for the DOE (Randolph, in prep.). The third study component is represented by the fifth paper following in this session. The fourth study component is described in a paper in Session III (Delineation of Benthic Impact and Recovery at Two Produced Water Discharge Sites in Inshore Louisiana).

The results of the study component on the economic impacts of offshore and coastal discharge requirements on present and future operations in the Gulf of Mexico have been previously presented (ICF Resources, Inc., 1995). This study component characterized the potential impacts of alternative treatment and discharge regulations on reserves and production in Gulf of Mexico coastal, territorial, and outer continental shelf waters, quantifying the impacts of both recent regulatory changes and possible more stringent requirements. The treatment technologies capable of meeting these requirements were characterized in terms of cost, performance, and applicability to coastal and offshore situations. As part of this analysis, an extensive database was constructed that included oil and gas production forecasts by field, data on existing platforms, and the current treatment methods in place for produced water treatment and disposal on offshore facilities. This work provides the first comprehensive evaluation of the impacts of alternative regulatory requirements for produced water management and disposal in coastal and offshore areas of the Gulf of Mexico.

REFERENCES

Ham, J. D., and Hochheiser, H. W. 1994. Environmental R&D Program aims to ease compliance. *The American Oil & Gas Reporter* December:110-12 and 117.

ICF Resources, Inc. 1995. Assessment of Economic Impacts of Offshore and Coastal Discharge Requirements on Present and Future Operations in the Gulf of Mexico. Draft report prepared for the U.S. Department of Energy. 74 pp.

Randolph, T. In prep. Produced Sand Characterization and Washing Equipment Evaluation. In: Continental Shelf Associates, Inc.. In prep. Environmental Field Sampling and Analysis of NORM, Heavy Metals, and Organics: Environmental and Economic Assessment of Discharges from Gulf of Mexico Region Oil and Gas Operations. Prepared for the U.S. Department of Energy.

Smith, B. 1995a. U.S. Department of Energy oil and gas environmental research programs. *The La. Environmental Professional* 12(1):1-40.

Smith, B. 1995b. U.S. Department of Energy Produced Water Research. In: Proceedings of the Fifth Annual Produced Water Seminar. Sponsored by the Texas Chapter of the American Filtration Society, Houston, Texas.

U.S. Department of Energy. 1993. The Domestic Natural Gas and Oil Initiative. Washington, D.C. 36 pp.

The results from the first two study components are presented in the [illegible] papers following in this session. Produced sand data are provided [illegible] (Randolph, in prep.). [illegible]

[illegible]

32

ASSESSING THE POTENTIAL FOR ENHANCED BIOACCUMULATION OF HEAVY METALS FROM PRODUCED WATER DISCHARGES TO THE GULF OF MEXICO

John H. Trefry, Robert P. Trocine, Kelly L. Naito, and Simone Metz

Division of Marine and Environmental Systems
Florida Institute of Technology
Melbourne, Florida 32901

INTRODUCTION

More than 140 million m^3 of produced water per year are discharged to surface water from more than 3000 production platforms in the northern Gulf of Mexico (Reilly et al., 1991; LeBlanc, 1994). Concentrations of heavy metals in this produced water can be enriched by factors of 10 to greater than 10,000 relative to ambient seawater (Boesch and Rabalais, 1989; Ray and Engelhardt, 1992). Concern for the fate and effects of such discharges continues to prompt a variety of research and monitoring programs.

The study described here was designed to assess whether organisms inhabiting areas adjacent to platforms that discharge produced water show enhanced bioaccumulation of heavy metals compared with organisms from non-discharging platforms. Such studies are limited in number because research efforts have focused more on laboratory toxicity studies to test the effects of produced water on organisms (Jacobs et al., 1992; Schiff et al., 1992). For example, Jacobs et al. (1992) showed that 50% of the populations of *Artemia salina* (brown shrimp) and *Salmo gairdneri* (rainbow trout) survived exposure to mixtures of 5:1 and 9:1 produced water:freshwater, respectively. In acute toxicity tests that concentrated on single elements, Jacobs et al. (1992) determined EC_{50} levels of 24,000 µg As/L, 11,600 µg Cd/L and 3,000 µg Pb/L. However, direct comparison of these acute toxicity results with data obtained in the field are difficult because concentrations in water surrounding the platform are rarely as high as encountered in laboratory studies. Furthermore, such elevated metal concentrations would persist for only for a brief time at the discharge site. For example, dispersion models predict a 1000-fold dilution of produced water within 20-30 m of the discharge point (Brandsma and Smith, 1996). To complement laboratory studies, field measurements of metal accumulation by organisms provide a longer-term integrator of metal uptake and facilitate more detailed ecological risk assessment.

Produced Water2: Environmental Issues and Mitigation Technologies
Edited by M. Reed and S. Johnsen, Plenum Press, New York, 1996

In this paper, we present concentrations of As, Ba, Cd, Cu, Fe, Hg, Mn, Ni, Pb, V and Zn in clams (*Chama macerophylla*), oysters (*Crassostrea virginica*), red snapper fish (*Lutjanus campechanus*) and gray triggerfish (*Balistes capriscus*) collected near discharging and non-discharging platforms. Our results are described within the context of the following: (1) concentrations of metals in produced water discharges versus ambient seawater, (2) concentrations of metals in organisms from reference versus discharge sites and (3) accumulation of metals in sediments adjacent to oil and gas platforms.

METHODS

Samples for metal analysis were collected during June through August, 1993, at nine sites in the Gulf of Mexico (Figure 1). Water depths at the various platforms ranged from 6-121 m (Figure 1) and surface water salinities ranged from 7-35 °/₀₀. Produced water was collected directly from the platform in acid-washed, low density polyethylene bottles. Ambient water samples were collected at distances of 20 and 2000 m away from the production platform by divers and filtered through acid washed, 0.4 μm, polycarbonate filters. Both types of water samples were acidified with concentrated Ultrex II HNO_3 to a pH <2. Clams (*Chama macerophylla*) and oysters (*Crassostrea virginica*) were sampled directly from the platform and frozen in polyethylene bags until return to the laboratory. Samples of red snapper (*Lutjanus campechanus*) and gray triggerfish (*Balistes capriscus*) were collected in traps and by hook and line. Sediment samples were collected by diver or with a grab sampler. The top 2 cm layer of sediment was frozen in polyethylene bottles.

Metal concentrations were determined by one or more of the following techniques, depending on sample type and metal levels: (1) atomic absorption spectrophotometry (AAS), using either flame atomization with a Perkin-Elmer (PE) model 4000 instrument (Cu, Fe, Mn, Ni, V, Zn) or heated graphite atomization using a PE model 5100 or 4000 system (As, Cd, Cu, Mn, Ni, Pb, V), (2) hydride generation in tandem with the PE model 4000 AAS using flame atomization (As), (3) cold-vapor AAS using a Laboratory Data Control system (Hg), (4) inductively coupled plasma-mass spectrometry using a PE ELAN 5000 (Ba, Cd

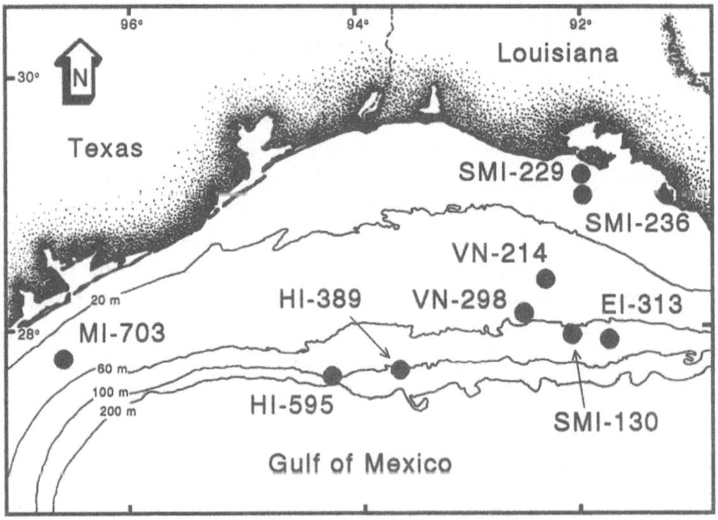

Figure 1. Map showing sampling sites at production platforms in the Gulf of Mexico.

and Pb) and (5) instrumental neutron activation analysis (Ba) using the research and training reactor at the University of Florida. Produced water samples were analyzed directly following appropriate dilution with distilled-deionized water. Ambient water samples were analyzed directly for Ba and Hg, and following preconcentration with Chelex 100 for Cd, Cu, Fe, Ni, Pb, V and Zn (Sturgeon et al., 1980) and after preconcentration by reductive precipitation for Mn (Nakashima et al., 1988).

Soft tissue from the bivalves was digested using concentrated HNO_3. Muscle tissue (fillet) was carefully excised from fish specimens and digested in HNO_3. Sediment was completely digested using a mixture of HF-HNO_3-$HClO_4$. Concentrations of total organic carbon (TOC) were determined with good agreement by both the wet digestion method of Fredericks and Sackett (1970) using an OI Corporation model 524C T.O.C. analyzer and by high-temperature combustion using a Shimadzu model TOC-500 instrument.

Standard reference materials (SRMs) were analyzed for each sample type and included the following: (1) NASS-2 seawater from the National Research Council (NRC) of Canada, (2) TORT-1 lobster hepatopancreas (NRC) and oyster tissue (SRM-1566) from the U.S. National Institute of Standards and Technology, and (3) BCSS-1 marine sediment from the NRC. All values obtained were within acceptable limits for the particular SRM as established by the certifying agency. Many replicate samples were analyzed to determine our analytical and sample precision. Results for precision varied as a function of metal concentration and typically ranged from <1 to 5% with values as high as 20% for very low metal levels.

HEAVY METALS IN PRODUCED AND AMBIENT WATER

The potential for enhanced bioaccumulation of heavy metals from seawater surrounding a platform that is discharging produced water is influenced by the concentration of a metal in the produced water relative to ambient seawater, the rates of discharge and mixing of the produced water with seawater, and the chemical forms of the metal in the produced water/seawater mixture. Concentrations of heavy metals in produced water from this study in the Gulf of Mexico were variable and generally much higher than in ambient seawater (Table 1). Highest enrichment factors for produced water versus ambient water were found for Ba, Fe and Mn with average values ≥10,000 (Figure 2). Levels of Pb and Zn in the produced water were >100-fold enriched relative to ambient water. In contrast, produced water samples from this study were depleted with Cu and V compared with ambient water (Table 1). For Hg, Cd and Pb, three metals of particular environmental concern, enrichment factors averaged about 3, 30 and 200, respectively.

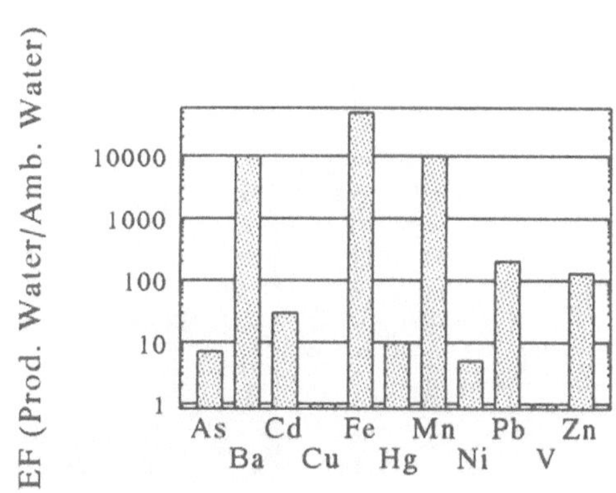

Figure 2. Graphs showing the enrichment factor (EF) for different metals in produced water relative to ambient water.

Table 1. Concentrations (mean or range) of total dissolved solids (TDS), total organic carbon (TOC) and selected heavy metals in produced water (PW) and ambient water (AW) from the Gulf of Mexico

	TDS (g/L)	TOC (mg/L)	As (µg/L)	Ba (mg/L)	Cd (µg/L)	Cu (µg/L)	Fe (mg/L)	Hg (µg/L)	Mn (mg/L)	Ni (µg/L)	Pb (µg/L)	Zn (mg/L)
PW, This Study (n=37)	90-176	120-800	0.5-31	81-342	<0.05-1.0	<0.2	10-37	<0.01-0.2	1-7	<1-7	<0.1-28	0.01-3.6
AW, This study (n=25)	7-35	1-2	1.0	0.01- 0.09	0.020	0.2-1.7	<0.001	<0.01	0.0005	0.25-1.6	0.02-0.05	0.005
Mean Enrichment Factor [1]	6	~400	7	10,000	30	<1	50,000	3	10,000	5	200	130
SMI-236-PW [2]	90	700	0.8	163	0.18	<0.2	12	0.11	1.2	2	0.3	0.02
SMI-236-AW	7	–	1.1	0.088	0.027	1.4	<0.002	<0.01	0.0002	1.2	0.04	0.005
HI-595-PW [3]	175	760	25	320	0.63	<0.2	35	0.03	7	<1	20	3.0
HI-595-AW	34	–	1.6	0.015	0.012	0.22	<0.0001	–	0.0004	0.26	0.03	0.006
Typical PW[4]	–	<0.1-1500	5-300	0.1-2000	0.1-200	0.4-1500	–	0.05-2	–	0.4-1700	1.5-8800	0.1-69

Concentrations of V in produced water from this study were all < 1 µg/L with an average of 1.2 µg/L for ambient water.

[1] Enrichment factors were calculated by dividing the concentration in produced water by the concentration in ambient water.

[2] South Marsh Island, Block 236 (SMI-236); for PW, n=8 and for AW, n=3.

[3] High Island, Block 595 (HI-595); for PW, n=8 and for AW, n=3.

[4] From Neff and Sauer (1995).

Concentrations of metals in produced water from our study were all within the range of values presented by Neff and Sauer (1995) in their summary of produced water composition, with the exception of very low Cu levels in our samples (Table 1). Maximum concentrations for As, Ba, Hg and Zn determined for produced water from our Gulf of Mexico study were 6-19 times lower than maximum values reported by Neff and Sauer (1995, Table 1). In even sharper contrast, our maximum values for Cd, Ni and Pb in produced water from the Gulf of Mexico were 200-300 times lower than maximum levels from other studies (Neff and Sauer, 1995, Table 1).

Discharge rates for produced water also influence local metal loadings and the dose to which organisms in the area of a specific platform are exposed. Rabalais et al. (1992) report discharge rates of 590-23,000 m^3/day for produced water along the coast of Louisiana. Present regulations limit discharges to <4000 m^3/day for each platform in the Gulf of Mexico. We have obtained daily loadings of 874 m^3 from the platform at HI-595 and 2085 m^3 for SMI-236. This difference in discharge rates for these two platforms yielded calculated metal loadings from each site that agreed within a factor of 3, despite much higher metal concentrations for produced water from HI-595 (Table 1). One notable exception to this similarity was a 47 times higher loading of Pb from the HI-595 platform versus the SMI-236 platform.

Overall, dissolved metal levels in produced water are controlled by a complex interplay of metals with organic matter, chloride, sulfide, pH, temperature, redox state and other variables (Barth, 1991; Helgeson et al., 1993). The behavior and subsequent bioaccumulation of heavy metals following mixing of produced water with seawater is partly a function of the chemical form of the metal of interest in solution. In the suite of samples analyzed for this study, a general trend of higher metal levels in fluids with higher total dissolved solids (TDS) was observed. For example, concentrations of most metals, and especially Ba, Cd, Fe and Pb, were higher in the more saline produced water from platform HI-595 relative to produced water from SMI-236 (Table 1). Considering all produced water from this study, the best trend for TDS versus metal concentrations was observed for Fe ($r^2 = 0.85$) and most likely reflects the stabilizing effect of chloride ion complexation on dissolved Fe (Templeton, 1960; Heinrich and Seward, 1990). Levels of TOC were enhanced in all produced water samples at 120-800 mg/L and, although no direct strong correlations were observed between concentrations of metals and TOC, organic matter complexation of metals certainly helps support elevated metal levels in produced water. Once discharged, Fe and Ba in the produced water most likely precipitate rapidly as oxide and sulfate phases, respectively, and co-precipitate other metals. In contrast, metals complexed by organic matter may stay dissolved in seawater for longer periods of time. Additional studies are necessary to evaluate these processes.

HEAVY METALS IN MOLLUSKS

Metal concentrations were determined in three separate samples of the clam *Chama macerophylla* from each of five discharging platforms and one reference platform (Table 2 and Figure 3). No statistically significant differences ($\alpha = 0.05$) were observed among the sites for concentrations of Ba, Fe, Pb, V or Zn. This statistical result is influenced by similar median values, some large standard deviations and a small sample population. However, it does suggest that produced water discharges are not significantly enhancing bioaccumulation of these metals in clams from the Gulf of Mexico.

Differences in As levels observed for clams from different sites seem to be more dependent on local concentrations of dissolved As in ambient seawater than on discharges of produced water. Clams from three offshore sites, including reference site HI-389 at 121 m,

Table 2. Concentrations (dry weight; mean ± 1 standard deviation) of heavy metals in clams (*Chama macerophylla*) and oysters (*Crassostrea virginica*) from discharging and non-discharging platforms in the Gulf of Mexico

Site	As (µg/g)	Ba (µg/g)	Cd (µg/g)	Cu (µg/g)	Fe (µg/g)	Hg (µg/g)	Mn (µg/g)	Ni (µg/g)	Pb (µg/g)	V (µg/g)	Zn (µg/g)
Clams											
HI-389-R	43±11	22±13	7±2	4.8±0.4	60±10	0.051	7±2	9±1	0.2±0.1	1.6±0.3	170±50
EI-313-D	47±8	13±1	14±6	5.2±1.1	80±30	0.065	11±3	8±1	0.3±0.1	2.8±0.6	200±95
HI-595-D	41±9	43±24	7±3	6.0±0.6	210±100	0.046	9±3	6±2	0.7±0.5	2.2±0.6	145±70
MI-703-D	29±3	14±15	4±1	6.8±1.3	110±60	0.110	9±3	11±5	0.2±0.1	2.3±0.2	100±30
SMI-130-D	26±8	14±2	5±1	11±3	180±40	0.054	19±10	4±2	0.2±0.04	3.0±2.5	90±10
VN-214-D	24±6	20±20	5±1	6.2±0.4	70±20	0.051	15±2	5±1	0.2±0.1	4.1±1.1	180±90
Oysters											
SMI-229-R	10±1	12±6	5±1	98±90	320±170	0.13	28±13	1.0±0.3	0.4±0.2	1.4±0.1	1760±90
SMI-236-D	11±1	11±4	7±2	140±30	480±210	0.06	18±5	1.0±0.1	0.6±0.02	1.6±0.3	1810±300
GOM*	8±3	—	6±2	250±120	370±140	0.05±0.02	18±8	2.2±0.6	0.6±0.5	—	1880±730
GOM**	10±7	—	4±2	150±100	290±160	0.13±0.11	14±7	1.8±1.6	0.6±0.9	—	2150±1580

R = reference (non-discharging) site. D = discharging site. Locations include: High Island 389, Eugene Island 313, High Island 595, Matagorda Island 703, South Marsh Island 130, Vermilion 214, South Marsh Island 229 and South Marsh Island 236.

*GOM = Gulf of Mexico oysters from Presley et al. (1990) for 3 comparable Louisiana locations (n = 27): Joseph Harbor Bayou, Vermilion Bay, and Atchafalaya Bay.

**GOM = Gulf of Mexico oysters from Presley et al. (1990) for 50-69 locations (n = 485).

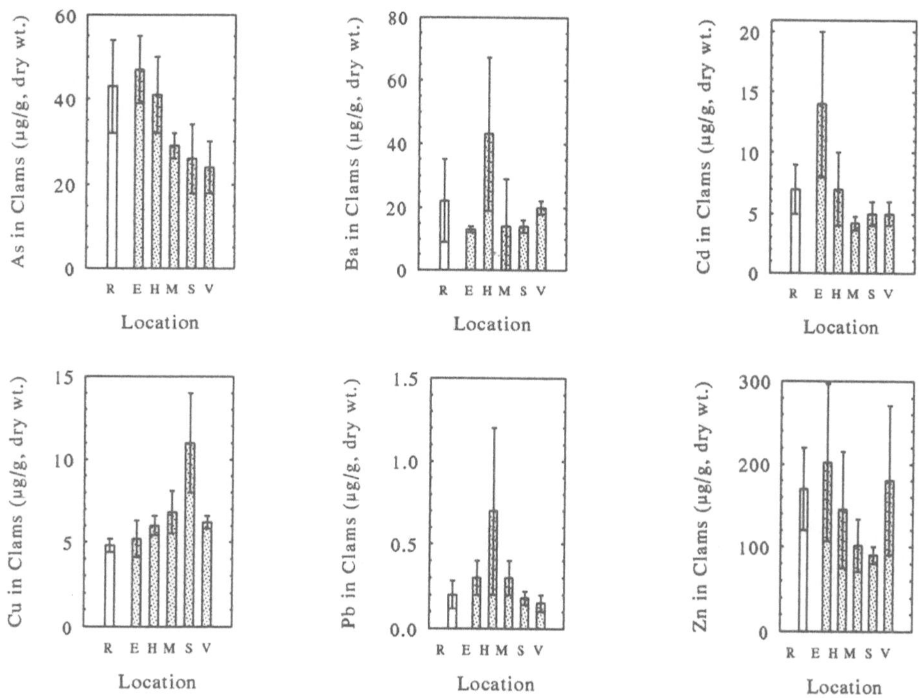

Figure 3. Graphs showing concentrations of As, Ba, Cd, Cu, Pb and Zn in clams (*Chama macerophylla*) collected from oil platforms in the Gulf of Mexico. Locations include Reference site (R) at High Island 389 and the following discharge sites: E = Eugene Island 313, H = High Island 595, M = Matagorda Island 703, S = South Marsh Island 130 and V = Vermilion 214.

and discharge sites HI-595 and EI-313 at 122 and 72 m, respectively (Figure 1) averaged 43 ± 9 μg As/g (dry weight) relative to values of 26 ± 6 μg As/g (dry weight) for clams from three discharging platforms in shallower water sites MI-703 at 51 m, SMI-130 at 66 m and VN-214 at 39 m (Figure 3). Dissolved As values for surface seawater from offshore site HI-595 averaged 1.58 ± 0.06 μg/L relative to values of 1.09 ± 0.09 μg/L for nearshore site SMI-130. Using data from sites HI-595 and SMI-130, we calculated tissue enrichment factors (EF_{tissue}) relative to seawater using Equation 1.

$$EF_{tissue} = [\mu g \text{ metal/g tissue wet weight}] / [\mu g \text{ metal/g water}] \quad (1)$$

Despite some of the potential variability in water content for tissue samples, wet tissue weights are used in Equation 1 to provide a more direct relationship between the seawater and a normal wet tissue sample. Collectively, the *Chama* samples averaged 83.0 ± 2.7% water (n = 21). Using the specific water content for each sample and ambient dissolved As concentration for each site, the EF_{tissue} values for sites HI-595 and SMI-130 were $10^{3.59}$ and $10^{3.70}$, respectively. Using tissue data from the three offshore sites versus the three shallower sites, combined with available water data (Table 1), values for EF_{tissue} are $10^{3.72}$ and $10^{3.65}$, respectively. Within the limits of the data set, the results suggest that As uptake by *Chama* is influenced by As levels in ambient seawater. This observation is supported by a maximum As enrichment of <20 in produced water from site HI-595 relative to ambient seawater.

For Cd, Cu and Ni in clams, concentrations for each element were notably higher at one site (Cd at EI-313, Cu at SMI-130 and Ni at MI-703, Table 2). Similarly, concentrations of Hg averaged 0.053 ± 0.007 µg/g dry weight for five locations with one higher value for site MI-703. The Fe values for *Chama* were variable but low enough to be sure that these platform dwellers did not contain any significant amount of sediment within their gut that could account for higher metal levels. Such sediment distortion of tissue metal levels would be most conspicuous for Ni and V in these samples, and none is observed (Table 2). Thus, we have no clear explanation for the above deviations, but note that such variations have been previously reported for bivalves (e.g. Presley et al., 1990).

Our mean values for Pb show that *Chama* from platform HI-595 contained 2 to almost 5 times more Pb than those from the other sites (Figure 3). Although this result is not statistically significant and may also be due to natural variability, we should note that Pb loading from HI-595 at 17.5 g Pb/day is 28 times higher than at SMI-130. In a similar fashion, Ba concentrations for *Chama* from platform HI-595 are also the highest for this study (Figure 3). This trend is consistent with loadings of about 700 kg Ba/day from the platform at HI-595, three times more than the Ba discharge at site SMI-130. More samples of organisms from platforms discharging produced water with high metal levels and high discharge rates may help to better identify worse-case scenarios for enhanced metal bioaccumulation.

At the two nearshore sites (SMI-229 and SMI-236), *Crassostrea virginica*, rather than *Chama*, was the more abundant bivalve. No significant differences were observed for any metals in samples of *Crassostrea virginica* from reference site SMI-229 versus the discharging platform SMI-236 (Table 2, Figure 4a). Furthermore, metal levels from our study were in good agreement with results obtained by Presley et al. (1990) for oysters from coastal Louisiana and throughout the Gulf of Mexico (Table 2).

In addition to showing similarities in metal levels between the reference and discharge sites, Figure 4a also shows a greater than three order-of-magnitude range in actual concentrations of the various metals in oyster tissue. These actual concentrations can also

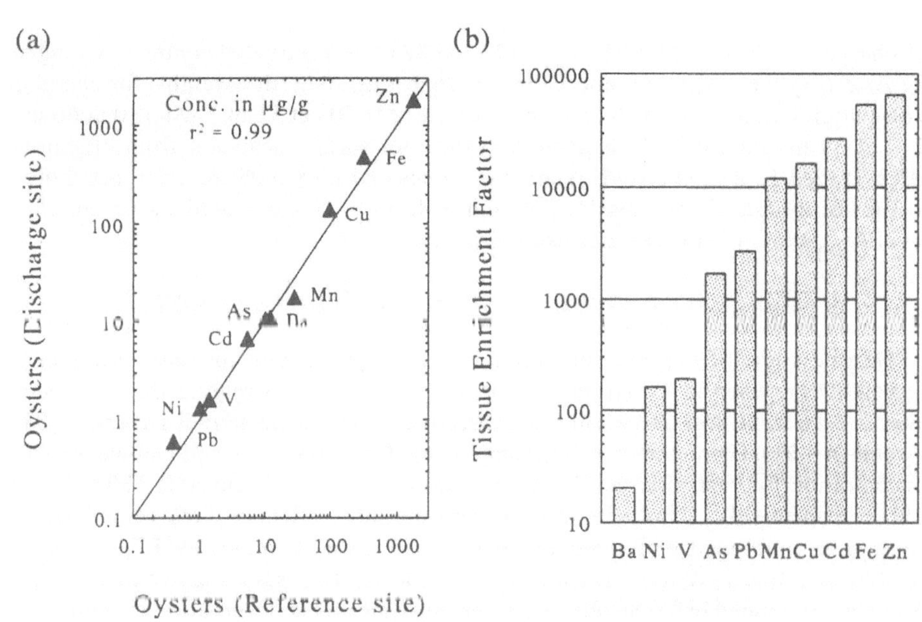

Figure 4. Graphs showing (a) metal concentrations in oysters (*Crassostrea virginica*) from discharge versus reference sites and (b) tissue enrichment factors for metals in oysters.

Table 3. Concentrations (dry weight; mean ± 1 standard deviation) of heavy metals in red snapper (*Lutjanus campechanus*) and triggerfish (*Balistes capriscus*) from discharging and non-discharging platforms in the Gulf of Mexico

Site	As (µg/g)	Ba (µg/g)	Cd (µg/g)	Cu (µg/g)	Fe (µg/g)	Hg (µg/g)	Mn (µg/g)	Ni (µg/g)	Pb (µg/g)	V (µg/g)	Zn (µg/g)
Red Snapper											
VN-298-R	3.1	0.06	<0.01	0.8	9±2	0.72	—	1.4	0.1	0.86	13.3
	±1.8	±0.03		±0.1		±0.19		±0.1	±0.05	±0.23	±0.7
VN-214-D	1.1	0.1	0.03	0.8	12±5	0.37	0.4	1.2	<0.1	0.52	14.6
	±0.6	±0.1	±0.01	±0.1		±0.20	±0.1	±0.1		±0.03	±0.4
SMI-130-D	0.8	0.05	—	0.8	7±1	0.28	0.2	1.4	<0.1	0.70	13.2
	±0.4	±0.05		±0.1		±0.03		±0.1		±0.07	±0.7
Triggerfish											
VN-298-R	83	0.07	<0.01	1.2	8±2	0.47	—	1.6	0.13	1.0	24
	±26	±0.02		±0.3		±0.31		±0.1	±0.07	±0.3	±8
VN-214-D	24	0.06	<0.03	1.0	8±1	1.1	1.1	1.5	0.07	0.7	19
	± 9	±0.04		±0.1		±0.3	±0.4	±0.06	±0.05	±0.1	±1
SMI-130-D	33	0.6	<0.1	1.2	9±3	0.44	1.1	1.6	0.09	0.8	23
	±13	±0.5		±0.2		±0.52	±0.7	±0.13	±0.06	±0.1	±3

R = reference (non-discharging) site; D = discharging site. Locations include: Vermilion 298, Vermilion 214 and South Marsh Island 130; n=5 for each site.

be viewed from the perspective of their enrichment relative to seawater. When the tissue data are converted to EF_{tissue} using Equation 1, values greater than 10,000 are calculated for Mn, Cu, Cd, Fe and Zn relative to levels of about 100 or less for Ni, V and Ba (Figure 4b). Essential micronutrients such as Cu, Mn, Fe and Zn can be very efficiently removed from seawater by marine organisms relative to non-essential elements such as As, Pb and Ba. These data provide a frame of reference for comparing metal concentrations in oysters from future monitoring efforts as well as for establishing universal trends in values for EF_{tissue}.

HEAVY METALS IN FISH

Metal concentrations also were determined in 5-8 separate samples of two species of fish, the red snapper (*Lutjanus campechanus*) and the gray triggerfish (*Balistes capriscus*) from each of two discharging platforms and one reference platform. No statistical differences ($\alpha = 0.05$) were observed among the sites for concentrations of Ba, Cd, Cu, Fe, Mn, Ni, Pb, V or Zn for both species of fish (Table 3). Although this statistical result is influenced by similar median values, some large standard deviations and a small sample population, the results in Table 3 and on Figures 5 and 6 certainly support similarities in metal levels for these reference and discharging sites.

Overall, Ba levels in fish fillets from this study were low with mean concentrations of <0.1 μg/g (dry weight). Samples of red snapper from site SMI-130 averaged 0.012 μg Ba/g, wet weight (wet tissue = 77±1 % water) and had EF_{tissue} values of about 1 (Figure 7),

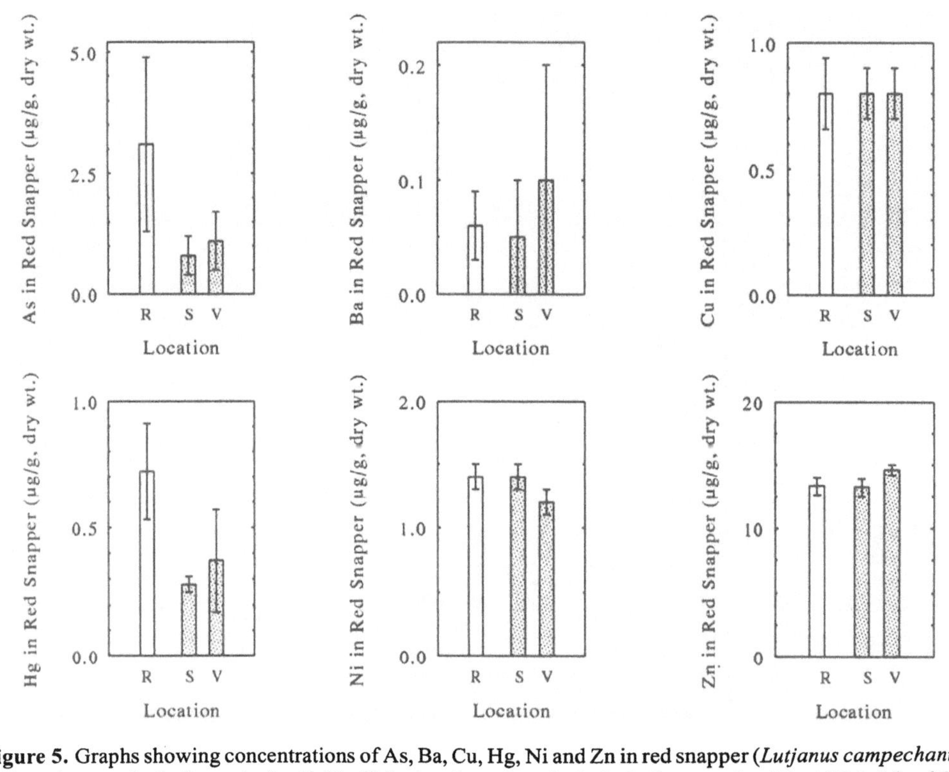

Figure 5. Graphs showing concentrations of As, Ba, Cu, Hg, Ni and Zn in red snapper (*Lutjanus campechanus*) collected near oil platforms in the Gulf of Mexico. Locations include Reference site (R) at High Island 389 and the following discharge sites: S = South Marsh Island 130 and V = Vermilion 214.

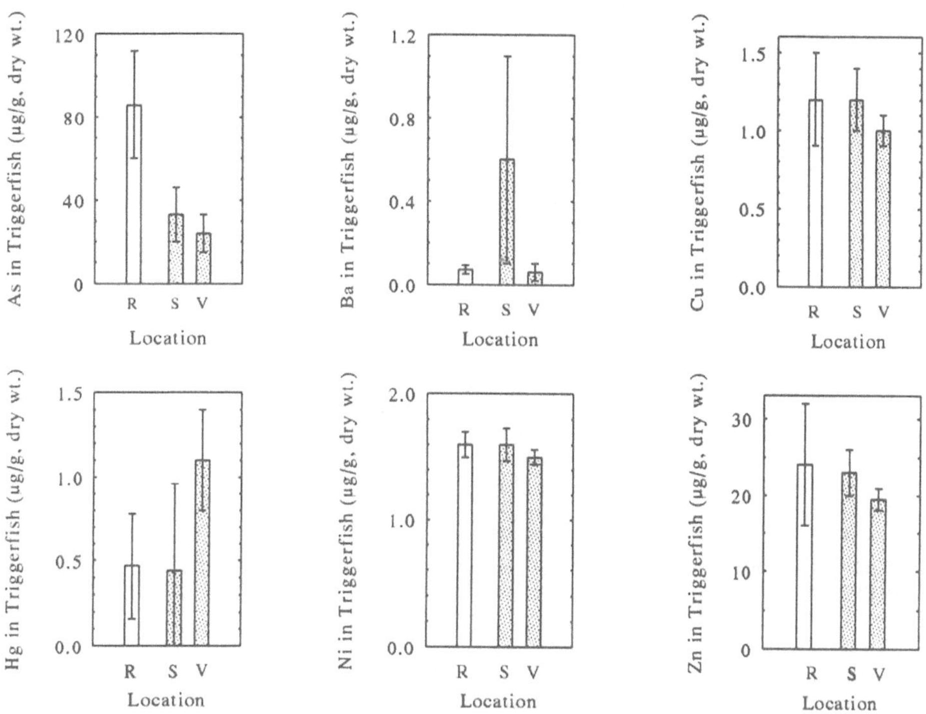

Figure 6. Graphs showing concentrations of As, Ba, Cu, Hg, Ni and Zn in triggerfish (*Balistes capriscus*) collected near oil platform in the Gulf of Mexico. Locations include Reference site (R) at High Island 389 and the following discharge sites: S = South Marsh Island 130 and V = Vermilion 214.

suggesting that no bioaccumulation of Ba above seawater levels occurs in red snapper. In contrast, about 25% of the samples had >0.025 µg Ba/g wet weight with four samples of triggerfish from the same nearshore, discharge site (SMI-130) with >0.070 µg Ba/g wet weight. The randomness of values in the Ba-enriched tissues may be related to incorporation of minute particle of barite as has been noted by Jenkins et al. (1989) and it would be valuable to identify the mode of tissue incorporation when enhanced Ba levels are determined.

Higher levels of As are observed in fish from the reference site on Sonnier Bank (VN-298 at a water depth of 20 m) relative to adjacent, open water locations at SMI-130 (water depth of 66 m) and VN-214 (water depth of 39 m; Figures 5 and 6). Observed As enrichment in fish collected from Sonnier Bank may be related to different food sources or even to movement of outer continental shelf waters into this area via advection or upwelling (Naito, 1994). Whatever the cause for this difference, the results once again show that local environmental conditions may have a greater influence on As levels in biota than produced water discharges. A similar trend to that found for As was observed for Hg in red snapper (Figure 5), but not triggerfish (Figure 6).

Enrichment factors presented for fish tissue from red snapper (Figure 7) are 10-20 times lower for As, Ba, Pb, Cu and Fe, 100 times lower for Zn and 300 times lower for Mn relative to values calculated for *Crassostrea virginica* (Figure 4). The micronutrient elements (especially Cu, Fe and Zn) are actively regulated in fish tissue, leading to the very uniform values shown in Table 3, and lower tissue enrichment factors. In sharp contrast, the degree of Hg bioaccumulation was >3000 in red snapper (Figure 7) and almost 20 times higher than in the oyster. Bioaccumulation of Hg in fish is favored because most of the Hg stored by fish

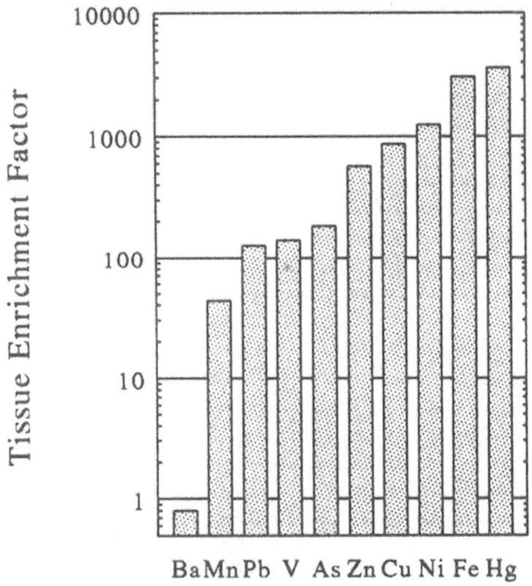

Figure 7. Graph showing tissue enrichment factors for metals in red snapper (*Lutjanus campechanus*) from site SMI-130.

is in the form of methyl-mercury (Fergusson, 1990). Once again, both concentrations and tissue enrichment factors provide a frame of reference for comparing metal concentrations in fish from future monitoring efforts as well as for establishing universal trends in values for EF_{tissue}.

HEAVY METALS IN SEDIMENTS

Sediment samples were collected at distances of 20 and 2000 m from production platforms. To assess the presence of contaminant levels of metals in these sediments, concentrations of each metal were plotted versus Al to normalize metal values and thereby correct for differences in grain size and mineralogy. Fine-grained, clay-rich sediments naturally contain higher levels of Al and trace metals. As fine-grained sediments introduced from riverine sources are mixed with metal-poor carbonates and coarse-grained sediments, concentrations of Al and metals are diluted. Thus, for a suite of uncontaminated sediments with a mixture of grain sizes and mineralogy, a simple straight line, such as that shown for Ni versus Al in Figure 8 is observed. This trend suggests that no enhancement of Ni was

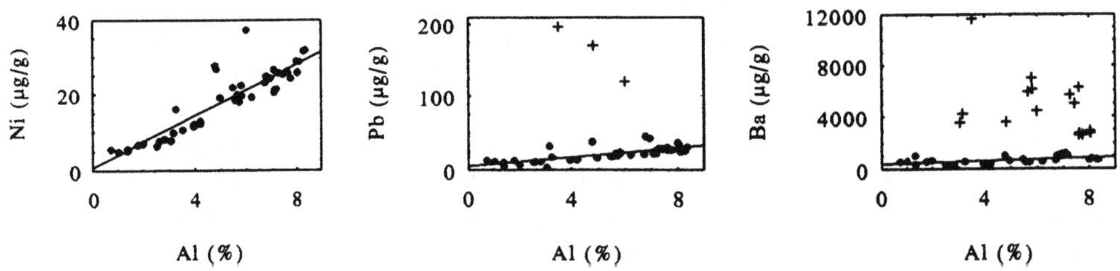

Figure 8. Scatter plots showing sediment concentrations of Ni, Pb and Ba plotted versus Al. The line on each graph identifies the natural metal/Al relationship for the northwestern Gulf of Mexico. Points marked with a (+) symbol are most likely influenced by anthropogenic inputs.

Table 4. Concentrations of heavy metals in sediments near oil platforms discharging produced water to the Gulf of Mexico

	Distance (m)	Al (%)	As (μg/g)	Ba (μg/g)	Cd (μg/g)	Cu (μg/g)	Fe (%)	Hg (μg/g)	Ni (μg/g)	Pb (μg/g)	V (μg/g)	Zn (μg/g)
SMI-236	20	7.42	7.7	963	0.24	19.7	3.59	0.044	26.9	36.1	115	268
SMI-236	2000	7.69	8.4	738	0.16	18.3	3.71	0.046	27.1	24.5	112	113
HI-595	20	3.25	4.4	6460	0.80	9.7	1.68	0.120	9.4	77.9	31	281
HI-595	2000	5.78	8.0	6350	0.12	12.6	2.64	0.056	20.6	23.5	87	96
Miss. R. Suspended Sediment	—	8.12	—	569	0.70	33.0	4.19	—	41.0	36.0	132	160

observed in sediments from these sites due to petroleum-related activities. Similar good correlations were observed for Cd, Fe, Ni and Hg indicating that these metals are generally present at close to natural levels near the drilling platforms studied. Data for Cd and Hg from the 20-m location near HI-595 are at the upper limits of natural concentrations and may suggest some small degree of contamination (Table 4).

Plots of Pb and Ba versus Al (Figure 8) show some large deviations from the natural metal/Al ratio. These anomalies indicate significant anthropogenic inputs of Pb and especially Ba. The influence of offshore drilling on heavy metals in sediments has been most notable for Ba (Trocine and Trefry, 1983; Boothe and Presley, 1987). Sources of Ba include drilling fluids wherein $BaSO_4$ is used as a weighting agent for pressure control and from produced waters that have high Ba concentrations naturally, due to its enhanced solubility at elevated chloride and low sulfate levels. Lead tends to be a more ubiquitous contaminant related to a variety of platform activities (Figure 8). Table 4 shows that concentrations of Pb in sediments from SMI-236 are comparable with values for incoming Mississippi River suspended sediment; however, a considerable excess was identified for HI-595. The HI-595 site also has the highest Pb loading; however, this enhancement is unlikely to be solely related to produced water.

CONCLUSIONS

Concentrations of Hg, Cd and Pb, three metals of particular environmental concern, were enriched in produced water from this study in the Gulf of Mexico by factors of about 3, 30 and 200, respectively, relative to ambient seawater. Concentrations of Ba, Fe and Mn were enriched by factors of ≥10,000. In contrast, produced water samples from this study were depleted with Cu and V relative to ambient water. Typical dilution factors for produced water upon admixture with seawater are ≥1000 at distances of 20-30 m from the discharge point at release rates of ≤4000 m^3/day (Brandsma and Smith, 1996). Thus, the produced water evaluated here most likely has a relatively limited opportunity for enhancing normal bioaccumulation of metals in resident organisms. No statistically significant differences ($\alpha = 0.05$) in concentrations of Ba, Fe, Pb, V or Zn were observed in clams (*Chama macerophylla*) from reference or discharge sites; and only one unexplained exception in clam levels was observed for Cd, Cu, Ni and Hg. However, absolute values for Ba and Pb in *Chama* were 2-5 times higher near one platform with much higher daily loadings of these metals than at other sites. Differences in levels of As observed for clams from different sites seem to be more dependent on local concentrations of dissolved As in ambient seawater than on discharges of produced water. No significant differences were observed for any metals in samples of *Crassostrea virginica* from a reference versus the discharge site. Furthermore, metal levels from our study were in good agreement with results obtained by Presley et al. (1990) for oysters from coastal Louisiana and throughout the Gulf of Mexico. In addition, no statistical differences ($\alpha = 0.05$) were observed among the sites for concentrations of Ba, Cd, Cu, Fe, Mn, Ni, Pb, V or Zn for red snapper (*Lutjanus campechanus*) and gray triggerfish (*Balistes capriscus*). Sediment in discharge areas from the Gulf of Mexico often had elevated levels of Pb and Ba relative to Al; however, this enhancement is unlikely to be solely related to produced water.

ACKNOWLEDGMENT

We thank our collaborators at Continental Shelf Associates, Inc. for their assistance during this study with special appreciation for the efforts of and discussions with Bruce

Graham and Al Hart. We also thank Annette Bernard of Florida Institute of Technology for her tireless efforts in preparing the manuscript. The study was supported by the U.S. Department of Energy, Contract Number DE-AC22-92MT92001.

REFERENCES

Barth, T., 1991. Organic acids and inorganic ions in waters from petroleum reservoirs, Norwegian continental shelf: a multivariate statistical analysis and comparison with American reservoir formation waters. *Appl. Geochem.*, 6:1-15.

Boesch, D.F., and Rabalais, N.N., 1989. Produced waters in sensitive coastal habitats: an analysis of impacts, Central Coastal Gulf of Mexico. *OCS Report/MMS 89-0031*, U.S. Department of the Interior, Minerals Management Service, Gulf of Mexico OCS Regional Office, New Orleans, Louisiana.

Boothe, P.N., and Presley, B.J., 1987. The effects of exploratory petroleum drilling in the Northwest Gulf of Mexico on trace metal concentrations in near rig sediments and organisms. *Environ. Geol. Water Sci.*, 9:173-182.

Brandsma, M.G. and Smith, J.P., 1996. Dispersion modeling perspectives on the environmental fate of produced water discharges. *Produced Water 1995*, pp. xxx.

Fergusson, J.E., 1990. *The Heavy Elements: Chemistry, Environmental Impact and Health Effects*, Pergamon Press, Oxford.

Fredericks, A.D., and Sackett, W.M., 1970. Organic carbon in the Gulf of Mexico. *J. Geophys. Res.*, 75:2199-206.

Heinrich, C.A., and Seward, T.M., 1990. A spectrophotometric study of aqueous iron (II) chloride complexing from 25 to 200°C. *Geochim. Cosmochim. Acta*, 54:2207-2221.

Helgeson, H.C., Know, A.M., Owens, C.E., and Shock, E.L., 1993. Petroleum oil field waters and authigenic mineral assemblages: Are they in metastable equilibrium in hydrocarbon reservoirs? *Geochim. Cosmochim. Acta*, 57:3295-3339.

Jacobs, R.P.W.M, Grant, R.O.H., Kwant, J., Marquenie, J.M., and Mentzer, E., 1992. The composition of produced water from Shell operated oil and gas production in the North Sea. In: *Produced Water*, J.P. Ray and F.R. Engelhardt (Eds.), Plenum Press, New York, pp. 13-22.

Jenkins, K.D., Ball, J.W., Burchard, J.M., Vivit, D.V., and Barks, J.H., 1989. Sediment deposition, biological accumulation and subcellular distribution of barium following the drilling of an exploratory well. In: *Drilling Wastes*, F.R. Engelhardt, J.P. Ray and A.H. Gillam (Eds.), Elsevier Applied Science, London, pp. 587-608.

LeBlanc, L., 1994. Industry marches into deepwater: the offshore years (1954-1994). *Offshore*, April, 36-56.

Naito, K.L., 1994. Bioaccumulation of As, Ba, Cu and Zn in fish and bivalves collected near produced water outfalls in the Gulf of Mexico. M.S. Thesis, Florida Institute of Technology, Melbourne.

Nakashima, S., Sturgeon, R.E., Willie, S.N., and Berman, S.S., 1988. Determination of trace elements in sea water by graphite-furnace atomic absorption spectrometry after preconcentration by tetrahydroborate reductive precipitation. *Anal. Chim. Acta*, 207:291-299.

Presley, B.J., Taylor, R.J., and Boothe, P.N., 1990. Trace metals in Gulf of Mexico oysters. *Sci. Total Environ.* 97/98:551-593.

Rabalais, N.N., McKee, B.A., Reed, P.J., and Means, J.C., 1992. Fate and effects of produced water discharges in coastal Louisiana, Gulf of Mexico, USA. In: *Produced Water*, J.P. Ray and F.R. Engelhardt (Eds.), Plenum Press, New York, pp. 355-369.

Ray, J.P., and Engelhardt, F.R., 1992. *Produced Water.* Plenum Press, New York.

Reilly, W.K., O'Farrell, T., and Rubin, M.R., 1991. Development document for 1991 proposed effluent limitation guidelines and new source performance standards for the offshore subcategory of the oil and gas extraction and point source category, *U.S. Environmental Protection Agency*, Washington, D.C.

Schiff, K.C., Reish, D.J., Anderson, J.W., and Bay, S.M., 1992. A comparative evaluation of produced water toxicity. In: *Produced Water*, J.P. Ray and F.R. Engelhardt (Eds.), Plenum Press, New York, pp. 199-208.

Sturgeon, R.E., Berman, S.S., Desaulniers, A., and Russell, D.S., 1980. Pre-concentration of trace metals from sea-water for determination by graphite furnace atomic- absorption spectrometry. *Talanta* 27:85-94.

Templeton, C.C., 1960. Solubility of barium sulfate in sodium chloride solutions from 25° to 95°C. *J. Chem. Eng. Data.*, 5:514-516.

Trocine, R.P., and Trefry, J.H., 1983. Particulate metal tracers of petroleum drilling mud dispersions in the marine environment. *Environ. Sci. Tech.* 17:507-512.

AN ECOLOGICAL RISK ASSESSMENT FOR POLYCYCLIC AROMATIC HYDROCARBONS IN PRODUCED WATER DISCHARGES TO THE WESTERN GULF OF MEXICO

Jerry M. Neff and Theodor C. Sauer, Jr.

Battelle Ocean Sciences Laboratory
397 Washington Street
Duxbury, Massachusetts 02332

INTRODUCTION

Produced water represents the largest volume waste stream in the entire exploration and production process on most offshore platforms (Stephenson, 1991). The amount of treated produced water discharged from a single platform usually is less than 1.5 million liters/day (≈ 400,000 barrels/day), whereas discharges from large treatment facilities that process produced water from several platforms may be as high as 25 million L/day (Menzie, 1982). The total volume of treated produced water discharged to U.S. waters of the Gulf of Mexico in 1991 was approximately 549 million L/day (Rabalais et al., 1991), approximately 65 percent of which was generated in and discharged to coastal (state) waters of Louisiana and Texas (Boesch and Rabalais, 1989). In 1991, an average of 438 million L/day of treated produced water was discharged to the entire North Sea (Stephenson et al., 1994). The Statfjord oil field alone discharged an average of 30.9 million L/day of treated produced water to the Norwegian sector of the North Sea (Ynnesdal and Furuholt, 1994). The daily rate of produced water discharge to the North Sea is expected to increase to about 955 million L/day in 1998.

Produced water contains a large number of inorganic and organic chemicals that have been dissolved or dispersed from the geologic formations in which the produced water resided for millions of years. Petroleum hydrocarbons usually represent 10 to 65 percent of the total organic matter in produced water. Much of the remainder of the organic chemicals in produced water is in the form of low molecular weight organic acids, such as acetic, propionic, and butyric acids, which are thought to have been derived from hydrous pyrolysis of hydrocarbons in the hydrocarbon-bearing formation (Borgund and Barth, 1994). Low molecular weight organic acids are readily biodegraded and are not considered toxic. Polycyclic aromatic hydrocarbons (PAHs; also called polynuclear aromatic hydrocarbons) are the organic chemicals of greatest environmental concern in produced water, because of their toxicity and persistence in the marine environment (Neff, 1987). Concentrations of total PAHs in treated produced water of the types discharged to the ocean usually range from about 70 to 3,000 μg/L. The most abundant PAHs in produced water are the more water-soluble 2- and 3-ring PAHs (naphthalene, fluorene, phenanthrene, and dibenzothiophene) and their alkylated homologues (Neff et al., 1987). High molecular weight PAHs, such as the carcinogen, benzo(a)pyrene, almost never are present at greater than ultratrace (< 1 μg/L) concentrations.

The objective of this paper is to evaluate the ecological risk of PAHs discharged to open ocean waters in treated oil well produced water. This risk assessment is based on the results of an investigation supported by the U.S. Department of Energy dealing with the composition and fate of chemicals in treated produced water discharged from four production platforms to outer continental shelf waters of the western Gulf of Mexico. Concentrations of individual and total PAHs in ambient water, sediments, and whole tissues of marine animals are compared to published concentrations in these media equivalent to the highest no observable effects concentrations.

MATERIALS AND METHODS

Sample Collection

As part of an offshore production platform monitoring study sponsored by the U.S. Department of Energy, produced water and environmental samples were collected during the summer of 1993 from four oil production platforms on the continental shelf off Louisiana. Reference seawater and sediment samples were collected at four sites without platforms or produced water discharges at water depths and distances from shore similar to those of the study sites.

Treated produced water samples were collected directly from the effluent pipe just before the discharge point. Several samples were collected from each production platform. Samples for hydrocarbon analysis were collected into and stored in 1- or 2-L pre-cleaned amber glass bottles that were sealed with Teflon-lined caps. Samples were stored chilled at ≈ 4°C and shipped immediately by air to the analytical laboratory.

Ambient seawater samples were collected at a depth of approximately 1 m below the sea surface approximately 2,000 m down-current from the four production platforms. Additional ambient water samples were collected at the four reference sites. Seawater samples were handled the same as produced water samples. Sediment samples were collected with several types of grab samplers. The surficial layer (upper 2 cm) of undisturbed sediments in the samplers were collected into precleaned wide-mouth glass jars, sealed with Teflon-line caps, and stored frozen until analysis.

Bivalve molluscs were collected from legs of the production platforms by divers. Fish and portunid crabs were collected by otter trawl at distances of 200 to 2,000 m from the production platforms. The biological specimens were frozen and shipped by air to the analytical laboratory. Whole soft tissues of bivalves and edible muscle tissues of crabs and fish were analyzed.

Sample Analysis

PAHs in produced water, ambient water, sediment, and tissue samples were analyzed by capillary column gas chromatography with quantification by mass spectrometry in the selected ion monitoring mode (modified EPA Method 8270). Details of the extraction, cleanup, and analytical methods are provided by Sauer and Boehm (1991) and Page et al.

Table 1. Characteristics of four produced water discharges to the western Gulf of Mexico

Platform	Water Depth (m)	Produced Water Discharge (L/day)	PAHs in Produced Water (μg/L)
SMI236	6.1	2,353,000	158
VR214	38.7	1,351,000	75.4
SMI130	65.5	3,600,000	578
HI595	122	838,000	36.9

(1995). A total of 41 parent PAHs and alkyl-PAH congener groups were quantified, including naphthalene, fluorene, dibenzothiophene, phenanthrene, fluoranthene/pyrene, and chrysene and their C_1 through C_3 or C_4 alkyl homologue groups, acenaphthylene, acenaphthene, biphenyl, anthracene, benz[a]anthracene, benzo[b]fluoranthene, benzo[k]fluoranthene, benzo[e]pyrene, benzo[a]pyrene, perylene, indeno[1,2,3-cd]pyrene, dibenz[a,h]anthracene, and benzo[ghi]perylene.

RESULTS AND DISCUSSION

Characteristics of the Produced Water Discharges

The four produced water discharges monitored in this project are located in the western Gulf of Mexico south of the state of Louisiana. Water depth at the four production platforms ranges from 6.1 to 122 m (Table 1). All four are large-volume discharges, ranging from 838,000 to 3,600,000 L/day. The combined discharges from these four platforms represent about 7 percent of the total volume of produced water discharged to US federal outer continental shelf waters of the western Gulf of Mexico (Boesch and Rabalais, 1989). The mean concentration of total PAHs in the four discharges ranges from 36.9 to 578 μg/L. These concentrations are in the low to middle part of the range of PAH concentrations (80 to 1,860 μg/L) reported previously for produced water discharged to the western Gulf of Mexico (Armstrong et al., 1979; Neff et al., 1989; Means et al., 1989; Boesch et al., 1989; Rabalais et al., 1991). The mass discharge rates (concentration X discharge volume) for total PAHs from the four produced water discharges range from 31 to 2,080 g/day.

The PAHs in produced water from the four offshore production platforms are dominated by naphthalene and alkyl naphthalenes, with lower concentrations of other two- and three-ring PAHs and their alkyl homologues (Table 2). Only trace amounts of four-, five-, and six-ring PAHs are present in the produced water samples. The most abundant of these higher molecular weight PAHs are alkyl fluoranthenes/pyrenes and chrysenes. Acenaphthylene, acenaphthene, benz[a]anthracene, indeno[1,2,3-cd]pyrene, and dibenz[a,h]anthracene were not present in any of the produced water samples at a concentration above the detection limit of the analytical method (≈ 1 ng/L).

This PAH composition is typical of produced water that has been treated effectively to remove dispersed oil droplets (Neff, 1987). The PAH assemblage is dominated by the more soluble, lower molecular weight two- and three-ring PAHs. As is typical of oil, alkyl PAHs are more abundant than the parent compounds and phenanthrene is much more abundant than anthracene (Neff, 1979).

Table 2. Mean concentrations (n = 3) of individual and total PAHs in treated produced water from four production platforms discharging to outer continental shelf waters of the western Gulf of Mexico. Concentrations are in ng/L

Compound	SMI236	SMI130	VR214	HI595
Naphthalene	90,000	57,000	54,000	19,000
C_1-Naphthalenes	70,000	74,700	34,000	14,000
C_2-Naphthalenes	40,000	94,000	17,000	7,800
C_3-Naphthalenes	16,000	92,000	9,300	4,900
C_4-Naphthalenes	7,000	59,000	4,500	2,100
Biphenyl	9,800	7,500	4,000	780
Fluorene	850	3,100	330	150
C_1-Fluorenes	1,400	9,500	370	300
C_2-Fluorenes	1,900	17,000	710	500
C_3-Fluorenes	1,800	19,000	800	610
Dibenzothiophene	197	4,900	103	220
C_1-Dibenzothiophenes	437	14,000	300	490
C_2.Dibenzothiophenes	553	26,700	320	960
C_3-Dibenzothiophenes	500	27,000	370	1,000
Phenanthrene	1,000	9,500	330	250
Anthracene	48	500	34	0.0
C_1-Phenanthrenes	1,400	27,000	430	500
C_2-Phenanthrenes	1,700	34,000	640	730
C_3-Phenanthrenes	1,200	24,700	520	670
C_4-Phenanthrenes	550	11,700	480	330
Fluoranthene	40	0.0	20	4.0
Pyrene	41	337	24	2.6
C_1-Fluoranthenes/Pyrenes	217	2,700	94	54
C_2-Fluoranthene/Pyrenes	197	4,700	63	82
C_3-Fluoranthene/Pyrenes	220	5,700	88	117
Chrysene	11	860	3.5	3.5
C_1-Chrysenes	72	2,900	16	31
C_2.Chrysenes	117	4,000	32	75
C_3Chrysenes	143	3,600	42	97
C_4-Chrysenes	137	2,700	0.0	70
Banzo[b]fluoranthene	0.0	26	3.6	0.0
Benzo[k]fluoranthene	0.0	76	0.0	0.0
Benzo[e]pyrene	0.0	0.0	2.6	0.0
Benzo[a]pyrene	14	121	2.3	12
Perylene	260	2,200	120	82
Benzo[ghi]perylene	0.0	32	0.0	0.0
Total PAHs	158,000	578,000	75,500	36,900

PAHs in Ambient Water

PAHs are diluted rapidly when produced water is discharged to the ocean. Samples of surface water collected approximately 2,000 m down-current from the four produced water discharges are diluted by 600- to nearly 5,000-fold compared to concentrations in the undiluted produced water (Table 3). The lowest dilution is for the lowest volume, lowest concentration produced water discharge (from Platform HI595) and the greatest dilution is for the highest volume, highest concentration discharge (from Platform SMI130). The reason for this is that the concentrations of total PAHs in ambient water down-current from the discharges are close to the background concentration for total PAHs in the northwestern Gulf of Mexico. Ambient water samples from four reference sites in the western Gulf of Mexico away from produced water discharges contain 0.01 to 1.28 μg/L total PAHs. Thus, concen-

Table 3. Initial dilution of PAH from produced water discharged to the ocean from four platforms in the Gulf of Mexico

Platform	PAH in Produced Water (μg/L)	PAH in Ambient Water (μg/L)*	Dilution (PW/Amb)
SMI236	158	0.071	2,225
VR214	75.4	0.060	1,257
SMI130	578	0.118	4,898
HI595	37.0	0.058	638

* Ambient water samples collected approximately 2,000 m down-current from the discharge

trations of PAHs in ambient surface waters of the western Gulf of Mexico are not notably higher near offshore produced water discharges than elsewhere in the offshore environment.

Concentrations of total PAHs (sum of 14 parent PAHs) in waters of Chesapeake Bay, USA are in the range of 0.0025 to 0.034 μg/L (Ko and Baker, 1994; Dickhut and Gustafson, 1995). By comparison, the total concentration of these 14 PAHs in the most heavily contaminated reference water sample in this study is 0.069 μg/L. The concentration of total PAHs (the same analytes as listed in Table 2) in surface waters of areas of Prince William Sound not affected by the Exxon Valdez oil spill of 1989 was in the range of 0.015 to 0.025 μg/L (Neff and Burns, 1995). Thus, outer continental shelf waters of the northwestern Gulf of Mexico are slightly contaminated with PAHs. Undoubtedly, produced water discharges are just one of several sources of these aqueous PAHs. Other sources include emissions of fuel and exhaust from motor vessel traffic, coastal and river runoff, natural oil seeps, and aerial deposition of PAHs from combustion sources (Neff, 1979).

The U.S. Environmental Protection Agency has published acute and chronic water quality criteria for protection of marine life for only four individual PAHs and for total PAHs. The published or estimated chronic criterion values are 250 to nearly 3,600 times higher than concentrations of individual and total PAHs measured in ambient water samples approximately 2,000 m down current from the produced water discharges (Table 3).

Produced water discharged to open ocean waters usually is diluted by more than 100-fold within 100 m of the discharge. For example, produced water discharged from offshore platforms to the Bass Strait, Australia, was diluted 220-fold at 100 m down-current of the discharge (Terrens and Tait, 1993). Brandsma (this publication) predicted dilutions of approximately 100-fold at 100 m from high-volume produced water discharges based on the Offshore Operators Committee model. The concentrations of PAHs in 100-fold dilutions of produced water from the four offshore produced water discharges are well below EPA

Table 4. Comparison of concentrations of selected PAHs in ambient water at produced water discharge sites and US EPA marine chronic water quality criteria. Concentrations are in μg/L

Compound	Range of Concentrations in Ambient Water	US EPA Chronic Criterion	Safety Factor (Crit./Amb.)
Acanaphthene	ND	710	∞
Naphthalene	ND - 0.066	235*	3560
Phenanthrene	ND - 0.007	4.6	657
Fluoranthene	ND - 0.017	16	941
Total PAHs	0.058 - 0.118	30*	254 - 517

* Estimated based on an acute/chronic ratio of 10.

Table 5. Comparison of concentrations of selected PAHs in a 100-fold dilution of produced water from the four discharges and US EPA marine chronic water quality criteria. Concentrations are in μg/L

Compound	Range of Concentrations in the 100-Fold Dilution	US EPA Chronic Criterion	Safety Factor (Crit/PW Conc)
Acenaphthene	0	710	∞
Naphthalene	0.19-0.95	235*	247-1237
Phenanthrene	0.002-0.10	4.6	46-2300
Fluoranthene	0-0.0004	16	40,000-∞
Total PAHs	0.37-5.78	30*	5.2-81

* Estimated based on an acute/chronic ratio of 10.

chronic criterion values (Table 5). Safety factors (criterion value/concentration in 100-fold dilution) range from 5.2 (total PAHs) to infinity (acenaphthene and fluoranthene).

The few PAHs for which water quality criteria are available are representative of the toxicity range of PAHs of the types found in produced water. For low molecular weight PAHs, the toxic concentration in water appears to lie between 20 and 70 percent of the saturated concentration (Abernethy et al., 1986). The toxic concentration for less soluble, higher molecular weight PAHs is at or in excess of the saturation concentration. Thus, the only PAHs from produced water that can reach lethal concentrations in receiving waters are the two- and three-ring PAHs. These PAHs have aqueous solubilities in the range of about 30 mg/L (naphthalene) to about 0.3 mg/L (methylphenanthrene) (Mackay et al., 1992b). Chronically toxic concentrations will range from about 6 to to 600 μg/L. Therefore, individual and total PAH do not reach concentrations that are potentially harmful to marine organisms in the near-field and far-field of these four high volume produced water discharges to the open ocean.

PAHs in Sediments

Concentrations of total PAHs are higher in surficial sediments collected 20 m down-current from the four produced water discharges than in sediments collected 2,000 m away (Table 6). PAH concentrations in surficial sediments 20 m from the discharges range from 214 to 2,485 ng/g dry wt. There is no clear relationship between water depth, produced water discharge rate, and concentration of total PAHs in the produced water (Table 1) on the one hand and the concentration of total PAHs in sediments near the discharges (Table 6) on the other. However, lowest PAH concentrations are present in sediments adjacent to platform HI595 which is in the deepest water and has the lowest discharge rate and the lowest concentration of PAHs in its produced water. Other factors that affect the concentrations of PAHs in surficial sediments near produced water discharges include circulation/mixing of the surface waters, sediment grain size and total organic carbon concentration, and the oxidation/reduction state of the surficial sediments (Neff et al., 1989). The most important factors other than mass discharge rate of PAHs in the produced water effluent are sediment grain size and total organic carbon concentration. Fine-grained sediments containing high concentrations of total organic carbon have a much higher capacity for binding and retaining nonpolar organic contaminants, including PAHs, than sandy sediments containing little organic matter (Karickhoff et al., 1979; Voice and Webber, 1983). Although anoxic events are frequent in bottom water and sediments along the coast of Louisiana and Texas, all or most of the platform discharges monitored in this study are located offshore of the most severe bottom water hypoxia (Rabalais et al., 1985).

Table 6. Concentrations of total PAHs in surficial sediments adjacent to production platforms discharging produced water to outer continental shelf waters of the western Gulf of Mexico. Concentrations are in ng/g dry weight

Platform	Water Depth	20 m Down-Current	2,000 m Down-Current
SMI236	6.1	510 - 1,220	236 - 388
VR214	38.7	374 - 2,485	142 - 324
SMI130	65.5	713 - 1,110	147 - 168
HI595	122	214 - 419	70 - 92
Reference Sites	36 - 64	10 - 242	—

Sediments 2,000 m from the four produced water discharges contain essentially background concentrations of total PAHs. PAH concentrations in these sediments (70 to 388 ng/g) are comparable to or only slightly higher than PAH concentrations (10 to 242 ng/g) in surficial sediments from four reference sites away from produced water discharges. Brooks et al. (1990) reported a background concentration of total PAH in sediments offshore Matagorda, Texas (south Texas outer continental shelf), of 29 ± 28 ng/g; nearshore sediments contained a mean of 96 ± 112 ng/g total PAHs. These concentrations generally are lower than concentrations reported here from 2,000 m from offshore produced water discharges and from reference sites. Neff et al., (1989, 1992) reported background concentrations of total PAHs in sediments 1,000 m from a produced water discharge in 8 m of water off Louisiana of 19 to 156 ng/g, comparable to background concentrations in sediments reported here. Thus, background PAH concentrations may be higher in surficial sediments from the outer continental shelf off Louisiana and north Texas than off south Texas. The higher background concentration of PAHs in sediments from the continental shelf of the northwestern Gulf of Mexico may be derived from the massive outflow of the Mississippi River, which is the dominant source of surficial sediments along the coasts of Louisiana and north Texas, or from long-range transport of dissolved and particle-bound PAHs from operational discharges from the large number of oil and gas platforms off Louisiana.

PAHs and other nonpolar organic chemicals partition between the sorbed phase of sediments and sediment pore water according to their relative affinities for the sediment organic carbon and water phases of the sediment (Karickhoff et al., 1979). Therefore, a fraction of the PAHs associated with surficial sediments is bioavailable and may adversely affect benthic fauna if its concentration in the pore water phase exceeds some critical concentration. Long and Morgan (1990) and Long et al. (1995) have gathered data on the concentrations of chemical contaminants in sediments that are associated with adverse effects in freshwater and marine organisms. From these data, they have developed effects level low (ERL) and effects level median (ERM) concentrations that can serve as threshold concentrations of contaminants in sediments, above which adverse effects in benthic organisms may occur. Concentrations of chemicals in sediments below the ERL represent a minimal effects range, a range intended to estimate conditions in which effects rarely would be observed (Long et al., 1995).

Values for the ERL for those PAHs that have been evaluated to date are higher, with one exception, than concentrations of individual and total PAHs measured in surficial sediments 20 m down-current from the four produced water discharges (Table 7). One sediment sample from platform VR214 contains 1,040 ng/g total low molecular weight PAHs (2- and 2-ring PAHs), nearly twice the ERL value for low molecular weight PAHs. A few sediment samples also contain concentrations of one or more of fluorene, phenanthrene, total high molecular weight PAHs, or total PAHs that approach, but do not exceed the correspond-

Table 7. Comparison of concentrations of PAHs measured in sediments within 20m of 4 produced water discharges and effects range low (ER-L) threshold concentrations in sediments for protection of marine life (Long et al., 1995). Concentrations are in ng/g dry weight

Compound	Concentration Range in Sediments	ERL	Safety Factor (ER-L/Sed)
Naphthalene	0 - 5.3	160	30 - ∞
Methylnaphthalenes	2.3 - 19	70	3.7 - 30
Fluorene	0.54 - 16	19	1.2 - 35
Phenanthrene	2.6 - 140	240	1.7 - 92
Low MW PAHs	79 - 1,040	552	0.53 - 7.0
Fluoranthene	4.5 - 150	600	4.0 - 133
Pyrene	6.2 - 120	665	5.5 - 107
Benz(a)anthracene	1.3 - 80	261	3.3 - 201
Chrysene	3.4 - 82	384	4.7 - 113
Benzo(a)pyrene	1.2 - 150	430	2.9 - 358
Dibenz(a,h)anthracene	0.32 - 21	63.4	3.0 - 198
High MW PAHs	144 - 1,446	1,700	1.2 - 11.8
Total PAHs	214 - 2,485	4,022	1.6 - 18.9

ing ERL. All other concentrations of individual and total PAHs in sediments 20 m from the four produced water discharges are well below the corresponding ERL values.

These results indicate that PAHs in sediments under and within a short distance from the produced water discharge of these platforms may be present at concentrations that could be toxic to some sensitive marine organisms. Generally, impacts of production discharges from offshore platforms, when they can be detected at all are restricted to the benthic environment in the immediate vicinity of the platform (Neff, 1987; Neff et al., 1992; Kingston, 1992). Adverse effects in benthic ecosystems near oil platforms usually are attributable to accumulation in the sediments of petroleum hydrocarbons, particularly PAHs. The severity of biological effects often are correlated with concentrations of petroleum PAHs in the sediments, or to severity of hypoxia in surficial sediments. The hypoxia, in turn, is caused by biodegradation of petroleum hydrocarbons in the sediments (Kingston 1987, 1992). In the present investigation, accumulation of PAHs in sediments near the four offshore produced water discharges was to marginally toxic concentrations and effects on the benthos probably were minimal or undetectable, particularly at the deeper water platforms.

PAHs in Tissues of Marine Animals

Marine invertebrates and fish collected near the four platform discharges monitored in this investigation contain concentrations of total PAHs in their soft or edible tissues ranging from less than 0.1 μg/kg dry wt. to 12,160 μg/kg (Neff and Sauer, this publication). With several exceptions, fish muscle contains the lowest concentrations, followed by muscle tissue of decapod crustaceans; highest concentrations are in whole soft tissues of bivalve molluscs.

Considerable progress has been made in recent years in developing quantitative structure activity relationships (QSARs) that relate physical/chemical properties of non-specific toxicants, particularly nonpolar organic chemicals, to biological end points, such as acute or chronic toxicity and bioaccumulation (McCarty, 1986; Abernethy et al., 1988; Warne et al., 1991; Connell and Markwell, 1992; Mackay et al., 1992a; McCarty et al., 1992; Van Leeuwen et al., 1992; Sijm et al., 1993). McCarty et al. (1992) showed, that for nonpolar organic chemicals that are non-specific toxicants (act primarily by physical accumulation in membrane lipids causing generalized narcotic effects), the critical body residue (CBR) of the chemical in the whole tissues of the animal is approximately 4.4 mmol/kg (95% C.I. = 3.7 to 5.2 mmol/kg). That is, adverse biological responses (e.g., death) occur when whole tissue residues of a nonspecific narcotic chemical reach 4.4 mM.

Because the mode and site of action of non-specific toxicants is the same, their toxicities can be considered to be additive. That is, the sum of the concentrations of all the non-specific organic toxicants in the tissues of an experimental animal will equal approximately 4.4 mmol/kg at the point where toxic responses are observed.

Several classes of nonpolar or slightly polar organic chemicals behave as non-specific toxicants in aquatic animals. Parent (unmetabolized) PAHs behave as non-specific toxicants. However, some of their metabolites may be highly reactive and bind, often irreversibly, to tissue macromolecules. Most acute and chronic effects of PAHs measured in aquatic animals can be attributed to non-specific toxicity and not to specific effects of metabolites.

Van Leeuwen et al. (1992) used these concepts to develop a more conservative (protective) toxicological end point, the estimated average concentration of the chemical in tissues causing harmful responses in 5 percent of the organisms in an aquatic community during chronic exposure (the HC5). HC5s are inversely related to log K_{ow}s. Values for HC5 in tissues of aquatic organisms range from 0.0673 mmol/kg for chemicals with a log K_{ow} of 1.5 to 0.03 mmol/kg for chemicals with a log K_{ow} of 6.0.

Concentrations of several PAHs spanning the molecular weight/log K_{ow} range from naphthalene (log K_{ow} 3.37) to benzo[a]pyrene (log K_{ow} 6.04) in tissues of marine animals from the vicinity of the four platform discharges were compared to HC5 values estimated according to the methods of Van Leeuwen et al. (1992) (Table 8). All measured tissue residues are at least 200 times lower than the HC5 values, defined as the concentration of a chemical in tissues that may be harmful to 5 percent of the organisms in a marine community. The safety factor (HC5/concentration in tissues) increases with increasing molecular weight of PAHs, despite the fact that HC5 values decrease with increasing log K_{ow} and molecular weight. This is because tissue residues of low molecular weight PAHs are higher than those of high molecular weight PAHs, reflecting the composition of the PAH assemblage in produced water and petroleum. Effects of tissue residues of nonpolar narcotic chemicals are additive. However, even the highest concentration of total PAHs in bivalve molluscs from the platform legs (12,160 μg/g dry wt.) is approximately half the HC5 for benzo[a]pyrene. Therefore, tissue residues of individual and total PAHs in marine animals near the four offshore production platforms are well below concentrations that might be toxic to marine animals.

Table 8. Comparison of concentrations of selected PAHs in tissues of marine animals near produced water discharges and the estimated tissue concentrations hazardous to 5% of an aquatic community (Van Leeuwen et al., 1992). Concentrations are in ng/g dry weight

Compound	Concentration Range in Tissues	HC5	Safety Factor (HC5/Tiss)
Naphthalene	ND - 170	38,450	226
Phenanthrene	ND - 89	25,650	288
Pyrene	ND - 62	24,800	400
Chrysene	ND - 63	22,850	363
Benzo(a)pyrene	ND - 2.6	23,450	9,019

ND, Not detected, detection limit about 0.10 ng/g dry wt. per analyte.

CONCLUSIONS

Four offshore production platforms discharge 838,000 to 3,600,000 L/day of treated produced water to the outer continental shelf of the northwestern Gulf of Mexico off Louisiana, USA. The produced water discharges contain 37 to 578 μg/L total PAHs; the PAH assemblage in the produced water is dominated by alkylated 2- and 3-ring PAHs. Background concentrations of total PAHs in ambient water and sediments of the northwestern Gulf of Mexico appear to be in the range of 0.01 to about 100 μg/L and 10 to about 250 ng/g dry wt., respectively. Near the four offshore produced water discharges, concentrations of total PAHs in the ambient water are elevated only slightly, if at all. Concentrations of total PAHs are elevated by 2- to 10-fold in surficial sediments within 20 m of the produced water discharges, but reach background concentrations within 2,000 m of the discharges. Thus, the slightly elevated concentrations of PAHs in the marine environment of the continental shelf of the northwestern Gulf of Mexico probably are not due to produced water discharges alone.

Measured and estimated concentrations of individual and total PAHs in the ambient water down-current from the four produced water discharges are well below US EPA chronic water quality criteria for PAH in marine waters. Concentrations of individual and total PAHs in sediments near the four produced water discharges are, with a few exceptions, well below threshold concentrations (ERL) that are considered safe for benthic marine animals. The concentration of total low molecular weight PAHs in one sediment sample is twice as high as the ERL value. Concentrations of individual and total PAHs in tissues of marine animals from the vicinity of the four produced water discharges are below critical body residues for PAHs associated with possible adverse effects in 5 percent of the biological community (the HC5).

These results indicate that marine organisms living in the water column in the vicinity of the four produced water discharges are at mimimal risk of harm from PAHs in the produced water. Benthic biological communities living in sediments near some platform discharges may be at slight risk of harm from PAHs derived from the produced water discharges that have accumulated in the surficial sediments. However, adverse effects are likely to be restricted to a very small radius around the platforms. These conclusions are consistent with observations in offshore waters in both the western Gulf of Mexico and the North Sea that effects of produced water discharges, when they can be observed at all, are restricted to the benthos, where petroleum hydrocarbons have accumulated to high concentrations, causing direct toxicity or depletion of oxygen in the sediments.

ACKNOWLEDGMENTS

The work reported in this paper was supported by a contract to Continental Shelf Associates, Inc., from the U.S. Department of Energy.

REFERENCES

Abernethy, S.G., Bobra, A.M., Shiu, W.Y., Wells, P.G., and Mackay, D. 1986. Acute lethal toxicity of hydrocarbons and chlorinated hydrocarbons to two planktonic crustaceans: the key role of organism-water partitioning. Aquat. Toxicol. 8:163-174.

Abernethy, S.G., Mackay, D., and McCarthy, L.S. 1988. "Volume fraction" correlation for narcosis in aquatic organisms: the key role of partitioning. Environ. Toxicol. Chem. 7:469-481.

Armstrong, H.W., Fucik, K., Anderson, J.W., and Neff, J.M. 1979. Effects of oilfield brine effluent on sediments and benthic organisms in Trinity Bay, Texas. Mar. Environ. Res. 2:55-69.

Boesch, D.F. and Rabalais, N.N. 1989. Produced water discharges in the northwestern Gulf of Mexico region. Pages 13-30 *In*: D.F. Boesch and N.N. Rabalais, Eds., An Analysis of Impacts. Produced Waters in Sensitive Coastal Habitats. Central Gulf of Mexico. OCS Study MMS 89-0031. U.S. Dept. of the Interior, Minerals Management Service, Gulf of Mexico OCS Regional Office, New Orleans, LA.

Boesch, D.F., Rabalais, N.N., Milan, C.S., Henry, C.B., Means, J.C., Gambrell, R.P., and Overton, E.B. 1989. Field assessments. Pages 31-116 *In*: D.F. Boesch and N.N. Rabalais, Eds., An Analysis of Impacts. Produced Waters in Sensitive Coastal Habitats. Central Gulf of Mexico. OCS Study MMS 89-0031. U.S. Dept. of the Interior, Minerals Management Service, Gulf of Mexico OCS Regional Office, New Orleans, LA.

Borgund, A.E. and Barth, T. 1994. Generation of short-chain organic acids from crude oil by hydrous pyrolysis. Org. Geochem. 21:943-952.

Brandsma, M.G. 1996. Dispersion modelling perspectives on the environmental fate of produced water discharges. (in press, this publication).

Brooks, J.M., Kennicutt, M.C., Wade, T.L., Hart, A.D., Denoux, G.J., and McDonald, T.J. 1990. Hydrocarbon distributions around a shallow water multiwell platform. Environ. Sci. Technol. 24:1079-1085.

Connell, D. and Markwell, R. 1992. Mechanism and prediction of nonspecific toxicity to fish using bioconcentration characteristics. Ecotoxicol. Environ. Saf. 24:247-265.

Dickhut, R.M. and Gustafson, K.E. 1995. Atmospheric inputs of selected polycyclic aromatic hydrocarbons and polychlorinated biphenyls to southern Chesapeake Bay. Mar. Pollut. Bull. 30:385-396.

Karickhoff, S.W., Brown, D.S., and Scott, T.A. 1979. Sorption of hydrophobic pollutants on natural sediments and soils. Wat. Res. 13:241-248.

Kingston, P.F. 1987. Field effects of platform discharges on benthic macrofauna. Phil. Trans. Roy. Soc. London B316:545-565.

Kingston, P.F. 1992. The impact of offshore oil production installations on the benthos of the North Sea. ICES J. Mar. Sci. 49:127-143.

Ko, F.-C. and Baker, J.E. 1994. Partitioning of hydrophobic organic contaminants to resuspended sediments and plankton in the mesohaline Chesapeake Bay. Mar. Chem. 49:171-188.

Long, E.R., MacDonald, D.D., Smith, S.L., and Calder, F.D. 1995. Incidence of adverse biological effects within ranges of chemical concentrations in marine and estuarine sediments. Environ. Manage. 19:81-97.

Long, E.R., and Morgan, L.G. 1990. The potential for biological effects of sediment-sorbed contaminants tested in the National Status and Trends Program. NOAA Tech. Memo NOS OMA 52. U.S. National Oceanic and Atmospheric Administration, Seattle, WA. 175 pp.

Mackay, D., Puig, H., and McCarty, L.S. 1992a. An equation describing the time course and variability in uptake and toxicity of narcotic chemicals in fish. Environ. Toxicol. Chem. 11:941-951.

Mackay, D., Shiu, W.Y., and Ma, K.C. 1992b. Illsutrated Handbook of Physical-Chemical Properties and Environmental Fate for Organic Chemicals. Polynuclear Aromatic Hydrocarbons, Polychlorinated Dioxins, and Dibenzofurans. Lewis Publishers, Boca Raton, FL.595 pp.

McCarty, L.S. 1986. The relationship between aquatic toxicity QSARs and bioconcentration for some organic chemicals. Environ. Toxicol. Chem. 5:1071-1080.

McCarty, L.S., Mackay, D., Smith, A.D., Ozburn, G.W., and Dixon, D.G. 1992. Residue-based interpretation of toxicity and bioconcentration QSARs from aquatic bioassays: neutral narcotic organics. Environ. Toxicol. Chem. 11:917-930.

Means, J.C., McMillin, D.J., and Milan, C.S. 1989. Characterization of produced water. Pages 97-110 *In*: D.F. Boesch and N.N. Rabalais, Eds., Environmental Impact of Produced Water Discharges in Coastal Louisiana. Report to Mid-continent Oil and Gas Association, New Orleans, LA.

Menzie, C.A. 1982. The environmental implications of offshore oil and gas activities. Environ. Sci. Technol. 16:454A-472A.

Neff, J.M. 1979. Polycyclic Aromatic Hydrocarbons in the Aquatic Environment. Sources, Fates, and Biological Effects. Applied Science Publishers, Barking, Essex, England. 262 pp.

Neff, J.M. 1987. Biological effects of drilling fluids, drill cuttings and produced waters. Pages 469-538 *In*: D.F. Boesch and N.N. Rabalais, Eds., Long-Term Environmental Effects of Offshore Oil and Gas Development. Elsevier Applied Science Publishers, London.

Neff, J.M. and Burns, W.A. 1995. Estimation of polycyclic aromatic hydrocarbon concentrations in the water column based on tissue residues in mussels and salmon: an equilibrium partitioning approach. Environ. Toxicol. Chem. (in press).

Neff, J.M., Rabalais, N.N., and Boesch, D.F. 1987. Offshore oil and gas development activities potentially causing long-term environmental effects. Pages 149-174 *In*: D.F. Boesch and N.N. Rabalais, Eds.,

Long-Term Environmental Effects of Offshore Oil and Gas Development. Elsevier Applied Science Publishers, London.

Neff, J.M. and Sauer, T.C., Jr. 1995. Aromatic hydrocarbons in produced water: bioaccumulation and trophic transfer in marine food webs. (in press, this publication).

Neff, J.M., Sauer, T.C., Jr., and Maciolek, N. 1989. Fate and Effects of Produced Water Discharges in Nearhore Marine Waters. API Publication No. 4472. American Petroleum Institute, Washington, DC. 300 pp.

Neff, J.M., Sauer, T.C., Jr., and Maciolek, N. 1992. Composition, fate and effects of produced water discharges to nearshore marine waters. Pages 371-385 *In*: J.P. Ray, and F.R. Engelhardt, Eds., Produced Water. Technological/Environmental Issues and Solutions. Plenum Press, New York.

Page, D.S., Gilfillan, E.S., Boehm, P.D., and Harner, E.J., 1995. Shoreline ecology program for Prince William Sound, Alaska following the Exxon Valdez oil spill: part 1—Study design and methods. Pages 263-295 *In*: P.G. Wells, J.N. Butler, and J.S. Hughes, Eds., Exxon Valdez Oil Spill: Fate and Effects in Alaskan Waters. ASTM STP 1219. American Society for Testing and Materials, Philadelphia, PA.

Rabalais, N.N., Dagg, M.J., and Boesch, D.F. 1985. Nationwide Review of Oxygen Depletion and Eutrophication in Estuarine and Coastal Waters: Gulf of Mexico (Alabama, Louisiana, Texas). Final Report to U.S.D.C. National Oceanic and Atmospheric Administration, National Ocean Survey, Ocean Assessment Division, Rockville, MD. 60 pp.

Rabalais, N.N., McKee, B.A., Reed, D.J., and Means, J.C. 1991. Fate and effects of nearshore discharges of OCS produced waters. Vol. 1: Executive Summary. Vol. 2: Technical Report. Vol. 3. Appendices. OCS Studies MMS 91-004, MMS 91-005, and MMS 91-006. U.S. Dept. of the Interior, Minerals Management Service, Gulf of Mexico OCS Regional Office, New Orleans, LA.

Sauer, T.C. and Boehm, P., 1991. Use of defensible analytical chemical measurements for oil spill natural resource damage assessment. Pages 363-369 *In*: 1991 Oil Spill Conference. Prevention, Behavior, Control, Cleanup. American Petroleum Institute, Washington, DC.

Sijm, D.T.H.M., Schipper, M., and Opperhuizen, A. 1993. Toxicokinetics of halogenated benzenes in fish: lethal body burden as a toxicological end point. Environ. Toxicol. Chem. 12:1117-1127.

Stephenson, M.T., Ayers, R.C., Bickford, L.J., Caudle, D.D., Cline, J.T., Cranmer, G., Duff, A., Garland, E., Herenius, T.A. Jacobs, R.P.W.M., Inglesfield, C. Norris, G., Petersen, J.D., and Read, A.D. 1994. North sea produced water: fate and effects in the marine environment. Report No. 2.62/204. E&P Forum, London, England. 48 pp.

Terrens, G.W. and Tait, R.D. 1993. Effects on the marine environment of produced formation water discharges from Esso/BHPP's Bass Strait platforms. Esso Environmental Report: 933/1. Esso Australia Ltd., Melbourne, Australia.

Van Leeuwen, C.J., Van Der Zandt, P.T.J.,Aldenberg, T., Verhaar, H.J.M, and Hermens J.L.M. 1992. Application of QSARs, extrapolation and equilibrium partitioning in aquatic effects assessment. I. Narcotic industrial pollutants. Environ. Toxicol. Chem. 11:267-282.

Voice, T.C. and Weber, W.J., Jr. 1983. Sorption of hydrophobic compounds by sediments, soils and suspended solids - I. Theory and background. Wat. Res. 17:1433-1441.

Warne, M.St.J., Connell, D.W., and Hawker, D.W. 1991. Comparison of the critical concentration and critical volume hypotheses to model non-specific toxicity of individual compounds. Toxicol. 66:187-195.

Ynnesdal, H. and Furuholt, E. 1994. Environmental effects of produced water from large fields in the North Sea. An overview. Society of Petroleum Engineers Paper 27150. Annual Symposium of the Society of Petroleum Engineers, Djarkata, Indonesia.

34

NATURALLY OCCURRING RADIOACTIVE MATERIALS ASSOCIATED WITH OFFSHORE PRODUCED WATER DISCHARGES IN THE GULF OF MEXICO

Alan D. Hart,[1] Bruce D. Graham,[1] David A. Gettleson,[1]
David L. Demorest,[2] and Brent W. Smith[3]

[1] Continental Shelf Associates, Inc.
759 Parkway Street
Jupiter, Florida 33477
[2] CORE Laboratories
420 West First Street
Casper, Wyoming 82601
[3] U.S. Department of Energy
Metairie Site Office
900 Commerce Road East
New Orleans, Louisiana 70123

INTRODUCTION

During production of oil and gas, fossil water is generally pumped from the reservoir in the formation, and following treatment, this produced water is commonly discharged from platforms located on the outer continental shelf in the Gulf of Mexico. Gulf of Mexico produced waters characteristically have elevated salinities compared to the receiving ambient water and often contain elevated concentrations of inorganic and organic constituents. Among the constituents of some Gulf of Mexico produced waters are naturally occurring radioactive materials (NORM), which is potentially bioaccumulated by marine organisms living around produced water outfalls. Meinhold and Hamilton (1992) identified biological uptake of radium by marine organisms as a potential human health concern because people ingesting radium in food may have increased cancer risk. Concern is therefore growing over the potential bioaccumulation and bioavailability of NORM in indigenous finfish and shellfish species available for human consumption. Because of this concern, the U.S. Environmental Protection Agency is considering additional regulations based on the potential adverse impacts of platform-associated produced water discharges containing NORM on indigenous fauna inhabiting the Gulf of Mexico outer continental shelf. In response to these potential new regulations, the U.S. Department

Produced Water2: Environmental Issues and Mitigation Technologies
Edited by M. Reed and S. Johnsen, Plenum Press, New York, 1996

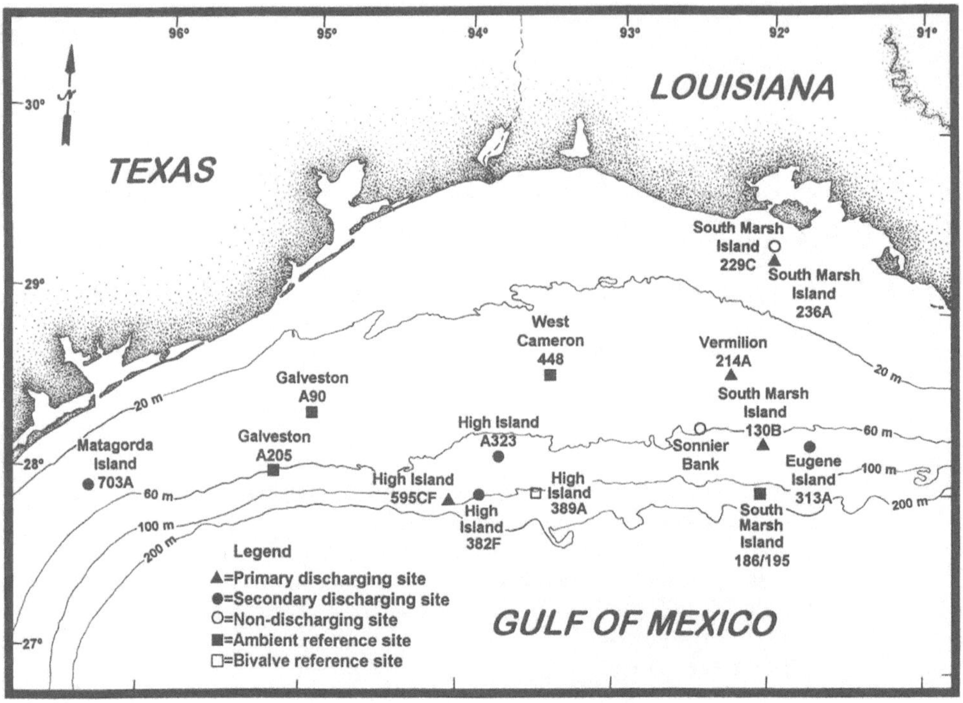

Figure 1. Geographic locations of study sites.

of Energy has funded a project to assess the environmental and human health risks associated with produced water discharges from offshore and coastal oil and gas operations in the Gulf of Mexico region (Smith, 1995). Relatively little information is available about the fate of radioisotopes in produced water discharges from offshore platforms, and reliable data on bioaccumulation of radium isotopes by marine animals are sparse. This study is part of the larger U.S. Department of Energy project, and its objective is to gather data to characterize the extent of radionuclide contamination in water, sediment, and biota at produced water and produced sand discharge points. Data collected during this study will eventually be used in a human-health risk assessment of NORM from produced water discharges.

The primary focus of this paper is selected NORM components (^{226}Ra, ^{228}Ra, and ^{210}Pb) in produced water, discharge plume water, ambient water, and biological tissue samples collected in the vicinity of Gulf of Mexico platforms and at reference sites. Sampling was conducted at 15 sites in the northwestern Gulf of Mexico (Figure 1). These sites were classified in four categories—primary discharging sites, secondary discharging sites, non-discharging sites, and ambient reference sites. There was an additional site where reference bivalve collections were made. Discharging sites were designated as primary because preliminary information indicated that high loadings of NORM occurred at these sites. More extensive sampling was performed at these primary discharging sites compared to the secondary discharging sites. Study sites were distributed over a range of water depths and oceanographic conditions (Table 1).

Table 1. Study site descriptions

Study Site		Water Depth (m)	Oceanographic Conditions
Primary Discharging Sites	South Marsh Island 236A	6.1	Offshore Louisiana; nearshore
	Vermilion 214A	38.7	Offshore Louisiana; midcontinental shelf
	South Marsh Island 130B	65.5	Offshore Louisiana; midouter continental shelf
	High Island 595CF	121.9	Offshore Texas; outer continental shelf
Secondary Discharging Sites	Matagorda Island 703A	50.9	Offshore Texas; midcontinental shelf
	High Island 323A	71.6	Offshore Texas; mid-outer continental shelf
	Eugene Island 313A	71.9	Offshore Louisiana; midouter continental shelf
	High Island 382F	103.9	Offshore Texas; outer continental shelf
Non-discharging Sites	South Marsh Island 229C	5.5	Offshore Louisiana; nearshore
	Vermilion 298 and 305 (Sonnier Bank)	20.1	Offshore Louisiana; midcontinental shelf
Ambient Reference Sites	West Cameron 448	36.6	Offshore Louisiana; midcontinental shelf
	Galveston 90	40.2	Offshore Texas; midcontinental shelf
	South Marsh Island 186/195	64.0	Offshore Louisiana; midouter continental shelf
	Galveston 205	64.0	Offshore Texas; mid-outer continental shelf
Bivalve Tissue Reference Site	High Island 389A	121.0	Offshore Texas; outer continental shelf

METHODOLOGY

After flushing for appoximately 10 min, produced water samples were collected into a sample container directly from a spigot that was located immediately prior to the waste stream being discharged overboard from the platforms. Ambient water samples were collected using a 5-L teflon-lined water sampling bottle, and the collected water was drained directly into the sample container from the water bottle. To collect samples of discharge plume water, rhodamine WT dye was injected into the produced water just prior to overboard discharge from the platform, and the center line of the plume was determined by fluorometric measurement using a Turner Design flow-through fluorometer. Plume samples were collected by pumping along the centerline of the plume. All water samples were stored at 4°C in non-contaminating polyethylene containers and preserved by the addition of nitric acid until the pH of the sample was less than 2.

Bivalve specimens were collected from the platform support structure in water depths of less than 15 to 20°m within a 100-m horizontal radius of the produced water discharge. Divers collected bivalve specimens by removing the whole specimen from the platform structure using a metal pry-bar. Specimens were taken intact to prevent incidental contamination during collection and subsequent processing. Crustaceans were collected in crab traps and by trawling. Fishes were collected using hook-and-line, traps, and trawls. All trawling was conducted within 1,000 m of the discharge outfalls. All tissue samples were stored frozen onboard the survey vessel.

In the laboratory, edible tissue was shucked from the shell of each bivalve specimen. Tails were removed from shrimp specimens, and the edible tissue was separated from the exoskeleton. Edible tissue was removed from the claws and body of crab specimens. Fish specimens were filleted to remove edible muscle tissue. Clean non-contaminating utensils were used for all dissection procedures. Bivalve and crustacean specimens from individual platforms were composited into one or more samples for analysis.

Radionuclide analyses of water samples were performed following standard protocols of Percival and Martin (1974) and U.S. Environmental Protection Agency (1980) with slight modifications to remove interferences known to occur in specific sample types following Demorest and Wallace (1992). Biological tissue samples were thoroughly homogenized and then dissolved in concentrated HNO_3. This solution was extracted with an equal volume of heptane to remove the long-chain fat molecules. The nitric fraction was then solubilized with consecutive digestions using dilute HNO_3, then HNO_3-H_2O_2, and finally HNO_3-$HClO_4$-HF digestion. After a Na_2SO_4 fusion to complete the destruction of the organic material, the samples were analyzed using procedures similar to those for water samples.

RESULTS AND DISCUSSION

The field effort was conducted from 9 June to 11 August 1993. At the four primary sites, samples of produced water, discharge plume water, and biological tissues were collected. At the four secondary discharging sites, produced water and biological tissue samples were collected. Ambient water and biological tissue samples were collected at the two non-discharging sites, and ambient water samples were collected at the four ambient reference sites, which were located at least 13 km from any active produced water outfall.

Three replicate produced water samples were collected at each of the primary and secondary discharging platform sites (Table 2). With the exception of High Island Block 595CF, the ^{226}Ra and ^{228}Ra activities were within the range reported by Stephensen and Supernaw (1990) for 42 Gulf of Mexico offshore produced water outfalls (4-584 pCi/L and 18-586 pCi/L, respectively). Continental Shelf Associates, Inc. (1993) reported ^{226}Ra and

Table 2. Mean (± standard deviation) activities of ^{226}Ra, ^{228}Ra, and ^{210}Pb in produced water and ambient water samples

Platform Location	^{226}Ra	^{228}Ra	^{210}Pb
South Marsh Island 236A (primary discharging site)	91 ± .13	239 ± 67	12.3 ± 4.3
Vermilion 214A (primary discharging site)	300 ± 157	228 ± 29	7.7 ± 4.7
South Marsh Island 130B (primary discharging site)	362 ± 43	164 ± 146	5.6 ± 5.5
High Island 595CF (primary discharging site)	1,494 ± 1,989	356 ± 19	12.5 ± 2.6
Matagorda Island 703A (secondary discharging site)	56 ± 3	69 ± 7	2.6 ± 0.5
High Island 323A (secondary discharging site)	112 ± 6	162 ± 12	5.2 ± 2.5
Eugene Island 313A (secondary discharging site)	270 ± 16	388 ± 15	13.8 ± 1.9
High Island 382F (secondary discharging site)	255 ± 43	600 ± 107	16.7 ± 4.1
West Cameron 448 (ambient reference site)	0.30 ± 0.17	<0.30	0.37 ± 0.15
Galveston 90 (ambient reference site)	0.07 ± 0.06	0.70 ± 0.66	0.23 ± 0.21
South Marsh Island 186/195 (ambient reference site)	0.13 ± 0.06	0.60 ± 0.53	0.03 ± 0.06
Galveston 205 (ambient reference site)	0.13 ± 0.06	0.93 ± 1.21	0.30 ± 0.26

^{228}Ra activities for two offshore Louisiana produced water outfalls in Eugene Island 189 and Ship Shoal 169, where ^{226}Ra activities were 225 and 111 pCi/L, and ^{228}Ra activities were 260 and 240 pCi/L, respectively. Mean ^{226}Ra, ^{228}Ra, and ^{210}Pb activities at the discharging platforms were several orders of magnitude greater than the activities in water samples collected at the ambient reference sites (Table 2). Although the activities of the three radionuclides in the produced water samples were variable, it was clear that produced water discharges are sources of ^{226}Ra, ^{228}Ra, and ^{210}Pb.

Additional produced water sampling was performed at the four primary discharging sites to investigate the variability of NORM activities of produced water at two time scales. After the initial produced water sampling, five additional produced water samples were collected sequentially at 12-h intervals to investigate short term variability of NORM in produced water. Up to five additional sets of replicate samples were also collected by the platform operators at monthly intervals to investigate longer term variability of NORM. With the exception of High Island 595CF, ^{226}Ra activities in the produced water collected at 12-h intervals were generally more consistent than samples collected at monthly intervals (Table 3). At High Island 595CF, the ^{226}Ra activities were variable at both time scales. ^{228}Ra and

Table 3. Summary of activities of ^{226}Ra, ^{228}Ra, and ^{210}Pb in produced water samples over time

Analyte	Time of Sampling	South Marsh Island 236A	Vermilion 214A	South Marsh Island 130B	High Island 595CF
^{226}Ra	Initial sampling	91 ± 13	300 ± 157	362 ± 43	1,494 ± 1,989
	12 h after initial sampling	70	400	261	490
	24 h after initial sampling	66	347	317	470
	36 h after initial sampling	47	441	110	83
	48 h after initial sampling	175	373	373	944
	60 h after initial sampling	68	344	229	734
	1 mo after initial sampling	39 ± 1	60 ± 8	196 ± 97	
	2 mo after initial sampling	65 ± 2	77 ± 11	315 ± 7	101 ± 11
	3 mo after initial sampling	38 ± 6	90 ± 25	265 ± 26	165 ± 259
	4 mo after initial sampling	292 ± 11	238 ± 71	287 ± 18	434 ± 13
	5 mo after initial sampling	321 ± 33		286 ± 12	
^{228}Ra	Initial sampling	239 ± 67	228 ± 29	164 ± 147	356 ± 19
	12 h after initial sampling	138	214	130	409
	24 h after initial sampling	307	193	113	362
	36 h after initial sampling	126	241	43	291
	48 h after initial sampling	187	263	187	220
	60 h after initial sampling	166	217 ±	279	257
	1 mo after initial sampling	119 ± 14	242 ± 6	193 ± 17	
	2 mo after initial sampling	277 ± 52	318 ± 11	188 ± 5	276 ± 30
	3 mo after initial sampling	126 ± 3	278 ± 26	156 ± 57	190 ± 297
	4 mo after initial sampling	274 ± 17	228 ± 63	196 ± 27	470 ± 55
	5 mo after initial sampling	306 ± 16		196 ± 23	
^{210}Pb	Initial sampling	12 ± 4	8 ± 5	6 ± 5	13 ± 3
	12 h after initial sampling	13	8	13	11
	24 h after initial sampling	16	5	4	15
	36 h after initial sampling	12	7	2	20
	48 h after initial sampling	16	11	5	23
	60 h after initial sampling	15	14	11	12
	1 mo after initial sampling	4 ± 1	6 ± 2	5 ± 2	
	2 mo after initial sampling	8 ± 2	7 ± 1	7 ± 0	4 ± 0
	3 mo after initial sampling	4 ± 1	6 ± 2	5 ± 1	8 ± 3
	4 mo after initial sampling	2 ± 3	4 ± 1	<2.1	9 ± 5
	5 mo after initial sampling	11 ± 9		<2.1	

Table 4. Mean ^{226}Ra activities (± standard deviation) in water samples

	Platform			
Water Sample Type	South Marsh Island 236A	Vermilion 214A	South Marsh Island 130B	High Island 595CF
Produced Water	91.17 ± 12.61	300.33 ± 156.87	362.33 ± 43.02	1494.00 ± 1988.78
Plume				
5 m	0.40 ± 0.17	0.18 ± 0.13	0.53 ± 0.21	0.93 ± 0.31
10 m	0.30 ± 0.10	0.70 ± 0.35	0.97 ± 0.32	0.87 ± 0.32
30 m	0.53 ± 0.12	0.45 ± 0.24	0.57 ± 0.31	0.47 ± 0.12
50 m	0.50 ± 0.20	0.33 ± 0.60	0.90 ± 0.26	0.45 ± 0.23
100 m	0.57 ± 0.15	0.33 ± 0.06	0.07 ± 0.06	0.23 ± 0.06
	0.20 ± 0.00	0.07 ± 0.06	0.10 ± 0.00	0.03 ± 0.06
>2,000 m	0.23 ± 0.12	0.07 ± 0.06	0.07 ± 0.06	0.03 ± 0.06
	0.07 ± 0.16	0.10 ± 0.00	0.10 ± 0.00	0.03 ± 0.06
Ambient Reference	0.30 ± 0.17	0.13 ± 0.06	0.13 ± 0.06	0.07 ± 0.06

^{210}Pb activities were not as variable as the ^{226}Ra activities at either time scale (Table 3). These data indicate that a single sampling of produced water at a platform may not be entirely representative of NORM levels over longer periods of time, which has implications for relating produced water activities to activities in organisms living in the vicinity of the outfall.

NORM was determined in water samples collected in the discharge plume at four platforms—South Marsh Island 236A, Vermilion 214A, South Marsh Island 130B, and High Island 595CF. These samples were collected at distances of 5, 10, 30, 50, and 100 m from the discharge point. In addition, ambient water samples were collected at three stations located greater than 2,000 m from each platform. All discharges were above the sea surface. These results are compared to those for the produced water and the ambient reference sites in Tables 4-6. Clearly, the activities of the radionuclides are quickly reduced after discharge of the produced water, approaching levels comparable to ambient conditions within short distances from the discharge point. At 5 m from the discharge point, ratios of produced water activity to plume water activity for ^{226}Ra and ^{228}Ra were 228:1-1668:1 and 162:1-759:1, respectively. The radium ratios observed at these four platforms were similar to those

Table 5. Mean ^{228}Ra activities (± standard deviation) in water samples

	Platform			
Water Sample Type	South Marsh Island 236A	Vermilion 214A	South Marsh Island 130B	High Island 595CF
Produced Water	238.67 ± 67.47	227.67 ± 28.54	164.00 ± 146.50	356.00 ± 19.31
Plume				
5 m	0.83 ± 0.49	0.30 ± 0.35	0.63 ± 0.60	2.20 ± 0.52
10 m	0.60 ± 0.56	1.87 ± 0.21	0.90 ± 0.66	1.93 ± 0.75
30 m	1.07 ± 1.22	0.93 ± 0.85	0.33 ± 0.29	2.03 ± 1.10
50 m	3.60 ± 4.94	0.50 ± 0.10	1.30 ± 0.20	1.93 ± 0.21
100 m	3.87 ± 2.25	1.60 ± 0.20	1.23 ± 0.40	1.07 ± 0.72
	3.20 ± 2.31	0.30 ± 0.52	1.67 ± 0.65	0.63 ± 0.45
>2,000 m	0.10 ± 0.17	0.60 ± 0.52	1.27 ± 1.22	0.53 ± 0.35
	1.60 ± 0.70	1.03 ± 0.95	0.13 ± 0.23	1.20 ± 1.32
Ambient Reference	<0.30	0.60 ± 0.53	0.93 ± 1.21	0.70 ± 0.66

Table 6. Mean ^{210}Pb activities (± standard deviation) in water samples

	Platform			
Water Sample Type	South Marsh Island 236A	Vermilion 214A	South Marsh Island 130B	High Island 595CF
Produced Water	12.27 ± 4.31	7.70 ± 4.70	5.60 ± 5.50	12.50 ± 2.60
Plume				
5 m	0.17 ± 0.15	0.10 ± 0.14	0.07 ± 0.12	0.53 ± 0.12
10 m	0.17 ± 0.15	0.23 ± 0.06	0.13 ± 0.12	0.80 ± 0.10
30 m	0.03 ± 0.06	0.10 ± 0.14	0.07 ± 0.06	0.70 ± 0.20
50 m	0.13 ± 0.12	0.33 ± 0.35	0.03 ± 0.06	0.53 ± 0.35
100 m	<0.10	0.80 ± 0.20	0.13 ± 0.12	0.17 ± 0.29
>2,000 m	0.33 ± 0.29	0.50 ± 0.10	0.13 ± 0.06	<0.20
	0.13 ± 0.15	0.13 ± 0.15	0.03 ± 0.06	<0.20
	0.30 ± 0.10	0.53 ± 0.21	0.07 ± 0.06	<0.20
Ambient Reference	0.37 ± 0.15	0.03 ± 0.06	0.30 ± 0.26	0.23 ± 0.21

observed by Continental Shelf Associates, Inc. (1993) at Eugene Island 189 and Ship Shoal 169. Produced water-to-plume water activity ratios were lower for ^{210}Pb, ranging from 24:1-80:1. The difference in produced water-to-plume water activity ratios between the radium isotopes and the lead isotope indicated that they may be behaving differently in the plume. Lead is commonly associated with fine-grained particles. When the radium isotopes enter the sulfate-rich seawater, they may form particles of radium sulfate just as barium does. White (1992) suggested that 90% of the radium in discharged produced water could be coprecipitated with barium.

The American oyster *Crassostrea virginica* was collected from the legs and supporting structure of a discharging platform located near the coast (Figure 1) and the jewelbox *Chama macerophylla* was collected from the legs at six offshore discharging platforms. Bivalves are excellent biological monitors of exposure to produced water discharges because

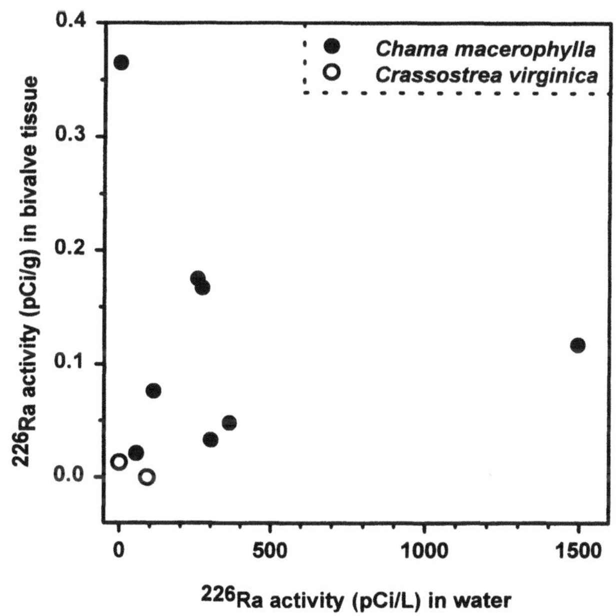

Figure 2. Comparison of ^{226}Ra activities in produced/ambient water and edible tissues of the bivalves *Chama macerophylla* and *Crassostrea virginica*. Produced water activities are plotted for discharging sites, and ambient water activities are plotted for non-discharging sites.

Table 7. Summary of radionuclide activities (pCi/g wet weight) in bivalve and crustacean edible tissues. Reference collections for *Chama* were made at High Island 389A; *Penaeus* and *Sicyonia* were collected at Sonnier Bank

Analyte	Platform	*Chama*	*Crassostrea*	*Callinectes*	*Penaeus*	*Sicyonia*
^{226}Ra	South Marsh Island 229C		0.013	<0.003	<0.0030.013	
	South Marsh Island 236A		<0.004	<0.005-0.015	<0.004	
	Sonnier Bank (High Island 389A)	0.365			<0.005-0.010	<0.004-0.007
	Vermilion 214A	0.033				<0.005-0.009
	South Marsh Island 130B	0.048			<0.004	<0.003
	High Island 595CF	0.118				
	Matagorda Island 703A	0.021			<0.003	
	High Island 323A	0.076			<0.004	
	Eugene Island 313A	0.167				<0.005-0.006
	High Island 382F	0.175			0.005	
^{228}Ra	South Marsh Island 229C		0.047	<0.011-0.047	<0.008-0.047	
	South Marsh Island 236A		<0.011	<0.014-0.037	<0.013-0.037	
	Sonnier Bank (High Island 389A)	<0.013			<0.018-0.046	<0.015-0.051
	Vermilion 214A	<0.014				<0.017-0.047
	South Marsh Island 130B	<0.009			<0.015	<0.017-0.105
	High Island 595CF	0.061				
	Matagorda Island 703A	<0.016			<0.008-0.159	
	High Island 323A	<0.066			0.027-0.099	
	Eugene Island 313A	<0.003				<0.017
	High Island 382F	0.081			<0.019-0.174	
^{210}Pb	South Marsh Island 229C		0.035	<0.011	<0.011-0.033	
	South Marsh Island 236A		0.024	<0.010-0.022	<0.009-0.014	
	Sonnier Bank (High Island 389A)	0.211			<0.010-0.034	0.041-0.070
	Vermilion 214A	0.075				<0.009-0.075
	South Marsh Island 130B	0.281			<0.009	<0.010
	High Island 595CF	0.416				
	Matagorda Island 703A	0.043			<0.010	
	High Island 323A	0.188			0.029-0.041	
	Eugene Island 313A	0.308				<0.009-0.035
	High Island 382F	0.454			0.018-0.046	

they attach to the platform structures and do not move. Bivalve specimens were collected at two other platforms in addition to the discharging platforms. *C. virginica* was collected at a platform in South Marsh Island Block 229, where produced water discharges do not occur. *C. macerophylla* was collected at a platform in High Island Block 389 because sufficient quantities of this bivalve were not available at Sonnier Bank. Although produced water is discharged at High Island 389A, the discharge is shunted to within 10 m of the seafloor. The water depth at this platform is greater than 120 m; therefore, the bivalves were not thought to be exposed to produced water discharges because of this shunting. The results of the NORM analyses of bivalve tissues are presented in Table 7. Mean activities in bivalve tissues exposed to near sea surface produced water discharges did not appear to be different from those from reference sites. Relationships between mean radionuclide activities in produced water (or ambient water for reference sites) and in bivalve tissues were not apparent (Figures

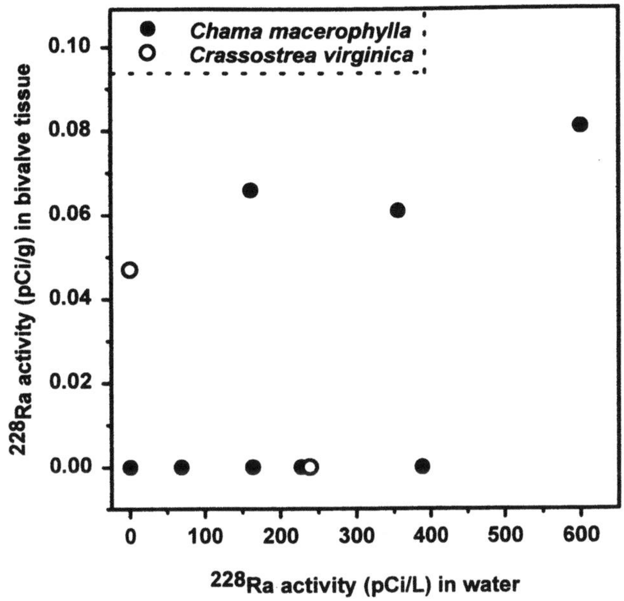

Figure 3. Comparison of ^{228}Ra activities in produced/ambient water and edible tissues of the bivalves *Chama macerophylla* and *Crassostrea virginica*. Produced water activities are plotted for discharging sites, and ambient water activities are plotted for non-discharging sites.

2-4). Similarly, mean radionuclide loadings from produced water discharges were not related to radionuclide activities in bivalve tissues (Figures 5-7).

Callinectes (blue crab) were collected in traps and trawls around the shallow water study sites. NORM activities were low in the edible tissue of these crab specimens and the data did not indicate greater levels in specimens collected around the discharging platform (South Marsh Island 236A) compared to the reference platform (South Marsh Island 229C). Penaeus (shrimp) were collected in trawls around the study sites. Radioisotope activities were low; activities at the discharging platforms were comparable with those observed at

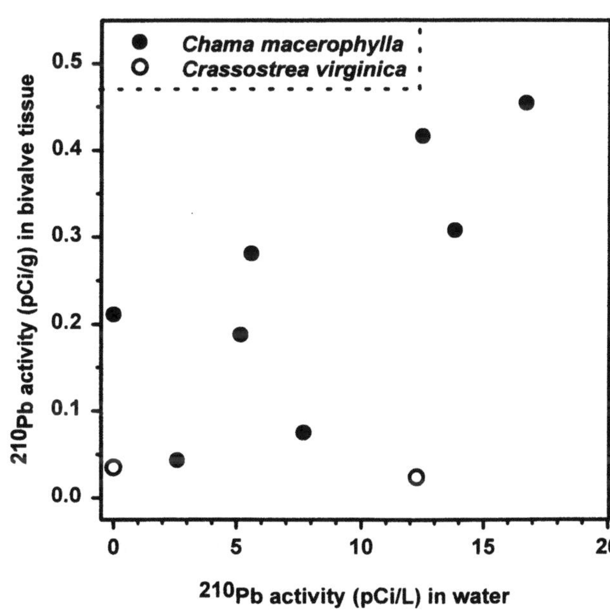

Figure 4. Comparison of ^{210}Pb activities in produced/ambient water and edible tissues of the bivalves *Chama macerophylla* and *Crassostrea virginica*. Produced water activities are plotted for discharging sites, and ambient water activities are plotted for non-discharging sites.

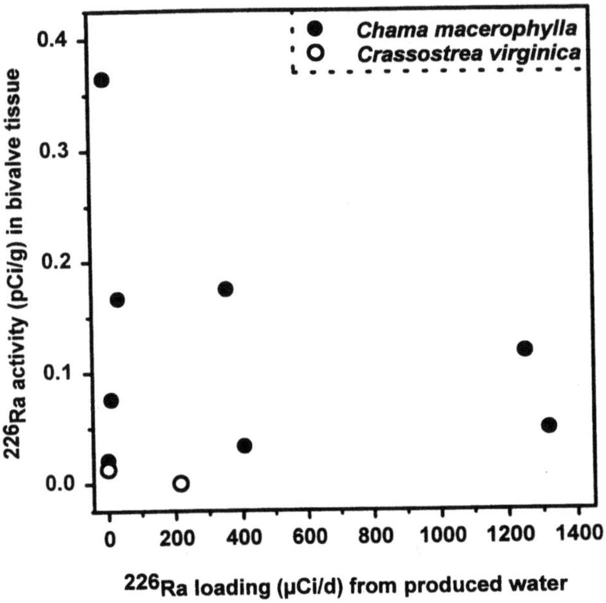

Figure 5. Comparison of ^{226}Ra produced water loadings and ^{226}Ra activities in edible tissues of the bivalves *Chama macerophylla* and *Crassostrea virginica.* Produced water loadings (activity × discharge rate) are plotted for discharging sites. Produced water loadings at non-discharging sites were zero.

the shallow water and offshore (Sonnier Bank) reference sites (Table 7). Similar results were observed for Sicyonia (rock shrimp), which were collected by trawling at the deeper study sites (Table 7).

NORM activities were also determined in the flesh of fish species living around the platforms. *Lutjanus campechanus* (red snapper) and *Balistes capriscus* (gray triggerfish) are associated with the platform structure, living among the legs, pilings, and cross-members of the platform. *Lagodon rhomboides* (pinfish) were collected in trawls over the seafloor in the vicinity of the platforms. The activities of the three radioisotopes were consistently low at

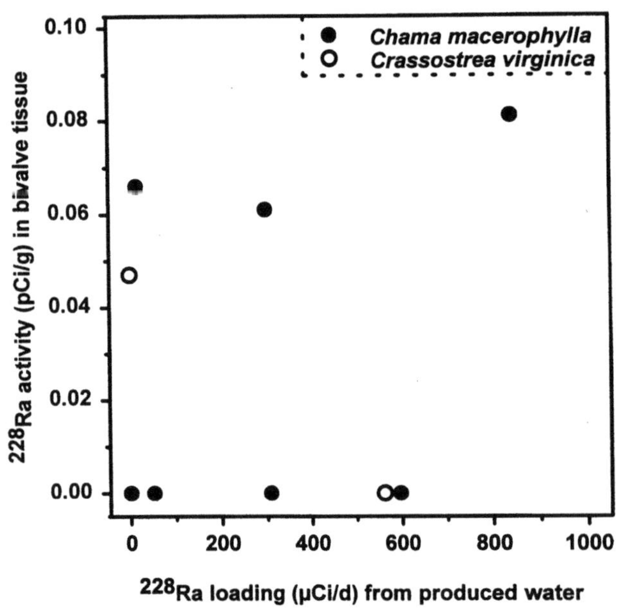

Figure 6. Comparison of ^{228}Ra produced water loadings and ^{228}Ra activities in edible tissues of the bivalves *Chama macerophylla* and *Crassostrea virginica.* Produced water loadings (activity × discharge rate) are plotted for discharging sites. Produced water loadings at non-discharging sites were zero.

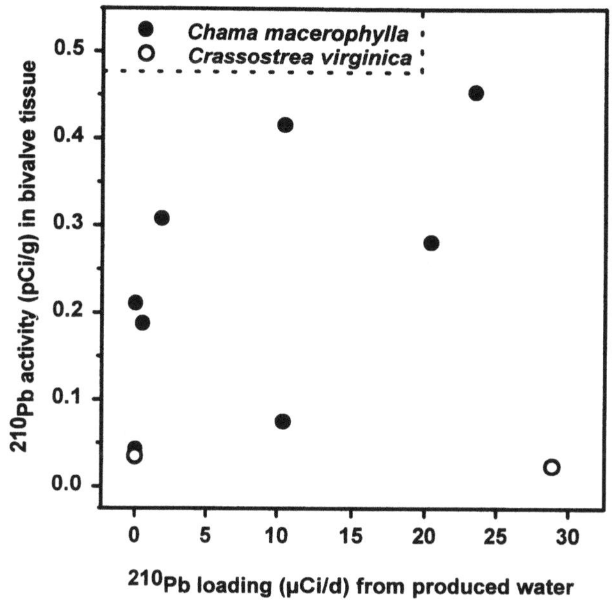

Figure 7. Comparison of ^{210}Pb produced water loadings and ^{210}Pb activities in edible tissues of the bivalves *Chama macerophylla* and *Crassostrea virginica*. Produced water loadings (activity × discharge rate) are plotted for discharging sites. Produced water loadings at non-discharging sites were zero.

Table 8. Summary of radionuclide activities (pCi/g wet weight) in edible tissue of three fish species

Analyte	Platform	Red Snapper (*Lutjanus campechanus*)	Gray Triggerfish (*Balistes capriscus*)	Pinfish (*Lagodon rhomboides*)
^{226}Ra	Sonnier Bank	<0.004-0.005	<0.004-0.008	
	Vermilion 214A	<0.005-0.008	<0.005-0.006	<0.002
	South Marsh Island 130B	<0.005-0.008	<0.004-0.011	<0.005-0.008
	High Island 595CF		<0.004-0.011	<0.002
	Matagorda Island 703A	<0.003	<0.004-0.007	0.005-0.008
	High Island 323A	<0.005	0.006	<0.003
	Eugene Island 313A	<0.003	<0.005-0.009	<0.005
	High Island 382F		<0.006	<0.003
^{228}Ra	Sonnier Bank	<0.013-0.028	<0.020-0.154	
	Vermilion 214A	<0.017	0.009-0.052	<0.014
	South Marsh Island 130B	<0.017-0.14	<0.015-0.025	0.037-0.045
	High Island 595CF		<0.013	0.039
	Matagorda Island 703A	<0.018	0.045-0.137	<0.010-0.020
	High Island 323A	<0.017-0.146	<0.014-0.018	<0.015
	Eugene Island 313A	<0.013-0.032	<0.017-0.051	0.017-0.094
	High Island 382F		0.049	<0.018-0.068
^{210}Pb	Sonnier Bank	<0.006-0.025	<0.010-0.041	
	Vermilion 214A	<0.010-0.026	<0.010-0.023	<0.010
	South Marsh Island 130B	<0.010	0.009-0.043	<0.008-0.025
	High Island 595CF		<0.009-0.014	<0.010
	Matagorda Island 703A	<0.010	<0.006-0.025	<0.007-0.008
	High Island 323A	<0.010	<0.009-0.032	<0.010
	Eugene Island 313A	0.011-0.028	<0.010-0.016	<0.010
	High Island 382F		<0.009	<0.010-0.018

all sampling sites (Table 8). For *L. campechanus* and *B. capriscus*, activities at discharging sites were comparable to those at Sonnier Bank, which was the offshore reference site. Although *L. rhomboides* were not collected at Sonnier Bank, activities in the fillets of this species were similar to those of the other two species.

In summary, these data indicate that produced water discharges are a potential source of NORM in offshore regions of the Gulf of Mexico. Although NORM activities in produced water are variable among platforms and over time, these greater produced water activities are rapidly diluted after entering the receiving waters. The data presented indicate that NORM is not accumulating in the edible tissues of organisms living in the vicinity of offshore produced water discharges.

SUMMARY

From 9 June to 11 August 1993, ten offshore production platforms were sampled to investigate the fate and environmental effects of naturally occurring radioactive materials (NORM) associated with produced water discharged into the northwestern Gulf of Mexico. Samples of produced water, discharge plume water, and ambient water were collected at discharging platforms to investigate the fate of NORM. For comparison, ambient water samples were collected at four reference sites. Fishes, crustaceans, and mollusks were collected at the offshore platforms and at three reference sites to examine potential bioaccumulation and bioavailability of NORM to marine organisms living in the vicinity of offshore produced water outfalls.

Produced water discharges were observed to be a potential source of NORM in offshore regions of the Gulf of Mexico. Although NORM activities in produced water varied among platforms and over time, these greater produced water activities are rapidly reduced after entering the receiving waters. NORM does not appear to be accumulating in the edible tissues of organisms living in the vicinity of offshore produced water discharges.

ACKNOWLEDGMENTS

This work is part of a U.S. Department of Energy-sponsored study entitled *Environmental and Economic Assessment of Discharges from Gulf of Mexico Region Oil and Gas Operations* (Contract No. DE-AC22-92MT92001). The authors would like to gratefully acknowledge the guidance of Dr. Brent W. Smith, Contracting Officers Technical Representative. The authors would like to thank operators of the platforms where sampling was performed—Chevron, Shell, Mobil, OXY, ARCO, and Texaco— for their cooperation.

REFERENCES

Continental Shelf Associates, Inc., 1993. Measurements of naturally occurring radioactive materials at two offshore production platforms in the northern Gulf of Mexico. A final report prepared for the American Petroleum Institute, Washington, DC.

Demorest, D.L., and E.S. Wallace, 1992. Radiochemical determination of NORM in produced water utilizing wet chemistry separation followed by radiochemical analysis. In: *Produced Water*, Plenum Press, New York, NY.

Meinhold, A.F., and L.D. Hamilton, 1992. Radium concentration factors and their use in health and environmental risk assessment. In: *Produced Water*, Plenum Press, New York, NY.

Percival, D.R., and D.B. Martin, 1974. Sequential determination of radium-226, radium228, actinium-227, and thorium isotopes in environmental and process waste samples. *Anal. Chem.* 46.

Smith, B.W., 1995. U. S. Department of Energy oil and gas environmental research programs. *The La. Envir. Prof.* 12(1):1-39.

Stephensen, M.T., and I.R. Supernaw, 1990. Offshore Operators Committee 44 platform study radionuclide analysis results. Offshore Operators Committee, New Orleans, LA.

U.S. Environmental Protection Agency, 1980. Prescribed methods for measurement of radioactivity in drinking water. EPA 600/4-80-032.

White, G. J., 1992. Naturally occurring radioactive materials (NORM) in oil and gas industry equipment and wastes, A literature review. A report to the U.S. Department of Energy, Contract No. DE-AC07-76ID01570.

35

DISTRIBUTION OF FINFISH CAUGHT NEAR OILFIELD STRUCTURES ALONG COASTAL LOUISIANA AND TEXAS

Fred E. Schultz,[1] Stephen E. Steimle,[1] Maureen M. Mulino,[1]
John C. Francis,[2] and Donna H. Redmann[3]

[1] Steimle & Associates, Inc.
Metairie, Louisiana
[2] University of New Orleans
New Orleans, Louisiana
[3] Louisiana State University
Baton Rouge, Louisiana

INTRODUCTION

Purpose

The purpose of this study was to gather and assemble information that will provide a synthesis of seafood catch, distribution and consumption patterns for the Gulf of Mexico (GOM) region. The study is part of a U.S. Department of Energy (DOE)-sponsored project entitled "Environmental and Economic Assessment of Discharges from GOM Region Oil and Gas Operations." Continental Shelf Associates was the prime contractor on the work.

The results provide a description of market availability of finfish species caught in onshore and offshore waters near oilfield operations in Louisiana and Texas. The results also provide an estimate of the amount of finfish harvested by recreational and commercial fishermen which becomes food locally and out-of-state through wholesalers/processors, retail outlets and restaurants. The data are provided for input into a human risk analysis to be conducted by Brookhaven National Laboratory.

Objectives and Limitations of the Study

The overall objective for the study was to identify the edible seafood catch, distribution and consumption patterns in the GOM region. The limitations of this study include the following:

- The primary target population for this study was fishermen who had the ability to fish near produced water discharge locations.

Produced Water2: Environmental Issues and Mitigation Technologies
Edited by M. Reed and S. Johnsen, Plenum Press, New York, 1996

- The majority of the surveys of fishermen and wholesalers/processors were conducted during the period May through November 1993, representing the peak fishing activity in Louisiana and Texas. Surveys of retailers were conducted during the period March through August 1993. Restaurant surveys were collected from July to December 1993, and oystermen surveys were conducted during April and May 1994. Intercept surveys were conducted during daylight hours, excluding night fishermen.
- Only restaurants in Louisiana and Texas that served fresh seafood were included in this study.
- For the purpose of identifying the distribution patterns of seafood, seafood processing plants and seafood wholesalers were combined into one survey. Some operations function both as processing plants and as wholesale outlets.
- Produced water discharges are not always evident to the recreational or commercial fisherman, therefore, the survey questioned only the nearness of the fisherman to oilfield structures. This assumes that all oilfield structures have produced water discharges, a very conservative assumption.

HISTORICAL PERSPECTIVE

Rabalais *et al.* (1991) estimated the amount of produced water discharged into the Outer Continental Shelf (OCS) and inshore coastal waters of Texas and Louisiana at 563 million L/day (3.45 million bbls/day). A map of discharge locations in inshore and offshore Louisiana and Texas is given in Figure 1. This map was plotted by Steimle & Associates, Inc. using U.S. Minerals Management Service (MMS) data as of 20 August 1992 for offshore discharge points. The inshore data was obtained from the Texas Railroad Commission and Louisiana Department of Environmental Quality (LDEQ) as of mid-1992.

Produced waters contain oil, heavy metals, radionuclides, treating chemicals, salt and dissolved oxygen (Stephenson, 1992). These constituents, which vary widely in concentration among produced waters from place to place, have the potential to affect the distribution and survival of aquatic organisms. Accumulation of these constituents by species commonly used for human consumption may also result in health risks.

Hydrocarbons from produced water discharges have been shown to accumulate in sediments. Boesch and Rabalais (1989a) found moderately elevated sediment hydrocarbon levels a maximum of 500 m (1,640 ft) from inshore Louisiana discharges. St. Pe' (1990) found elevated hydrocarbons and Ra^{226} in sediments within 300 m (1,000 ft) of the discharge. Steimle & Associates (1991) found during sampling at 36 produced water locations throughout coastal Louisiana, that 14 sites showed sediment contamination above background levels up to 300 m (1,000 ft) from the discharge. The accumulation of constituents in the sediments is known to affect species diversity in benthic communities close to shallow water discharges (Middleditch, 1984; Neff, 1987). Neff *et al.* (1989) found elevated total petroleum hydrocarbon concentrations less than 300 m (1,000 ft) from a shallow water discharge and less than 100 m (328 ft) from a deep water discharge. Neff *et al.* (1989) concluded that benthic communities within 20 m (66 ft) of both discharges were influenced by sediment contamination.

Kraemer & Reid (1984) in geochemistry studies showed a direct relationship between the salinity of produced water and Ra content. Hanan (1981) documented a doubling of the Ra^{226} concentration in bottom sediments near a 3.5 million L/day (22,000 bbls/day) produced water discharge when compared to a nearby control site in a fresh marsh habitat off the mouth of the Mississippi River. In conjunction with Hanan's work, Landa and Reid (1983) demonstrated that the Ra^{226} associated with the sediments from the same study site was

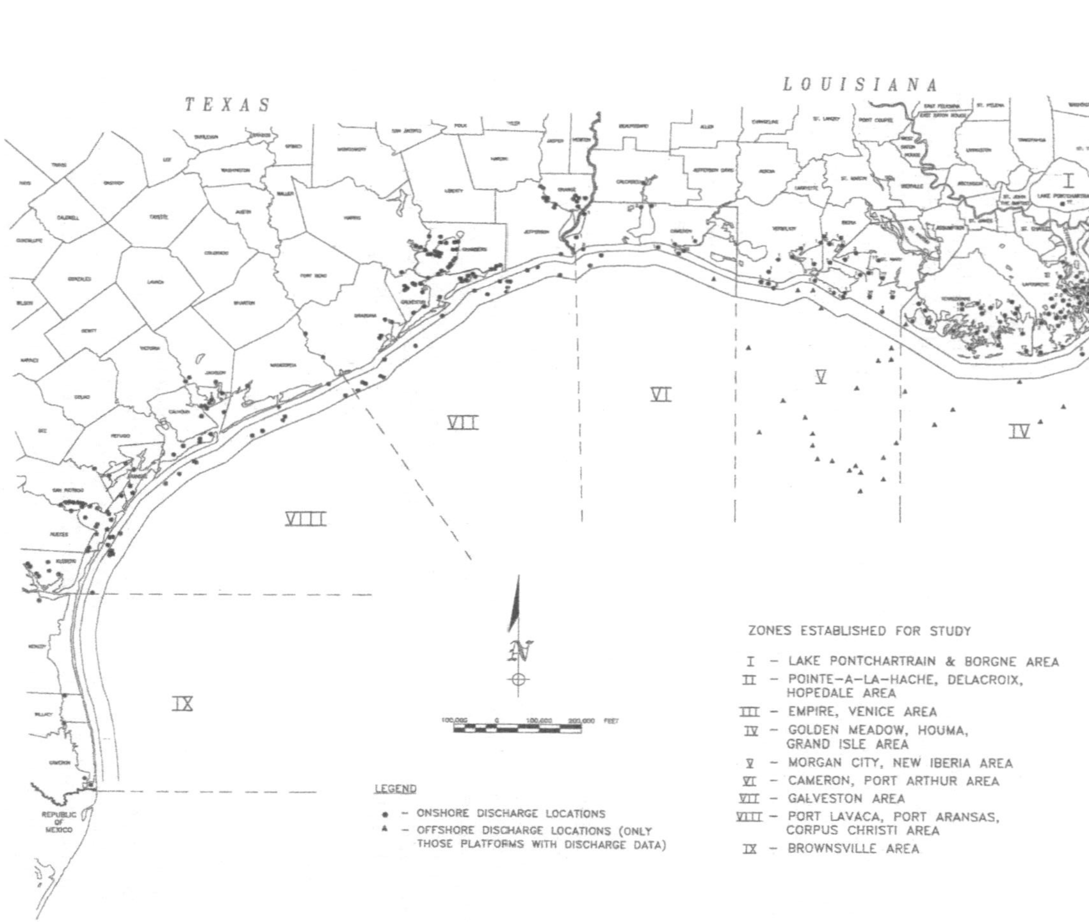

ZONE	ONSHORE* NUMBER OF DISCHARGES	OFFSHORE NUMBER OF PLATFORMS
I	6	0
II	31	49
III	124	513
IV	90	881
V	121	929
VI	20	461
VII	105	264
VIII	96	157
IX	5	10
TOTAL	598	3264

* – INCLUDES INLAND WATER BODIES

Figure 1. Discharge and platforms in the OCS and coastal zones of Texas and Louisiana.

associated with the mineral fraction of the sediments, not the organic fraction. This demonstrated that the organic matter is unlikely to provide a pathway into the food chain.

Neff (1988) reviewed the primarily offshore literature on the bioaccumulation potential of oil field wastes including produced waters. He found that produced water related hydrocarbons were bioavailable and rapidly accumulated, but they were not persistent in animal tissue and did not generally magnify in marine food webs. A limited amount of data also exists regarding the availability and personal consumption of seafood along the Gulf Coast. Species of seafood which are more often found in proximity to oilfield structures and/or potential discharge sites become exposed to contaminants and therefore may present a greater risk if consumed by humans. It may also be reasonable to assume that human populations residing near readily available sources of fresh seafood are more likely to consume it than populations where fresh seafood is less available.

Fisheries data show that 4.7 billion kg (5.23 million tons) of fish and shellfish were landed in the U.S. in 1993. This total has increased slightly over the past few years due to increased landings reported for the state of Alaska. The total for 1993 represents record commercial landings for both Alaska and the U.S. Louisiana landings for that year were approximately 12% of the U.S. total, while Texas landings represented 0.9% of that total. Fisheries landing data for 1993 for Louisiana show that the most common commercial saltwater finfish species landed was mullet with sheepshead and black drum following in succession (Usie, pers. comm. 1994, National Marine Fisheries Service [NMFS]). In Texas, the yellowfin tuna was listed as the most common commercial saltwater finfish species landed in 1992 with nearly 635,600 kg (1.4 million lb) caught. Red snapper was the next most common with 408,600 kg (0.9 million lb) landed (Hightower, pers. comm. 1994, Texas Parks and Wildlife Department [TPWD]).

Several families of fish are typically associated with offshore oilfield structures. Two of these families (i.e., Serranidae, Lutjanidae) include the commercially important groupers and snappers. Commercial landings of grouper and snapper in Louisiana for 1993 totaled 128,366 kg (283,000 lb) and 896,752 kg (1,977,000 lb), respectively. This represents 1.12% and 7.81% of the total Louisiana fish landings (excluding menhaden) for 1993. In Texas, the grouper and snapper species totaled 60,781 kg (134,000 lb) and 481,443 kg (1,061,400 lb), respectively, representing 2.71% and 21.51% of the total fish catch for 1992.

One of the most effective methods of determining the species-specific amounts of shellfish and finfish caught by residents of coastal areas is by conducting surveys. The most recent published survey information, covering the 1991 fishing season (NMFS, 1992), shows that 1991 marine recreational anglers caught spotted seatrout 15% of the time they fished in the GOM. Other frequently caught fish included white grunt, gray snapper, yellowtail snapper, black seabass, Spanish mackerel and sheepshead. The 1991 survey also showed that anglers primarily fished in private/rental boats (64-67%) in inshore waters (43%) catching 9.1-11.3 fish/trip (NMFS, 1992).

Seafood consumption surveys have been performed. Dellenbarger *et al.* (1993) performed a survey of residents in south Louisiana over the Lenten period of 1993. The results of their study show that 70% of the respondents consumed seafood during the five day period. The authors concluded that for the five day period respondents consumed an average meal of 60 g (2.1 oz) of finfish and shellfish on a per capita basis (i.e., 16 g [0.6 oz] of freshwater finfish, 13 g [0.4 oz] of saltwater finfish and 31 g [1.1 oz] of shellfish).

Anderson and Rice (1992) completed a survey of fish and shellfish consumption patterns for the greater New Orleans area. Their study consisted of telephone interviews of residents regarding their seafood consumption habits for the week prior to the interview. Their results showed that 61.2% of the respondents ate seafood the week prior to the interview. Shrimp was the most popular seafood with catfish, speckled trout, crab and other saltwater fish next in order.

The U.S. annual per capita consumption of fresh and frozen commercial fish and shellfish has risen nearly 2.5 times since 1908 when data were first recorded. The average annual per capita consumption of fresh and frozen seafood for the past ten years was 4.5 kg (9.9 lb) (NMFS, 1994).

There is a dearth of information regarding the distribution of seafood from the source (fishermen) through the various outlets (wholesaler, retailer, etc.) to the consumer. One purpose of this study was to obtain some of these data as well as to ascertain the amount of seafood potentially exposed to oilfield discharges.

METHODOLOGY

Intercept/personal interviews, mail questionnaires and telephone questionnaires were used to survey the following groups in the states of Louisiana and Texas: commercial finfishermen; recreational fishermen; processing plants and wholesalers; restaurants; and retailers. The recreational and commercial fishermen were surveyed by intercept and telephone surveys.

Processors and wholesalers were surveyed by personal interview and by mail. Lists of processors and wholesalers for Louisiana and Texas were provided by the Seafood Division of the Louisiana State Department of Natural Resources (LDNR) and the TPWD. Retailers were surveyed by site visits.

Restaurants were surveyed by mail. The Louisiana Restaurant Association and the Texas Restaurant Association provided assistance in accomplishing these surveys.

Types of Surveys Used

Both primary and secondary modes of surveying were used. Each primary survey mode was selected because it was the best and most efficient means of acquiring data from that particular group. For example, the intercept interview is the most labor intensive survey approach but it is accepted as the most effective for commercial and recreational fishermen (Dillman, 1978; Ary *et al.*,1990; Guthrie *et al.*, 1990; U.S. Environmental Protection Agency (EPA), Office of Water, 1992; Pollock *et al.*, 1994).

Development of Survey Forms

To meet the objectives of the study, seven questionnaires were developed. The questionnaires were constructed to minimize the time required to collect the information and a panel of experts were used to check the content, validity and reliability of the surveys. Technicians were trained to administer the surveys and the surveys were field tested twice before finalization.

Fishermen Surveys (Recreational and Commercial)

The primary mode for surveying both commercial and recreational fishermen was the intercept/personal interview, while the telephone interview served as a secondary mode.

Location of Seafood Catch. For the data analysis the ports were coded to match the zones (see Figure 1). To determine what section of the region(s) they fished, the respondents were asked to indicate the location in relation to the shoreline: inshore, 0 to 4.8 km (zero to three mi) offshore, 4.8 to 16 km (three to 10 mi) offshore and greater than 16 km (10 mi) offshore. For these locations, approximate percentage of catch was requested.

To determine how close fishing was conducted near potential produced water discharges, fishermen were asked to estimate distance from oilfield structures i.e., within 300 m (1,000 ft), from 300 m to 0.8 km (1,000 ft to 0.5 mi), from 0.8 km to 1.6 km (0.5 to 1.0 mi) and greater than 1.6 km (1.0 mi).

Species Caught. The species selected for the surveys were the ones most commonly caught and of commercial value along the GOM coast according to NMFS landing data (Usie, pers. comm. 1994, NMFS). Finfish species included amberjack, black drum, croaker, dolphin, flounder, grouper, kingfish, king mackerel, redfish (recreational only), red snapper, shark, sheepshead, Spanish mackerel, speckled trout, tuna and vermilion snapper. An identification guide with fish illustrated in color was shown to the respondents to enable them to correctly identify species.

Distribution Patterns. To determine the distribution of seafood catch through the population, both commercial and recreational fishermen were asked to estimate by species the percentage given away to others for consumption and the percentage kept for personal consumption. Commercial fishermen were asked to specify what percentage of the listed species were sold and to whom (i.e., wholesalers or processing plant, retailers or restaurants, consumers—friends/relatives and others).

Processing Plants and Wholesalers Survey

The survey design for the processors and wholesalers concentrated on identifying the sources and quantities of seafood species, as well as the distribution patterns of the seafood handled. The primary mode used to survey processing plants and wholesalers in Louisiana and Texas was a mail survey. The secondary mode utilized was the personal interview method.

Only processing plants that process seafood for human consumption were included in the study database. The surveys were designed to determine the source, quantities and distribution of saltwater seafood sold in 1992.

Sources and Distribution. To determine the sources of seafood to processing plants and wholesalers, the respondents were ask to estimate the percentage of saltwater finfish which were bought from the following sources: 1) in-state commercial fishermen; 2) in-state wholesalers/dealers; 3) in-state processing plants; 4) out-of-state sources; and 5) out-of-country sources.

To further define the origin of seafood catch, the respondents were asked to identify the fishing zones and ports of their commercial fishermen suppliers. If commercial fishermen were suppliers of their seafood, the respondents were asked to estimate the relative percentages of the fishermen fishing inshore (i.e., inside the barrier islands) and offshore (i.e., outside the barrier islands). To determine whether the fishing sites were near oilfield structures, the respondents were also asked to estimate the percentage of suppliers who fished inshore and the percentage fishing near inshore oilfield structures. Similarly, they were asked the percentage of suppliers who fish offshore and the percentage fishing near offshore oilfield structures. To determine importation and exportation patterns of seafood handled, the survey addressed the distribution of the end product.

Retail Seafood Market Onsite Survey

The retail seafood market surveys were performed by onsite visits. Finfish species were identified from specimens lying under glass countertops. To determine the distribution of the seafood, the surveyors asked the retailers if they shipped seafood to other areas of the state and if they shipped out-of-state.

Restaurant Survey

Seafood Served. Restaurant surveys were mailed to a set number of members in the Louisiana and Texas Restaurant Associations. The restaurant surveys were designed not only to provide data on seafood consumption in restaurants, but to infer population consumption preferences in the areas of the restaurants. The respondents were asked to provide information concerning the source, quantities, and method of preparation of seafood served/sold by their restaurant for the time period of 1 January through 31 December 1992.

Respondents were asked approximately how many seafood dishes were served per week for the year 1992. Of the seafood dishes served, they were asked to estimate the percentage of shellfish, finfish and/or a combination of shellfish and finfish for the year 1992. Data was gathered on the seafood dishes and parts eaten (e.g. meat, bone, skin).

Sampling Plan

The sampling procedure was designed to provide a stratified random sample of the various subgroups of the population. Within each zone (see Figure 1) survey information was obtained from recreational fishermen, commercial fishermen, processors and wholesalers, retail seafood markets and restaurants concerning finfish consumption and use.

DATA QUALITY ANALYSIS

The objective of the data quality analysis was to determine whether the information obtained in the several surveys was consistent both within and among surveys.

The data quality analysis focused only on finfish information from the ten surveys because finfish species information was the only information common to all surveys.

The analysis, which employed standard analysis of variance procedures described in Sokal & Rohlf (1994), indicated that there was consistent reporting of finfish information among the several surveys. This observation suggests, for example, that finfish catches reported by commercial and recreational fishermen are representative of finfish availability in retail markets and restaurants.

RESULTS

Recreational Fishermen Surveys

Of the 245,952 individuals holding Louisiana saltwater sport fishing licenses in 1993 issued by LDWF, 524 individuals (0.21%) were interviewed whereas 381 (0.06%) of the 600,000 Texas license holders were interviewed. Louisiana and Texas recreational fishermen were asked questions regarding their fishing and seafood consumption habits as well as some personal information. The fishermen responded that they primarily finfished (96%) compared to 2.2% for crabbing, 1.9% for shrimping and 0.3% for oystering.

The majority of fishermen respondents fished in a private boat in either inshore waters (55%) or offshore waters (34%). The mean number of fishing trips in the past three months was 10.9. Zone VIII was the most popular area fished (14.1%) followed closely by Zone VII

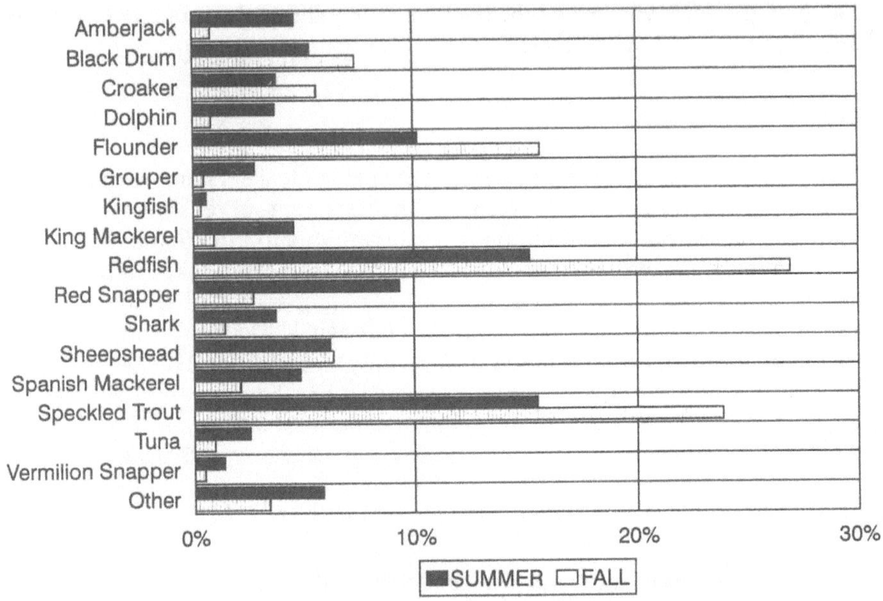

Figure 2. Frequency distribution of finfish species caught in summer and fall seasons in Louisiana.

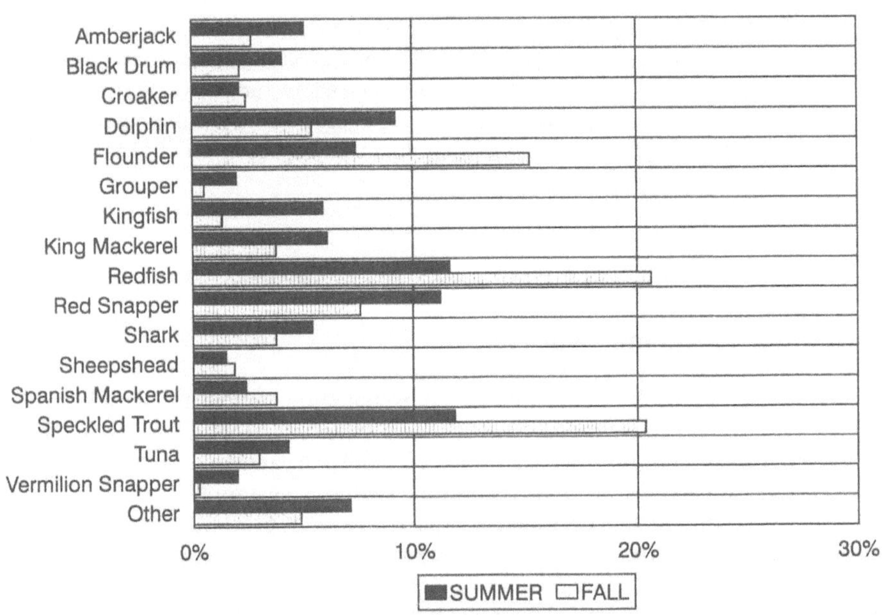

Figure 3. Frequency distribution of finfish species caught in summer and fall seasons in Texas.

(13.9%) and Zone IV (13.2%) Figure 1. Fishermen reported they primarily fished inshore (55%), although fishing offshore greater than 16 km (10 mi) was also common (38%). A large percentage of fishermen fished within 300 m (1,000 ft) of an oilfield structure (36%), although slightly more (38%) fished more than 1.6 km (1.0 mi) from an oilfield structure.

Recreational fishermen typically kept over 70% of the finfish they caught for their own personal consumption. In Louisiana and Texas, speckled trout, redfish, flounder and red snapper were the finfish species most commonly caught (see Figure 2 and Figure 3).

Commercial Fishermen Surveys

In 1993, commercial fishing licenses issued by Louisiana were held by 33,036 individuals. This is inclusive only of shrimpers, crabbers, oystermen and finfishermen. Of these, 146 (0.44%) were interviewed for this survey. Texas commercial licenses numbered 4,336 in 1993 (TPWD). Of these, 72 (1.66%) were interviewed for this survey. The Louisiana and Texas commercial fishermen profiled in this survey fishes shrimp (77%), finfish (12%), crabs (8%) and oysters (4%). Zone III was the most fished area (20%) although the other zones were also frequently fished. The number of days fished in the last three months averaged 45. Frequency of fishing location was largely inshore (39%) with frequent fishing also reported 4.8 to 16 km (3 to 10 mi) offshore (21%) and greater than 16 km (10 mi) offshore (28%). Of all trips, 33% were within 300 m (1,000 ft) of an oilfield structure: 24% were between 300 m and 0.8 km (1,000 ft and 0.5 mi) of an oilfield structure; 21% were between 0.8 and 1.6 km (0.5 and 1.0 mi) and 22% were more than 1.6 km (1.0 mi) mi from an oilfield structure.

The great majority of the seafood was sold, however some small amounts were given away or kept for personal consumption. The predominant species sold included shrimp, oysters, black drum, croaker, dolphin, flounder, king mackerel, shark, sheepshead, Spanish mackerel, speckled trout and tuna. The predominate species which were given away were blue crabs and speckled trout. The predominate species which were kept for personal consumption were blue crab, amberjack, croaker, flounder, king mackerel, shark and speckled trout.

Wholesalers and Processors Survey

The total number of wholesalers/processors in Louisiana in 1993 dealing in saltwater species was 300 according to numbers obtained from NMFS. Questionnaires were sent to all of these wholesalers/processors and 74 (24.7%) responses were returned. The average quantity of saltwater finfish received by the Louisiana wholesalers/processors was 307,600 kg (678,141 lb) of whole fish and 31,720 kg (69,930 lb) of fish parts.

The TPWD reported 420 wholesalers/processors dealing in saltwater species in Texas in 1993. Questionnaires were sent to all of these wholesales/processors and 21 (5.0%) completed surveys were received. The average quantity of saltwater finfish received by the Texas wholesalers/processors was 24,479 kg (53,967 lb) of whole fish and 14,742 kg (32,500 lb) of fish parts.

Among respondents to the Louisiana and Texas wholesalers and processors survey, 5% act as a processor only, 53% operate as a wholesaler/dealer only and 42% function as both wholesalers and processors. Their typical source of seafood was in-state commercial fisherman and in-state wholesalers/dealers with generally greater than 25% of the finfish received from these groups. Approximately 21% of finfish were obtained from other in-state processors. Approximately 9% of finfish were obtained from out-of-state sources. Imports from other countries provided 12% of the finfish.

The wholesalers/processors indicated that their sources of seafood fished in both inshore and offshore locations. Of those sources fishing inshore, 52% of them fished near an oilfield structure. Of those sources fishing offshore, 68% fished near an oilfield structure. Fishing Zone IV provided 33% of the seafood received at the wholesaler/processor with Zone VII contributing 15% and Zone VIII supplying 12%.

Over twenty percent of black drum, sheepshead, tuna, vermilion snapper and speckled trout were sold to out-of-state sources. Thirty-one percent of tuna, 30% of black drum and 29% of red snapper were sold to other in-state wholesalers/dealers. Fifty-four percent of flounder and 50% of speckled trout and red snapper were sold to in-state retailers/restaurants. Few species of finfish were sold to in-state processing plants.

Retail Seafood Market Survey

In Louisiana, 226 retail outlets were surveyed and in Texas, 115 outlets were surveyed.

In Louisiana and Texas finfish were sold as whole fish (gutted) and fillets, and may or may not have head and skin removed. Croaker, king mackerel, sheepshead, flounder, black drum, kingfish and Spanish mackerel were sold as whole fish with the skin left on more than 60% of the time. Amberjack and shark were usually sold as fillets with no skin, 50% and 55%, respectively. Approximately 42% of tuna, 27% of dolphin, 20% of shark, 19% of red snapper, and 12% of speckled trout was available fillets with skin. Less than 10% of black drum, flounder, grouper, Spanish mackerel and sheepshead was available in this form. Both tuna (8%) and grouper (13%) were available in parts (steaks). Most finfish were available unprepared although some of the speckled trout (12%), black drum (9%), flounder (3%), red snapper (3%) and sheepshead (3%) were available prepared.

In Louisiana and Texas, shipping of seafood products to other parts of the state occurred in 28% of the markets. About 18% of the respondents reported that they shipped seafood out-of-state.

Restaurant Survey

Three thousand survey questionnaires were sent to members of the Louisiana and Texas Restaurant Associations who serve seafood (1,500 each state) with a return of 5.1%.

The results of the Louisiana and Texas Restaurant Mail Survey showed an average number of 889 seafood dishes were served each week in 1992, of which 48% consisted of shellfish, 45% involved finfish and 31% contained both shellfish and finfish. The most common finfish species served were red snapper, flounder, tuna and speckled trout.

Texas restaurant survey respondents indicated that they import nearly twice as much seafood as the Louisiana restaurant survey respondents, 62% versus 31%, respectively. Collectively though, wholesalers/processors provided about 88% of the seafood served at restaurants. Other seafood sources were retailers (42%), commercial fishermen (37%) and other sources (14%).

DISCUSSION

Intrasurvey Comparisons

Recreational Fishermen. All of the fishermen (both Texas and Louisiana) surveyed fished from a boat in inshore waters rather than onshore fishermen who would not be able to fish in proximity to oilfield structures. Recreational fishermen reported they preferred to keep their catch instead of giving it away. The respondents indicated that they infrequently served their catch in their household, but when they did, it was usually filleted, skinned and fried. Louisiana recreational fishermen reported fishing in close proximity to oilfield structures more often than the Texas respondents. This may be due to the large number of well canals and oilfield structures in the marshes and bayous of south Louisiana where many of the fishing trips occurred.

Commercial Fishermen. Commercial fishermen depend primarily on their catch for their livelihood. Commercial shrimpers comprised 77% of the commercial fishermen interviewed. Shrimpers were more accessible for survey since their hours of operating were more predictable than the other types of commercial fishermen. Shrimp unloading locations were more numerous and easier to access. Commercial fishermen, although more likely to fish inshore waters, were more likely to fish near oilfield structures which they believed attracted organisms. The fishermen reported that they consume seafood on a regular basis in their household, but sell the vast majority of their catch to wholesalers/processors or restaurants/retailers.

Wholesalers and Processors. Good response was obtained from the wholesalers/processors for both Louisiana and Texas. Their primary source of seafood was from in-state commercial fishermen with very little seafood coming in from other countries. They indicated that their offshore suppliers fish in proximity to oilfield structures a large percentage of their time. Finfish species associated with oilfield structures and wellheads would be more frequently caught under these conditions.

Louisiana wholesalers/processors handle significantly more finfish and shellfish then their Texas counterparts. Louisiana respondents handled an average of 12 times more finfish and almost six times more shrimp. This is in agreement with the landings data which show Louisiana landings nearly twice as great as Texas landings.

Retail Seafood Markets. Retail seafood outlets in Louisiana and Texas primarily offer and serve their seafood in similar forms. Outlets typically have uncooked portions of shellfish and finfish. Among prepared seafood, Louisiana retailers were more likely to offer prepared hard shell blue crabs, while Texas retailers commonly prepared their peeled and headed shrimp. The finfish species were well represented by both states' retailers, although Texas outlets offered more finfish species in parts, typically "steaks" or cross-sectional pieces of fish meat. Louisiana retailers are more likely to ship fresh and prepared seafood to other parts of the state and out-of-state than the Texas retailers.

Restaurants. Respondents to the restaurant survey for both Louisiana and Texas reported similar trends and tendencies in seafood preparation and service. For each state shellfish (particularly shrimp) dishes were more popular than finfish or a combination of the two. Speckled trout and red snapper were the finfish species most often served.

Restaurants in Louisiana and Texas depend on wholesalers/processors to provide them with their seafood. More than 60% of the Texas restaurant respondents depend on out-of-state sources for their seafood. This is a very large number in view of the fact that Louisiana was *not* considered an out-of-state source for purposes of this survey. Since Louisiana has the largest fisheries landings for the continental U.S. and is adjacent to Texas, it is not surprising that Texas would import some of Louisiana's seafood.

Statistical Findings. Data quality analyses were conducted to determine whether survey responses concerning finfish species were consistent both within and among surveys.

The analyses focused only on finfish information from the ten surveys. This was a necessary choice because finfish species information was the only information common to all ten surveys. The results of analysis of variance indicate consistency of reporting both within and among surveys.

Intersurvey Comparisons

Location of Seafood Catch. The data indicate that the recreational and commercial fishermen for both Louisiana and Texas choose to fish near an oilfield structure on approximately 30% of their fishing trips. Louisiana and Texas commercial fishermen and Louisiana recreational fishermen more commonly fish near oilfield structures than Texas recreational fishermen. This may be attributable to location and accessibility of the Louisiana oilfield structures and to the life cycle and migratory habits of the finfish.

The Texas and Louisiana wholesalers/processors report that significant amounts of their seafood comes from waters near oilfield structures, particularly from suppliers who fish offshore. Louisiana wholesalers/processors report their suppliers fishing near oilfield structures because of a large number of oilfield structures along the Louisiana coast.

Source of Seafood. Information obtained from wholesalers/processors indicates that a large amount of the seafood was supplied by in-state commercial fishermen and other in-state wholesalers/processors. For the purpose of this study, Louisiana and Texas were not considered out-of-state in either states' survey. Therefore, large amounts of seafood from the neighboring state could be received by a wholesaler/processor and significantly affect the reported source of seafood. The data show only a small percentage of the seafood comes from out-of-state and out-of-country sources.

Products from wholesalers/processors represent a major source of seafood to restaurants. The data indicates that restaurants attempt to provide fresh seafood to their customers and the acquisition of seafood from local or in-state sources helps ensure that fresh seafood will be served.

Distribution Patterns. As one would expect, the recreational fishermen from Louisiana and Texas reported that they typically kept their seafood catch. Commercial fishermen most likely sell their seafood catch to wholesalers/processors. Wholesalers/processors were more likely to purchase the entire catch of the commercial fishermen, eliminating the commercial fishermen's need to transport and sell his catch to various outlets.

Wholesalers/processors usually sell their products to restaurants and retailers. Since they do not necessarily change the form of the seafood prior to sale, they serve as an intermediate step in transferring seafood from fishermen to consumer. The large volume of seafood handled by the wholesaler/processor suggests that they are able to supply seafood to various parts of the region.

Proximity to Oilfield Structures

Some finfish species are more frequently observed in close proximity to oilfield structures than are other species. One might reasonably expect, therefore, that Louisiana and Texas recreational fishermen would report catching some species in close proximity to oilfield structures more frequently than other species. This expectation was indeed realized. Two distances relative to oilfield structures were particularly significant: 1) closer than 300 m (1,000 ft); and 2) farther away than 0.8 km (0.5 mi.). Finfish species whose reported catch frequencies were at least 50% greater in close proximity to oilfield structures included amberjack, grouper, king mackerel, red snapper, shark and vermilion snapper. Finfish species whose reported catch frequencies were at least 50% greater at distances of one mi or more from oilfield structures included flounder, redfish and speckled trout.

CONCLUSIONS

Over the two state area surveyed, a large percentage of seafood was harvested commercially and recreationally near inshore and offshore oilfield structures. Thirty six percent of the recreational and 33% of the commercial fishermen that responded to the survey reported fishing within 300 m (1,000 ft) of an oilfield structure. This was particularly true in the large expanses of Louisiana estuaries from which was harvested much of Louisiana's landings, the second largest landings in the U.S. Seafood harvested by commercial fishermen —primarily shrimp fishermen—was most often sold to wholesaler/ processors. These middlemen ship the seafood with little value added to retail outlets and restaurants.

Retail outlets provide mostly uncooked fresh seafood to the public. Restaurants which serve seafood do so as a large percentage of their total meals served. Fresh seafood served to consumers was fried. Stock was often used in the preparation of seafood dishes.

Seafood harvested by recreational fishermen was usually kept for home consumption and prepared fried. Most of the seafood harvested locally, both commercially and recreationally, remains in the local population. Very little of the fresh seafood sold locally was imported from out-of-state or out-of-country. The species of finfish most frequently consumed and distributed included flounder, sheepshead, speckled trout and red snapper. With the exception of the flounder, these species are commonly associated with oilfield structures.

ACKNOWLEDGMENTS

The authors would like to acknowledge the U.S. Department of Energy (DOE) (DE-AC22-92MT92001) without whose support this work would not have been possible. Thanks are also extended to Dr. Brent Smith, Project Officer and Dr. Gerry Ham, Director of Metairie Site Office of DOE. Appreciation is extended to Fisheries Information and Seafood Harvest Inquiry Enterprise (FISHIE) for having performed the surveys.

REFERENCES

Anderson, A. C. and J. C. Rice. 1992. A survey of fish and shellfish consumption by residents of the greater New Orleans area. Final Report, LA DEQ Contract: 24400-91-18, New Orleans: School of Public Health and Tropical Medicine, Tulane University.

Ary, D., L. C. Jacobs, and A. Razavieh. 1990. Introduction to research in education. Fort Worth: Holt, Rinehard, & Winston, Inc.

Boesch, D. F. and N. N. Rabalais, editors. 1989. Environmental impact of produced water discharges in coastal Louisiana. Report to LA Div. Mid-Continent Oil and Gas Association. LUMCON, Chauvin, LA.

Dellenbarger, L., A. Schupp, and B. Kanjilal. 1993. Seafood consumption in coastal Louisiana. Louisiana Department of Environmental Quality, Office of Water Resources, Baton Rouge, LA.

Dillman, D. A. 1978. Mail and telephone surveys: The total design method. New York: Wiley-Interscience Publication.

Guthrie, D., J. M. Hoenig, M. Holliday, C. M. Jones, M. J. Mills, S. A. Moberly, K. H. Pollock, and D. R. Talhelm. 1990. Creel and angler surveys in fisheries management. American Fisheries Society Symposium 12. Bethesda, MD.

Hanan, M. A. 1981. Geochemistry and mobility in sediments of radium from oilfield brines - Grand Bay, Plaquemines Parish, Louisiana: M. S. thesis, Univ. of New Orleans.

Kraemer, T. F. and D. F. Reid. 1984. The occurrence and behavior of radium in saline formation water of the U.S. Gulf coast region. *Isotope Geosci.* 2:153-174.

Landa, E. R. and D. F. Reid. 1983. Sorption of radium-226 from oil-production brine by sediment and soils. *Environ. Geol.* 5:1-8.

Middleditch, B. S. 1984. Ecological effects of produced water discharges from offshore oil and gas production platforms. Final Report on API Project No. 248. American Petroleum Institute, Washington, D.C.

National Marine Fisheries Service. 1992. Marine recreational fishery statistics survey, Atlantic and Gulf coasts, 1990-1991. U.S. Department of Commerce, National Oceanic and Atmospheric Administration, National Marine Fisheries Service. Current Fisheries Statistics No. 9204.

National Marine Fisheries Service. 1994. Fisheries of the United States, 1993. U.S. Dept. of Commerce, National Oceanic and Atmospheric Administration, National Marine Fisheries Service. U.S. Government Printing Office, Washington, D.C. Current Fishery Statistics No. 9300.

Neff, J. M. 1987. Biological effects of drilling fluids, drill cuttings and produced waters. Pages 469-538 in D. F. Boesch and N. N. Rabalais (eds.), Long-Term Environmental Effects of Offshore Oil and Gas Development, Elsevier Applied Science Publishers Ltd., London.

Neff, J. M. 1988. Bioaccumulation and biomagnification of chemicals from oil well drilling and production wastes in marine food webs: A review. Final Report to American Petroleum Institute, Washington, D.C.

Neff, J. M., T. C. Sauer, and N. Maciolek. 1989. Fate and effects of produced water discharges in nearshore marine waters. American Petroleum Inst. Publ. No. 4472, American Petroleum Institute, Washington, D.C.

Pollock, K. H., C. M. Jones, and T. L. Brown. 1994. Angler survey methods and their applications in fisheries management. American Fisheries Society Special Publication 25. Bethesda, MD.

Rabalais, N. N., B. A. McKee, D. J. Reed, and J. C. Means. 1991. Fate and effects of nearshore discharges of OCS produced waters. Vol. II: Technical Report. OCS Study/MMS 91-0005. U.S. Dept. of the Interior Minerals Management Service, Gulf of Mexico OCS Regional Office, New Orleans, LA.

Sokal, R. R. and F. J. Rohlf. 1994. Biometry. San Francisco: W. H. Freeman.

Steimle & Associates. 1991. Produced water impacts on Louisiana wetlands. American Petroleum Inst. Publ. No. 4517, American Petroleum Institute, Washington, D.C.

Stephenson, M. T. 1992. A survey of produced water studies. In Produced Water: Technological/Environmental issues and solutions, eds. J. P. Ray and F. R. Engelhardt, 1-11. New York: Plenum Publishing Corp.

St. Pe', K. M. 1990. An assessment of produced water impacts to low energy, brackish water systems in Southeast Louisiana. Louisiana Department of Environmental Quality, Water Pollution Control Division, Baton Rouge, LA.

U.S. Environmental Protection Agency, Office of Water. 1992. Consumption surveys for fish and shellfish a review and analysis of survey methods. U.S. Environmental Protection Agency, Washington, D.C. EPA/822R-92-001.

West, P. C., J. M. Fly, R. Marans, and F. Larkin. 1989. Michigan sport anglers fish consumption survey. Natural Resource Sociology Research Lab Technical Report #1, Univ. of Michigan, Ann Arbor, MI.

36

RISK ASSESSMENT FOR PRODUCED WATER DISCHARGES TO OPEN BAYS IN LOUISIANA

Anne F. Meinhold, Seymour Holtzman, and Michael P. DePhillips

Department of Applied Science
Brookhaven National Laboratory, Upton, New York

INTRODUCTION

Potential human health and environmental impacts from discharge of produced water to the Gulf of Mexico concern regulators at the State and Federal levels, environmental interest groups, industry and the public. Current regulations in the United States require or propose a zero discharge limit for coastal facilities based primarily on studies performed in low energy, poorly flushed environments. Produced water discharges in coastal Louisiana, however, include a number located in open bays, where potential and impacts are likely to be larger than the minimal impacts associated with offshore discharges, but smaller than those demonstrated in low-energy canal environments.

This paper summarizes results of a conservative screening-level health and ecological assessment for contaminants discharged in produced water to open bays in Louisiana, and reports results of a probabilistic human health risk assessment for radium and lead. The initial human health and ecological risk assessments consisted of conservative screening analyses that identified potentially important contaminants and excluded others from further consideration. A more quantitative probabilistic risk assessment was completed for the human health effects of the two contaminants identified in this screen: radium and lead. This work is part of a series of studies on the health and ecological risks from discharges of produced water to the Gulf of Mexico, supported by the United States Department of Energy (USDOE).

RISK ASSESSMENT

Risk assessment can be defined as a process of estimating magnitudes and probabilities of potential adverse effects on human health or the environment. Risk assessments provide risk managers with the scientific information needed to balance the degree of risk permitted, against competing risks and the costs of risk reduction.

A human health risk assessment for a pollutant describes its discharge, transport and fate in the environment, and the resulting human exposure. Human-health risks are then calculated using data and models that relate exposures to health effects.

Produced Water2: Environmental Issues and Mitigation Technologies
Edited by M. Reed and S. Johnsen, Plenum Press, New York, 1996

The United States Environmental Protection Agency (USEPA) considers excess individual lifetime cancer mortality risks less than 1 x 10^{-4} (one in ten thousand) to 1 x 10^{-6} (one in one million) to be acceptable (Federal Register, 1991). No similar standard "acceptable risk" value is available for toxic effects — intakes are usually compared to a chemical-specific reference dose to determine if toxic effects are expected.

With some modifications and the addition of important uncertainties, the paradigm developed for assessment of human health risks is being applied to estimation of risks to the environment. The receptors or values of concern in an ecological risk assessment may range from individual organisms to entire ecosystems and fundamental ecological processes. Ecological risk assessments may be more qualitative than human health assessments because of the many sources of uncertainty in assessing risks to ecological values (USEPA, 1992).

A tiered approach to human health and ecological risk assessment is logical and cost-effective. In a tiered approach the initial analysis is a conservative (i.e. worst case) screening step, designed to screen out contaminants and pathways that are not of concern.

If conservative models and assumptions yield small estimated risks no further analyses are needed. If a conservative analysis suggests that risks may be high, a more detailed and realistic assessment is performed. The state-of-the-science now involves use of a probabilistic approach that explicitly considers uncertainties.

Monte Carlo analysis, a commonly used tool in probabilistic risk assessment, was used here (Crystal Ball®, Decisioneering, Inc., Denver, Colorado). In a Monte Carlo analysis, a sample from the distribution of an input parameter is placed into a simulation to interact in a model with samples from other input parameters.

RECEPTORS AND EXPOSURE PATHWAYS

Important receptors for radium discharged in produced water are recreational fishermen and their families. Recreational fishermen may fish close to a platform, return often to the same fishing spot, and derive a large percentage of their diet from fish caught near a platform. Ingestion of contaminated fish is expected to be the most important exposure route for people. Children of recreational fishermen fishing near produced water platforms were considered the subpopulation at risk for exposure to lead.

In the screening assessment, potential ecological receptors included recreationally and commercially important finfish and shellfish species living close to platforms. Screening assessments for other potential receptors will be reported elsewhere. The exposure pathway addressed was direct exposure of biota to contaminants in water.

The State of Louisiana has identified a standard acute toxicity mixing zone of 50 feet and a standard chronic and human health zone of 200 feet from produced water discharges. In the risk assessment reported here, these distances were used because they imply a risk management decision about the "acceptable" location for potential impacts.

CONTINUING OPEN BAY DISCHARGES

Characterization

Louisiana regulations (Title 33, March 20, 1991) required the termination of all produced water discharges to natural or man-made water bodies located in intermediate, brackish or saline marsh areas after January 1, 1995, unless the discharge(s) were authorized in an approved schedule for elimination or effluent limitation compliance. A variance through January, 1997 was granted (12/16/94) for permitted discharges located in open waters, and

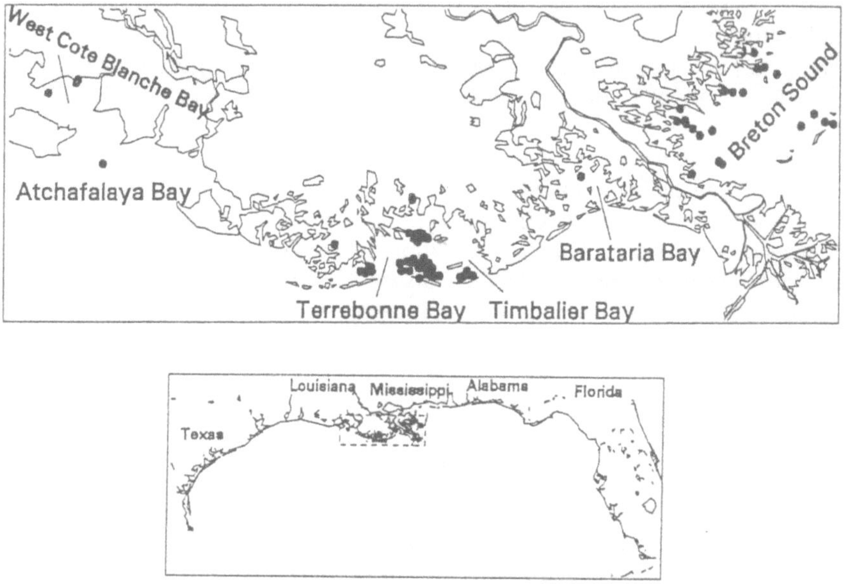

Figure 1. Assumed continuing discharges in open Louisiana bays.

at least 1 mile from any shoreline, in Chandeleur Sound, Breton Sound, Barataria Bay, Caminada Bay, Timbalier Bay, Terrebonne Bay, East Cote Blanche Bay, West Cote Blanche Bay or Vermillion Bay.

The Louisiana Department of Environmental Quality (LDEQ) identified produced water discharges in open bay areas that may qualify for this variance. In August, 1994, a telephone survey of these operators was conducted to determine if they would take advantage of an extension of the phase-out rule for coastal Louisiana produced water discharges. Most operators indicated that they would continue to discharge through 1997, if allowed. Figure 1 shows the locations of the 72 assumed continuing discharges in open Louisiana bays. Table 1 summarizes data describing platform depths and discharge rates abstracted from LDEQ permit files.

Dilution

The USEPA surface water transport model CORMIX (Cornell Mixing Zone Expert System Model; Doneker and Jirka, 1990) was used to estimate the dilution expected at 200 feet (61 m) from open bay discharges. A depth of 8 feet (2.4 m) and a range of discharge rates (100 - 37,500 bbl/day) was modeled.

Table 1. Platform depths, discharge rates and dilution factors

	Depth (feet)[1]	Discharge[2] (bbl/day)	Dilution Factor[3]
Arithmetic mean	9.1	4,500	62
Standard deviation	2.3	7,200	52
Minimum	4.0	1	8
Maximum	18.0	37,000	750

[1] 29 data points available
[2] 62 data points available
[3] Modeled dilution factor at 200 feet.

Because of the shallow depth, the model was run using an unstratified scenario with a surface and bottom water density of 1005 kg/m^3. These values were derived from temperature and salinity data published in literature reviewed by USEPA (USEPA, 1995a). A produced water discharge density of 1020 kg/m^3 was derived from USEPA's review of produced water effluent density estimates, and an ambient velocity of 0.05 m/s was used (USEPA, 1995a).

CORMIX forces a submerged single port discharge to be in the bottom 1/3 of the water column. This is unrealistic for produced water discharges, normally discharged on or close to the surface. The model can be adjusted to make the projections more accurate by creating a mirror image using a stratified water column and inverting the ambient densities (Avanti Corporation, 1993). Sensitivity analyses using this approach did not produce significantly different results.

Model results are presented in terms of the expected dilution factor at 200 feet, where the dilution factor (DF) is the concentration of the discharge (100 %) divided by the concentration of effluent in the water at 200 feet. A relationship between discharge rate and dilution factor was derived from model results (Figure 2):

$$\text{DILUTION FACTOR} = 7315.6 * (\text{DISCHARGE})^{-0.6473} \; (R{=}0.95) \tag{1}$$

This relationship fits well at high discharge rates, but tends to underestimate dilution at lower discharge rates (Figure 2). The derived relationship was applied to the distribution of discharge rates to produce a distribution of dilution factors (Table 1).

SCREENING ASSESSMENT

Human Health Risk

A conservative screening human health risk assessment was done for metals, organic compounds and radionuclides in continuing open bay discharges using the USEPA Superfund approach(USEPA, 1989).

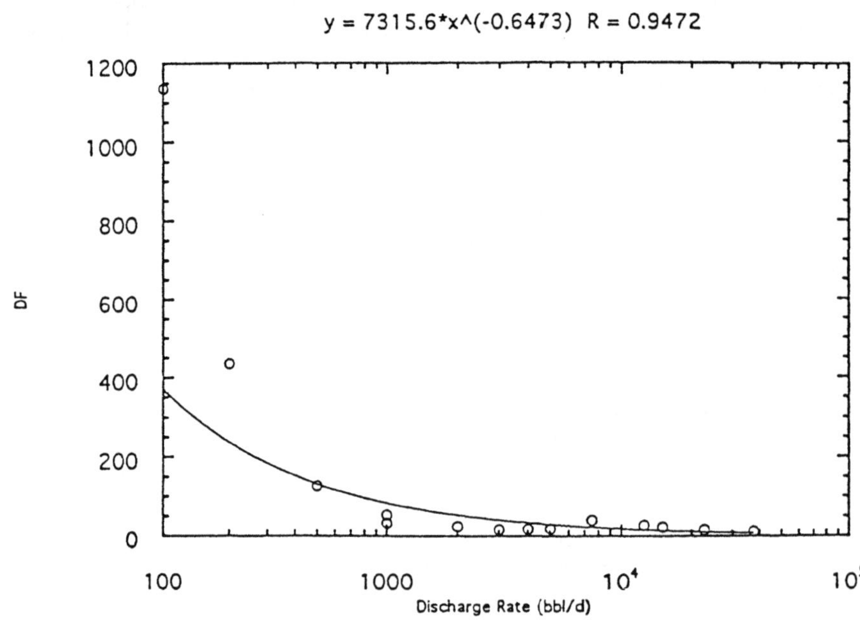

Figure 2. Relationship between discharge rate and modeled dilution factor at 200 feet from discharge.

Worst-case water concentrations were estimated, using the highest recorded (in LDEQ permit files) concentration of a contaminant for any of the assumed continuing discharges at any time, and assuming a DF of 20. Concentrations in fish were estimated using the bioaccumulation factor method (USEPA, 1989; Strenge and Peterson, 1989). Exposures were assumed to continue over a lifetime (70 years), and the intake rate used was the USEPA (1990) 95th percentile value of 132 g/day. Hazard quotients and individual lifetime fatal cancer risks were calculated (USEPA, 1989). Hazard quotients less than one and risk estimates less than 1×10^{-4} suggest that a contaminant is not of concern in terms of human health risk. The details of this screening analysis will be reported elsewhere.

This conservative screening analysis served to eliminate contaminants that do not warrant further time and attention. Contaminants eliminated from further consideration were arsenic, chromium, copper, silver, naphthalene, phenol, toluene and xylenes.

Contaminants of concern included antimony, cadmium, lead, mercury, nickel, zinc, benzene and radium. Because of the conservative nature of this screening analysis, no important effect on human health can be assumed. Screening hazard quotients for antimony, cadmium, nickel and zinc exceeded one by less than an order of magnitude. The cancer risk estimate for benzene exceeded 1×10^{-4} by less than an order of magnitude (6.4×10^{-4}). A more realistic and quantitative assessment using effluent concentration distributions and predicted dilutions for the entire range of discharges is expected to predict few hazard quotients greater than one, and a median risk for benzene of less than 1×10^{-4}. This analysis is being done.

Risk from ingestion of radium in fish exceeded 1×10^{-4} by more than an order of magnitude (7.8×10^{-3}). Contaminants that exceeded hazard quotients by more than an order of magnitude were lead and mercury. Comparison of the few mercury concentrations in effluent reported above the detection limit suggested that mercury in the continuing discharges was similar to that expected in the Gulf of Mexico. Mercury was not analyzed further. Radium and lead were analyzed quantitatively in a probabilistic risk assessment.

Ecological Risk

Radium. Radium concentrations reported in LDEQ permit files were used to assess potential ecological effects. Worst-case water concentrations were estimated for the highest reported concentrations in open bay discharges, using a conservative DF of 20. Predicted water concentrations were compared to screening dose-rate factors (IAEA, 1988) that relate radiation exposure of an organism to a unit concentration of the radionuclide in the water in which the organism lives. Estimated doses were compared to published reference dose rates (NCRP, 1991).

No estimated doses exceeded the National Council on Radiation Protection and Measurements (NCRP) reference limit of 10 mGy/day. Several estimated doses exceeded the NCRP suggested screening level for detailed assessment (2.4 mGy/day). Additional quantitative assessments could be performed to assess the extent to which the NCRP screening level of 2.4 mGy/day is likely to be exceeded. Because of the conservative nature of this analysis, it can be concluded that no effects are expected on aquatic organisms from radium discharged in produced water to open bays in Louisiana.

Individual Contaminants. Worst-case water column concentrations of measured contaminants in continuing open bay effluents (Section 5.1) were compared to USEPA and Louisiana water quality criteria. This conservative analysis served to eliminate contaminants that do not warrant further time and attention.

Contaminants eliminated from further consideration included arsenic, chromium, benzene, naphthalene and toluene. Worst-case water concentrations exceeded acute water quality standards for copper, lead, nickel, silver and zinc. Chronic water quality criteria were exceeded for antimony, cadmium, copper, lead, mercury, nickel, silver, zinc and phenol.

No important effect on aquatic biota can be assumed because of the conservative nature of the screening analysis. Water quality standards were exceeded by less than an order of magnitude for cadmium, silver, zinc, and phenol. A more realistic and quantitative assessment using predicted dilutions for the entire range of discharges and effluent concentration distributions is expected to predict few exceedances for these contaminants and is being done.

Water quality standards for copper, lead, mercury, and nickel were exceeded by more than an order of magnitude. These contaminants are being assessed in a quantitative risk assessment that includes estimated distributions for: DFs, effluent concentration, and dose-response relationships.

PROBABILISTIC RISK ASSESSMENT: RADIUM AND LEAD CONCENTRATIONS IN WATER AND FISH

Concentration in Water

Measured concentrations of radium and lead in open bay produced water discharges, reported in LDEQ permit files, are summarized in Table 2. The largest lead concentration reported in permit files (800,000 μg/l) was several orders of magnitude larger than maximum values reported in other studies (Stephenson, 1992; Middleditch, 1984) and was not included in the data set for the risk assessment. Many of the lead concentrations in produced water were reported as "less than (<)" the detection limit. The detection limit for lead ranged from 50 to 125 μg/l. These values were replaced by one-half the value of the reported detection limit.

To estimate ambient water concentrations, the distribution of radium and lead concentrations reported for open bay discharges was modified by the distribution of DFs. All radium, and 38% of lead was assumed to remain in solution based on calculations performed by LDEQ (USEPA, 1995). Table 2 gives estimated lead and radium concentrations in the water column at 200 feet.

Concentration in Fish

Fish Near Platforms. Radium and lead concentrations in fish near produced water discharges were estimated using the bioaccumulation factor (BAF) method:

Table 2. Radium and lead concentrations in open bay produced water discharges, and estimated concentrations in water and fish at 200 feet

	Effluent[1]			Ambient Water[2]			Fish[3]		
	^{226}Ra	^{228}Ra	Lead	^{226}Ra	^{228}Ra	Lead	^{226}Ra	^{228}Ra	Lead
Arithmetic mean	191.4	250.0	284.6	5.5	6.9	3.6	0.2	0.2	0.5
sd deviation	122.4	163.6	616.2	5.5	6.7	5.6	0.2	0.27	0.8
Minimum	0.0	0.0	25.0	0.0	0.0	0.03	0.0	0.0	0.0
Maximum	592.0	560.0	2,280.0	51.2	65.1	78.0	2.6	3.5	12.8

[1]Measured in effluent: pCi/l radium, μg/l lead.
[2]Modeled concentrations in water: pCi/l radium, μg/l lead.
[3]Modeled concentrations in fish: pCi/g radium, μg/g lead.

$$C_{fish} = BAF \times C_{water} / 1000 \text{ (g/l)} \tag{2}$$

where:

C_{fish} = contaminant concentration in fish (pCi/g; μg/g)
BAF = bioaccumulation factor
C_{water} = contaminant concentration in water (pCi/l; μg/l))

A radium BAF distribution, derived from data collected in coastal Louisiana (Meinhold and Hamilton, 1992), was used to estimate concentrations in fish. This distribution is lognormal, has a range of 2 to 100, a mean of 30.4 and a standard deviation of 28. Table 2 gives the estimated distributions for radium concentrations in fish.

A distribution for a lead BAF was developed from published estimates. In a report prepared for USEPA, Avanti Corporation (1993) cited a range of 10 to 100 for bioaccumulation of lead. IAEA (1982) presented a default BAF of 300 for lead in seawater. A triangular distribution for BAF of lead ranging from 10 to 300, with a most likely value of 100 was used in this analysis.

Fish Away from Platforms. For comparison, concentrations of radium and lead in fish caught in the Gulf of Mexico away from platforms (and associated health risks) were estimated. Radium concentrations in fish not associated with platforms were assumed to be uniformly distributed, with a range of 0 to 0.01 pCi/g (Meinhold et al., 1995).

Distributions of lead in fish not associated with platforms were derived from measured concentrations of lead in whole fishes at two Environmental Monitoring and Assessment Program (EMAP) sites on the coast of Louisiana (USEPA, 1995b). These measurements may under- or overestimate background concentrations because the samples were of whole fish rather than edible fillets. Background concentrations were assumed to be lognormally distributed with an arithmetic mean value of 0.05 μg/g (standard deviation: 0.06; range: 0.01 - 0.28). Although the data used in deriving this distribution have been funded wholly or in part by the USEPA through its EMAP-Estuaries Program, it has not been subjected to Agency review, and therefore does not necessarily reflect the views of the Agency and no official endorsement should be inferred.

PROBABILISTIC RISK ASSESSMENT: FISH INGESTION RATES

Ingestion rates were derived from reported data on meals per week for recreational fishermen in Louisiana, collected in a survey of recreational and commercial fishermen (Steimle & Associates, Inc., in preparation). The ingestion rate distribution for recreational fishermen and their families was derived as follows:

$$FI = \frac{M \times MS}{7d \times week^{1}} \tag{3}$$

where:

FI = derived ingestion rate (g/day)
M = meals per week (assumed lognormal distribution: arithmetic mean 1.8; sd 1.3; range 0-15; derived from Steimle & Associates Inc.
MS = meal size (150 g/meal; USEPA, 1989b).

Table 3. Derived intake distributions for fish caught near open bay platforms

	Intake (g/day)	
	Recreational Fishermen and Families	Children
Arithmetic mean	38.4	16.6
Median	31.5	13.6
Standard deviation	26.4	11.6
Minimum	3.3	1.3
Maximum	228.6	115.7
95th percentile	89.5	38.5

This distribution (Table 3) was used to estimate exposures of recreational fishermen and their families to radium in recreationally caught fish.

The populations with highest susceptibility to adverse health effects from lead intakes are infants and young children. USEPA (1990) reported data for intake rates of seafood by the population consuming seafood, obtained in a survey conducted over a period of one year (1973-1974). For juveniles (0-9 years of age) the rate of ingestion of seafood was approximately 43% that of the general population. The intake rate distribution derived for recreational fishermen and their families was multiplied by a factor of 0.43 to estimate the rate of juvenile ingestion of fish (Table 3). This intake rate distribution was used in the risk assessment for lead.

PROBABILISTIC RISK ASSESSMENT FOR RADIUM

Exposure Period

Exposure periods (i.e., number of years fishermen catch and eat fish close to an offshore produced water discharge) may vary from several years to a large part of a lifetime. The exposure period for recreational fishermen was assumed to be a triangular distribution, ranging from 5 to 65 years, with the most frequent value set at 20 years.

Intake

Daily ^{226}Ra and ^{228}Ra ingestion rates during the exposure period were calculated as:

$$RI = FI \times [Ra]_{fishes} \quad (4)$$

where:

RI = radium intake (pCi/day)
FI = intake of fish (g/day) for recreational fishermen and their families
$[Ra]_{fishes}$ = concentration of radium in fishes (pCi/g)

Dose-Response Assessment

Current practice in radiation protection assumes there is a cancer risk associated with even small doses of radiation. Risk factors, derived from epidemiological data are extrapolated down to low doses to describe the cancer risk associated with small exposures. Risk factor distributions were derived from USEPA values (Federal Register, 1991) by assuming

Table 4. Risk factor distribution for ^{226}Ra and ^{228}Ra (lognormal distributions)

	Individual lifetime fatal cancer risk per pCi/day	
	^{226}Ra	^{228}Ra
Arithmetic mean	1.5×10^{-6}	1.0×10^{-6}
Lower 90% confidence limit	9.4×10^{-7}	4.7×10^{-7}
Upper 90% confidence limit	2.2×10^{-6}	1.9×10^{-6}

that the USEPA values represent the upper 90% upper confidence limit of a lognormal distribution (Meinhold et al., 1995). Table 4 shows the distribution of risk factors used.

The risk factor was modified for each exposure period by adding 10 years for radium retention (Meinhold et al., 1995).

$$RF(EP) = \frac{[EP + 10]}{70 \text{ years}} \times URF(70) \quad (5)$$

where:

RF(EP) = risk factor as a function of exposure period EP (lifetime risk per pCi/day)
EP = exposure period (years)
URF(70) = USEPA unit risk factor for lifetime exposure (lifetime risk per pCi/day)

Risk Characterization and Results

Individual lifetime fatal cancer risks were calculated as:

$$ILR = RI \times RF(EP) \quad (6)$$

where:

ILR = individual lifetime fatal cancer risk
RI = daily radium intake during the exposure period (pCi/day)
RF(EP) = risk factor modified by exposure period (lifetime risk per pCi/day)

Individual lifetime risks were calculated separately for ^{226}Ra and ^{228}Ra and then summed. Results are presented in Table 5. Median individual lifetime fatal cancer risks were 9.8×10^{-6}, and 95th percentile risks were 5.4×10^{-5}. Median risks from ingestion of fish caught away from open bay discharges were 3.8×10^{-7} , and 95th percentile risks were 1.3×10^{-6}.

Table 5. Risk for ingestion of radium in fishes

	Individual Lifetime Fatal Cancer Risk	
	Fish Near Platforms	Fish Away From Platforms
Mean	1.7×10^{-5}	5.0×10^{-7}
Median	9.8×10^{-6}	3.8×10^{-7}
Standard deviation	2.3×10^{-5}	4.4×10^{-7}
5th percentile	1.8×10^{-6}	8.1×10^{-8}
95th percentile	5.4×10^{-5}	1.3×10^{-6}

PROBABILISTIC RISK ASSESSMENT FOR LEAD

Intake

Background Intake. Lead is ubiquitous in the environment, and children, in particular are exposed to lead through a number of pathways. Important sources of lead exposure to children include food, drinking water, air, soil and dust. Exposures from specific sources are added to background exposures experienced by children and increase the probability of exceeding blood lead levels of concern identified by USEPA (Section 9.3). This analysis used the assumed age-specific background intakes for children ages one-half to 7 years described in USEPA (1994).

Recreational Fishing. Lead intake was estimated for children eating fish caught near platforms, and away from platforms. Distributions of lead intake in recreationally caught fish were calculated as:

$$PbI = FI \times [Pb]_{fishes} \tag{7}$$

where:

PbI	=	lead intake (μg/day)
FI	=	intake of fish (g/day) for children of recreational fishermen
$[Pb]_{fishes}$	=	concentration of lead in fish (μg/g)

Intake estimates were divided into groups (27 groups for fish caught near platforms, 13 groups for fish caught away from platforms) and the midpoint of the intake range for each group was used to represent the intake of lead ingested in recreationally caught fish. Daily lead ingestion rates in food were calculated for each year of life to age 7 by adding the background intake for that age (USEPA, 1994) to the estimated intake from recreationally caught fish. This approach slightly overestimates lead intake in food because recreationally caught fish would actually replace a small amount of lead in fish and meat obtained from other sources.

Dose-Response Assessment

Lead exposure can affect a number of systems, including the brain, hematopoietic system, cardiovascular system and the developing fetus (Derosa *et al.*, 1991). Extensive data are available to link low-level lead exposure of young children to deficits in neurobehavioral-cognitive performance (Rosen, 1995). Federal agencies and advisory groups including USEPA (USEPA, 1986, 1990a), have defined a level of concern for children as a blood lead level ≥10 μg/dl (Rosen, 1995; USEPA, 1994). USEPA has developed a biokinetic/uptake model for lead (UBK Model; USEPA, 1994) that relates intake in food, air, water and soil to the probability of exceeding a blood lead level of 10 μg/dl (BL>10). This analysis used this probability as the metric for risk from ingestion of lead in fish.

Risk Characterization

The UBK model (USEPA, 1994) was used to estimate the blood lead concentration and the probability of BL>10 for each level of intake of recreationally caught fish. All other UBK model parameters reflected USEPA (1994) estimates of average background intakes.

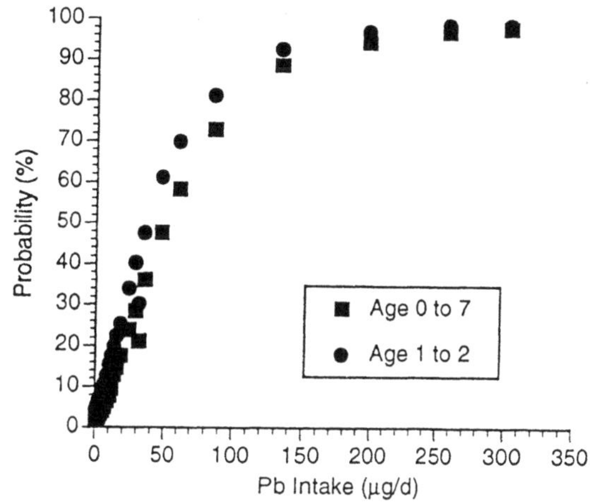

Figure 3. Relationship between intake of lead in recreationally caught fish and probability of exceeding 10 μg/dl blood lead for two age groups.

Blood lead levels were estimated for two age groups: age 1-2 years when they are at their maximum level for a given intake; and averaged over age 0 to 7 years. Figure 3 shows the relationship between the intake of lead in recreationally caught fish and the probability of BL>10. For comparison, background intakes of lead are associated with a probability of BL>10 of 1.56% for age 0-7 years and of 4.42% for age 1-2 years.

The total risk (that is, the probability of BL>10 across all predicted intake rates) was calculated as:

$$TP = \Sigma\, P(PbI) * P(BL>10 \mid PbI) \qquad (8)$$

where:

TP	=	total probability (%) of exceeding a blood lead level of 10 μg/dl
P (PbI)	=	probability (%) of a given lead intake in recreationally caught fish
P (BL>10 ∣ PbI)	=	probability (%) of exceeding a blood lead level of 10 μg/dl for a given intake of lead in fish

Results

The frequency distribution for the probability of BL>10 is given in Figure 4 for fish caught near platforms and fish caught away from platforms. Risks for most intake rates in the distribution for fish caught near platforms only slightly exceeded the background risks of 1.6% and 4.4 % for 0-7 years and 1-2 years, respectively. Fifty-two percent of the predicted intake rates at platform sites had risks of BL>10 less than 2.7% and 6.4% for age 0-7 and 1-2 years, respectively. Seventy-four percent of the predicted intake rates at platform sites had risks of BL>10 less than 5% and 10.6% for age 0-7 and 1-2 years, respectively. Figure 5 shows the cumulative frequency distribution of the risks for ingestion of fish caught near platforms.

Table 6 shows the total probability of BL>10 for fish caught near platforms, fish caught away from platforms and background intakes. Risk from ingestion of fish caught away from platforms only slightly exceeded risks from background intake of lead. The total probability of BL>10 from intake of lead in fish caught near platforms was approximately 3-5 times higher than risks from background intakes of lead.

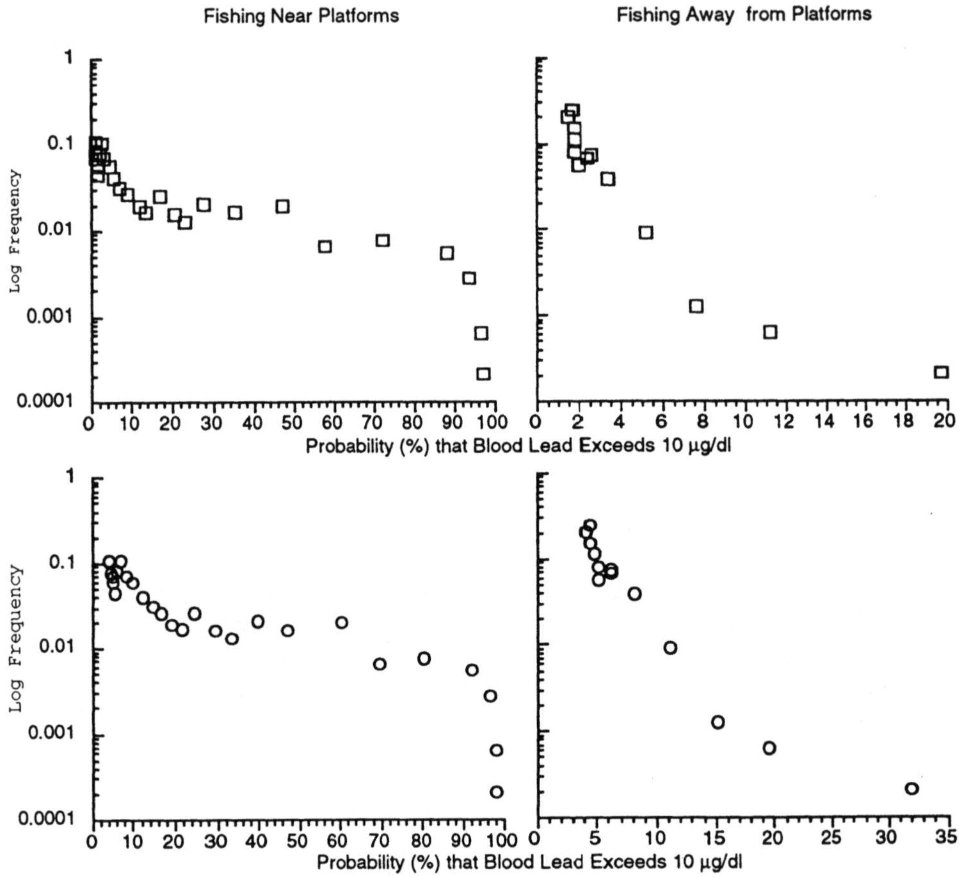

Figure 4. Frequency distribution of the probability of exceeding 10 μg/dl blood lead from ingesting fish caught near platforms and away from platforms. (□ = 0 to 7 years; O = 1-2 years; note differences in x axes).

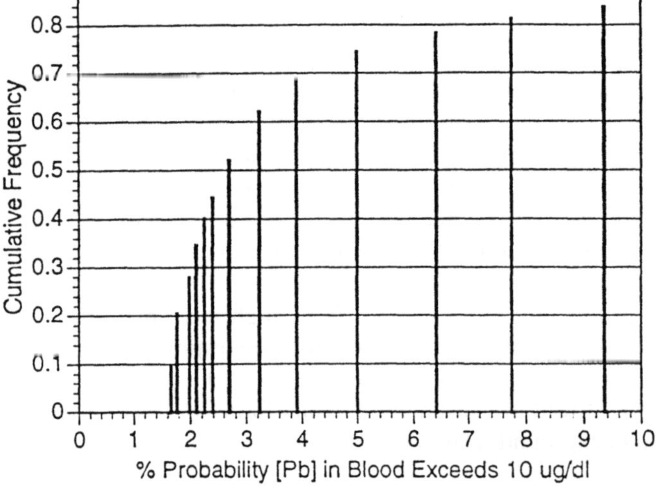

Figure 5. Cumulative frequency distribution of the probability of exceeding 10 μg/dl blood lead from ingest-ing fish caught near platforms (□ = 0 to 7 years; O = 1-2 years; 70-80% of values represented, up to 10% probability).

Table 6. Total probability (%) of exceeding a blood lead level of 10 μg/dl

	0-7 years	1-2 years
Fish Near Platforms	7.8	14.3
Fish Away From Platforms	2.0	5.2
Background	1.6	4.4

UNCERTAINTIES

Uncertainties associated with this analysis include: uncertainty in modeling of radium and lead discharge concentrations, dilution and bioaccumulation; uncertainty in ingestion rate distribution; and uncertainty in dose-response functions. These uncertainties were included in the probabilistic risk assessments by describing each of the relevant variables as a distribution in the Monte Carlo analysis. Because of a lack of knowledge and data, a number of major conservative assumptions remain. Because risks from radium ingestion in fish are predicted to be small (less than 1×10^{-4} individual lifetime fatal cancer risk), these remaining conservatisms are not of concern for interpretation of the results for radium. Based on the analysis presented for lead, however, risks for children ingesting a large amount of fish caught near some produced water platforms may be of concern.

Two major conservative assumptions that remain in the analysis were investigated in a sensitivity analysis:

1. Discharge concentrations of lead were based on a very limited data set (15 values), with 8 values reported as "less than" the detection limit. The detection limit varied from 50 to 125 μg/l. A standard approach of using 1/2 the reported detection limit was used in the analysis. The true concentration in the effluent is unknown and may be much lower than suggested by the reported detection limits.
2. The transport model used to predict lead concentrations in water underestimates dilution at low discharge rates.

A sensitivity analysis was performed to assess the importance of these assumptions to the estimates of risk from lead in ingested fish. Additional analyses were done by either replacing "less than" values with a high estimate of background lead in sea water (1.0 μg/l); or assuming that the water concentrations of lead were overestimated by a factor of 10. The results of the sensitivity analyses are summarized in Table 7.

Table 7. Sensitivity analysis: importance of "less than" values and overestimation of lead concentrations in water on estimating the total probability (%) of exceeding a blood lead level of 10 μg/dl for ingestion of fish caught near platforms

	Probability blood lead > 10μg/dl	
Scenario	0-7 years	1-2 years
Base case[1]	7.8	14.3
Replace "less than, < "values with 1 μg/l	6.7	11.7
Lead concentrations in water divided by 10	2.1	5.2

[1] base case is the risk assessment reported in the text, "less than, <" value replaced by 1/2 their value and water concentrations estimated using the model described in the text

Results of the sensitivity analysis suggest that the two major conservatisms remaining in the assessment have an important effect on the estimate of total risk from lead ingestion in fish. Replacing "less than, <" values with an estimate of background lead concentrations reduced the total risk by a factor of approximately 1.2. Total risks were reduced by a factor of 3-4 (similar to background risks) when lead water concentrations were assumed to be overestimated by a factor of 10.

SUMMARY AND CONCLUSIONS

The tiered approach to risk assessment is a cost-effective way to provide risk managers with information needed to make decisions. A screening assessment for human health risks eliminated a number of contaminants from further consideration. Quantitative assessments were performed for two contaminants of concern identified using this screen: radium and lead.

An ecological screening assessment concluded that radium discharged to open bays in produced water is not expected to affect aquatic fish or shellfish. Contaminants of concern identified by the screening ecological risk analysis were copper, lead, mercury, and nickel. These contaminants are being assessed in a quantitative risk assessment.

Median individual lifetime fatal cancer risks for continuing open bay discharges of radium were 2.2×10^{-7}, and 95th percentile risks were 1.9×10^{-5}. These results suggest that ingestion of radium in fish caught near open bay produced water platforms does not present an important risk to human health.

Risks from ingestion of lead in fish caught near platforms exceed background risks and risks from ingestion of lead in fish caught away from platforms. Most predicted intake rates are associated with risks that only slightly exceed the risk from background intakes. Because of the conservatisms embedded in the analysis (assumptions concerning "less than" effluent concentrations, underestimate of dilution at low discharges rates) the risk from ingestion of lead discharged from open bay discharges in Louisiana appears to be small, although exposures to children eating a large amount of fish caught near the few platforms discharging measurable amounts of lead may be of concern.

ACKNOWLEDGMENTS

This work was supported by the United States Department of Energy, Office of Fossil Energy, through the Metairie Site Office with guidance from Dr. Brent Smith as project officer. Dr. Leonard D. Hamilton, Dr. Samuel Morris and Dr. Michael Rowe reviewed drafts of this paper and made many useful suggestions. The primary contractor for the field study mentioned in the text was Continental Shelf Associates, Inc. (CSA) in Jupiter, Florida. Steimle & Associates, Inc., in Metairie, Louisiana performed the recreational fishermen survey used in the analyses presented here. Work Performed Under Contract No. DE-AC02-76CH00016 with the U.S. Department of Energy

REFERENCES

Avanti Corporation, 1993. *Ocean Discharge Criteria Evaluation For The NPDES General Permit For the Western Gulf of Mexico OCS.* prepared for the United States Environmental Protection Agency, Region VI, Dallas, Texas, prepared by Avanti Corporation Vienna, Virginia.

Derosa, C.T., H. Choudhury, and W.B. Peirano, 1991. An integrated exposure/pharmacokinetic based approach to the assessment of complex mixtures, lead: a case study. *Toxicology and Industrial Health*, Vol. 7, No. 4: 231-248.

Doneker, R.L., and G. H. Jirka, 1990. *Expert System for Hydrodynamic Mixing Zone Analysis of Conventional and Toxic Submerged Single Port Discharges (CORMIX 1)*. EPA/600/3-90/012, United States Environmental Protection Agency, Athens, Ga.

Federal Register. 1991. Environmental Protection Agency, national primary drinking water regulations; radionuclides. *40 CFR* Parts 141,142, 56:138:33050.

IAEA, 1982. *Generic Models and Parameters for Assessing the Environmental Transfer of Radionuclides from Routine Releases: Exposures of Critical Groups*. Safety Series No. 57, International Atomic Energy Agency, Vienna.

IAEA, 1988. *Assessing the Impact of Deep Sea Disposal of Low Level Radioactive Waste on Living Marine Resources*. Technical Reports Series No. 288, International Atomic Energy Agency, Vienna.

Meinhold, A.F., and L.D. Hamilton, 1992. Radium concentration factors and their use in health and environmental risk assessment. pp. 293-302, *in: Produced Water*, J.P. Ray and F.R. Engelhart, eds., Plenum Press, New York.

Meinhold, A.F., S. Holtzman and L. D. Hamilton, 1995. Human health risk assessment for radium discharged offshore in produced water, *Proceedings of the SPE/EPA Exploration and Production Environmental Conference*, Houston Tx.

Middleditch, B.S., 1984. *Ecological Effects of Produced Water Discharges From Offshore Oil and Gas Production Platforms*. American Petroleum Institute, Washington, D.C.

NCRP, 1991 *Effects of Ionizing Radiation on Aquatic Organisms*. NCRP Report No. 109, National Council on Radiation Protection and Measurements, Bethesda, Md.

Rosen, J. F., 1995. Adverse health effects of lead at low exposure levels: trends in management of childhood lead poisoning. *Toxicology* 97:11-17.

Steimle & Associates, Inc. in preparation. *Synthesis of Seafood Catch, Distribution, And Consumption Patterns in the Gulf of Mexico Region*. Draft Report prepared for the United States Department of Energy, New Orleans, La.

Stephenson, M.T., 1992. A Survey of Produced Water Studies, in: J.P. Ray and F.R. Engelhardt, eds., *Produced Water Technological/Environmental Issues and Solutions*. Plenum Press, New York.

Strenge, D.L. and S.R. Peterson, 1989. *Chemical Data Bases for the Multimedia Environmental Pollutant Assessment System (MEPAS): Version 1*. PNL-7145, Pacific Northwest Laboratory, Richland, Washington.

USEPA, 1986. *Air Quality Criteria for Lead*. Office of Research and Development, Office of Health and Environmental Assessment, Environmental Criteria and Assessment Office, United States Environmental Protection Agency, Research Triangle Park, N.C., EPA 600/8-83-028AF, BF, DF. EPA/602/8-83/028A.

USEPA, 1989a. *Risk Assessment Guidance for Superfund, Volume I, Human Health Evaluation Manual (Part A) Interim Final*. Office of Emergency and Remedial Response, EPA/540/1-89/002, United States Environmental Protection Agency, Washington, D.C.

USEPA, 1989b. *Assessing Human Health Risks From Chemically Contaminated Fish and Shellfish: A Guidance Manual*. Office of Marine and Estuarine Protection, United States Environmental Protection Agency, Washington, D.C., EPA-503/8-89-002.

USEPA, 1990. *Exposure Factors Handbook*. Office of Health and Environmental Assessment, United States Environmental Protection Agency, Washington, D.C., EPA/600/8-89/043.

USEPA, 1992. *Framework for Ecological Risk Assessment*. Office of Research and Development, United States Environmental Protection Agency, Washington, D.C., EPA/630/R-92/001.

USEPA, 1994. *Guidance Manual for the Integrated Exposure Uptake Biokinetic Model for Lead in Children*. Office of Solid Waste and Emergency Response, United States Environmental Protection Agency, Washington, D.C., EPA/540/R-93/081, PB93-963510.

USEPA, 1995a. *Water Quality Benefits Analysis for the Proposed Effluent Guidelines for the Coastal Subcategory of the Oil and Gas Extraction Industry*. EPA821-R-95-001, Office of Water, United States Environmental Protection Agency, Washington, D.C.

USEPA, 1995b. *Data from the EMAP-Estuaries Program Level database, Louisianian Province, 1991-1992*. United States Environmental Protection Agency, Office of Research and Development, Narragansett, Rhode Island.

Section VII

TECHNOLOGY FOR MITIGATION

37

SESSION SUMMARY

Kjell Lohne

STATOIL

Technologies for separation of suspended oil from produced water have improved dramatically over the last 10 - 15 years. Until the mid 80's oilfield produced water treatment systems, both onshore and offshore, were generally dominated by various forms of plate separators and induced gas flotation systems. The now well-known de-oiling hydrocyclone was first developed at the University of Southampton in the early 80's. This technology has since been further improved and is now available in various models from a number of different equipment suppliers. Although other technologies have also been further developed and refined in this period and are used in many cases, the de-oiling hydrocyclone generally represents more efficient and compact technology and is usually the preferred alternative, in particular on offshore installations. Centrifugation is also used to some extent, and on onshore facilities various forms of media filtration are sometimes used.

Now in the mid 90's, a gradually increasing focus is being directed on technologies for removal of components other than suspended oil, such as dissolved organics and heavy metals. There is also an increasing interest in technologies that can represent an alternative to treatment and surface disposal, such as produced water reinjection, downhole water shut off techniques and downhole water separation.

Produced water reinjection, which for many years was rarely seen on other than onshore facilities, has now been explored for offshore use by many oil companies and research institutions. The paper presented at this seminar by *Bakke* on the experience from produced water reinjection at BP's *Ula* field in the North Sea attracted considerable interest. The discussion of this issue in the plenary session centered around the economical aspects and the potential operational problems associated with produced water reinjection, such as loss of injectivity, reservoir souring, and scaling and corrosion both topside and downhole. It was pointed out that although BP consider they have few such problems and have reduced the operating cost due to lower consumption of seawater in the case of Ula, reinjection as a method for produced water disposal will in many cases not represent an economically attractive option. As both topsides and reservoir aspects vary substantially from field to field, no universal conclusions can be drawn and reinjection needs to be evaluated on a case by case basis. It was also pointed out that reinjection in some cases would result in an increase of energy consumption, with a corresponding increase in atmospheric pollution.

Reinjection, nevertheless, represents the only viable disposal option for many on-shore oilfields. Extensive pretreatment is often required, as was exemplified by *Huff* in his valued presentation of a fully hydrocyclone-based system for removal of both suspended oil

Produced Water2: Environmental Issues and Mitigation Technologies
Edited by M. Reed and S. Johnsen, Plenum Press, New York, 1996

and fine solids from produced water prior to reinjection at the *Lisburne* field at the North Slope of Alaska.

Removal of *dissolved organic components and heavy metals* from produced water requires much more sophisticated process technologies than those used for separation of oil droplets, and is very rarely seen offshore. An interesting paper was however presented at the seminar by *Buffagni* on the composition and treatment of produced waters from gas fields by gravity separation followed by activated carbon filtration in the Adriatic Sea.

On some onshore fields where produced water is reused for steam generation or other similar purposes, or where the water has to be discharged to fresh water surface recipients, technologies for dissolved components removal are also in use, at a considerable treatment cost. In his presentation *Khandan* reported on the experimental results from biological oxidation of produced water in New Mexico. Although the reported trials demonstrate that it is technically possible to biodegrade petroleum hydrocarbons in high salinity produced waters, the energy and hydraulic retention times required are very high.

As is correctly pointed out in the paper on dissolved component removal presented by *Turnbull*, these technologies are likely to attract more interest should the trend to use toxicity based effluent legislation continue. However, with the technologies that are available today, the removal of dissolved components from produced water can generally only be achieved at a very high capital and operating cost. In offshore areas with good recipient conditions, such as the North Sea, it is in my opinion very difficult to justify such high costs with the current understanding of environmental effects of these discharges.

The use of membrane technology to remove dispersed and dissolved components from produced water has been extensively researched and experimentally tested over the last 5 - 10 years. In the paper presented at the seminar by *Fraser*, further experimental results were presented on the use of NF and UF membranes for these purposes in Canada. These results support my general impression that there is still a long way to go before membrane technology can be successfully applied, within sound economical limits, for treatment of produced water, be it onshore or offshore. Too low flux rates, fouling problems, high energy requirements and low operational availability are just some of the problems and disadvantages associated with this technology that still remain to be resolved.

Reinjection as an alternative to disposal to surface waters has already been mentioned. *Downhole water separation* based on hydrocyclone technology and downhole injection and lift pumps is another concept currently attracting considerable interest, with research projects ongoing in both Canada and Norway. At the seminar *Lie-Andersen* presented an interesting paper on the preliminary results of the Norwegian project. To me it is clear that there are still a few practical problems with this concept that need to be resolved, such as monitoring and control, and aspects related to maintenance requirements. However, if the necessary improvements could be achieved, the concept would effectively reduce the amount of produced water reaching the surface, and as such is certainly very interesting for some applications.

As the chairman for this session of the seminar, I would like to take this opportunity to again thank those who contributed to a successful session with excellent papers and presentations, and those who contributed by their participation in the interesting plenary discussions.

38

AIR STRIPPING COMBINED WITH FID DETECTION FOR OIL-IN-WATER ANALYSIS

Arne Lund Kvernheim,[1] Kristin Eitrem Landmark,[1] Hanne M. Øren,[2] and Ingolf Caspari[3]

[1] SINTEF Oslo, Norway
[2] SAGA Petroleum AS, Stavanger, Norway
[3] Gregersen and Johnson AS, Stavanger, Norway

INTRODUCTION

Oil-in-water is analysed offshore in the North Sea for two reasons:

1. The National State Pollution Control Authority's and the Oslo - Paris Commission (OSPAR) has set the target for discharge of non polar aliphatic hydrocarbons to 40 mg/l based on IR spectrophotometric detection and freon (CFC 113) extraction. This discharge limit should be reported as a monthly average based on daily measurements.
2. Offshore analysis based on a number of different principles off-line and on-line are requested to test the efficiency of oil/water separation facilities offshore.

The main driving force to develop alternative analytical methods for oil-in-water is the ozone depleting properties of freon. Norway, together with 50 other nations, has signed the Montreal Protocol which means that the production, import, export and use of these substances are strictly regulated and that these substances will be phased out as soon as alternatives are available. For the analysis of oil-in-water a global exemption for 1996 and 1997 has been issued. This exemption use is discussed each year at the United Nations Environmental Programme (UNEP) - Technology and Economic Assessment Panel (TEAP) meetings. As yet, no international accepted alternatives are available. Because of this the EU program "Standards, Measurement and Testing", OSPAR and several national environmental protection boards has specifically drawn the attention to this area for development of a new standard.

In Norway the main use of oil and grease analytical methods is for offshore analysis. The high fraction of volatile components, especially in condensate and gas field discharges, prohibits the use of gravimetric methods (like the US-EPA 1664), or methods where a large fraction of the analyte may be lost during sample preparation and analysis. Research on SPE techniques combined with Supercritical Fluid Extraction (SFE) and IR detection has been conducted for the Norwegian State Pollution Control Agency (SPAC), the Swedish Environmental Protection Board (SNV) and STATOIL (Jensen, 1993; Jensen, 1994a; Jensen, 1994b). Further research for a standard method is presently conducted for SPAC. Presently, partners within EU are sought for a proposal within the "Standards Measurement and Testing" programme.

In order to use data from hydrocarbon analysis for feed-back control of the separation process offshore, rapid methods, or on-line methods would be beneficial. Such data could be used to tune the separators and to optimise the use of oil/water separation chemicals. So far most on-line methods has been based on: 1) Automated extraction and analysis or; 2) Direct optical methods in the water-phase. Unattended operation of these methods offshore has met limited success mainly because of vibration problems or fowling of optical surfaces.

SINTEF has been working on a new analytical concept in collaboration with SAGA Petroleum as and Gregersen & Johnson AS. This system has no moving parts and does not involve spectroscopy. The principle involved is air stripping combined with direct flame ionisation detection (FID). It is a fact that air stripping has been discussed as an efficient technique for removal of hydrocarbons from effluent waters. This has tested, in the laboratory, for produced water offshore (Fang and Lin, 1988) and for discharge water from land based oil refineries (Al-Muzaini et al., 1994). In this paper a commercial available instrument from Hartmann & Braun has been used to test this analytical principle using an off-line mode of operation. The instrument is also designed for on-line analysis in explosion hazardous environments.

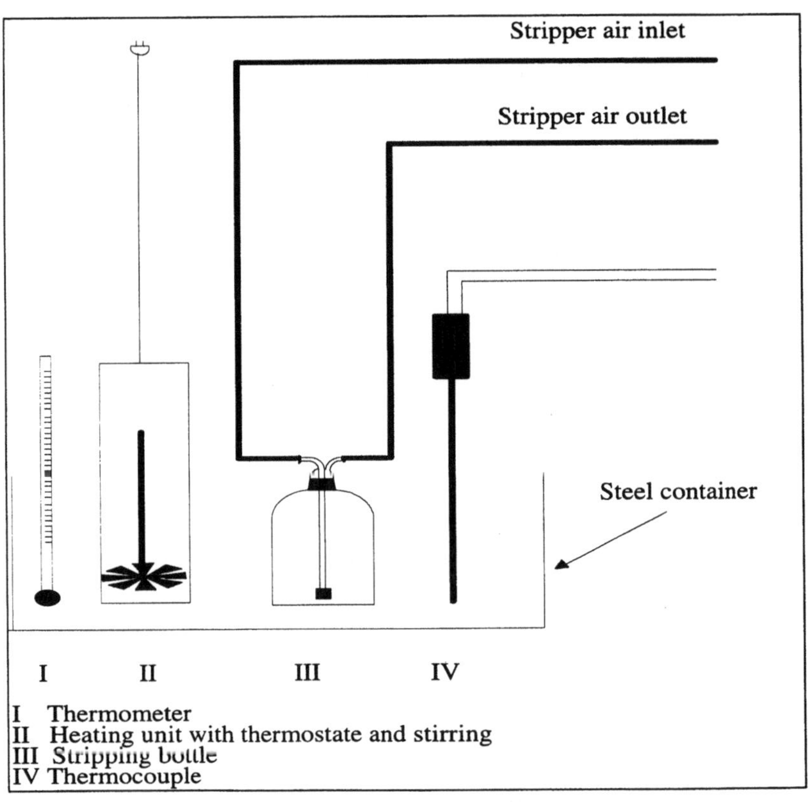

Figure 1. Sketch of the manual stripper unit.

EXPERIMENTAL

Instrumentation

The following instrument was made available by Mannesmann Hartmann & Braun AG in Germany, through the Norwegian Agent Gregersen & Johnson AS.

- COMPUR catalyzer K-83-5 U - for removal of trace amounts of hydrocarbons in the air supply.
- M & C Mikroprozessorgesteuerter temperarturregler mit Selbstoptimierung TYP 2040 - for temperature regulation and control of the COMPUR catalyzer.
- COMPUR Stripper - for air stripping of flowing water samples.
- COMPUR FID - for detection of volatile organic compounds.

To test the suitability of the analytical principles involved, a manual stripper unit was designed and used throughout this project. The air inlet was fitted with a glass-sinter end to obtain a better air bubble diffusion. The stripped compounds from the manual unit were fed into the transfer line to the FID. A drawing of this unit is found in Figure 1. The main advantage by using this manual stripper unit was that the analysis could take place in the same bottle in which the solution was made or sampled. A Intermed Heto OI TT 620 heating unit with thermostat and stirring was used in the water bath.

An IBM compatible computer (Acer 915V) with AAC-2 PC software version 7.XX and an INTAB analogue/digital interface from INTAB Interface-teknik AB was used to log the FID signal and the temperature during air stripping of the samples.

Air flows of 80 ml/min and a temperature of 25°C in the water bath were used for all experiments of oil-in-water. The signal from the FID were registrated after 10 min of airstripping.

Measurements by the Standard Method

The Norwegian Standard NS 9803 were used on the samples from the North Sea and on the laboratory made samples. This method includes freon extraction, alumina column removal of polar components and IR detection in the C-H stretch region. Samples of drainage water and produced water was analysed by SAGA Petroleum AS.

Samples

A stabilised crude oil from SAGA Petroleum as was used in this project, hereafter called SAGA oil. Information about the composition of the oil was given by SAGA Petroleum as and found in Table 1.

Because the use of air stripping methods will be dependent on the specific oil composition in the sample, no effort was made to test the method with different crude or

Table 1. Composition of crude oil used in the project

Boiling point range °C	Approximate n-alkane range	Composition in weight %
<90	< C7	10
90 - 165	C7 - C10	15
165 - 225	C10 - C12	10
225 - 320	C12 - C18	20
320 - 375	C18 - C22	10
> 375	> C22	35

oils. If the method do work with one oil composition, any oil in water mixture can be analysed provided a calibration solution can be made and that the volatility of the oil are sufficient for a FID signal to be monitored. No effort was made to analyse the composition of the compounds that was stripped from the waterphase.

No solvent could be used for making these samples. The 1 litre bottles were washed and heated to 550 °C to get rid of possible contaminants. Syringes were used to withdraw the SAGA oil from the bottle. The syringe was weighted before and after transferring the oil to the sample bottle. The difference was used as a measure of the added amount of oil. The bottles were made up to 1 litre using distillate water and closed with Teflon lined screw caps.

In addition to the samples made in the laboratory, 40 samples of drainage water and 13 samples of produced water were provided by SAGA Petroleum as. The SAGA oil was used as a calibration standard in both cases for the air stripping method. The data for oil-in-water for replicates of the received samples were given by SAGA Petroleum as measured offshore.

RESULTS AND DISCUSSION

Method Validation

The values obtained by air stripping were tested for linearity, repeatability and accuracy compared with added amounts of SAGA oil. The validation was performed at six

Table 2. Repeatability of SAGA oil on three levels (1, 50 and 500 ppm)

Added amount of SAGA oil	μl	Added amount of SAGA oil, ppm
1	0.64	1.47
1	0.50	1.98
1	0.90	2.27
1	0.89	1.22
Average	0.73	1.74
% RSD	26.8	27.5
10	8.28	16.44
10	8.29	18.63
50	41.24	124.3
50	41.12	138.0
50	41.24	135.9
50	41.10	120.6
Average	41.18	129.7
% RSD	0.18	6.61
100	83.11	285.25
500	466.85	1383
500	460.19	1473
500	462.01	1473
500	461.82	1376
Average	462.72	1426
% RSD	0.62	3.77
1000	877.3	2306

concentration levels with 4-6 replicates at the 1.0; 50 and 500 ppm levels and one analysis at the 10; 100 and 1000 ppm levels.

In Table 2, the results are listed. Four replicate samples of the oil at three levels were measured for estimation of standard deviation. Relative standard deviation (RSD) was found to be: 27.5% at 1.0 ppm; 6.6% at 50 ppm and 3.8% at 500 ppm concentration levels respectively. Relative standard deviation was found to decrease with increasing concentrations up to 500 ppm.

Accuracy (recovery) was not measured at 1.0 ppm due to the low repeatability in added amounts of oil. Accuracy was estimated to 84% and 112% on the 50 ppm and 500 ppm concentration levels respectively. These numbers should be compared with the accuracy of 70-80% and repeatability of 9-10% RSD on the 50 ppm and 500 ppm levels obtained with the NS 9803 method for the same oil.

The linearity of the method in the 1-1000 ppm range were tested by plotting added amounts of oil versus the air stripper signal and calculating the correlation coefficient. In Figure 2 the correlation is displayed.

The regression coefficient was found to be 0.997. The correlation as seen in Figure 2 is highly influenced by the sample point at about 900 ppm. A non-linear fit may be used to obtain even higher correlation coefficients in the 0-1000 ppm range. The same correlation coefficient was obtained when results from the NS 9803 method were plotted versus the air stripper signal.

An obvious problem related to these samples was that the oil is not soluble in water. In addition to the components soluble in the water phase, some oil was found as "lumps" dispersed in the water and some of the oil got stuck on the glass walls. These effects are expected to have an influence on the analyses especially at high and low levels of oil when using the air stripping technique with the manual stripper unit.

The robustness of the method was tested at 20 ppm oil levels in a few experiments using additions of methanol, acetic acid and sodium chloride. The effect of pH adjustments and temperature were also tested.

If temperature variations is expected, this parameter has to be included in the calibration model. This is because the mass transfer from the water to air increases with increasing water temperature.

The effect of addition of volatiles like methanol and acetic acid will be to increase the FID signal. In the work discussed in this paper, a 10 min delay was used before the signal was registrated, and the effect of methanol and acetic acid additions were small unless added in g/l amounts. In on-line applications, absorbents for polar components in the transfer line may be needed if fluctuations in the methanol or acetic acid content is expected. In the acetic

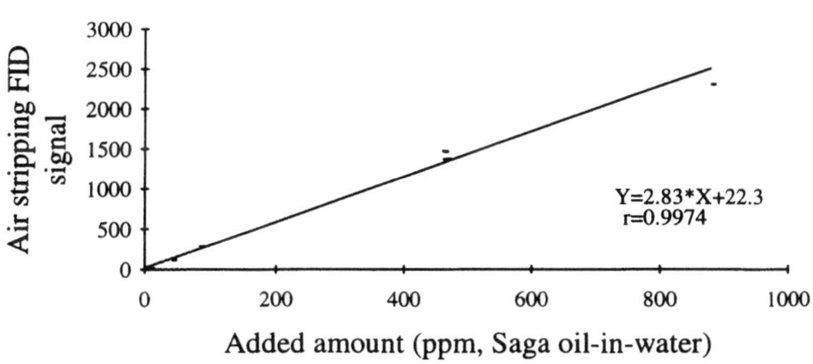

Figure 2. Correlation between added amounts of oil and air stripping FID for SAGA oil.

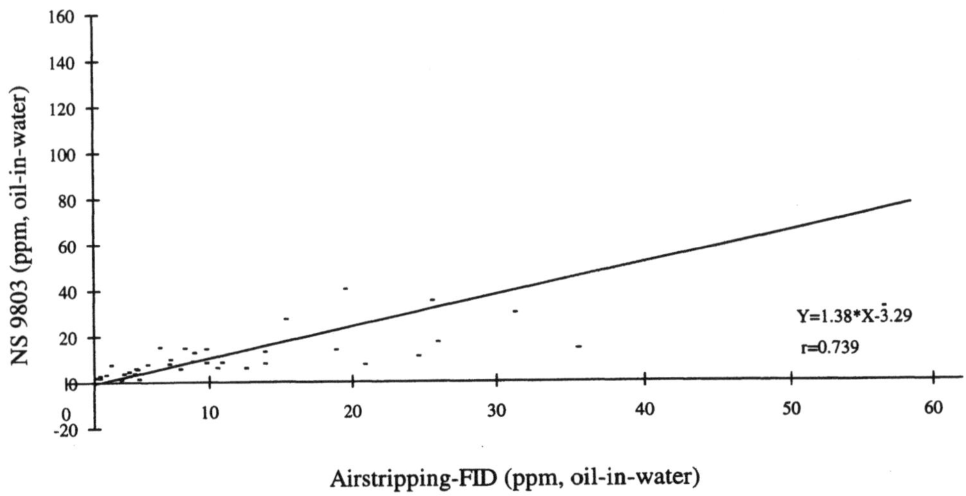

Figure 3. Correlation with NS 9803 for 40 samples of drainage water.

acid case, the addition of a base like sodium hydroxyde, will decrease the air stripper FID signal because the "free" acid will not longer be present.

The effect of salt addition (and other ionic substances like NaOH) is expected to be a "salting out" effect. Higher density of the water phase, gives phase separation and higher tendencies for wall adherence. Thus a lower air stripper FID signal is expected.

Only a few experiments were completed to verify these assumptions. One of the most important tasks in on-line implementation of this analysing system will be robustness testing and interferences detection and elimination. These tests must be conducted for each discharge effluent, taking into account the compounds present and their range of variation.

Analysis of Drainage and Produced Water

Offshore samples of drainage water were analysed using a four point calibration curve. SAGA oil in water were used for calibration as discussed above. A total of 40 samples of drainage water have been analysed with the air stripping FID method. The aim of the analysis was to compare the results with NS 9803.

For about 20 of 40 samples deviations of less than 30 % between NS 9803 and air stripping were found. For 30 of the drainage water samples deviations less than 50% difference were found. Six samples showed a deviation between 50-100% and five samples deviated more than 100%. In Figure 3 a comparison of all the data are made. The correlation coefficient was measured to 0.739.

Air stripping FID is likely to overestimate the oil content at low ppm values compared to NS 9803. The five samples which deviate more than 100% contained 1-14 ppm using the NS 9803 method and 4-36 ppm with air stripping FID. The probable cause for this deviation is that other constituents in the water phase, like polar acids, contributes to the signal, or that the actual composition of the hydrocarbons in the water is different from the ones in SAGA oil. At higher oil-in-water levels, the air stripping FID method gave a minor underestimation compared with NS 9803. Some of these differences could probably be eliminated with a non-linear calibration model. The probable cause for this deviation is higher amounts of dispersed oil which is less accessible to volatilisation by air stripping.

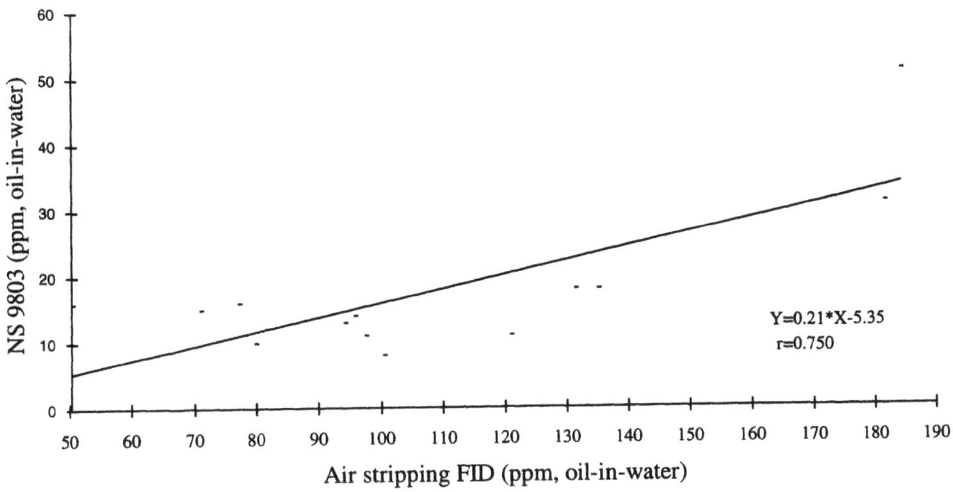

Figure 4. Correlation with NS 9803 for 13 different samples of produced water.

When comparing the NS method and the air stripping data one should take into account that the samples analysed with the two methods were not identical, but sampled one after the other. Another point to make is that both methods is subjected to uncertainty. If this uncertainty were known, error bars in both X and Y directions could be included.

Offshore samples of produced water were analysed in the same manner as the drainage samples. A total of 13 samples were analysed. The correlation coefficient between air stripping and the NS 9803 for 13 different samples of produced water was measured to 0.750 as displayed in Figure 4.

In the case of produced water, air stripping FID signals were significantly higher than results obtained with NS 9803 with an average factor of 7.

Two possible reasons could be thought of for this behaviour:

1. The calibration oil may not be representative for the produced water samples and therefore not suitable as calibration standard
2. The results were biased by interfering compounds.

In order to test the last hypothesis GC-MS analysis of selected produced water samples were performed. The aim was to reveal which type of compounds that contributed to these differences if any.

The analysis revealed that the samples, beside of the hydrocarbon content, contained mainly polar compounds such as phenols, benzoic acids and carboxylic acids. One of the produced water samples, was studied more in detail. Polar and volatile components were shown to contribute to the high ppm "oil-in-water" measured with air stripping FID. Addition of a base like NaOH prior to air stripping analysis, could probably eliminate the interfering effect from the acidic volatiles.

To test the hypothesis of interfering acidic substances, five new produced water samples were analysed in duplicate at pH 2 and pH 10 respectively. The idea was that the acidic compounds would be non-volatile in alkaline water.

Addition of a base like NaOH prior to air stripping analysis, reduced the air stripping FID signal by about 20-40% compared with acidic samples. It is therefore concluded that acidic interfering substances can be removed from the analysis by addition of a base.

The largest contribution to the high values obtained for the produced water samples is probably the calibration standard, giving erroneous results when comparing the results directly with added amounts of oil and with the drainage water samples results.

The normal way to calibrate the air stripping FID signal with the NS 9803 oil-in-water method on "real world samples" will be to do parallell analysis over a period of time. In addition to this, zero water and a stable one point calibration solution containing a soluble volatile compound (eg. an alcohol) should be used. In this way a correlation between a stable calibration solution and the oil-in-water content is established in such a way that during normal operation, the stable calibration solution is used. At abnormal situations, a manual sampling system could be activated, and manual analysis with the existing standard method performed.

When used in an on-line fashion with on-site calibration and correlation with the valid standard method, air stripping has the potential of becoming a rapid oil-in-water detection system. The signal output could be used both for separation process tuning purposes and for oil-in-water discharge level reports.

SUMMARY

The aim of the project was to demonstrate air stripping combined with FID detection as a possible principle for "solvent free" analysis of produced water and drainage water offshore. The principle was tested in the laboratory in an off-line mode. The results were compared with the Norwegian Standard for oil-in-water analysis based on freon extraction and IR detection.

The airstripping method, using a commercial available COMPUR system from Hartmann & Braun, was found to give linear responses in the 1-1000 ppm range using direct additions of oil-in-water, when used in an off-line mode. Correlation coefficients was better than 0.99.

Repeatability was measured at three levels. Four replicate samples were measured for estimation of standard deviation (RSD) and found to be: 27.5% at 1.0 ppm; 6.6% at 50 ppm and 3.8% at 500 ppm concentration levels respectively.

Accuracy was estimated to 84% and 112% on the 50 ppm and 500 ppm concentration levels respectively. These numbers should be compared with the accuracy of 70-80% and repeatability of 9-10% RSD on the 50 ppm and 500 ppm levels obtained with the NS 9803 method for the same oil.

When using "SAGA oil" as a calibration standard, 35 out of 40 drainage water samples gave deviations less than 100% between airstripping and the standard method. Correlation coefficients better than 0.7 was found for the complete data set.

Used on produced water samples, results 3-13 times above the amounts found by the standard method were noticed. About 20-40% of this deviation could possibly be contributed to interferences from acidic components, and could be corrected for by pH adjustments. Correlation coefficients of 0.7 was found for all 13 samples of produced water. The results indicated that the produced water samples were different from the drainage water samples and should be calibrated with a different calibration standard.

The conclusion is that airstripping is a very promising candidate for off-line and possibly on-line oil-in-water analysis offshore.

ACKNOWLEDGMENTS

The authors wish to thank Mannesmann Hartmann & Braun AG for making their instrument available for this work.

REFERENCES

Al-Muzaini, S., Khordagui, H. and Hamouda, M.F., 1994, Removal of VOCs from refinery and petrochemical wastewaters using dissolved air flotation, *Wat. Sci. Tech.*, 30 (3):79-90.

EPA method 1664: N-Hexane Extractable Material (HEM) and Silica Gel Treated N-Hexane Extractable Material (SGT-HEM) by Extraction and Gravimetry (Oil and Grease and Total Petroleum Hydrocarbons). EPA-821-B-94-004, Issued October 1994.

Fang, C.S. and Lin, J.H., 1988, Air Stripping for Treatment of Produced Water, *J. Petroleum Technology*, May 1988.

Jensen, L., 1993, Utvikling av Nordisk Standard for analyse av olje og fett i vann basert på superkritisk fluid ekstraksjon, Forprosjekt, del 1, SINTEF report STF27 F93095 to SFT (SPAC). In Norwegin.

Jensen, L., 1994a, Utvikling av Nordisk Standard for analyse av olje og fett i vann basert på superkritisk fluid ekstraksjon, Forprosjekt, del2, SINTEF report STF27 F94033 to SNV. In Norwegian.

Jensen, L., 1994b, Alternativer til superkritisk ekstraksjon for analyse av olje i vann, SINTEF report STF27 F94022 to STATOIL. In Norwegian.

Kvernheim, A.L. and Landmark, K., 1995, Air stripping combined with FID detection for oil-in-water analysis, SINTEF Report STF 27-A95015 to SAGA Petroleum as, ISBN 82-595-8912-5.

Norwegian Standard NS 4753, 1980, Vannundersøkelse - Bestemmelse av olje og fett - Infrarødspektrofotometrisk metode, Norges Standardiseringsforbund, Adapted from Svensk Standard SS 02 81 45 (1979). In Norwegian.

Norwegian Standard NS 9803, 1993, Vannundersøkelse - Bestemmelse av olje i vann - Infrarødspektrofotometrisk metode, Norges Standardiseringsforbund, Adapted from Svensk Standard SS 02 81 45, utgave 3, 1991. In Norwegian.

OLF Environmental Programme, Phase II, Summary report, March 1993.

Simms, K., Zaidi, A. and Bhargava, O., 1992, Protocol for determining oil and grease in produced waters, In: Produced water, Technological/Environmental Issues and Solutions, Proceedings from the 1992 International Produced Water Symposium in San Diego, CA, USA, Ray, J.P. and Engelhardt, F.R. (Eds). ISBN 0-306-44358-9. Plenum Press, NY, USA.

39

DISSOLVED COMPONENT REMOVAL FROM OILFIELD WATERS

R. W. Turnbull and S. J. Tulloch

Orkney Water Test Centre
Environment and Resource Technology Ltd.
Flotta, STROMNESS
Orkney KW16 3NP

INTRODUCTION

The removal of dissolved organic components from produced water is recognised as being a complex problem. This is primarily because the number (and nature) of components is vast and highly variable from field to field.

Several comprehensive studies have been carried out to investigate the suitability, and likely performance, of equipment designed to remove dissolved components from produced waters offshore. At least one major study, such as that prepared for the Norwegian Oil Industry Association (OLF), concluded that there was currently no technology commercially available, capable of removing or breaking down all the probable organic components of produced water, that could be cost effectively re-engineered for offshore usage. In addition it was concluded that whilst many of the technologies identified appear suitable for purpose, extensive pilot scale testing of the units was still required to validate performance and reliability. OWTC is presently carrying out a study to compare the performance of a range of commercial technologies for use on oilfield effluent waters.

At present operators in the North sea are required to reduced the dispersed component of produced water prior to discharge. Currently this limit stands at 40 mg/l. Analysis of the dissolved components present in produced waters reveal that the levels are considerably higher than the dispersed hydrocarbons. Increasing concern for the environmental effect of these components has meant that future legislation may restrict the 'total' oil discharges ie, the dispersed *and* dissolved components, In addition it is possible that toxicity based effluent limits may be introduced to regulate discharges, rather than the standard oil in water extraction techniques used at present.

In terms of environmental effect dissolved components can have on the toxicity of produced water, the impact can vary considerably depending on both the concentrations and types of components present. These can also vary depending on whether the field is an oil or gas producer. Commonly the dissolved components present in produced water consist of a range of aliphatic, aromatic and polar components. Typical concentrations from oil and

Produced Water2: Environmental Issues and Mitigation Technologies
Edited by M. Reed and S. Johnsen, Plenum Press, New York, 1996

Table 1. Dissolved components in produced water

	Concentration - Oil Field		Concentration - Gas Field	
Components	Typical - mg/l	Range - mg/l	Typical - mgl	Range - mg/l
Aliphatics <C5	1	0-6	1	0-6
Aliphatics >C5	5	0-30	10	0-60
Aromatics				
BTX	8	0-20	25	0 - >50
Naphthalenes	1.5	0-4	1.5	0-4
Polar Compounds				
Phenols	5	1-11	5	0-22
Fatty Acids	300	30-800	150	0-500

gas field waters are given in Table 1 (source: OLF Environmental Programme Project D01 Treatment of Produced Water).

It is acknowledged that in most offshore situations the volume of produced water being discharged has a limited effect on the surrounding environment due to the high mixing and dilution rate. Nevertheless in shallow water, low current or confined waterways the potential impact of produced water discharges is likely to increase. In such a situation it is likely that action will therefore have to be taken to control the impact. Since the treatment of produced water is aimed at improving the environmental 'quality' of the discharge, the effectiveness of any treatment system must be judged in terms of the environmental benefits achieved, rather than simply on the arbitrary reduction in concentration of specified components of the discharge.

Different methods for the removal of dissolved organics will remove different components, such as BTX's, naphthalenes, and phenols, with different efficiencies, resulting in discharged effluents with a potentially wide range of compositions. The potential environmental effects of different compositions is often difficult to predict with confidence, since the cumulative toxicity of the residual oil components will depend on knowledge of the concentrations of these components and on literature values for their toxicity. Consequently the real environmental benefits of different treatment methods are often difficult to quantify and compare.

The quantification of environmental benefits can however be rapidly and reliably. assessed by means of 'end-of-process' toxicity tests, in which test organisms can be exposed to the different mixes of components discharged from different treatment systems. This approach needs no assumptions about the effects or interactions of different components, and therefore provides a high degree of confidence in the estimation of treatment efficacy. If used in combination with chemical analysis, 'end-of-process' toxicity testing can also be used as a means of establishing a calibration standard for future treatment systems or for modifications to existing systems. The primary deliverable from 'end-of-process' toxicity testing is therefore a direct indication of the reduction in toxicity achieved by a particular treatment process, which can be compared with the measured reductions in component concentrations in order to determine the extent to which actual match assumed benefits.

As detailed earlier Orkney Water Test Centre is presently conducting a 12 month project to assess the performance of various technologies designed to remove dissolved organic components from oilfield waste waters. This paper presents information on some of the competing technology types available today for such applications, and considers what treatment applications they may be used in within the oil and gas industry. Some of the more common applications requiring effluent treatment are given in Table 2.

Table 2. Oil and gas industry effluent discharges

Effluent Source	Typical Dissolved Components	Typical Volumes (m3/day)
Produced Water	Aliphatics, Aromatics, Phenols	> 10,000
Oil Field	Fatty Acids	< 30,000
Produced Water	Aliphatics, Aromatics, Phenols	
Gas Field	Fatty Acids, (alcohols/gycols, amines)	< 100
Gas Terminal Onshore	Aromatics, Alcohols/Glycols, Amines	< 10
Oil Terminal Onshore		
Ballast	Aliphatics, Aromatics, Phenols	> 25,000
	Fatty Acids	< 60,000
Desalter	Aliphatics, Aromatics, Phenols	> 100
	Fatty Acids	< 700

TECHNOLOGY TYPES

A considerable amount of published data exists on technologies designed for the removal of dissolved organic components. Of the different types available only stripping systems have been developed to the point where they could readily be installed offshore with a degree of confidence. Several pilot scale trials have also been carried out on stripping systems and their performance is now well understood.

This paper concentrates on technologies which have been identified as offering potential for the treatment of oilfield effluent waters, but which have, so far, not been tested extensively for this application. Specifically the technologies detailed in this paper are as follows:

- Adsorption Systems/Absorption (Solvent Extraction)
- Oxidative Systems - Wet Air -UV/Ozone/Peroxide
- Pervaporation Systems

A summary table detailing the removal efficiencies, and ultimate fate, of the dissolved components is given in Table 3. Brief technical details on each of the above systems, together with advantages and disadvantages of their use, are given below and are summarised in Table 4.

ADSORPTION/ABSORPTION (SOLVENT EXTRACTION) SYSTEM

Adsorption of dissolved organic components using activated carbon is a physical process in which the compounds adhere to a porous carbon particle's surface or are trapped within its pores. Such a system for removing dissolved organics from produced water would typically have a moving bed of granular activated carbon which could be continuously removed from the bed for regeneration by wet air oxidation or other means. There are some

Table 3. Typical removal capability and ultimate fate for dissolved components

Technology	Components Removed	Ultimate Fate
Absorption/Adsorption Systems	Variable depending on medium	Disposal or Recovery
Wet Air Oxidation	All organic components destroyed	H_20 and $C0_2$
UV/Ozone/Peroxide Oxidation	All organic components destroyed	H_20 and $C0_2$
Pervaporation Systems	Variable depending on membrane	Disposal or Recovery

Table 4. Brief technical information on dissolved component removal systems

Technology	Advantages	Disadvantages	Removal Efficiency
Absorption/Adsorption Systems	Easily understood	Regeneration or disposal of materials	Up to 100% for specific components
Wet Air Oxidation	Potential for high efficiency	Energy requirements Initiation Concentrations	Theoretically 100%
UV/Ozone/Peroxide/ Oxidation	Potential for high efficiency	Reactive chemicals Controllability	Theoretically 100%
Pervaporation Systems	Selective removal	Selective removal Membrane damage	Up to 100% for specific components

concerns about the ability of such a system to operate in the presence of dispersed oil and suspended solids. Again such systems are already in use in a range of onshore applications, though it does not appears to have yet been tested offshore. Modified carbon adsorbents, in pellet form, are also available which claim to offer improved removal efficiencies compared with traditional activated carbons. Polymer based absorbents are now commercially available and are finding application in the oil and gas industry for effluent treatment. The polymers can be tailored to remove either dispersed or dissolved components. A novel feature of one polymer absorbent is that the polymer particles contain a non-volatile extraction liquid within the pores of the polymer. The extractant absorbs the dissolved hydrocarbons during treatment and thus the process could be termed solvent extraction. Both activated carbon and polymer processes require regeneration once the adsorbent or absorbent reaches capacity. Typically steam stripping is used. This obviously transfer to volatile hydrocarbons to the vapour phase which may result in the requirement for further downstream treatment. Absorbent and adsorbent processes offer relatively simple treatment for effluents with efficient removal of volatile hydrocarbons. They do not however result in complete removal of destruction of the hydrocarbons, but simply a transfer to another phase.

OXIDATION SYSTEMS

Wet Air

In a wet air oxidation system, organic and inorganic contaminants are oxidised whilst remaining in the aqueous phase at elevated temperatures (typically 150-300°C and 15-150 bar). The presence of the liquid modifies the oxidation reactions, enabling them to occur at considerably lower temperatures than during open flame combustion. Where a high degree of oxidation occurs (at higher temperatures and pressures), hydrocarbons are broken down to carbon dioxide and water. At lower degrees of oxidation easily degraded low molecular weight organics such as acetic acid are formed. The process is already used in industrial, municipal and refinery discharge application. Wet air systems offer the advantage that all hydrocarbons present are removed by destruction $C0_2$ and H_20. Potentially, therefore, they offer a single stage treatment for effluents with destruction of both the dispersed and dissolved components. The utility requirement for the systems is, however, relatively high for high volume throughputs.

Uv/Peroxide/Ozone Oxidation Systems

Oxidative systems are widely used for pollution control in a variety of applications. The systems offer a similar end result to wet air oxidation in that they are designed to break down the contaminants into carbon dioxide and water, the difference being that a catalyst is used to promote the reaction rather than high temperature and pressures. Systems are also available which use chemical oxidising agents, such as hydrogen peroxide (in the presence of UV light), to break down the process contaminants. As with the wet air oxidation systems catalytic oxidation systems potentially offer single stage treatment through removal of both dispersed and dissolved hydrocarbons. Catalytic systems can also operate with much lower levels of hydrocarbon present in effluent, whereas wet air systems tend to require a high organic loading in the effluent to promote the reaction. The systems do have a relatively high utility requirement. In particular the use of peroxide may not be attractive for use offshore. In some countries ozone is now considered as suspected carcinogen which may restrict utilisation.

PERVAPORATION SYSTEM

Pervaporation is a membrane process in which a phase change takes place across the membrane. Unlike other membrane filtration processes that use feed side hydraulic pressure as the driving force for mass transfer, the difference between feed side vapour pressure and permeate side partial pressure is what drives the separation. The selective polymer absorbs the desired species, which is then desorbed by evaporation on the permeate side of the membrane. An elegant feature of the system is that the trace contaminant (oil) passes through the membrane whilst most of the water does not. The composite membrane material can be designed to have specific separation properties depending on the desired separation. Systems are now commercially available for the removal of VOC's and other organics from waste water streams. Pervaporation systems are virtually untried in oil and gas field situations so their operability and efficiency is largely unknown. If their performance is subject to the same constraints as traditional membrane systems the it is likely that such systems will require careful operation if they are to operate successfully.

CONCLUSIONS

At present there are a few applications in the offshore oil and gas industry where equipment for removing dissolved organic components has been actively tested. Use of the technology onshore is, however, meeting with increasing interest, particularly at refineries. Legislative requirements are currently centred around the removal of dispersed oil from oilfield waste waters. Should this be changed to reflect removal of both the dispersed and dissolved components then it is likely that there will be a further increase in interest for these technology types.

Several onshore facilities in both the US and UK are known to be using toxicity based effluent analysis, rather than 'traditional' oil and grease analysis, to reflect effluent water quality. Should this trend continue then this again is likely to result in increased demand for the use of such technology. The performance of many of these technology types has still to be assessed through comprehensive testing to assess both fitness for purpose and cost effectiveness. Potentially some of the technologies offer 'single step' processing for effluent streams both offshore and onshore. It is hoped that the work carried out at OWTC, and elsewhere, will provide better understanding of how these are likely to perform in the field.

REPORT ON THE START-UP OF A UNIQUE HYDROCYCLONE-BASED SYSTEM FOR TREATING PRODUCED WATER

Craig Dotson,[1] Richard Huff,[1] and Steven R. Haskin[2]

[1] ARCO Alaska, Inc.
P.O. Box 100360
Anchorage, Alaska 99510-0360
[2] Krebs Engineers

INTRODUCTION

On February 22, 1995, ARCO Alaska and Krebs Engineers successfully commissioned the world's first combined solid/liquid and liquid/liquid hydrocyclone-based produced water treatment system. This hydrocyclone system was designed to treat produced water as it flows from either the slug catcher or the electrostatic treaters and can handle a combined flow up to 224,000 BPD. This is the first time that hydrocyclones have been incorporated in such a way that solids, as well as trace oil, are removed from the same flow stream.

This system initially removes coarse solids (larger than 80μ) from the produced water flow stream. The flow then continues to the de-oiler hydrocyclones, removing the oil and lightly washing the fine solids. This mild scrub effectively increases the solids specific gravity. The oil free (less than 40 ppm oil) produced water then travels to a fine particle de-sander and undergoes its final stage of cleaning (removing all solids 20μ or larger). All produced water is recovered, processed, and reinjected into the Cretaceous disposal wells at the Lisburne Production Center, or pumped to Point McIntyre Drill Sites 1 and 2 for waterflood purposes. Initial testing has shown that the system performance has met or surpassed all specifications.

Lisburne Production Center

The Lisburne Production Center (LPC) is an independent oil processing facility designed to separate produced fluids (oil, gas, and water). The LPC, located at Prudhoe Bay, Alaska, started production on December 15, 1986. This plant, under its original design, processed oil at a nominal rate of 100 MBOPD, gas at 400 MMSCFPD, produced water injection at 20 MBWPD, and generated 4,800 STBD of Natural Gas Liquids (NGL) that were commingled with treated oil prior to shipment. In 1993, the LPC Expansion Project

Produced Water2: Environmental Issues and Mitigation Technologies
Edited by M. Reed and S. Johnsen, Plenum Press, New York, 1996

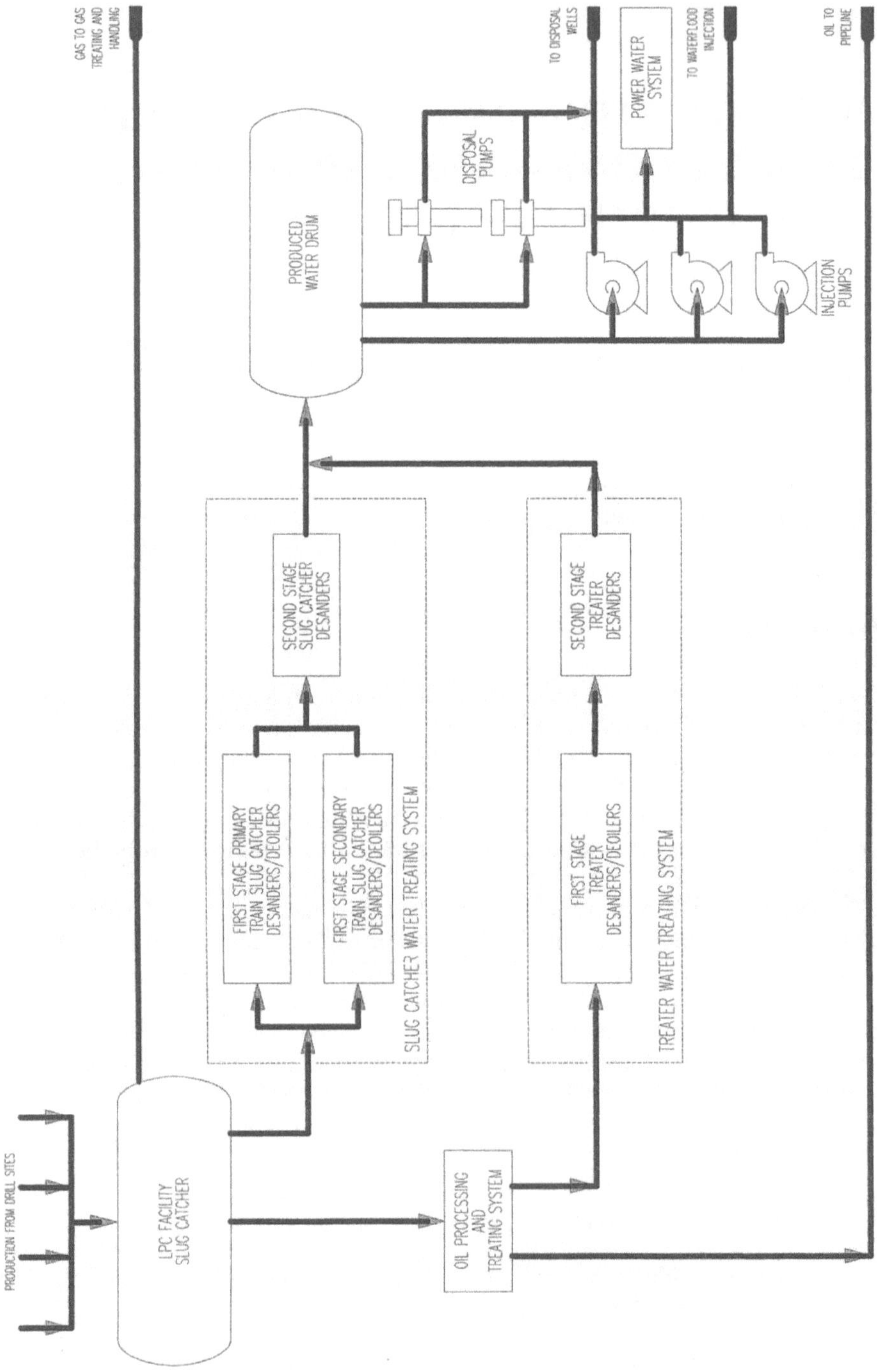

Figure 1. Produced water handling: general description.

upgraded the LPC enabling the facility to receive fluid from other oil fields. The LPC increased its fluid capacity to the nominal oil rate of 180,000 STBD, and added water treating facilities to handle 224,000 BPD of produced water, with a injection pump capacity of 160,000 BPD. The LPC currently receives untreated crude oil from the fields at Lisburne, Point McIntyre, West Beach, North Prudhoe, Niakuk, and Alapah. The production fluids are a mixture of oil, gas, and water. Pipeline specifications require crude oil to comply with water removal to a maximum of 0.35 percent Basic Sediment and Water (BS&W), and gas removal to an atmospheric vapor pressure of 145°F (62.7°C). The recovered gas is compressed and treated by flowing the crude oil through three separation stages at pressure drops of 450 to 545; 115; and 6 to 21 psig (3,103 to 3,758; 793; and 41 to 145 kPa), respectively. Triethylene glycol contactors dehydrate the gas to a dew point of -40°F (-40°C) at 1,500 psig (10,342.1 kPa). The gas then flows to the depropanizer, where processing recovers heavy components such as NGL's. The LPC uses the remaining gas for fuel gas consumption, or it is compressed to 4,700 psig (32,405.4 kPa) and used for reinjection into the producing formations.

Water is removed from the produced liquids in two stages consisting of the high pressure separator followed by the treater system (includes treater flash drums and electrostatic treaters). The crude oil is heated to the optimum oil treating temperature in the interstage crude oil heaters to assist water removal. Water is collected and treated to remove oil and sand. After removing the hydrocarbons, the water is either injected into disposal wells or used for waterflood injection. Currently, the LPC injects produced water into the Cretaceous formations through two injection wells on the pad.

Currently, there are 100+ oil wells capable of flowing into the LPC from eight drill sites. The prevailing daily averages (1995) are:

Oil:	180,000 STBD	Water:	20,000 BPD
Gas:	400 Million SCFD	NGL:	4,500 STBD

North Slope of Alaska

The remoteness of the Alaskan North Slope creates major logistical challenges. Since 1968, large oil production modules and other oil field facilities, weighing thousands of tons, have been barged to Alaska's North Slope from various fabrication sites worldwide. The trip takes several weeks, traveling thousands of miles through tempestuous weather elements. ARCO plans these sealifts for summer delivery, usually pin-pointing a four to six week period, when the frigid ice pack moves offshore. With the completion of major oil field facilities at Prudhoe Bay, Kuparuk, and Endicott, and the advent of smaller truckable oil production modules being built in Alaska, future sealifts will be smaller and less frequent.

THE LPC EXPANSION PROJECT DESIGN

Oil Handling

The ARCO LPC examined its separation train's capacity, to determine the impact of additional crude oil production upon oil and water separation performances. The LPC oil train's original design included a high pressure (HP) slug catcher, two crude heaters, a treater flash drum, a pair of parallel operating treaters, a surge drum, three shipping pumps, and two crude coolers. A capacity analysis included the fabrication of a dynamic model of the existing HP separator, that simulated the behavior of the vessel under the anticipated slug flow streams from Point McIntyre.

This effort covered the traditional level response analysis of the vessel, and investigated the water separation response during the pipeline slug flow. The simulation evaluated the effect of different production flowline configurations and operating temperatures on the separation capacity. The study's conclusions resulted in the final design components listed below:

- The addition of one 24-inch common line approximately ten miles (16 km) long, with provisions to add an additional 24-inch common line from the Point McIntyre Field.
- Installation of another interstage crude heater to improve water separation performances.
- Required additional oil/water separation capacity. Converted the existing produced water drum to a second treater flash drum. Retrofit one of the treater flash drums with water draws to achieve the required freewater knockout. The treaters were divided to form independent trains.

Produced Water Treatment Design

Based on the recommended waterflood scenario, LPC added equipment to handle and treat predicted integrated produced water forecasts for Point McIntrye/Lisburne/Niakuk. Treatment consisted of removing suspended oil using hydrocyclone technology, removing solids using hydrocyclone-based separators, and biological mitigation using biocide injection. The recommended water quality specifications (used as design criteria) are: <50 milligrams per liter (mg/l) of suspended oil in water, <10 mg/l suspended solids in water, solid removal of particles greater than 20μ in size, and 1000 counts/ml maximum bacteria. The water treating system's basic concept is to treat the slug catcher stream separately and then combine with the effluent from the treater system. The design flow rate of the slug catcher water treating system is 160 MBPD. The design flow rate of the treater water treating system is 60 MBPD.

Key Considerations in Produced Water Treatment Design

The Lisburne Field is a limestone formation. Early tests indicated that produced water in this type of formation has a tendency to form $CaCO_3$ scale, especially when the field is operated at less than 60 psig (414 kPa). The LPC's original design operated the treater system and produced water drum system above 85 psig (586 kPa). Field experience indicates very little $CaCO_3$ forms in a produced water system operating at that level. The LPC Produced Water Expansion project incorporated the original design criteria for all produced water used in waterflood injection water. Operations and maintenance (O&M) costs decreased due to LPC's low amount of $CaCO_3$ scale. Insignificant costs were incurred for chemical inhibitors. In addition, booster pumps were not required to charge the large water injection pump, and there have been no failures in piping, control valves, pumps, or wells due to scale formation.

Hydrogen Sulfide (H_2S) was discovered during the development drilling program for the Lisburne Reservoir. The LPC was originally designed to NACE Standard MR-01-75. LPC was designed, assuming the inlet crude, gas, and water stream would have 600 ppm of H_2S. Corrosive Resistance Alloys (CRA) were used in all process areas that would be in contact with H_2S, produced water, or wet gas. 316L material was used in process areas that had temperatures of less than 160°F (71°C). Duplex 2205 material was used in process areas where the temperature could potentially rise above 160°F (71°C). The slug catcher produced water trains were designed with 316L materials. The heater treater hydrocyclone system and piping were designed with Duplex 2205 materials. No corrosion inhibitors are presently being injected at the LPC plant.

The LPC evaluated criteria on handling solids and minimizing the potential growth of Sulfite Reducing Bacteria (SRB). The system was designed to remove the solids as quickly as possible, thereby reducing the potential of SRB growth. This design decreases the overall system downtime, because of fewer equipment failures and phenomenal injection well performances. Slop/divert tanks were designed to pump directly into the disposal injection line, minimizing SRB contamination in the process vessels and waterflood injection fluids.

The LPC's produced water expansion design amplified the kinetic and other potential energy available within the system. The pressure drop between the different process vessels became an important factor in the placement of the hydrocyclones, control valves, and pumps. This design operated the system at a higher pressure, meeting process specifications, and minimizing the energy required to pump or compress fluids prior to injection or sales delivery.

HYDROCYCLONE SEPARATION TECHNOLOGY

All LPC's De-sanders and de-oilers are hydrocyclone-based separators; their components and operation are comparable. In the de-oilers, the heavy phase (water) represents 98% of the flow; while with De-sanders, the heavy phase (solids) is comparatively small. Therefore, liquid/liquid de-oilers operate in a similar, albeit, opposite manner compared to solid/liquid de-sanders. Also, de-oilers and de-sanders operate flooded; this means the air core typical of solid/liquid cyclones is not present. The reject orifice in the liquid/liquid de-oiler is comparable to the apex of the solids/liquid de-sander, and the tailpipe in the de-oiler is comparable to the vortex finder in the de-sander.

The spiral path traveled by the incoming feed water converts its straight line (linear) velocity to tangential velocity. This action develops centrifugal force that propels the heavier weight (higher specific gravity) water to the wall of each hydrocyclone, along its tailpipe into the underflow chamber, and out the underflow nozzle.

Feed movement along the cyclone and tailpipe extension further increases tangential velocity, centrifugal force, and residence time, thereby improving separation. Lighter weight oil concentrates in the center of the hydrocyclone, and (due to backpressure on the underflow) flows as an oil core axially in the opposite direction (of the water) and through the reject orifice where it travels into the overflow chamber and out its nozzle.

The core stabilizing shield enhances separation over conventional designs by stabilizing the oil core in the inlet area, resulting in higher efficiency and better response to gas entrainment. Oil content of the overflow stream is typically 1% to 3%, depending on the feed concentration. Underflow is usually very clean water, often with less than 40 ppm of oil content. Underflow specification for the LPC is 50 ppm of oil content. Factors affecting oil removal efficiency include oil droplet size, temperature, oil concentration, pressure differential across the hydrocyclones (flowrate), and specific gravity difference. In turn, these factors influence de-oiler design variables such as inlet area, cyclone diameter, tailpipe diameter, and length.

- Droplet size - Maintaining mean oil droplet size and promoting coalescence of smaller droplets to larger droplets has the greatest impact on de-oiler performance. To minimize turbulence leading to droplet shearing, the LPC placed its de-oilers upstream of the pressure reducing or flow control valves, piping turns, and pumps.
- Temperature - Based on the LPC crude oil characteristics, the water temperature of 150°F (65.5°C) promotes separation efficiencies of at least 90 - 95%.

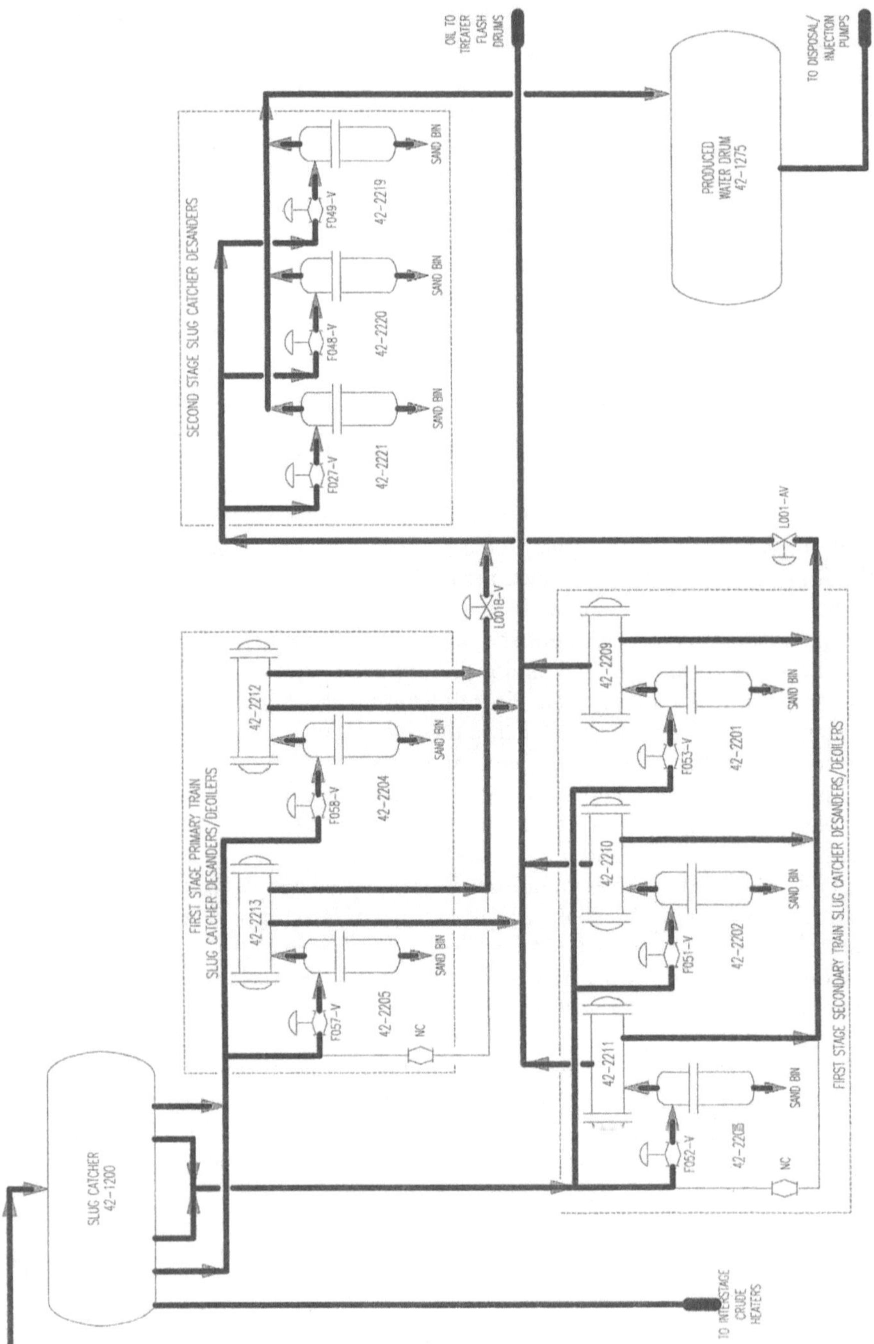

Figure 2. Slug catcher water treating: general description.

- Oil concentration - Oil concentration increases efficiency due to the expanded chance of oil particle collision and coalescence to larger particle size. It also improves efficiency needed to meet specified discharge requirements.
- Pressure differential across hydrocyclones (flowrate) - The LPC increased the efficiency with flowrate, resulting in a higher inlet velocity that outweighs the slight decrease in residence time caused at higher rates.
- Specific gravity difference - The greater the difference in weight or specific gravity of the materials being separated, the easier the separation.

De-oiler flow split or reject ratio is the percentage of the feed that exists in the overflow. For LPC units it is 2%, as in most produced water applications. Reject ratios over 2% are normally avoided to minimize waste stream disposal. This ratio must be at least as great as the feed oil concentration. The pressure ratio controls the reject ratio. This is the difference between feed pressure and overflow pressure, divided by the difference between feed pressure and underflow pressure. Since the underflow is back-pressured to force the oil core axially out the cyclone overflow, the underflow is always greater than the overflow pressure, and the pressure ratio is always greater than one.

Cyclonic-type separators use centrifugal force to cause separation of solids from liquids or liquids from other liquids of different miscibility's or gravity's. In order to sustain sufficient centrifugal force, the separation of the oil and sand from the water requires the inlet feed of each unit to enter tangentially at a specified minimum velocity. The greater the differences in specific gravity of the materials entering the unit, the more effective the unit will be in accomplishing their separation. A control system ensures each de-sander and de-oiler operates within the required pressure differential across hydrocyclone (flowrate) range. Accomplishing high separation efficiency also depends upon avoiding high inlet stream turbulence that could potentially form an emulsion between the oil and water or cause shearing of the oil droplets. For this reason, the de-oilers are located upstream of the slug catcher level control valves, whose high pressure drop of 59 psig (407 kPa) would otherwise introduce turbulence into the de-sanders. Extensive testing proved that a de-oiler's efficiency is not significantly reduced by locating a de-sander upstream.

SLUG CATCHER WATER TREATING SYSTEM

The current design divides the slug catcher produced water treating system into two trains. The primary train has two de-sander/de-oiler pairs that operate in parallel with each other; the secondary train has three de-sander/de-oiler pairs. Each train consists of de-sanders, de-oilers, and a slug catcher level control valve.

The first treatment step for each stream is desanding (coarser than 80μ). The slug catcher de-sanders have the same pressure rating as the slug catcher, and remove 95% of the particulates greater than 80μ. de-sanders are vertically mounted, single cone, solid/liquid, hydrocyclone-based separators. The design of the desanding operation prevents plugging of the de-oilers. The de-oilers are downstream of the first stage de-sanders, and once the streams have gone through the de-oilers, the system streams are combined and a second desanding stage (removing all solids 20μ or larger) takes place. When the two primary trains can no longer handle the slug catcher inlet flow, the overlapping split range level control system (operating control valves) senses the liquid level rise in the slug catcher, and causes the secondary train to begin operation. The sandy underflow from the de-sanders flows to a sand bin located beneath the module. de-sander water overflow continues to its de-oilers, where the remaining oil is separated and recycled to the treater flash drums. Water leaving the de-oilers flows to the second stage slug catcher de-sanders, where sandy underflow again

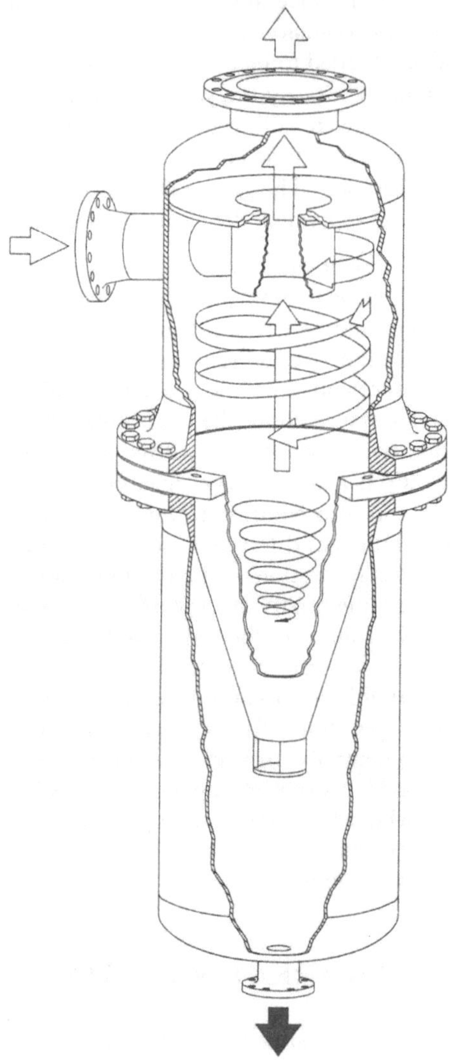

Figure 3. Coarse particle de-sander.

exits to sand bins below the module, and water overflow continues to the produced water drum.

First Stage Slug Catcher De-Sanders

de-sanders, designed for a non-turbulent feed entry, provide greater solid recovery than multiple slotted inlet designs, while reducing inlet wear. Fully replaceable liners (the internal cyclone shapes) reduce downtime and eliminate the need for complete de-sander replacement. The tapered cyclone cone enhances separation by generating greater centrifugal forces than conventional cylindrical separating chambers. Each de-sander has an accumulator, vertically mounted below the sandy water outlet of the de-sander unit. Size and capacity of the accumulators are based upon a nominal sand rate of 6.9 pounds (3.1 kg) per hour. The accumulator empties once every 12-hours and is a maximum of 1/3 full when it empties. The accumulators are approximately 12 inches (30.5 cm) in diameter and 3 feet (0.91 m) in length. The Distributed Control System (DCS) opens and closes the three valves located on each accumulator. An inlet valve from the de-sander is its source for sand. An outlet valve allows it to dump solids to the sand bin, and an additional valve provides flush water during the dumping operation.

First Stage Slug Catcher De-Oilers

Slug catcher de-oilers are horizontally-mounted, multi-cone, liquid/liquid, hydrocyclone-based separators. de-oiler vessel components include the feed, overflow, and underflow chambers with appropriate thermal relief, vent, drain, and inlet/exit nozzles. Internals consist of cyclone mounting and tailpipe stabilizer plates, core stabilizing shields, inlet heads, cyclones, and tailpipes. The internal cyclones or hydrocyclone assemblies (consisting of the shields, inlet heads, cyclones, and tailpipes) connect the three chambers and permit

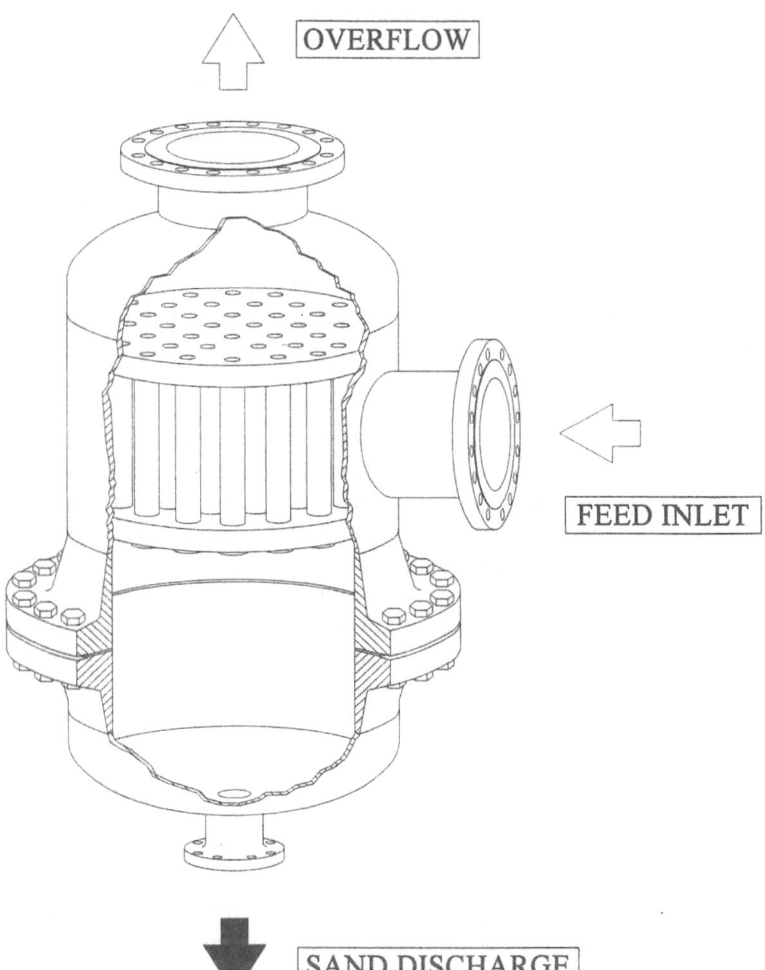

Figure 4. Fine particle de-sander.

oily water entering the feed chamber to enter each cyclone assembly through an opening in its inlet head (its cyclone inlet). The number of cyclone assemblies in each inlet chamber is based on the design separation requirements for the stream being processed.

Second Stage Slug Catcher De-Sanders

The first stage de-sanders remove the larger diameter particles from the produced water, upstream of the slug catcher de-oilers. The second stage de-sanders, located downstream of these de-oilers, can remove some of the finer sand particles from the produced water, prior to its flow to the produced water drum. These second stage units consist of three de-sanders functioning in parallel with each other.

Second stage slug catcher de-sanders are vertically mounted, multi-cone, solid/liquid, hydrocyclone-based separators. These de-sanders are larger in diameter than first stage de-sanders. They remove smaller solid particles from the slug catcher produced water effluent stream, after it flows from the slug catcher de-oilers. The second stage de-sanders are designed to remove 98% of 20μ particles.

Each second stage de-sander has multiple cyclones that are vertically positioned in a inlet section. This section is bound above by an overflow divider plate and below by an underflow divider plate. Flow to each de-sander enters the inlet section (at about vessel midpoint) where the flow splits and enters each of the 42 cyclones. In these cyclones, the centrifugal force generated throws the heavier solids to the outer wall of each cyclone. Clean water discharges out the top of each cyclone, through the overflow divider plate, into the upper section of the de-sander, and out the vessel overflow nozzle. Solids are discharged out the lower conical sections, through the underflow divider plate, and accumulate in an external in-line accumulator, where they are periodically removed by a produced water wash to sand bins for subsequent disposal. Each de-sander vessel has 8-inch (20 cm) blinded access ports to both the inlet and overflow sections of the vessel for inspection or cleanout. The underflow or discharge head section may be inspected by unbolting and separating it from the overflow section with the lifting lugs provided.

Each de-sander has an accumulator, vertically mounted below the sandy water outlet of the unit. Size and capacity of these accumulators are based on a nominal sand rate of 1.43 pounds (0.65 kg) per hour. The accumulator empties once every 12- hours and is 1/3 full when it empties. The accumulators are approximately 8-inches (20 cm) in diameter and 2-feet (61 cm) in length. The Distributed Control System (DCS) opens and closes the three valves that each accumulator has. An inlet valve from the desander is its source for sand. An outlet valve (rotating disc type) allows it to dump sand to the sand bin. The rotating disc outlet valve is a key factor to the successful operation of the accumulators. The Everlasting Valve™ is a self-lapping, self-cleaning valve that works extremely well in this accumulator application. This valve configuration has been used quite successfully in the mining and slurry handling industry for over eighty years. An additional valve provides flush water during the dumping operation.

TREATER WATER TREATING SYSTEM

After crude oil exits the slug catcher with entrained water, sand, and dissolved gases, it flows through an interstage crude oil heater to promote optimum temperatures for further separation. The heated stream is pressure reduced in treater flash drums, flashing the gases. The flashed gases flow to the intermediate stage compression. The oil, water, and sand mixture exiting the drums, flow to electrostatic treaters, where the oil separates and flows

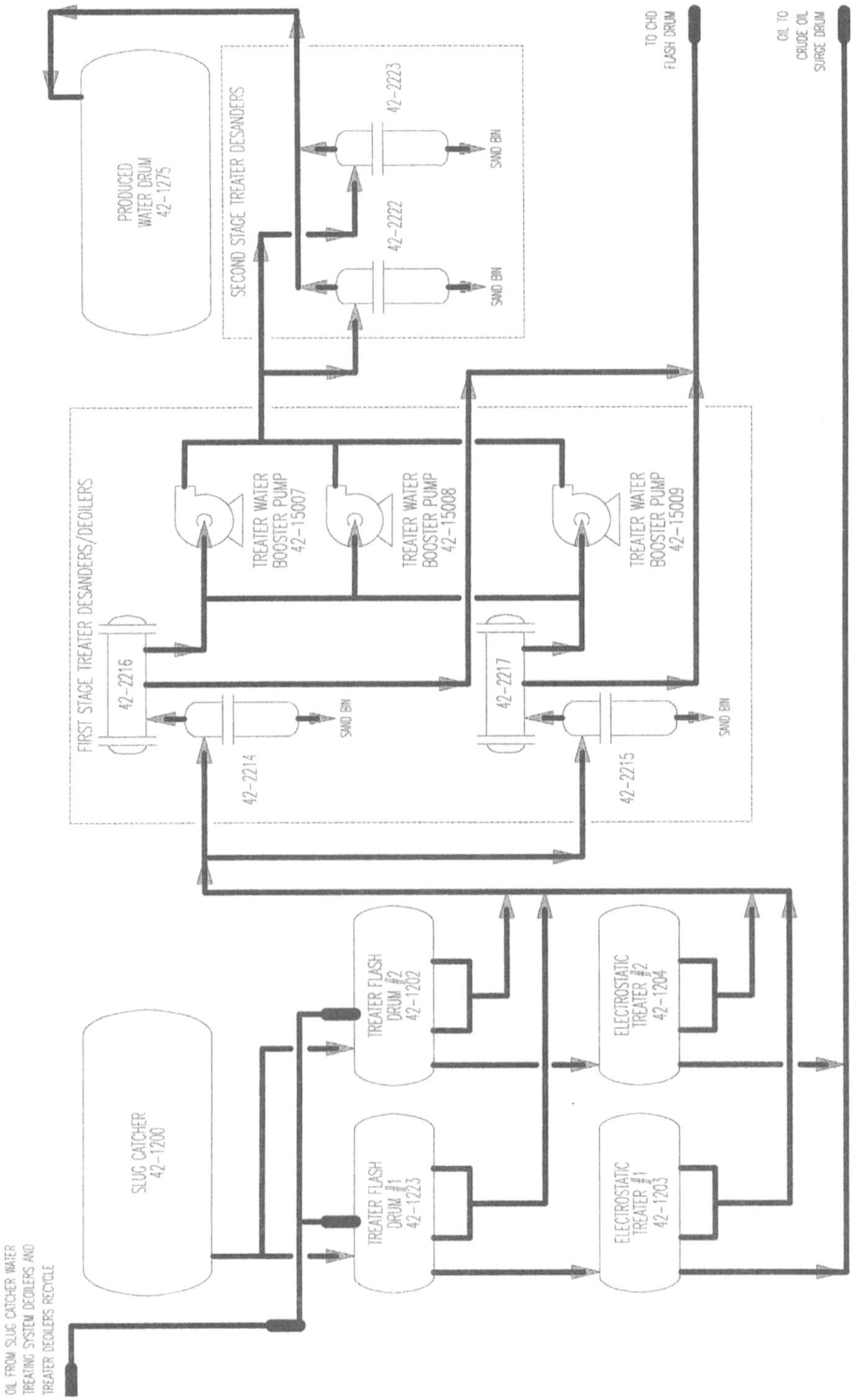

Figure 5. Treater water treating: general description.

to the crude oil surge drum. The water and sand effluents from the treater flash drums and the electrostatic treaters commingle and feed the treater water treating system.

The first stage consists of two parallel trains of de-sanders and de-oilers; each train includes a treater water booster pump. There is a spare treater water booster pump that serves either train. Inlet water to these de-sanders overflows to the de-oilers and is then pumped by the treater water booster pumps to the second stage de-sanders. Overflow water flows to the produced water drum. Sandy underflow from both the first and second stage de-sanders flows to a sand bin beneath the module. The reject oil (from the de-oilers) flows to the closed hydrocarbon drain (CHD) flash drum, and is recycled by pumping it back to the treater flash drums.

There is pressure differential measurement across the hydrocyclones on the treater de-sanders that can be used to detect plugging. The pressure differential measurement is also used as a flow meter to control the actuation/operation of the hydrocyclone trains. Also, inlet valves to the de-sanders are located in a vertical run of pipe to prevent sand accumulation in the valves. Because both the de-sanders and their associated de-oilers are hydrocyclone separators, they use centrifugal force to separate the oil and sand from the water. Since separation depends on centrifugal force, it is important to achieve a minimum tangential velocity of the fluids within the cyclone. To achieve this velocity, a pressure differential is maintained across the hydrocyclone in each unit at or above the design minimum for proper separation to occur. In addition to having the same minimum tangential velocity requirements as de-sanders, de-oilers are required to be located upstream of equipment causing turbulence that could emulsify the oil and water. This requirement necessitates de-oilers being placed upstream of outlet water control valves on the treater flash drums and electrostatic treaters. Level control valves feeding the treater water system could not be located downstream of the de-oilers, because the treater flash drums and electrostatic treaters interface levels need to be controlled independently. The turbulence across each control valve is minimized by designing a low pressure drop of 10 psig (69 kPa). To accomplish this design, the treater water booster pumps were specified with variable speed drives. Each pump speed is reset by a pressure controller across the de-oiler. Maintaining a constant pressure of 115 psig (793 kPa) at de-oiler inlets, ensures the pressure drop from the treater flash drums to the de-oilers remains constant. This pressure maintenance also ensures that the pump suction pressure does not drop below 60 psig (414 kPa), minimizing the potential of scale formation in the produced water system.

The de-sanders and de-oilers in this system are similar in design to those used in the slug catcher water treating system. The first stage, single cyclone desanders are designed to remove 95% of 80μ particles. The second stage multi-cone de-sanders have 32 cyclones in each vessel and are designed to remove 98% of 20μ particles. Each de-sander has an accumulator, vertically mounted below the sandy water outlet of the de-sander unit. The accumulators are similar to the slug catcher accumulators. The size and capacity of first stage accumulators are based upon a normal sand rate of 17.3 pounds (7.8 kg) per hour. The first stage accumulators are approximately 14-inches (35.6 cm) in diameter and 5.5-feet (1.7 m) in length. The size and capacity of second stage accumulators are based on a normal sand rate of 2.2 pounds (1 kg) per hour. The accumulators are approximately 8-inches (20 cm) in diameter and 2-feet (61 cm) in length. These accumulators empty once every 12-hours and are 1/3 full when they empty.

SAND COLLECTION SYSTEM

The sand collection system is designed to accumulate waste sand and grit removed from the produced water and some oily water streams. The sand is accumulated in specially designed bins. Sand bins permit free water to drain into individual outer compartments. A dewatering pump is used to force the water out of the compartment to the slop oil tank, where it's eventually reprocessed. Periodically, the sand bins are removed for sand disposal. The bin compartments are located inside the containment enclosure that rest on supports. A heating coil is provided under the compartments to prevent the water from freezing. The enclosure is ventilated to remove any gas that may break out of the sand or water. Water may leak out of the bin compartment during bin removal; however, the water drains to a containment sump and is pumped to the slop oil tank. There are seven sand bins. Each bin is placed inside a compartment for water accumulation. These bins each measure 48" W x 48" L x 54" H (1.2m W x 1.2m L x 1.4m H) and holds 1.9 cubic yards (1.45 cubic meters) of sand. The outer compartment holds about 140 gallons (530 l)of water. A gauge rod determines the sand level in each bin.

Dewatering Pumps

The dewatering pumps remove the accumulated water and some fine sand particles from the sand bin compartments. These pumps are designed with a high suction lift ability of over 12-feet (3.7m). There are seven pumps, one for each sand bin. The pumps use plant air at full pressure as a motive force. A full open/close solenoid operated valve is used in the air line to start/stop the pumps.

PRODUCED WATER INJECTION SYSTEM

The produced water, from the high pressure separator (or slug catcher) and the treater system (the treater flash drums and the electrostatic treaters), flows through various desanders and deoilers and gathers in the produced water drum. Any vapor flashing off the water is routed to the stock tank vapor (STV) compressor for recovery. The water is pumped by the water injection pumps from the produced water drum into the disposal wells or to the waterflood injection.

The LPC staged the three new injection pumps for disposal well injection. The pumps were designed to pump 1,667 gpm (6.31 m^3/min) at 1,881 psig (12,969 kPa) discharge pressure. The three injection pumps can be staged for waterflooding to deliver a combined rate of 5,000 gpm (18.9 m^3/min) at a discharge pressure of 2,665 psig (18,375 kPa). The produced water drum operates at 85 psig (586 kPa), maintaining the water pressure above where calcium carbonate scale would form.

A produced water study indicates there is a high potential for barium sulfate scale formation. This could happen if Point McIntyre produced water is mixed with produced waters from Lisburne and Prudhoe Bay, or by mixing it with Prudhoe Bay sea water. Point McIntyre's produced water has barium and the presence of sulfate in the other waters accounts for the potential scale formation. An inhibitor program is in place to mitigate scale formation.

SAND JETTING OPERATIONS

Sand jetting is an intermittent operation required to remove sand that accumulates in the process vessels. The sand jetting system is done on-line without causing any production impact. Vessels that are sand jetted on a weekly basis include the slug catcher, electrostatic treaters No. 1 and 2, the treater flash drums No. 1 and 2, and the produced water drum. Water for the sand jetting operation is supplied by the new injection pumps at a flow rate of 584 gpm (20 MBPD). The effluent from the vessels is desanded to remove 95% of particles 50μ and larger. The sand jet water is then pumped down the disposal well.

CONCLUSIONS

The ARCO LPC successfully completed the start-up and commissioning of the Lisburne Water Expansion Project with no major obstacles. The hydrocyclones and injection pumps were brought on-line without problem. The electrostatic treater system train was commissioned first. The possibility of solids plugging the high pressure desanders while commissioning the system was a vital start-up concern. The high pressure slug catcher had not been sand jetted since July 1993, when the LPC installed the new sand jet system inside the vessel. During start-up, a flush line was connected from the treated produced injection line to the high pressure catcher's primary water train header. This procedure allowed the operators to commission the high pressure slug de-sanders/de-oilers with clean produced water, that permitted the control system to be tuned before introducing solids to the hydrocyclone system. Overall, each set of de-sander/de-oilers has performed well. The LPC tested both treater hydrocyclone trains to a full rating of 60 MBPD. The slug catcher has operated one hydrocyclone bank to full rating with no problems indicated.

Design Considerations for Future Projects

- Install permanent flush lines from the injection line. This allows commissioning and control system tuning before introducing solids to the system.
- Use pressure differential across the deoilers to control operation of the de-sander/de-oiler trains. The first stage de-sander and de-oiler were designed to have the same operational ranges. Lisburne Production Center used strap-on sonic flow meters with limited or no-success due to gas entrainment in the fluids.
- The sand collection system can be improved. The sand accumulators tanks, along with the Everlasting Rotating Disc™ outlet valves, appear to have been successful.

ARCO designed the LPC and the expansion project with two directives in mind. One, the facility should be a safe and successful plant. Two, the O&M costs are to be competitive. A primary motivation was to create a unique, first class facility and manage it under a reasonable budget. That mission has been accomplished. The LPC is the world's first commissioned combined solid/liquid and liquid/liquid hydrocyclone-based produced water treatment system.

ACKNOWLEDGMENTS

- Conceptual Engineering staff and consultants:
  - Project Management Team
  - Technical Assist Team
  - John Cologgi (AAI)
  - James P. Meyer (AEPT)
- Preliminary Engineering Report staff and consultants:
  - Tom Gibbons (BPX)
  - Raytheon Engineers and Contractors
  - Krebs Engineers, Menlo Park, CA
  - Joe Mills, M&H Systems, Training Manual Project Leader

- Robert Brewster, LPC Operator, Prudhoe Bay, AK
- LPC Operations Staff
- LPC Maintenance Staff
- LPC Start Up Team

REFERENCES

1. ARCO Alaska, Inc., Project Management Team, Point McIntyre Development - Lisburne Facility Sharing; Conceptual Engineering Design Report, Vol. 1 Study Results, (June, 1990).
2. C. C. Paten & Associates, Applied Water Technology, Campbell Petroleum Series, Publisher, (June, 1986).
3. M&H Engineering Systems, Houston, Texas, ARCO Alaska, Inc., a Subsidiary of Atlantic Richfield Company, Lisburne Unit, Operator Training Manuals, Volumes I - III.
4. Raytheon Engineering, Point McIntyre Development - Lisburne Facility Sharing, LPC Modifications; Preliminary Engineering Report, Volume 1, (May, 1991).
5. Trans-Alaska Pipeline Directive, 1995-07, (February 6, 1995).

PRODUCED WATER REINJECTION (PWRI)

Experiences from the Ula Field

Siri Bakke,[1] Eilen A. Vik,[1] Hans Grüner,[2] and Thor A. Hjelmås[2]

[1] Aquateam - Norwegian Water Technology Centre A/S
P.O. Box 6326 Etterstad
N-0604 Oslo, Norway
[2] BP Norway Limited
P.O. Box 197
N-4033 Forus, Norway

INTRODUCTION

As the oil producing fields in the Norwegian sector mature, the volumes of water produced will increase. The standard offshore method for produced water disposal is to remove excess oil down to the legal specification which currently is 40 mg/l hydrocarbons. It is generally recognized that the discharged produced water contains levels of dissolved organic components, chemicals and inorganics (heavy metals etc.) that may also be harmful to the marine environment. The environmental risks associated with produced water discharges will increase with increasing water production, unless improved water treatment (including removal of dissolved organics) or produced water reinjection (PWRI) is implemented.

Produced water handling represents one of the more significant environmental challenges to the oil industry, not only from an environmental point of view, but also from a technical and cost view. Corrosion, erosion and scaling mechanisms may lead to high maintenance costs for produced water treatment equipment and extensive use of harmful chemicals. If the water treatment efficency needs to be improved, the treatment costs will increase accordingly. BP Norway (BPN) has accepted that PWRI may be the only way to reduce the produced water discharges, although considerable uncertainty about the costs of implementation and the consequences of reinjection still exist. BPN therefore completed a PWRI trial in one single well on the Ula field in 1994, and based on these results (Bakke et al., 1995) decided to continue with a full scale PWRI trial.

Globally, PWRI has been evaluated and practiced for many years; e.g. by BP at the Forties, Wytch Farm and Prudhoe Bay fields (Paige and Sweeney, 1993, Simpson and Paige, 1991 and Martins et al., 1994), the Maureen field (King, 1994) and others (Evans, 1994). In most cases the PWRI activity has been on individual wells and has not included mixing with seawater prior to injection. The experiences from these trials have been variable. In most cases some loss in injectivity has been seen, in some cases the problems observed have been more severe; i.e. acclerated reservoir souring and increased scaling have also been noted. Experiences from the BP fields stress the need for a better understanding of the mechanisms that influence the impact of PWRI. Industry acceptance of combined produced water/seawater reinjection as a viable option in field development concepts, and industry confidence in any predictive modelling capability will be greatly enhanced by a full scale implementation of such a scheme, with appropriate documentation and verification of models through the results observed. BPN's decision for initiating the full scale PWRI via the existing water injection system on Ula were based on:

- The need for upgrading the produced water treatment facility.
- An evaluation of reservoir characteristics and production/injection balance showing that the risk for not meeting production targets if PWRI was implemented, would be minimal during the next two years.
- The need to push forward the knowledge and technology for marginal field developments.
- The need to find a cost-effective solution to the potential environmental problems associated with the discharges of large volumes of produced water.

BPN believes that it is in the best interests of the industry to spread the experience and knowlege on PWRI gained within the company. Through the Norwegian Research Council's MUST (Environmental and Profitable Development of Small Fields) programme, BPN is carrying out the PWRI programme together with Aker Engineering, Aquateam, ResLab and Saga to build up an integrated service on produced water reinjection (PWRI). It is BPN's intention that all project participants will have full and open access to all the information related to this project. The results and the observations obtained during the first period of the full scale PWRI trial (January-August 1995) are presented in this paper.

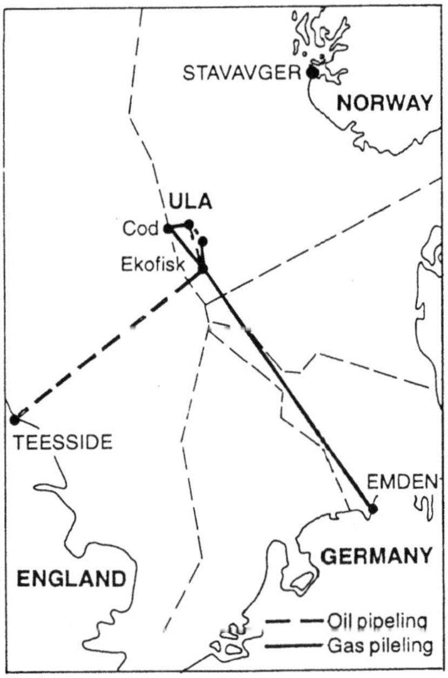

Figure 1. Location of the Ula field.

WATER PRODUCTION AND INJECTION HISTORY ON ULA

The Ula field is located in block 7/12 (BPN operator 57.5%) of the Norwegian sector of the North Sea, 280 km south-west of Stavanger, see Figure 1. Ula was discovered in 1976 and was sanctioned for development in 1982. The oil reservoir is 3345 m below the seabed in sandstones from the Jurassic age (160 million years). The water depth in the area is 71 meters. The oil production started in October 1986 and is today carried out from 9 production and 6 water injection wells. The reserves on Ula are 435 million barrels of oil, 167 billion ft^3 gas and 43 million barrels of condensate. The field lifetime is estimated to 20+ years. Figure 2 presents the oil and water production and water injection rates from 1986 to 1995.

Water Injection History

The seawater injection started in 1988 and has today exceeded approximately 62 million m^3 of water. Initially the seawater was treated by coarse filtration followed by direct fine filtration and deoxygenation. BPN's water injection strategy has altered significantly from 1986 till today. When the seawater treatment system at Ula was designed, it was considered essential that the injected seawater for reservoir pressure maintenance had to be fine filtered to prevent loss in injectivity. Water injection specifications (98% removal of particles > 2 μm), were derived from a combination of pore size distribution measurements and statistical analyses, in conjunction with linear core flood studies. The need for increased volumes of injection water, and the costs of the fine filters led to a study of injection water quality, both on the Ula field, as well as on other BP operated platforms in the North Sea (Paige and Sweeney, 1993). The water quality measurements were carried out both topside and downhole. The results showed that the fine filters operated efficiently and according to the specification, reducing the overall solids loading. The downhole samples showed, however, a dramatic rise in the solids loading, see Table 1. The filtered material from the downhole samples consisted predominantly of iron based products and sulphur. These results showed a minimum risk of bypassing the fine filters, which was performed in 1990. Subsequent measurements indicated that the fine filtration had a negligble effect on the downhole water quality and did not lead to any reduction in the injectivity.

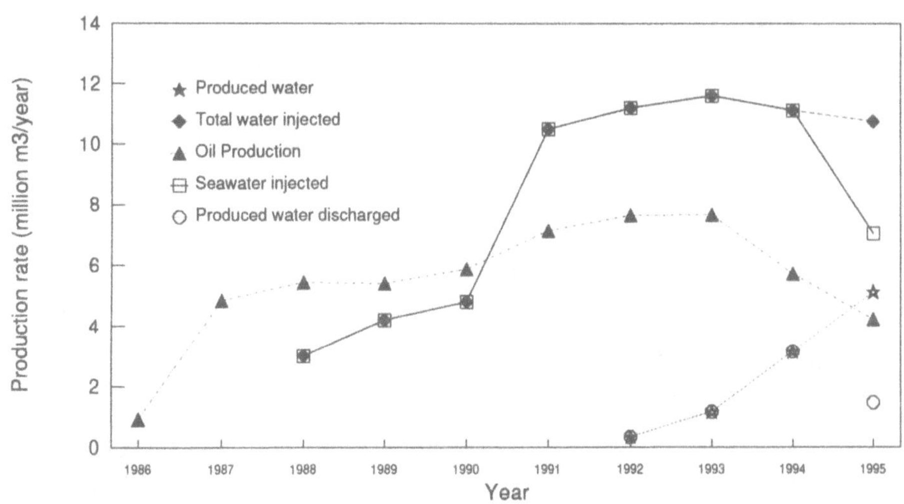

Figure 2. Oil and water production and water injection rates on Ula from 1986 to 1995.

Table 1. Test results obtained from an injection water quality study on Ula in 1989

Particulates in water	Water qualities studied				
	Upstream fine filters	Downstream fine filters	Wellhead injection header	Downhole well A05	Downhole well A03
# of particles per 500 μl					
> 2 μm	6,653	58	555	17,010	32,517
> 4 μm	900	2	25	3,327	3,874
> 8 μm	73	0	0	418	462
> 16 μm	5	0	0	54	56
Suspended Solids (mg/l)	3.5	0.1	0.2	6.0	7.5

Water Production History

The water production started in June 1992, and to date approximately 8 million m^3 water has been produced. Initially the produced water was treated in a gas flotation unit (GFU). As the produced water volume increased (see Figure 2), the capacity of the treatment plant became too low, already during the summer of 1993. The existing GFU became the limiting factor for the oil production, and at the highest water treatment capacity it was difficult to meet the existing regulation of 40 mg/l hydrocarbons in the produced water discharged. It was therefore decided to install a hydrocyclone based water treatment system, designed to treat all produced water to 20-30 mg/l hydrocarbones. It was further decided to evaluate the possibility for re-injecting the produced water.

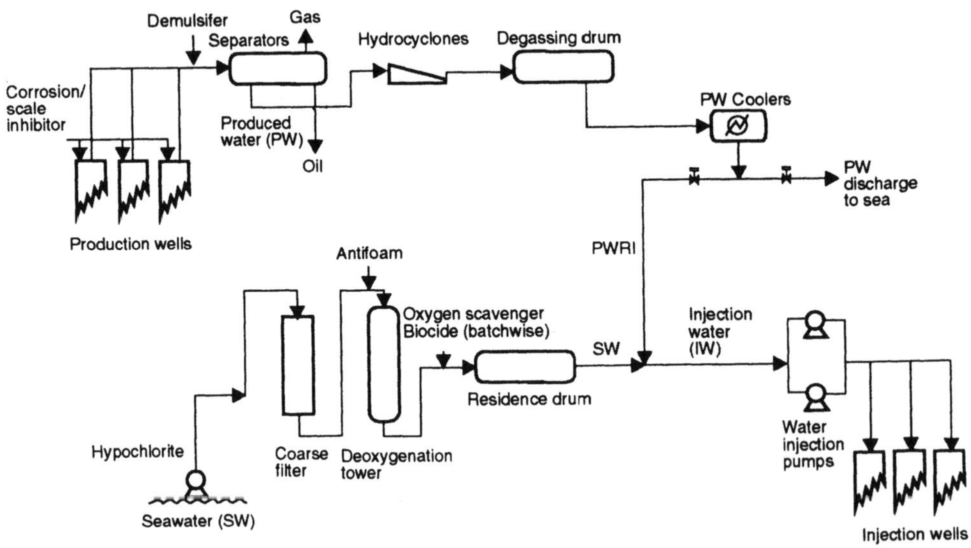

Figure 3. Schematic diagram of the process facilities used for the full scale PWRI trial on Ula.

FULL SCALE PWRI TRIAL ON ULA

Water Treatment

The full scale PWRI trial on Ula was initiated in February 1995. A sketch of the process facilities is presented in Figure 3. The produced water is handled in the separators, treated in the hydrocyclones and the degassing drum before it is cooled down to approximately 50°C. Downstream the coolers, the produced water (PW) is either mixed with seawater (SW) and reinjected (PWRI) or discharged. PWRI is, however, the normal mode of operation and disharge is only occuring if the water injection system is closed down, or the temperature of the PW exceeds 55°C. This temperature limit was set to avoid cavitation of the injection pumps. The SW is treated in the coarse filters, the deoxygenation tower and the residence drum before being mixed with PW prior to water injection.

Monitoring Programme

A monitoring programme is carried out to determine reservoir, process and environmental effects. The programme covers the following aspects:

- Injectivity
  - Well performance monitoring including continuous recording of water injection rates and pressures for the injection wells.
  - Laboratory core flooding experiments for measurements of permeability (injectivity). Different prediction models are used and the results compared to the full scale well performance. This work is described by van der Zwaag and Øyno (1995).
  - Water quality measurements with respect to oil and solids (particle distribution and suspended solids) loading.
- Scaling Potential
  - Determination of variation in the water quality (ion composition) of the injection water.
  - Prediction of potential scale formation and comparison of the model predictions with field experiences.
- Souring Potential
  - Measurements of bacteriological activity (nutrients and bacteria) in injection water for prediction of potential for H_2S formation.
  - Evaluation of models for prediction of H_2S production, and verification of results with field experiences.
- Corrosion Potential
  - Evaluation of the corrosion related changes in the injection water qualities.
  - Corrosion monitoring by use of ER-probes and weight loss coupons.
- Environmental Aspects
  - Determination of environmental related water quality parameters of produced water.
  - Monitoring environmental effect of the produced water (toxicity, biodegradation and bioaccumulation potential).
  - Prediction of environmental risk of produced water discharges, and the effect of various risk management options. This work is described by Øfjord et al. (1995).

Table 2. Water quality data parameters of water injected on Ula (week 10, 1995)

	Injection water	
Water quality parameter	Seawater (SW)	Mixed produced water/seawater (IW)*
Injectivity parameters		
Total solids[1] (mg/l)	n.d.	22
Suspended solids (mg/l)	4.2	6
Total number of particles (# per 500 μl)	15,000	155,000
Hydrocarbons (mg/l)	0	15
Temperature (°C)	18	31
Scaling and Corrosion parameters		
Calcium (mg Ca/l)	410	2,330
Barium (mg Ba/l)	0	5.3
Strontium (mg Sr/l)	7.7	96
Iron (mg Fe/l)	0.3	10.7
Chloride (mg Cl/l)	19,850	28,300
Sulphate (mg SO_4/l)	2,790	1,900
Bicarbonate (mg HCO_3/l)	140	192
Free carbondioxide (mg CO_2/l)	7.0	52
Oxygen (μg O_2/l)	80	15
pH	8.1	7.3
Souring parameters		
Acetic acid (mg/l)	0	1.2
Dissolved organic carbon (DOC) (mg C/l)	0.8	7.5
Total phosphorus (mg P/l)	0.03	0.1
Total nitrogen (mg N/l)	0.2	19
Sulphate reducing bacteria (SRB) (#/ml)	0	4.5
General anaerobic bacteria (GAB) (#/ml)	45	200

n.d. = not determined
* IW = 35% PW : 65% SW
[1] Includes oil

RESULTS AND DISCUSSION

Typical relevant water quality parameters of the injection water (IW) with respect to injectivity, scaling, souring and corrosion are summarised in Table 2 (data from week 10, 1995).

Water Treatment

The hydrocyclones were in place by the spring of 1994. The development in the hydrocarbon (oil) concentration in the produced water from January to August 1995 is illustrated in Figure 4. Initially the oil/water separation problems contributed to the large variability in the oil concentration in produced water. A number of actions were taken to improve the oil/water separation (e.g. optimising the use of demulsifier, modifications of the HP-separators and improving the water balance over the hydrocyclones).

A problem with the efficiency of the produced water coolers was observed during a reinjection function test in December 1994. The produced water coolers were designed to operate at a temperature interval of 48-55°C. The function test showed that scales (consisting of 23% asphaltenes, 24% iron sulfide and 27% zinc sulfide, as well as 26% other metal sulphides) were formed at the inlet of the coolers. A procedure was worked out for routinely

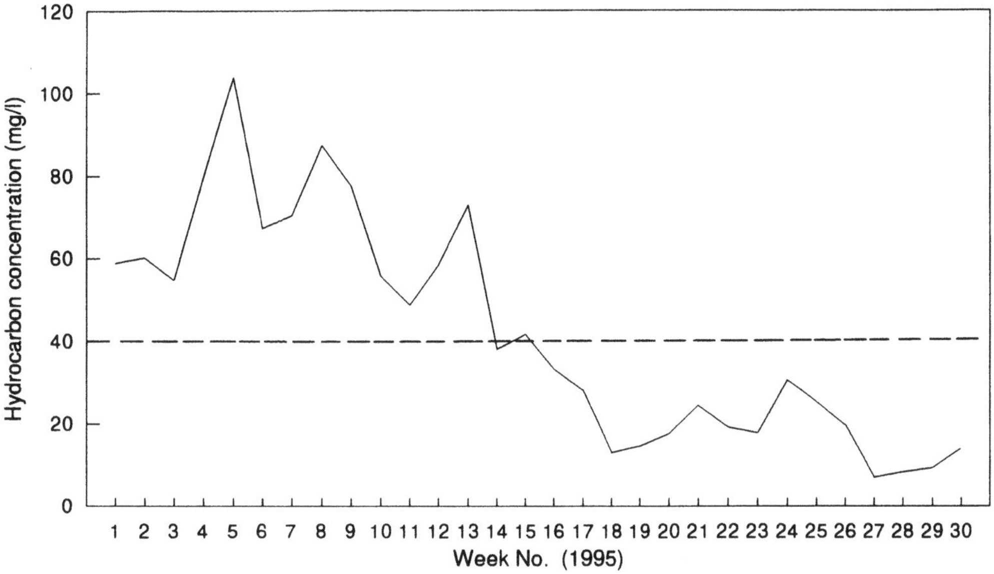

Figure 4. Development of the produced water hydrocarbon concentration during the first period of the full scale trial on Ula.

acid washing and mechanically cleaning of the coolers, before the full scale PWRI trial started. Figure 5 shows the temperature fluctuations in the produced water during the initial period of the trial. The coolers were mechanically cleaned up till May '95, and then acid cleaning was introduced. During certain well interventions it was observed that large amounts of particles plugged the inlet of the produced water coolers. To reduce these problems, a by-pass for the PW coolers was installed in week 13. Figure 5 illustrates that the fluctuations in the PW temperature still are occurring.

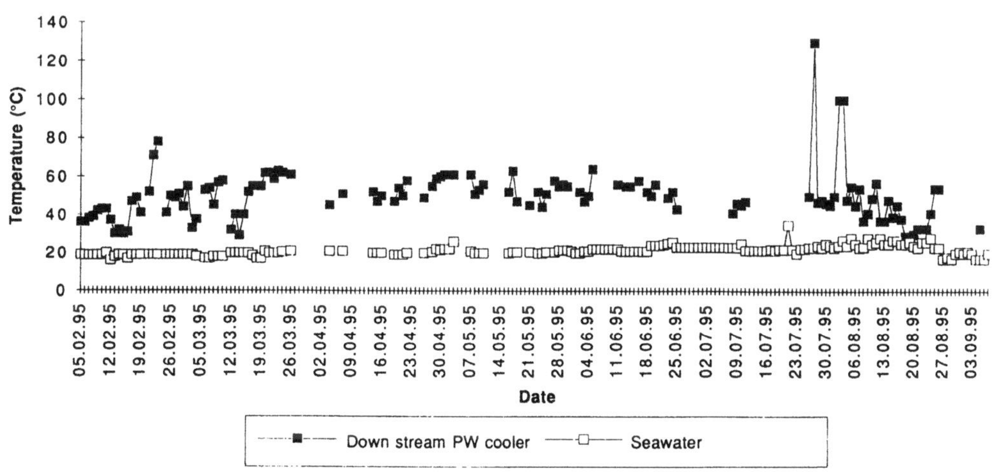

Figure 5. Temperature profiles of produced water monitored downstream the produced water coolers, and seawater monitored before injection during the first period of the full scale trial on Ula.

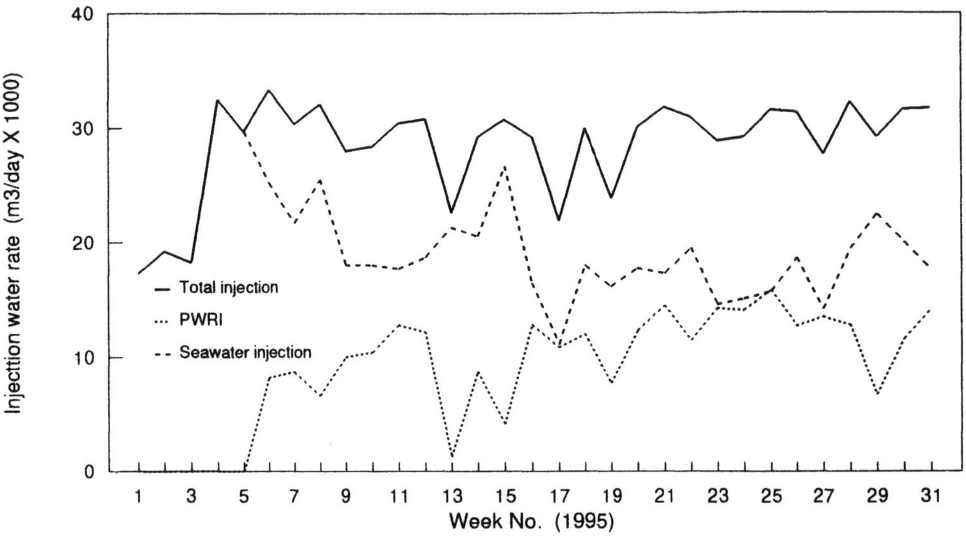

Figure 6. Water injection rates during the first period of the full scale trial on Ula.

Injectivity

When the full scale trial started the typical mixture of injection water (IW) was 35 % PW and 65% SW. Today the injection water is a mixture of approximatly 50 % PW and 50 % SW. The goal is on an annual basis to reinject 90 % of the PW produced. From the PWRI started to August 1, approximately 70% PW has been be reinjected. The water injection rates during the first period of the full scale PWRI trial is illustrated in Figure 6. The water injection rate is in the moment stabilized at a level of 30,000 m^3/day. The produced water reinjection (PWRI) started in week 5, and increased from 0 to around 15,000 m^3/day in week 34. The need for seawater to be injected was steadily reduced from 30,000 to, on an average of 15,000 m^3/day.

A typical development in the water injection rates and wellhead pressures as observed during the trial examplified for well A03A, is presented in Figure 7. This figure shows no obvious reduction in injectivity, illustrated as injection flowrate and pressure, during this initial period. The same observations are made for the other injection wells.

According to Martins et al. (1994) the water quality specification for PWRI at Prudhoe Bay were relaxed to less than 400 mg/l hydrocarbons and 15 mg/l solids (primary silt = suspended solids). The values given in Table 2 show that the Ula injection water is well within these limits.

Scaling Potential

Scale predictions of the produced water (PW) have shown that scale formation is likely to be a minor problem at temperatures of 60°C and less. Any significant quantities of scale would only be formed in the reservoir, well away from the injection wells. At high temperatures such as those in the reservoir, there would be a strong tendency to form calcium sulphate ($CaSO_4$) and some calcium carbonate ($CaCO_3$). At this stage of the production at Ula, the PW contains approximately 85% SW and 15% formation water. Increasing the

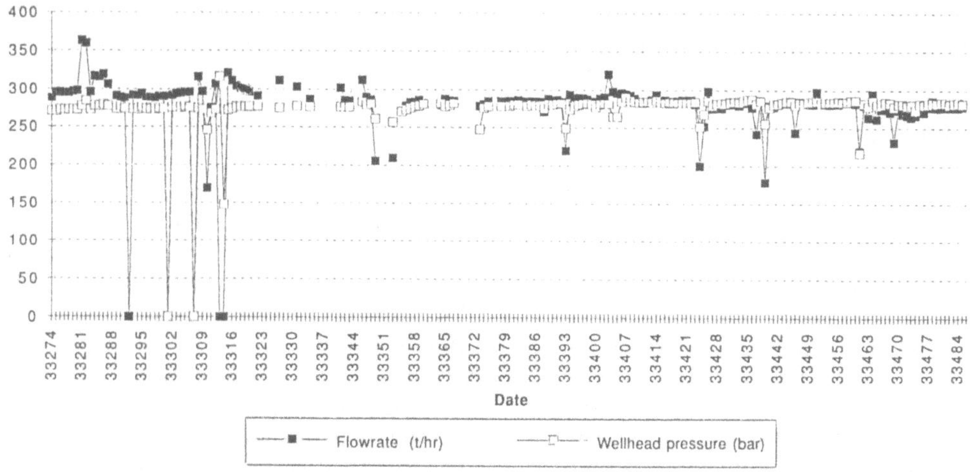

Figure 7. Water injection rate and wellhead pressure for well A03A during the first period of the full scale trial on Ula.

proportion of SW will reduce the potential to form both barium sulphate ($BaSO_4$), strontium sulphate ($SrSO_4$)and $CaCO_3$, while the tendency to deposit $CaSO_4$ will be slightly increased.

Throughout Ula's production history, $CaSO_4$ scaling was expected, but no problems have been observed. $BaSO_4$ and $SrSO_4$ scaling have been observed. Some $CaCO_3$ scale has also been observed in pressure drop areas. Scale inhibitor squeezing of production wells keeps, however, the problems under control. Iron scales are not predicted by the existing models, but water sampling test programmes have illustrated that ferric oxides and ironsulfides are formed topside. Other sulphide scales are also formed in the produced water coolers. Prediction of these scales cannot be performed with the existing scale predicting models.

Souring Potential

Based on the data given in Table 2 theoretical calculation of the H_2S-generation potential for the injection water has been performed (based on Sawyer and McCarthy, 1978) indicating low potential for H_2S-formation. The calculations indicated that there are excessive amounts of phosphorus and nitrogen nutrients as well as sulphate in the system, and that the limiting factor for biological activity is the amount of carbon. The predictions were based on the acetic acid concentration in PW, and dissolved organic carbon (DOC) content in SW, as the sources of carbon available. The content of acetic acid and even total amount of DOC in Ula's PW are, however, much lower than seen in other North Sea fields (Tibbets et al., 1993), as well as Prudhoe Bay (Martins et al, 1994), were several 100 mg/l have been reported.

Before the full scale PWRI trial was initiated on Ula, a souring study was conducted based on BP's H_2S forecasting model (similar to Ligthelm et al., 1991). The general conclusions were; 1) it is unlikely that souring will take place with seawater injection, 2) it is unlikely that sufficient H_2S would have been generated to reach the producing wells in sufficient quantities to be detected above the natural production levels, even if seawater souring is occuring; and 3) there is a limited potential for bacterial souring with PWRI. Further studies using this model will be performed with the existing water quality of IW.

The H_2S concentration in both SW and IW are regularly controlled and no increase has been observed throughout this test period.

Only limited amount of sulphate reducing bacteria (SRB's) and general anaerobic bacteria (GAB's) have been found (see Table 2), indicating a minimum bacteriological activity. No increases in bacterial counts have so far been observed.

Corrosion Potential

The water quality parameters that could influence the corrosion rates of carbon steel in the injection system are free carbondioxide, oxygen, free chlorine, pH and SRB, as well as residual oxygen scavenger. Some changes in the water quality are occuring when produced water is reinjected on Ula, but since the PW already contains 85% seawater, no dramatic changes have been seen. The fact that PW also contains some oil could influence the formation of a protective surface film on the pipe walls. The variation in water quality occurring when PW is not reinjected could negatively influence the protective surface film.

To date, no increase in the corrosion rates based on ER-probes, corrosion coupons and non-destructive testing (NTD) have been observed since the PWRI started. The average corrosion rate is approximately 0.1 mm/year, which is in the same order as previously found on Ula.

Environmental Aspects

The environmental related water quality data parameters for the produced water during this full scale PWRI trial are summarised in Table 3 (data from week 10, 1995). The

Table 3. Environmental related water quality parameters of produced water discharged (or reinjected) on Ula (week 10, 1995)

Environmental water quality parameter	Produced water (PW)
Ecotox	
Toxicity	
- *Skeletonema costatum* (algae), EC_{50} (%)	14
- *Acartia tonsa* (herbivore), LC_{50} (%)	14
Biodegradation potential in seawater, 28 days* (%)	93
Bioaccumulation potential	
(log P_{ow} - weighted average)	0.7
Organics	
Hydrocarbons (mg/l)	39
Dissolved organic carbon (DOC) (mg/l)	17.3
Acetic acid (mg/l)	5.5
Alkylbenzenes (BETX) (mg/l)	7.2
Aromatics (16 PAH) (mg/l)	0.3
Phenols (mg/l)	1.7
Heavy metals	
Cadmium (mg Cd/l)	< 0.01
Lead (mg Pb/l)	< 0.05
Copper (mg Cu/l)	< 0.05
Zink (mg Zn/l)	0.10
Tin (mg Sn/l)	< 0.05
Mercury (mg Hg/l)	< 0.001

* % biodegradation determined from DOC reduction and CO_2 production.
$E(L)C_{50}$ = 50% effect (lethal) concentration
BETX = benzene, ethylbenzene, toluene and xylene
PAH = polyaromatic hydrocarbons
P_{ow} = octanol/water partition coefficient determined by HPLC

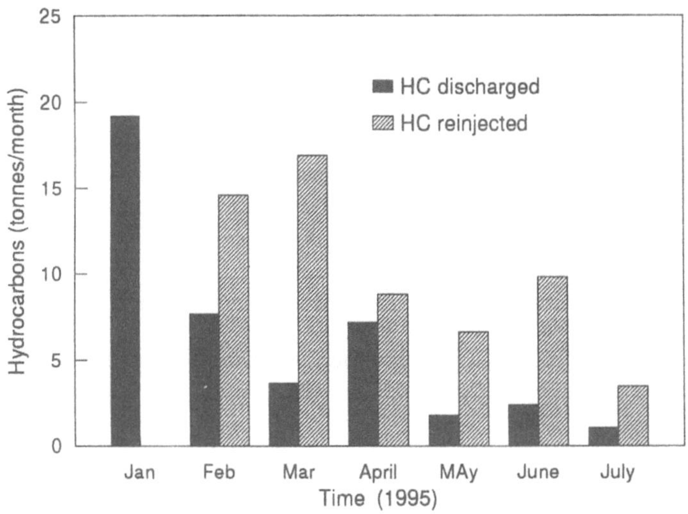

Figure 8. Amount of hydrocarbons discharged and reinjected during the first period of the full scale trial on Ula.

toxicity of produced water shows that when 14% PW is presented in the seawater, 50% of the algaes and herbivores will be killed or affected. At higher dilution factors the acute toxic effects are reduced and eliminated. The organics in the produced water was 93% biodegradable. The produced water is containing minimum amounts of bioaccumulating compounds.

The hydrocarbon concentration in the produced water has, after the hydrocyclones were installed, been reduced to 20-30 mg/l (see Figure 4). Figure 8 shows that the amount of hydrocarbons discharged have been significantly reduced as PWRI has been implemented. 38 tonnes of oil have been discharged and 68 tonnes reinjected in the first period of this trial. In the same period, 20 tonnes of the production/injection chemicals used have been discharged and 50 tonnes reinjected.

The concentration of the other dissolved organics in the PW, like polyaromatic hydrocarbons (PAH) and alkylated benzenes (BETX), are low compared to other typical North Sea fields (Tibbetts et al., 1992). The same is true for the heavy metals (see Table 3).

PRELIMINARY CONCLUSIONS

Based on the results from the first period (approximately 6 months) of the full scale produced water reinjection (PWRI) trial on Ula, the following conclusions can be drawn:

- The initial oil/water separation problems of the newly installed hydrocyclones were overcome during the first 6 months of operation.
- A scale consisting of asphaltenes and sulphides is formed on the inlet of the produced water coolers. There is a need to improve the remedial action.
- To date, no apparent reduction in injectivity, or increase in injection pressure have been observed. The monitoring programme will be continued to evaluate the potential for long term injectivity reductions due to PWRI.
- The scale prediction studies and field observations indicate that neither carbonate nor sulphate scales are likely to be formed in the near well-bore area.

- The souring prediction studies indicate low potential for H_2S formation. No increases in bacterial counts or H_2S have been observed. The results from the H_2S prediction model will be validated with results from long term field monitoring.
- No increases in the corrosion rate have been observed.
- The amount of hydrocarbons, dissolved organics, residual chemicals, and heavy metals in the produced water discharges to the marine environment, have been significantly reduced as the PWRI has been implemented.

ACKNOWLEDGMENTS

The financial and professional contribution from BP Norway Ltd are greatfully appreciated. The financial support from the Norwegian Research Council is also appreciated.

REFERENCES

Bakke, S., Vik, E.A., Øyno, L. , Grüner, H., Hjelmås, T.A., and Stranna, T.E., 1995. Reduction of offshore disharges by produced water reinjection - experiences from the single well trial at the Ula field. Paper presented at the 6th International Symposium on Oil Field Chemicals, Geilo, Norway, 19-22 March.

Evans, R.C., 1994. Developments in environmental protection related to produced water treatments and disposal (produced water reinjection). SPE 27179 paper presented at the Second International Conference on Health, Safety & Environment in Oil & Gas Exploration & Production, Jakarta, Indonesia, 25-27 January.

King, R.E., 1994. Re-injection of produced water in the Maureen field. Paper presented at the BICS International Conference on Produced Water Discharges, Aberdeen, UK, November 24.

Ligthelm, D.J., de Boer, R.B., Brint, J.F. and Schulte, W.M., 1991. Reservoir souring: an analytical model for H_2S generation and transportation in an oil reservoir owing to bacterial activity. SPE paper 23141 presented at the Offshore Europe Conference, Aberdeen, 3-6 September.

Martins, J.P., Murray, L.R., Clifford, P.J.G., McLelland, G., Hanna, M.F., and Sharp Jr., J.W., 1994. Long term performance of injection wells at Prudhoe Bay; the observed effects of thermal fracturing and produced water reinjection. SPE 28936 paper presented at the SPE 69th Annual Technical Conference and Exhibition, New Orleans, L.A., 25-28 September.

Paige, R.W., and Sweeney, F.M., 1993. Produced water re-injection - understanding the problems. Paper presented at the Water Management Offshore Conference.

Sawyer, C.N; McCarthy, P.L. 1978. Chemistry for environmental engineering. Third ed., McGraw Hill Company, New York, USA.

Simpson, A.J. and Paige, K.W., 1991. Advances in Forties field water injection. SPE paper 23140 presented at the Offshore Europe Conference, Aberdeen, 3-6 September.

Tibbetts, P.J.C., Buchanan, I.T., Gawel, L.J. and Large, R., 1992. A comprehensive determination of produced water composition. In: Produced Water, ed. by J.P. Ray and F.R. Engelhart, Plenum Press, New York.

van der Zwaag, C. and Øyno, L., 1995. Application and comparison of injectivity prediction models to estimate Ula water injector performance for PWRI. Poster to be presented at the 1995 International Seminar on Produced Water, Trondheim, Norway, 25-28 September.

Øfjord, G.D., Vik, E.A. and Bakke, S., 1995. Environmental risk management of produced water using the CHARM model - for Ula. Paper to be presented at the 1995 International Seminar on Produced Water, Trondheim, Norway, 25-28 September.

DOWN HOLE SEPARATION (DHS) TECHNOLOGY

J. F. Michelet

READ Process Engineering A.S
Ravnsborgvn. 56
P.O. Box 25
N-1364 Hvalstad

INTRODUCTION

Liquid separation is one of the heavy processes on an oil production facility. The Down Hole Separation (DHS) technology provides an efficient method to reduce the space occupied by process equipment on an oil rig.

READ Process Engineering is developing* a new separation process (DHS) designed to fit inside the casing at the bottom of the well. The DHS is designed to remove a large portion of the produced water from the oil. The separated water can be re-injected through slim branches at the bottom of the well.

The most obvious effect of a DHS, is the reduction of process equipment required topside. Since the total water production at the surface is reduced less and smaller process equipment is needed for the liquid separation.

Production with a DHS opens for new and favourable options in terms of production.

- The pressure drop in the well can be reduced, due to removal of the dense face.
- Selective re-injection can enhance the reservoir recovery and maintain the reservoir pressure.
- Dynamic control of the water-oil contact (WOC-control) through dual completion and re-injection of the separated water in the aquifer may reduce or eliminate water production due to water coning and thus improve oil recovery.

This paper discusses the effects a DHS may have on the production and the environment.

Traditionally, water and oil are separated at the topside in gravity separators. The required retention time and the flow rate determine the size of a gravity separator. In offshore production the gravity separator often occupies large portions of the available space and weight on the platform. A lot of research has been done to reduce the separation facilities. In the mid 80's operators started to use hydrocyclones and centrifuges to treat produced water before

* Patent pending

disposal. Hydrocyclones and centrifuges both make use of centrifugal force and therefore occupy a smaller area and have lower weight. The main difference between the two is that centrifuge operates with high speed rotating parts while hydrocyclones have no moving parts. Consequently centrifuges require more services (down time) than hydrocyclones.

Around 1990 operators started to test separation facilities with hydrocyclones as bulk separators. READ Process Engineering is developing a new separation system, Light Equipment Separation (LES), combining hydrocyclones and traditional separation equipment. The use of hydrocyclones in the separation process upstream and downstream the first stage separators are expected to reduce the space and weight occupied by separation equipment. in fact the separators may be replaced by small, compact hydrocyclone vessels. Production ships are subject to wave motion resulting in poor separation conditions for traditional gravity separator. However, hydrocyclones are not effected in the same manner and the use of the LES system on production ships would therefore improve the reliability and quality of the separation system. A major further step in limiting the separation process at the surface would be to move the liquid separation down hole in the production well. The liquid separation often limits the production from an oil field. By moving the separation down hole this production bottle neck can be removed and space can be made available for other activities. Hydrocyclones are suited for the down hole separation process due to their ability to fit inside a tubing and their high reliability. By separating and re-injecting the produced water down hole, the liquid density can be reduced and the Gas-liquid-ratio (GLR) increases. Thus, a well producing with a high water cut can improve the vertical flow performance radically by producing with a Down Hole Separator (DHS).

This paper discusses the feasibility and the production potential of moving the liquid separation down hole. The feasibility studies are related to well configuration and hydrocyclone arrangement.

COMPACT PROCESS EQUIPMENT

READ Process Engineering has a wide experience in the use of hydrocyclone.

READ has developed produced water hydrocyclones that removes oil from the produced water. A new generation of hydrocyclones is currently being developed. The new generation of hydrocyclones can operate at a higher oil concentration, making them ideal for bulk separation. A bulk hydrocyclone battery operating upstream the first stage separator can reduce the retention time in the first stage separator radically. In a case where a high viscosity emulsion of oil and water is being produced, a bulk hydrocyclone battery can break the emulsion and reduce the required retention time in the gravity separator to an acceptable level.

In addition to the oil/water hydrocyclones do Read Process Engineering also have a sand/liquid hydrocyclones that separates sand from water or oil. The solid liquid hydrocyclones is large and has a high flow capacity. The sand/liquid hydrocyclones is integrated with a sand cleaning unit, "READCLEAN". The sand cleaning unit uses READ Process Engineering's unique cleaning system to remove the oil from the sand particles. Tests have proven that the "READCLEAN" package has an excellent performance. Third party test has given following typical results:

- Oil contents on solids before washing: 26.8 % wt
- Oil contents on solids after washing: 0.52 % wt.

READ Process Engineering are therefore able to deliver a compact and highly efficient sand separation and cleaning system.

READ Process Engineering has specialised in developing compact and efficient process equipment. It is therefore natural that READ Process Engineering now introduce a separation process designed to operate at the bottom of a well (DHS).

DOWN HOLE SEPARATION (DHS) ARRANGEMENT

Gravity separation at the surface is simple but located at an undesirable location of the process train, namely at the topside. The separation conditions at the topside are unfavourable compared with the conditions down hole.

For an undersaturated oil the down hole separation conditions are as follows:

- Two phases only (above the bubble point)
- Maximum droplet size (minimum exposure to shear)

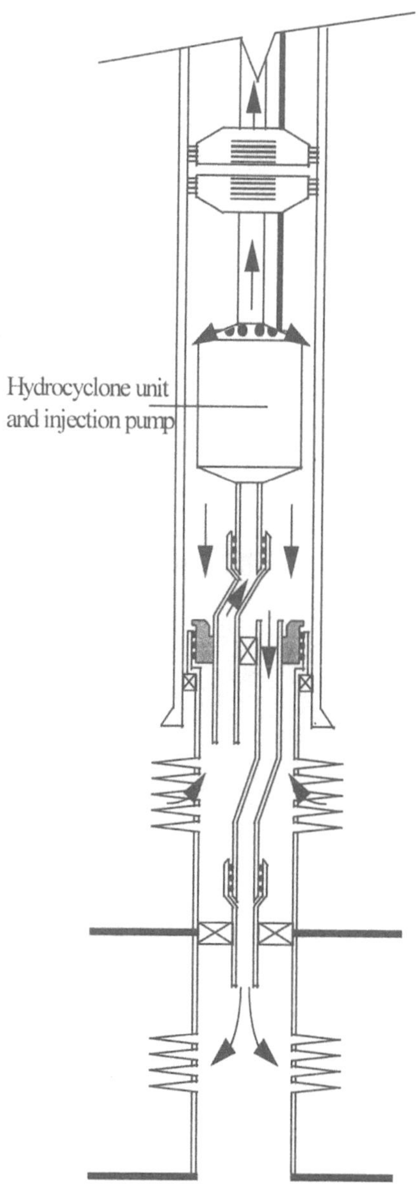

Figure 1. Separation unit and re-injection of water below the production zone.

- Higher temperature (minimum viscosity)
- Maximum dissolved gas (minimum viscosity)
- Asphaltines in soluble form.

Moving the separation process down hole will therefore improve the hydrocyclone efficiency. A hydrocyclone operating at down hole conditions can remove a larger portion of the water phase from the oil phase than a hydrocyclone operated at surface conditions, due to the improved down hole conditions.

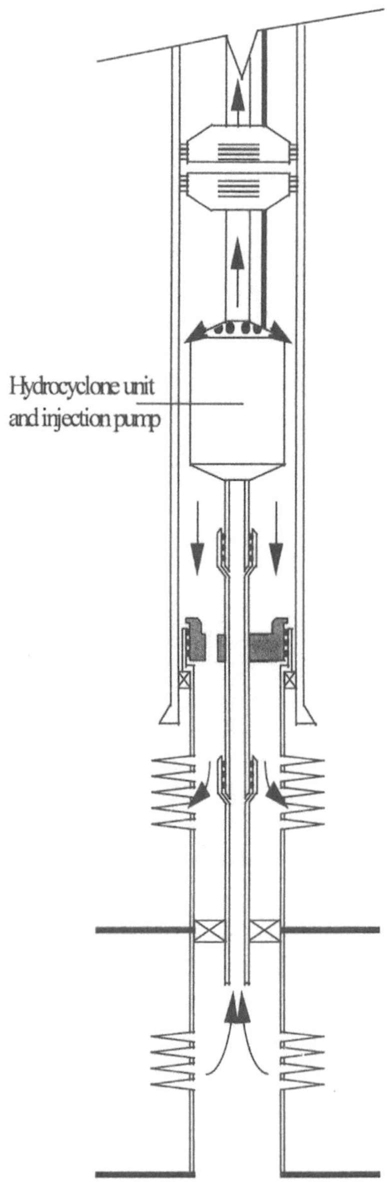

Figure 2. Separation unit and re-injection of water above the production zone.

A key to down hole equipment reliability is simplicity. In order to design a simple but still relatively efficient system one stage separation is required. A separation system with several separation stages, similar to the system used in LES, would require a number of down hole valves, a distribution network and monitoring equipment. It is therefore quite obvious that one stage separation, where the system can be controlled by adjusting the well head pressure, is more reliable than a multi stage system. READ Process Engineering has developed a bulk hydrocyclone with a high separation efficiency enabling us to design a one stage system with high efficiency. However, we continu our develupment work to futher improve the efficiency by adding one more stage of hydrocyclones.

The separation system and re-injection system are showed in Figure 1. The figure illustrates a system where the separated waters are re-injected in the aquifer below the production perforations. The ideal injection point may not, in a specific case, be below the production zone, but above. Figure 2 shows a second production case where the separated water is re-injected above the production perforations. Kjos et. al[2] presents a more detailed description of the well components and procedures for installation and removal of the equipment.

HYDROCYCLONE SEPARATION

Hydrocyclones are a conical device that make use of centrifugal force to separate the light phase from the heavier phase. Figure 3 illustrates how a hydrocyclone operates. It is important to note that there are no moving parts. The fluid enters the hydrocyclone in a tangential feed inlet, which directs the fluids against the wall of the cylindrical section. The centrifugal force creates a core of the light fluid in the centre of the hydrocyclone. This core is drawn off through a vortex finder into the overflow. The heavy phase accumulates on the cone wall and is rejected via the underflow. In the case of liquid separation oil is removed in the overflow and water goes to the underflow.

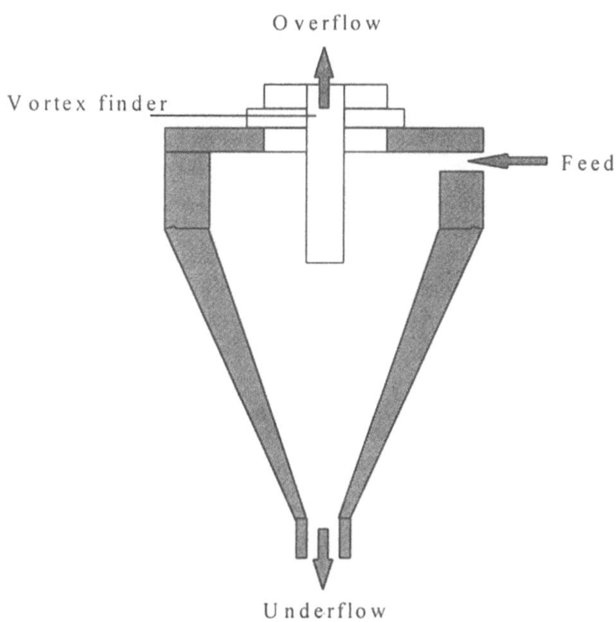

Figure 3. Cross section view of an hydrocyclone. The mixture of oil and water enters tangential at the feed. The light phase (oil) concentrates in the core of the hydrocyclone, due to the centrifugal forces created by the spin. The light phase is removed through the vortex finder and the heavy phase (water) moves down through the underflow.

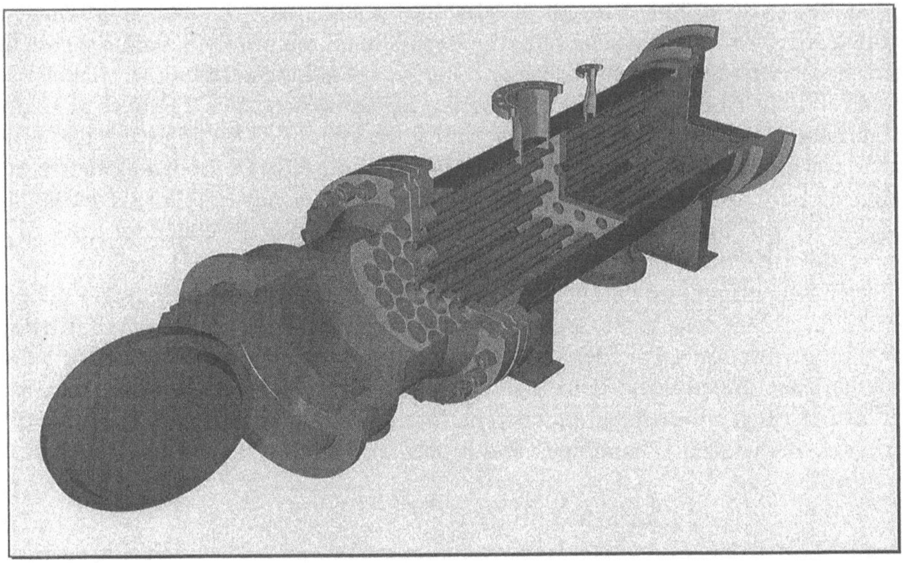

Figure 4. Cross section view of READ Process Engineering's hydrocylone battery. The battery consist of a number of hydrocylone in parallel. The design is used for top side separation of oil and water.

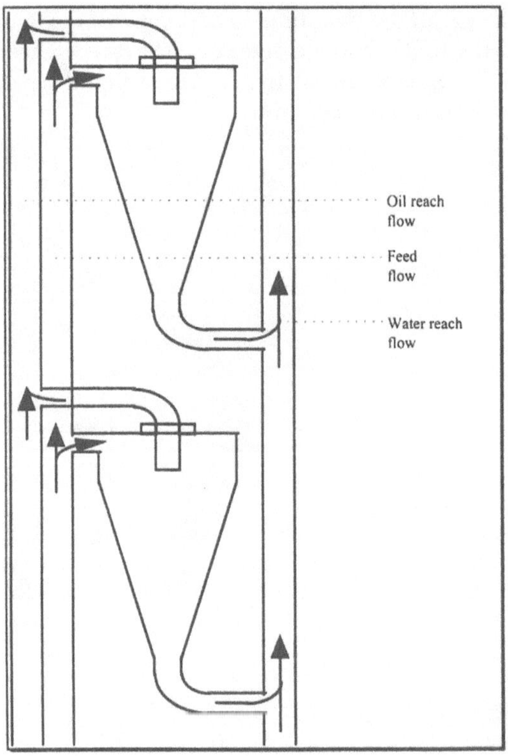

Figure 5. Hydrocyclones in a down hole configuration where the hydrocyclones are operating in parallel.

The flow capacity of one hydrocyclone is low compared with the total flow of a well. Hydrocyclones are therefore placed in parallel. In a topside design several hydrocyclones are placed in a common tank. The design is shoved in Figure 4. This design is not practical in a down hole system where the space is limited in the horizontal plane. The hydrocyclones in a down hole system are therefore arranged vertically. This design is illustrated in Figure 5. The numbers of hydrocyclones required depends on the total flow from the reservoir.

The physical separation conditions down hole are quite different from the conditions topside. The temperature and pressure drops as the fluid flows to the surface. Due to high temperature and pressure, the fluid viscosity is at a minimum at the bottom of the well. Low viscosity improves the separation and reduces the pressure loss in the hydrocyclones. High temperature and pressure at the bottom of the well also reduce the shear forces. A droplet collapses easier if the shear forces are small resulting in improved hydrocyclone efficiency. Hydrocyclones are used in the North Sea for produced water treatment. An analysis of the physical separation conditions topside and down hole indicates that hydrocyclones may be successfully used for separation down hole.

PRESSURE DROP DOWN STREAM THE DHS

Separation of oil and water down hole causes changes in the flow regime down stream the DHS. In almost all cases, the water phase is more dense than the oil phase. Removal of the heavy liquid phase, water, reduces the liquid density. Thus, the hydrostatic pressure drop in the vertical potion of the well is reduced. The effect depends on the density difference between oil and water and the amount of water present in the liquid phase before down hole separation.

Gas are often injected at the bottom of the well to create an artificial lift, i.e. gas lift, in order to reduce the pressure drop in a well. However the effect depends on the natural gas liquid ratio, GLR_{nat}. Studies of gas lift have shown that there is an optimum gas liquid ratio, GLR_{opt}, which creates a minimum pressure drop in the well. A GLR higher or smaller than the optimum will result in a higher pressure drop in the well. A well with a GLR_{nat} smaller than the GLR_{opt} can benefit from gas lift but a well with a gas liquid ratio greater than the optimum gas liquid ratio can not benefit from gas lift. The effects of down hole separation and gas lift are similar. Because the liquid rate is reduced and the gas rate is kept constant, the GLR is increased by using a down hole separator. In those cases where the GLR_{nat} is less than the GLR_{opt} can a down hole separator create an artificial lift, similar to gas lift.

The separation process down hole requires a pressure drop across separation facilities. A typical pressure drop across the down hole separator is eight bar. This pressure drop is insignificant compared with the pressure gained by using a down hole separator in those cases where the well is producing with a high water cut.

RESERVOIR BEHAVIOUR UPSTREAM THE DHS

The flow rate in the reservoir depends on the pressure drop in the reservoir, i.e. draw down. A large draw down results in a high flow velocity in the reservoir. The effect of an increased draw down depend on the permeability of the reservoir. A reservoir with a high permeability, i.e. high Production Index (PI), can increase the production rate with a large portion by increasing the draw down compered with a reservoir producing with a small PI.

As an example, the production from a well producing from a reservoir with a production index of 35 [Sm^3/Day/Bar] was simulated with and without a down hole separator for different water cuts. The well was 2700 m deep and had a natural GOR of 71 Sm^3/Sm^3.

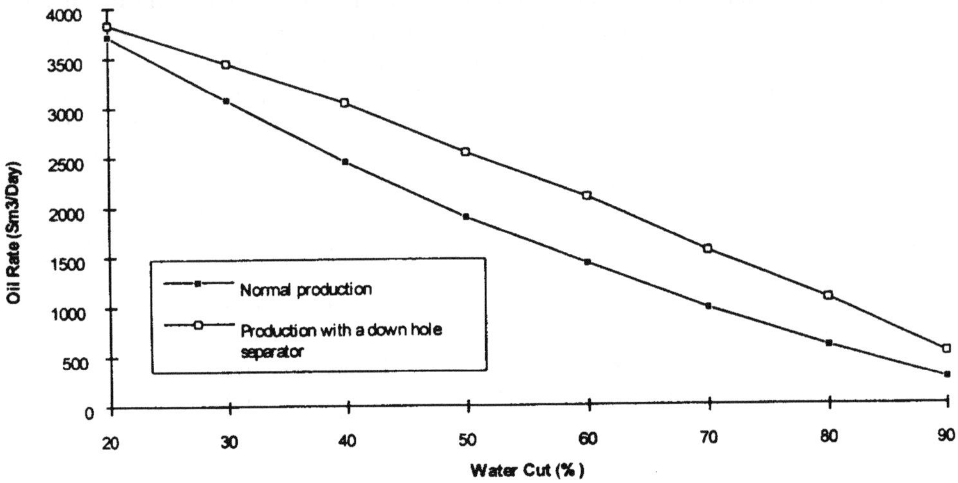

Figure 6. Simulations of well performance at different water cuts with and without a DHS. The simulations is based on data from the Statfjord formation. The Production Index used was 35 (Sm^3/day/Bar), well depth 2700 m and the GOR was 71 Sm^3/Sm^3.

The fluid data was based on data from the Statfjord formation in the North Sea. The results are presented in Figure 6 (Kjos et. al[1]).

When an oil reservoir is produced, the pressure around the well bore is reduced in order to allow the oil to flow. In a partly penetrated well the original water oil contact (WOC) represents a constant pressure line as long as no water is being produced. The WOC around the well bore will move towards the production well due to the reduced pressure around the well bore. The WOC around the well bore will take the form of a cone (Muskat et. al[4]). The height of the cone depends on the draw down and the density difference between oil and

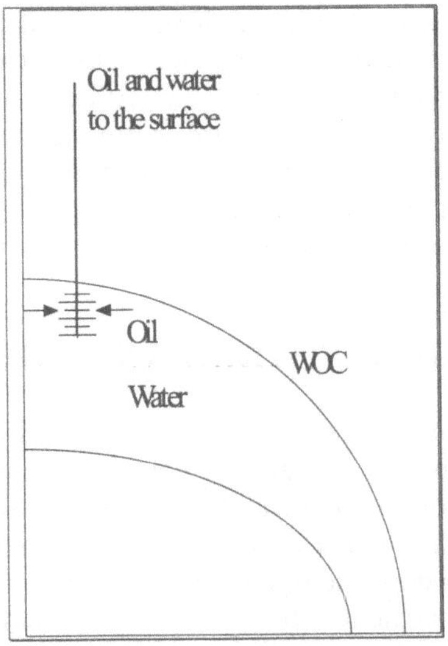

Figure 7. Production from a partial penetrating well. The figure illustrates how a cone develops around the well bore.

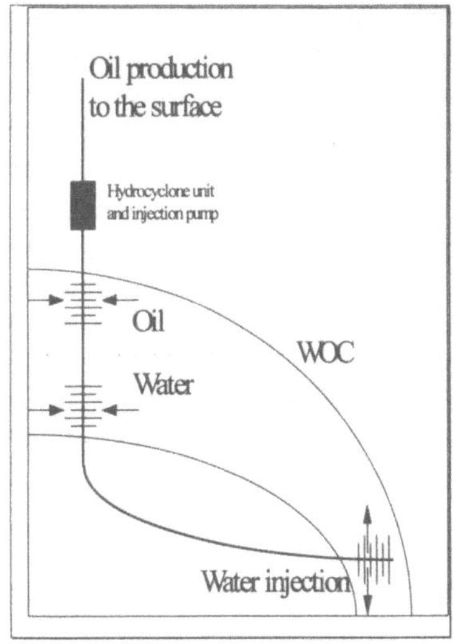

Figure 8. Production from dual perforation zone and re-injection of the produced water down into the aquifer. In the case illustrated the drag force from the aquifer production zone is stronger than the drag force from the oil zone production zone. Thus, a negative cone develops around the well bore.

water. A second perforation zone can be introduced in the aquifer, i.e. dual production zone, in order to decrease the cone height. The bottom production perforation will create a draw down that tends to drag the fluids downward, thus reducing the height of cone. If the drag produced from the bottom production perforation is larger than the drag from the top perforation a negative cone will develop. Figure 7 shows a traditional completed well, i.e. partial penetrated well, with a normal cone. Figure 8 shows a well producing from two intervals, one in the oil-zone and one in the aquifer, with a negative cone. Dual completion in the oil zone and in the aquifer may in principle balance the draw down above and below the WOC so that water coning is eliminated. Of course this would require individual control of the production rates, which may be difficult to achieve. However, for the purpose of this study, we assume that production rates may be varied individually. In a homogenous reservoir where the two production rates are balanced, see Figure 9, the bottom perforations will produce pure water and the upper perforations will produce pure oil. Of course in a

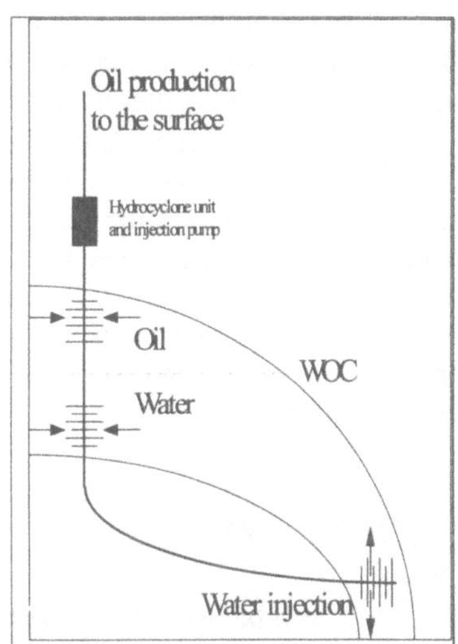

Figure 9. Production from dual perforation zone and re-injection of the produced water down into the aquifer. In the case illustrated the drag force from the aquifer production zone balance the drag force from the oil zone production zone. Thus, no cone develops around the well bore.

homogenous case there would be no need for a down hole separator. However, there are no well producing at homogenous conditions. The upper perforations will therefore, in a normal case, produce some water along with the oil. However, it may be feasible that the water produced through the lower perforation interval is re-injected directly into the aquifer without running through the separation unit. Thus, the water capacity is increased.

Kleppe et. al[3] showed that in a specific case the aquifer rate (lower perforation zone) must be between 1.0 and 1.5 of the oil zone rate (upper production zone) in order to avoid water coning. In the cases that Kleppe simulated was the two perforation intervals mirrored around the original WOC. The WOC moves upward towards the upper production zone towards the end of the production period, due to aquifer expansion and/or re-injection of water. Since the WOC moves upwards as the oil is produced the balancing aquifer rate (lower perforation zone) at the end of the production period is larger than the balancing aquifer rate at initial conditions.

POTENTIAL ENVIRONMENTAL BENEFITS

Water production may cause production problems in terms of corrosion and forming of hydrates in the pipeline. Normal procedure to avoid this problem is to inject corrosion and hydrate inhibitors. These chemicals are not only expensive but also often toxic. Separation down hole and re-injection of the produced water back into the formation reduces the problems with corrosion and hydrates. The need for corrosion and hydrates inhibitors are reduced or eliminated when the water is separated and re-injected down hole.

Because of strict regulation for produced water dumping, chemicals that flocculate the oil are often used to clean the produced water before disposal. Flocculators have the ability to increase the oil droplet's size. Large droplets are easier to separate than small. Because the oil droplets often are small at the surface, flocculators are used to improve separation at the surface. Down hole the oil droplets are larger and flocculators are not required to separate oil and water. Flocculators also pollute the environment and their use should therefore be minimised.

Sand particles in the flow may cause erosion. Especially are elements where the flow is restricted, e.g. at valves, bends, subject to high erosion rates. This erosion may cause accidents if the eroded elements fail before being repaired. Accidents can have tragic consequences on people and environment. Sand is removed with the water down hole and all elements down stream the separator is protected from sand erosion. Erosion on the process equipment topside is therefore reduced and safety is improved.

In a normal topside process the produced water contains a small portion of oil after the water treatment. This oil is usually dumped to the sea. As dumping regulations are getting more strict, the separation equipment required is getting more expensive. More and more operators are considering re-injection of the produced water back into the reservoir in order to meet future regulations. Production with a down hole separator and re-injection of the produced water eliminates the pollution from produced water. A down hole separator provides an environmentally friendly alternative production method that also is economically profitable.

CONCLUSIONS

1. Production with a DHS may extend field lifetime and increase oil recovery, due to improved well performance and reservoir flow conditions. A well producing at

conditions where the well normally would be shut down, due to high water production rate, can in many cases continue to produce with a DHS.
2. The use of a down hole separator reduces the required space for liquid separation equipment at the surface. Thus, more space is available for other activities on the surface or smaller production vessels can be used.
3. Hydrocyclones are well suited for separation down hole, due to their ability to fit inside a small tubing, high reliability (no moving parts) and their high efficiency.
4. The physical conditions down hole are favourable for separation compared with conditions at the surface. Thus, a hydrocyclone operating at bottom hole conditions is expected to have a higher efficiency than a hydrocyclone operating at the surface.
5. Hydrostatic pressure drop in the vertical portions of the well is reduced by removing the produced water down hole, due to water being more dense than oil.
6. In the cases where the GLR_{nat} is less than the GLR_{opt} a down hole separator can create an artificial lift similar to gas lift.
7. Water coning control is feasible by dual completions in oil zone and in aquifer, by balancing the withdrawal rates.
8. By re-injection of the produced water, we avoid bringing large volumes of water to the surface, reduce pressure decline and enhance the water displacement of oil.
9. The use of a DHS can reduce the use of toxic chemicals such as corrosion inhibitors, hydrates inhibitors and flocculators.
10. By re-injecting the produced water back into the reservoir no oil in the produced water is exposed to the open environment.

NOMENCLATURE

DHS	=	down hole separator
GLR	=	gas-liquid ratio, Sm^3/Sm^3
GLR_{nat}	=	natural gas-liquid ratio, Sm^3/Sm^3
GLR_{opt}	=	optimum gas-liquid ratio, Sm^3/Sm^3
GOR	=	gas-oil ratio, Sm^3/Sm^3

REFERENCES

1. Kjos, T., Ledsaak, G. M., Gunnarson, B., Michelet, J. F.,: "Downhole Separation and Injection of Produced Water", presented at the PEA Meeting nr. 95/1, Stavanger, Norway, March 9 1995.
2. Kjos, T., Sangesland, S., Michelet, J. F. and Kleppe, J.: "Down Hole Water-Oil Separation and Water Reinjection Through Well Branches", paper SPE 030518 (to be) presented at the 1995 SPE Annual Meeting, Dallas, Texas, 22-25 October 1995.
3. Kleppe, J. , Michelet, J. F., Sangesland, S. and Kjos, T.: "Dynamic Water Coning Control Through Dual Completion, Downhole Separation and Reinjection of Water", paper IATMI 950141 presented at Symposium on Production Optimization, Society of Indonesian Petroleum Engineers, Bandung, Indonesia, 24-26 July 1995.
4. Muskat, B. and Wyckoff, R. D.: "An Approximate Theory of Water-Coning in Oil Production", *Trans. AIME*, (1936)144-163.

EVALUATION OF NF/UF MEMBRANE TREATMENT IN DE-OILING PRODUCED WATER IN-SITU AT A HEAVY-OIL PRODUCTION FACILITY IN WESTERN CANADA

J. Fraser,[1] A. Zaidi,[1] M. Preston,[1] T. Liu,[1] and E. Doyle[2]

[1] Wastewater Technology Centre (operated by RockCliffe Research Management Inc.)
867 Lakeshore Road
Burlington, Ontario, Canada, L7R 4L7
[2] Delcan Corporation
2151 Harris Crescent
Burlington, Ontario, Canada, L7R 1G3

INTRODUCTION

Over the past three years, the Wastewater Technology Centre (WTC) in cooperation with governmental department agencies and some support from interested parties within the oil producing industry, have been evaluating the application of MF/UF/NF for treatment of produced water from heavy oil recovery operations.

The overall objective of the study has been to identify the most effective commercially available membrane(s), together with the best operating conditions with respect to the removal of oil and contaminants from produced water.

The project was originally conceived to involve three phases. At present, Phase I and Phase II are complete.

In Phase I a series of MF and UF organic and inorganic membrane coupons (flats) were evaluated at bench scale, both on-site at the Norcen heavy oil recovery operation in Alberta and at WTC's testing facilities in Burlington, Ontario. The Norcen site is a steam flood operation with a produced water flow of approximately 300m^3/day. A knock-out tank and two skim tanks in series are used for gravity separation of the oil. The produced water samples used for this phase of the study were taken form the final skim tank.

The specific focus of Phase I was:

- correlate membrane performance to membrane characteristics for the selected systems;

- define membrane fouling with respect to specific produced water characteristics; and
- through this knowledge select the protocols required for a more defined larger scale testing program.

Twenty-five polymeric (organic) and 8 inorganic membranes were tested for permeate production over time (flux), oil/water rejection and recovery of the two preceding parameters after cleaning. The produced water stream ranged in oil content from 205 mg/L to 350 mg/L and ranged in temperature from 65 to 80 °C over the test period.

From this work it was concluded that in general, oil/water separation performance was considered acceptable for all the units tested based on permeate oil concentrations of <10 mg/L. Flux rates varied greatly, ranging between 5 L/m^2/hr and 100 L/m^2/hr, after six hours of screening tests. The highest flux rates were associated with those membrane exhibiting the greatest degree of hydrophilic character and to a lesser extent with decreasing pore size for polymeric membranes. Inorganic membranes showed an opposite effect in that, larger pore sizes had a tendency for increased flux rates. In general, flux rates increased, and were found to be less variable with increasing temperature. It should be noted here, that although coupon testing is quite valid in assessing oil/water separation it really does not parallel commercial membrane systems in terms of flux comparison.

Fouling of the polymeric membranes was generally the result of dynamic fouling associated with pore plugging and deposition of foulants. There was no significant flux decline due to adsorption. Iron, silica and aluminum were detected on the surface and in the pores of several of the fouled membranes. A pH adjustment pretreatment was evaluated for control of fouling. The results demonstrated that increased flux rates were realized at pH 5 and pH 11 for all organic membranes with the exception of one, though a greater correlation remained between flux and membrane material.

The inorganic membrane manufacturers recommend that units with 300 000 MWCO is best suited for oil/water separations. In the Phase I study three Tech-Sep modules were tested with 10 000 MWCO, 50 000 MWCO and 300 000 MWCO respectively, and indeed increasing flux rates corresponded to increasing MWCO. However since these tests were not run in parallel and considering the potential for variation within the produced water characteristics, further confirmation of this relationship is required. Generally, it was noted that flux decline was less evident with inorganic modules as compared to the organic modules under these study conditions. Highest flux rates were observed at pH 11 for all inorganic membranes tested, an observation which may be associated with the softening of the produced water at the elevated pH.

Based on the Phase I study, testing protocols were formulated for Phase II, focussing on a limited number of selected modules to be tested on-site at the Amoco Canada, Wolfe Lake heavy crude production facility in Alberta.

For Phase II , the work was directed towards the:

- evaluation of nanofiltration flatsheets with respect to flux, and the rejection of contaminants, in produced water applications
- evaluation of flux and rejection performance of those modules that demonstrated the greatest potential in the Phase I testing; and
- investigation of specific fouling incidents that may occur during the field testing.

SITE FACILITIES

At the Wolfe Lake site, oil recovery is enhanced by stream injection. Produced water flows through the Amoco treatment system from which a large portion of it is reused, with

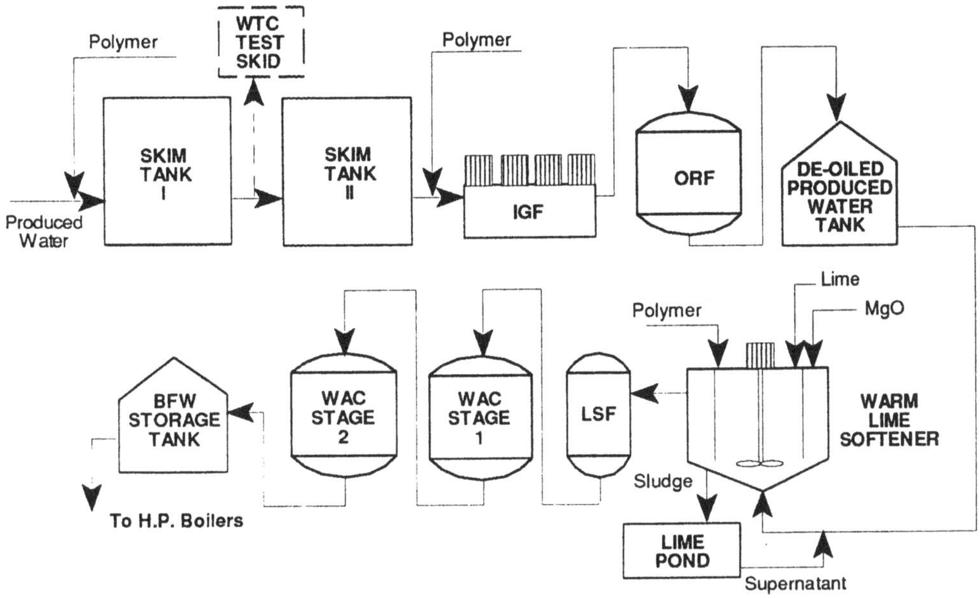

Figure 1. AMOCO produced water treatment train.

the remainder being deep well injected. Figure 1 shows a schematic of the treatment train, Table 1 illustrates the characteristics of the Wolfe Lake heavy crude and Table 2 shows the characteristics typical of the produced water stream at Wolfe Lake.

Deimulsifier agents are added to the produced water stream prior to Skim Tank 1. The WTC bench-scale pilot MF/UF testing unit withdrew a feed stream from the line transporting the aqueous fraction from Skim Tank 1 to Skim Tank 2.

The feeds for the flatsheet testing unit were sampled from the effluent of the de-oiled produced water tank, the effluent from the warm lime softener prior to the sand filter and from the discharge of the WAC Stage 2 unit.

Table 1. Characteristics of Amoco Wolfe Lake heavy crude oil

Parameter	Sample
Relative Density	0.985
API Gravity (degrees)	10.5 - 11.5
Pour Point (°C)	15
Kinematic Viscosity (cPs) @ 30°C	10,000
Kinematic Viscosity (cPs) @ 45 °C	120
Saturates (%)	15
Aromatics (%)	55
Asphaltenes (%)	15.6
Polars (%)	14.4
Water (volumetric fraction)	0
Sulphur (mass fraction)	3

Table 2. Produced water analysis at Wolfe Lake and Norcen heavy-oil field sites

Component	Units	Average Value Wolfe Lake	Average Value Norcen
Hardness	mg/L as $CaCO_3$	136	513
Calcium	mg/L	42.2	98
Magnesium	mg/L	7.4	67
Total Alkalinity	mg/L as $CaCO_3$	371.3	399.8
Bicarbonate	mg/L as $CaCO_3$	355.7	487.4
pH	units	7.7	8.1
Iron	mg/L	0.2	0.52
Chlorides	mg/L as Cl	2,763	7,200
Sulphites	mg/L as SO_4	87	<1
Sodium	mg/L	1,900	4,520
TDS	mg/L	5,732	12,263
Silica	mg/L as SiO_2	275	69
TOC	mg/L as C	141	58.8
Carbon Dioxide	mg/L as CO_2/L	20	
O & G	mg/L	40	41

EXPERIMENTAL PROGRAM

Nanofiltration Membrane Testing

The following seven NF flatsheet membranes were tested to identify optimal operating parameters and to derive the MWCO range best suited for full-scale implementation in treating produced water:

1. Desal 5 - proprietary thin film organic membrane manufactured by Desalination Systems Inc. with a 200-300 MWCO range and a maximum efficient operating temperature of 50°C.

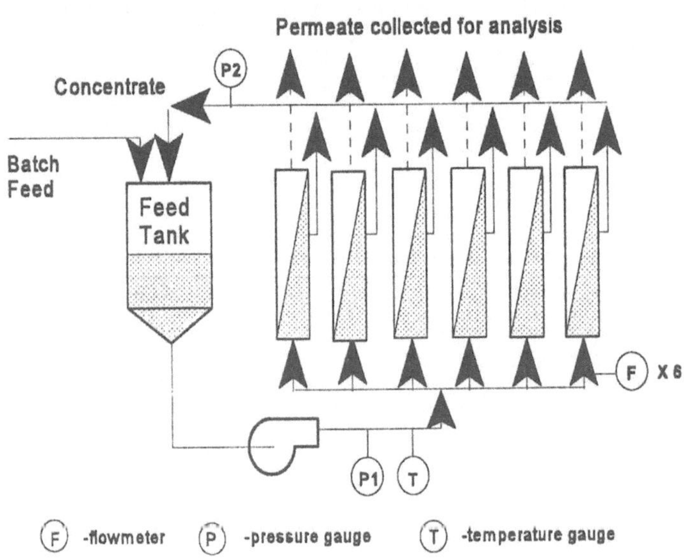

Figure 2. WTC six cell flatsheet test unit.

2. NTR 7410 - charged, thin film composite membrane with a maximum operating temperature of 50°C, and a NaCl rejection rate of 10%. It is manufactured by Nitto Denko.
3. NTR 7450 - similar in most aspects to NTR 7410 except in performance where it is defined as having a 50% NaCl rejection.
4. BW30 - thin film composite membrane designed for treatment of feed water with turbidity values of <1 NTU and a maximum temperature of 45°C, and is manufactured by Filmtec Corporation.
5. MPF 31 - manufactured by Membrane Products Kiryat Weizman Ltd., this membrane has a MWCO of 350 and has a maximum operating temperature of 70°C.
6. NF 40 - thin film composite with a MWCO of 300 and a maximum operating temperature of 45°C is manufactured by Filmtec Corporation.
7. PES 10 - polyether sulphone membrane designed for temperatures of up to 90°C. Manufactured by Hoechst Celanese Corporation, it has a nominal MWCO of 4000.

The 6.5 cm diameter circular membrane coupons were tested in parallel configuration. Figure 2 shows a schematic of the test unit.

Three feed waters were tested: 1) the effluent from the de-oiled produced water storage tank, typically at 85 °C; 2) the effluent from the warm lime softener in which both the silica and hardness have been reduced to <40 mg/L. The temperature of this stream is typically 70°C; and 3) the effluent from the weak acid cation unit in which the hardness has been further reduced to 1.0 mg/L or less. The temperature of this stream is generally between 50 - 60°C.

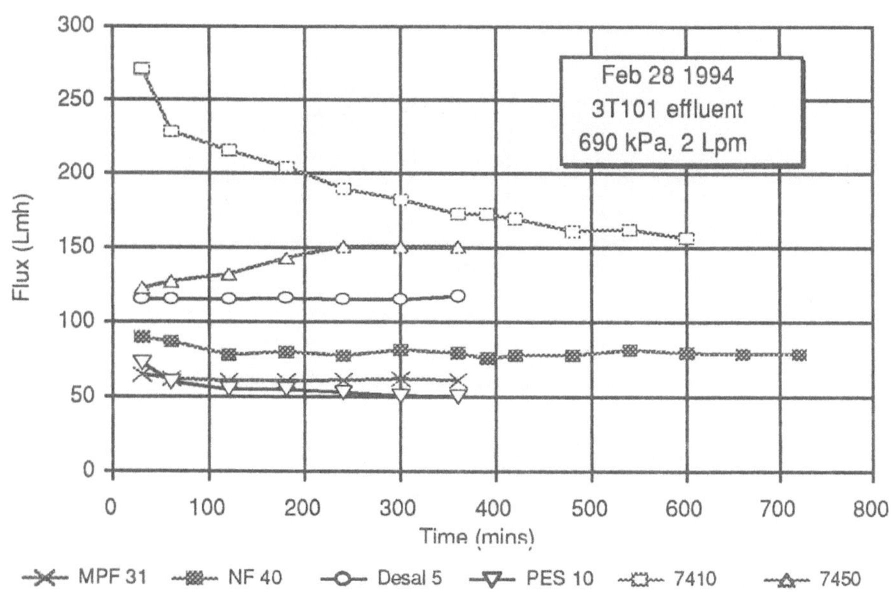

Figure 3. NF Flux on 3T101 (80 C vs. time).

RESULTS AND DISCUSSION

The relatively high temperatures of the feed streams caused substantial stress to the majority of the membranes tested.

The Desal 5 membrane was the only unit, that performed reasonably well for two of the three feed streams. The other units demonstrated flux rates that were substantially lower than their designed value in every case and were ineffective in rejecting silica and hardness to values taht are acceptable as boiler feedwater. Oil and grease levels, though reported as less than 40 mg/L (freon, gravimetric), were consistiently monitored at < 15 mg/L by Horiba OCMA22 IR. The NTR 7450 demonstrated the best flux performance, hwever exhibited significant physical stress under the test conditions. The desal-5 performed well with the two of the three feed streams. Flux, though not as high as the NTR 7450 showed minimal decline over the six hour test period. The Desal-5 showed the least effect of the relatively high temperature conditions under which the tests were conducted. Once more it should be noted that flux rates observed using flatsheets can differ substantially from those realized in engineered systems using the same membrane material.

Figures 3, 4 and 5 illustrate the flux experienced for specific feed streams of some of the NF membranes tested. Those membranes that delivered no measurable flux are not included. Table 3 presents the analyses of NF permeate solutions generated in this study from the treatment of the de-oiled Produced Water Tank Effluent (3T101).

Three feed waters were tested: 3T101- the effluent from the produced water storage tank, 3W102- the effluent from the warm-line softener and 3W103 - the effluent from the weak acid cation unit.

UF MEMBRANE MODULE TESTING

Two ultrafiltration membrane modules were selected for further testing based on experience gained in Phase I. The Desal G-80 spiral wound module and the Tech-Sep

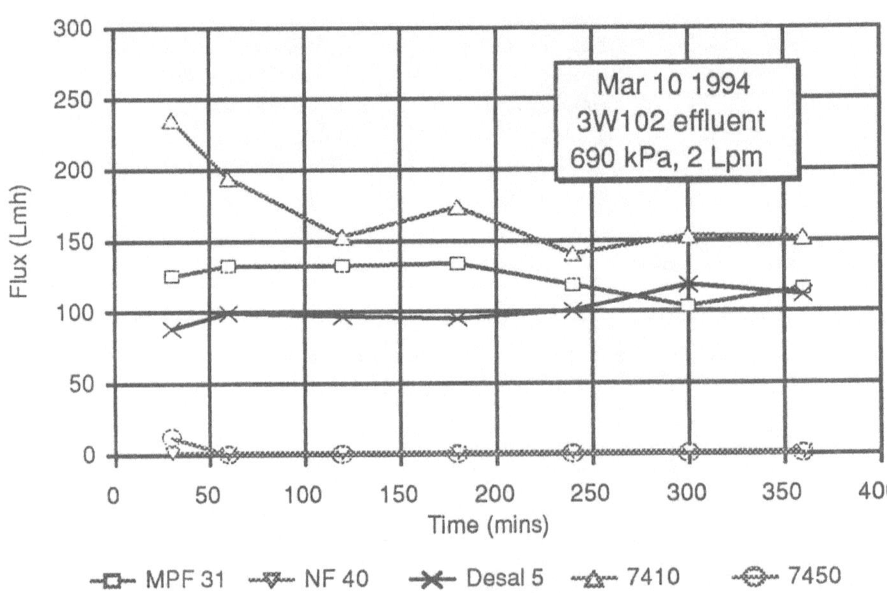

Figure 4. NF Flux on 3W102 (65.C vs. time).

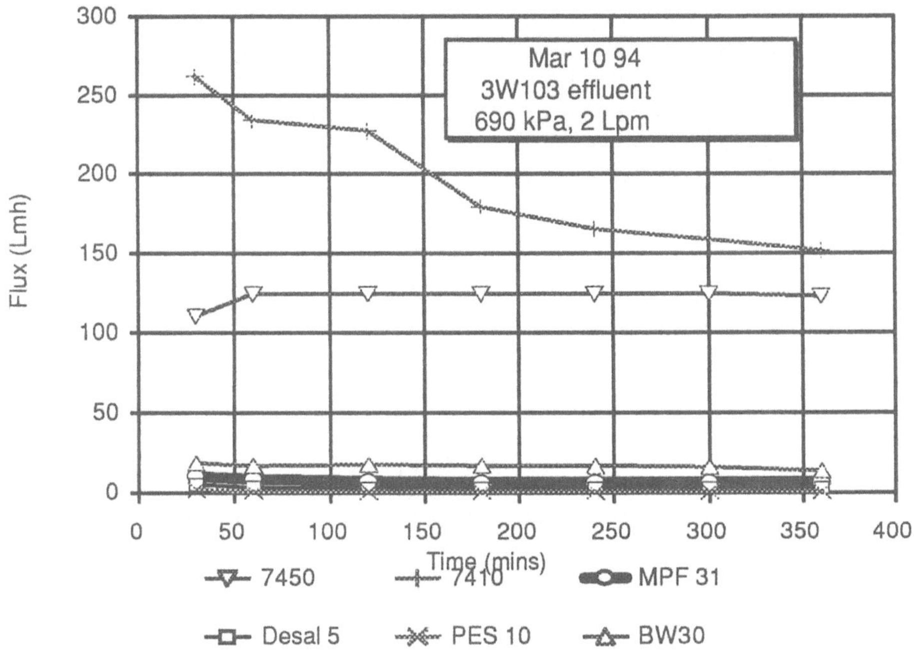

Figure 5. NF Flux on 3W103 (50 C vs. time).

Table 3. Feed 3T101 and NF flatsheet permeate concentrations

	TOC mg/L	TOC % rej	Tot a mg/L	pH	Carb mg/L	BiCar mg/L	TH mg/L	Si mg/L	O&G mgL
Feb 28									
3T101	176		381	8.3	0.8	463	135	131	<40
PES 10	158	10.27	323	8.4	8.2	377	111	130	<40
Mar 01									
3T101	178		400	8.5	20.6	446	137	128	<40
Desal 5	99.6	44.05	206	8.4	13.2	224	57.6	134	<40
Mar 02									
3T101	171		410	8.4	17.8	464	141	132	<40
NF40	136	20.47	280	8.4	9.2	323	85.3	144	<40
Feb 23									
3T101	x		355	8.4	13.9	404	120	x	<40
7410	x		313	8.4	17.8	345	103	x	<40
Feb 24									
3T101	245		386	8.1	<	471	132	133	<40
BW30	95.8	60.89	181	8.1	<	221	16.3	<	<40
7450	191	22.04	314	8.1	<	383	88.3	157	<40
MPF31	179	26.94	289	8.1	<	352	96.5	<	<40
Mar 04									
3T101	178		397	8.2	<	484	151	121	<40
MPF 31	133	25.53	310	8	<	378	109	198	<40

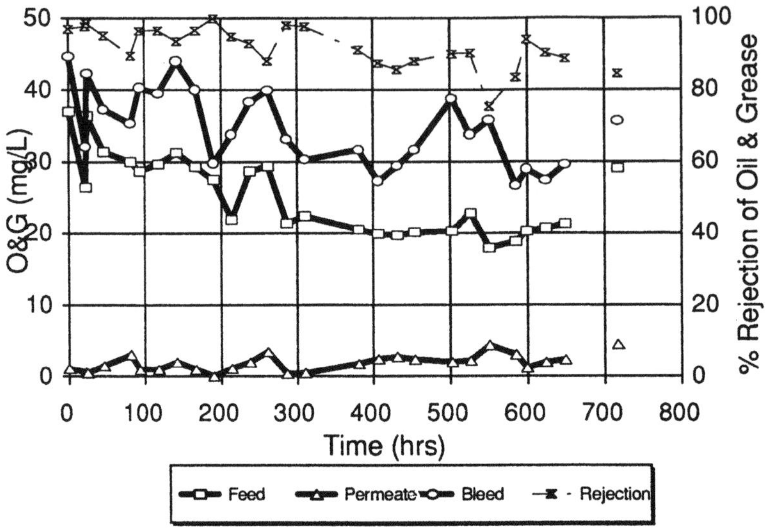

Figure 6. Desal G-80 total oil rejection.

SC-7M9 inorganic tubular module were both tested for extended periods on-site at Wolfe Lake.

Desal G-80 Module Description

The Desal G-80 is in a thin film composite, configured as a spiral wound membrane. It has a surface area of 8.92 m^2 and a defined pore size of 15 000 Dalton. The tolerable operating parameters for this module are:

- maximum pressure 4 200 kPa;
- maximum temperature 50 °C
- pH range 2 - 11

Desal G-80 Testing Conditions

During the test period the Desal G-80 was run with an inlet pressure of 345 kPa and an average outlet pressure of 330 kPa (the outlet pressure varied depending on the degree of fouling). Influent flow was 57 Lpm with the temperature averaging between 55 and 60 °C. The total oil concentration of the feed, as measured by an on-site Horiba OCMA - 220 analyser, ranged from 30 to 40 mg/L. The study period involved the continuous operation of the module over 700 hours, punctuated with 1-2 hour cleaning cycles every 100 to 150 hours as required.

Desal G-80 Test Results

Throughout the test period, the Desal G-80 consistently achieved rejection rates of 90% or greater, attaining permeate oil concentration values of <3 mg/L as illustrated in Figure 6.

The maximum flux attained was 60 Lm^2/hr falling to <10 Lm^2/hr over approximately 100 hours of continuous operation. There was a steady decline in flux with decreasing

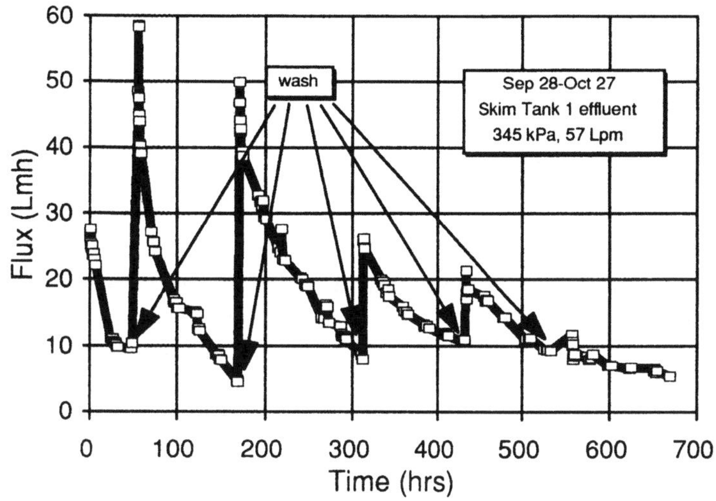

Figure 7. Desal G-80 flux performance (65 C vs. time).

recovery after each cleaning cycle indicating irreversible fouling of the membrane. Figure 7 represents the change in flux that occurred over the 700 hour test period.

Tech-Sep SC - 7/M9 Module Description

The Tech-Sep SC-7/M9 is a ZrO_2/TiO_2 tubular membrane module that consists of seven 1 200 mm long tubes, each having a diameter of 6 mm. The total membrane surface is 0.16 m^2 and has a defined porosity of 300 000 Dalton. Being relatively thermally inert, it

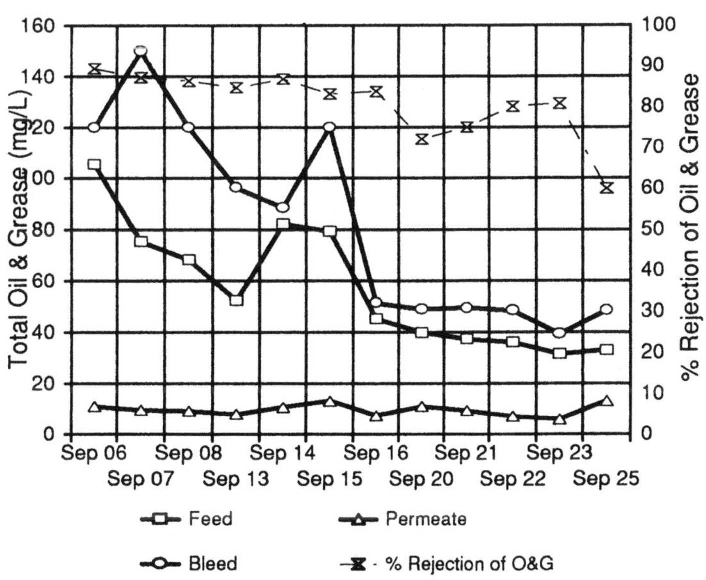

Figure 8. Tech-Sep SC-7/M9 total oil rejection.

can withstand temperatures of up to 350 °C. The tolerable operating parameters for this module are:

- maximum pressure 4 000 kPa;
- maximum temperature 350 °C; and a
- pH range 0 to 14.

Tech-Sep SC - 7/M9 Testing Conditions

The module was tested in runs lasting from 3 to 5 days, with routine cleaning cycles occurring every 30 to 40 hours. The inlet flow was 59 Lpm with a crossflow velocity of 5 m/s. The applied pressure was 240 kPa. The feed temperature varied greatly, between 25 and 75 °C with total oil concentrations ranging from 40 to 100 mg/L as measured by an on-site Horiba OCMA- 220 analyser.

Tech-Sep SC - 7/M9 (Test Results)

On average, the module attained a 60 to 90% rejection of total oil over the test period with permeate concentrations between 5 and 15 mg/L. Considering the wide variation in the feed stream oil concentrations, the total oil content of the permeate stream remained relatively uniform. Figure 8 displays the oil rejection performance of this module.

The flux performance ranged from a maximum of 500 Lm^2/hr to a minimum of 200 Lm^2/hr over the 50 hour run periods between cleaning cycles. Cleaning was effective in regenerating acceptable flux performance indicating that the fouling mechanism was quite different to that experienced with the Desal G-80 module. Figures 9 and 10 show the flux performance of the Tech-Sep module over the test period.

The Tech-Sep module was allowed to run continuously using the same parameters as defined in the test conditions over a period of 140 hours without cleaning. Starting with a flux level of approximately 400 Lm^2/hr a steady decline was observed, with acceptable flux values being maintained for 100 hours. This would suggest that this type of module has the potential of being productive in the treatment of produced water. Figure 11 illustrates the flux performance without any cleaning protocols being applied to the Tech-Sep module.

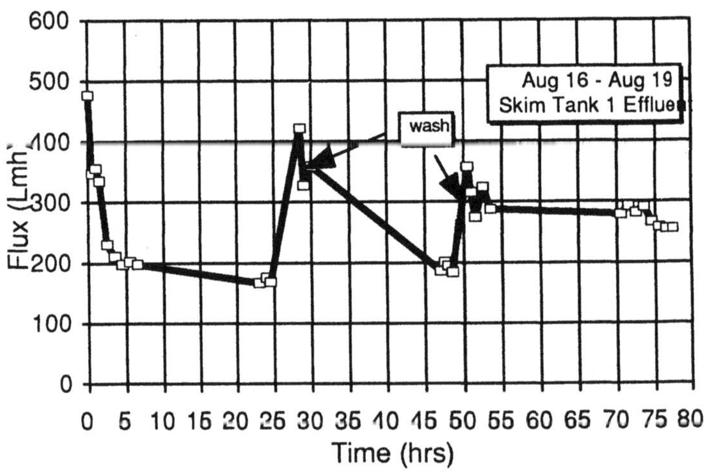

Figure 9. Tech-Sep module flux - period 1 (65 C vs. time).

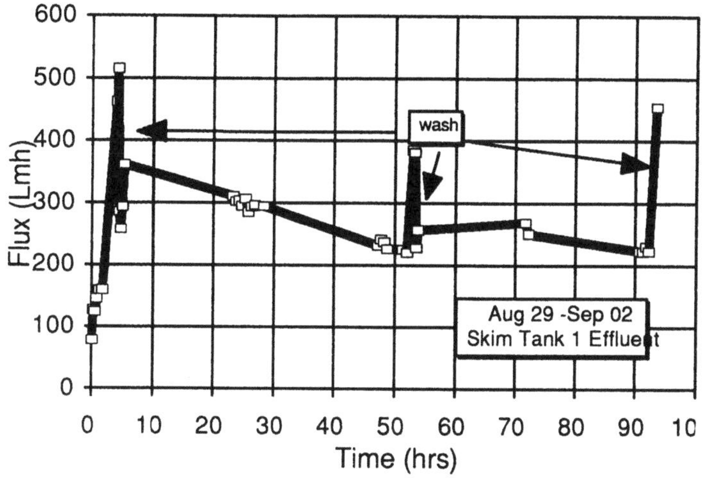

Figure 10. Tech-Sep module flux - period 2 (65 C vs. time).

Tech-Sep SC-7/M9 vs Desal G-80 Performance Comparison

In comparing the flux performances of the Desal G-80 and Tech-Sep SC-7/M9 modules respectively, it is obvious that there is a great difference in the absolute flux rates. There is also a significant difference in the flux recovery after cleaning between the two modules, which suggests that the degree of fouling is of much greater concern with the Desal G-80 as compared to the Tech-Sep module. However two important factors must be considered before conclusions are drawn: 1) the difference in porosities is significant and can perhaps be used to explain the differences in the respective flux values; 2) the operating times between cleaning events were quite different which may impact upon the flux recovery after cleaning, of the respective modules; and 3) the oil concentrations of the respective feed streams were substantially different. The lower porosity of the Desal G-80 module could

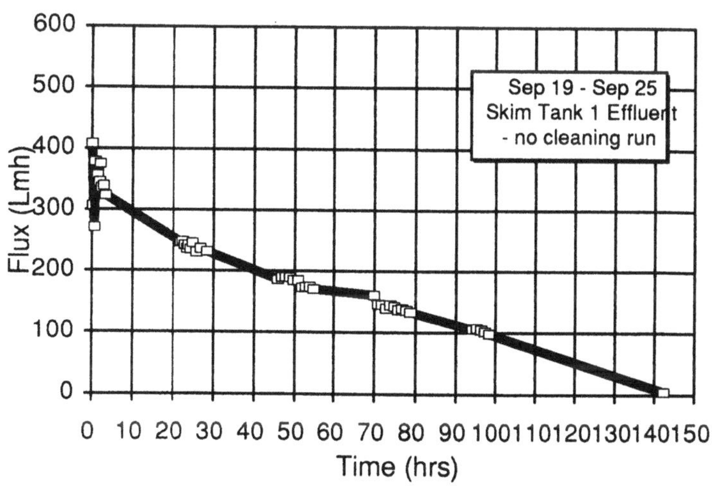

Figure 11. Tech-Sep module flux without cleaning (65 C vs. time).

Table 4. Module performance comparison

Component	Units	Average Value Influent	Average Value Desal G-80	Average Value Tech-Sep
Hardness	mg/L	152	140	124
pH		7.9	7.9	7.9
Silica	mg/ as SiO_2	204.2	111.6	180.0
Conductivity	mS	16.10	15.9	16.05
TOC	mg/L as C	141	152	164
Total O&G	mg/L	35.0	1.1	10.0
Emulsified O&G	mg/L	32.1	0.9	9.2

also explain the lower oil concentrations realized in the permeate of this unit as compared to that of the Tech-Sep permeate stream, although there is the question of the highly variable oil concentrations of the respective feed streams. Table 4 presents a comparison of the rejection performance of the two modules.

Upon examining the modules on completion of the test program, it was noted that the sealants that bind the Desal G-80 membrane had disintegrated under the testing conditions, most likely due to thermal rather than chemical stresses.

CONCLUSIONS

The result of the nanofiltration flatsheet screening demonstrated the difficulties in applying these types of materials to the relatively high temperatures of produced water feed streams. The choice of inorganic NF membranes is limited at this time though development work is very active. Tighter (lower MWCO) membranes would definitely be useful in lowering silica and hardness concentrations to levels that could be used for steam generation prividing there was no large compromise in flux, or an increased susceptibility to fouling.

The Desal G-80 polymeric membrane, though providing good separation characteristics, was not engineered for this type of application. The use of thermally stable binding and construction material and the development of a more effective cleaning protocol, would be required for further testing within this environment. The flux performance realized would not be of commercial interest in the context of produced water de-oiling.

The Tech-Sep SC-7/M9 inorganic membrane generally performed well and was not adversely affected by the relatively high temperature or harsh nature of the produced water. It also responded well to cleaning thereby extending its usefulness. A more effective rejection rate is desirable and therefore a tighter membrane with similar construction would improve its cost-effectiveness

ACKNOWLEDGMENTS

This work was funded primarily through the Canadian government's Panel of Energy Research and Development (PERD) program. Additional contributions were made by the Alberta Oil Sands Technology and Research Authority (AOSTRA) and by several major oil companies including, Amoco, ARCO, Chevron, Elf Aquitaine, Norcen, PanCanadian, Shell, Statoil and Texaco.

REFERENCES

Chen, A.S.C, Flynn, J.T., Cook, R.C., Casaday, A.L., 1991. Removal of oil and grease, and suspended solids from produced water with ceramic crossflow microfiltration. *SPE Production Engineering.*

Tech-Sep, France, 1991, Product Literature.

Desal, USA, 1992, Product Literature.

Zaidi, A., Kok, S., Simms, K., and Schmidt, J., 1991. Oil and suspended solids removal from produced water suing ultrafiltration and microfiltration. Presented at the UNITAR/UNDP Fifth International Conference on Heavy Oil and Tar Sands, Caracas, Venezuela, September, 1991.

Zaidi, A., Kok, S., 1990. Recent advances in the treatment of produced water from in-situ heavy oil recovery. Presented at the Second Canada-China Heavy Oil Technology Symposium, Beijing, October, 1990.

Zaidi, A., Simms, K., Kok, S., and Nelson, R., 1992. Recent advances in the application of membrane technology for the removal of oil and suspended solids from produced water. Presented at the International Produced Water Symposium, San Diego, California, 1992.

44

AGIP AND THE ENVIRONMENT

Disposal of Produced Water from Gas Fields Exploitation in the Adriatic Sea

M. Buffagni,[1] D. Giacca,[1] and C. Biancifiori[2]

[1] AGIP S.p.A. Drilling fluids and Field Operations Laboratories
Via Maritano 26, S.Donato Mil.se 20097, Milano, Italy
[2] AGIP S.p.A. Geodynamics and Environment
Via Emilia 1, S. Donato Mil.se 20097, Milano, Italy

INTRODUCTION

The exploitation of oil and gas fields is generally accompanied by the production of variable amounts of water associated with the reservoir hydrocarbons. This naturally occurring water could sometimes be mixed with waters injected purposely into the formation in order to keep the reservoir pressure constant. Furthermore, in the specific case of gas fields (generally characterized by lower amounts of water than oil fields though more difficult to be treated), there could be also condensed water resulting from pressure and temperature variations. The total amount of water generated during the life of a field may be very high, reaching volumes up to 10 times more than the volume of the hydrocarbons produced. This is particularly true in the case of oil fields and usually the situation worsens with the age of the field: older fields produce more water than newer fields.

The disposal of these large amounts of effluents represents a serious economic and, also, technical-operative problem for all oil companies, especially in the case of offshore activities; in fact, most waters can contain, at different concentration levels, many organic and inorganic compounds (Stephenson, 1992), deriving both from the formation waters themselves and from the chemicals added to the hydrocarbon in order to cope with the problems occurring during the production phase.

In most cases and in the majority of the countries these produced waters are discharged, on the basis of permits given by the Authorities, directly into the sea, after a treatment step carried out on the platforms by using especially-designed equipment and procedures aimed at minimizing the concentration of oil and all other undesired substances.

In the case of Italy, the Ministry of the Environment has very recently (July 28, 1994) issued a Decree which defines the conditions to be satisfied by the oil companies in order to obtain the authorization to discharge the water into the sea. This decree indicates not only the maximum allowable oil concentration in the waters, but also requires a long series of additional information, such as chemical and physical analyses of the waters to be disposed of, the ecotoxicological characterization of all the chemicals of possible use during the life of the field and the conditions of the receiving sea environment.

To fulfill all the requests introduced through this new decree, AGIP has been forced to further strengthen its control on the waters produced in its fields, particularly in the Adriatic Sea, where most of its activity is concentrated.

This report, therefore, describes the current situation in Italy, taking into account not only the legislative environment in which the oil companies have to operate, but also the results of the in-depth characterization campaigns performed by AGIP in the last years on the produced waters as well as the initiatives and actions undertaken to further improve the performances of the real treatment plants and, consequently, to minimize the potential impact of this category of wastes on the environment.

ITALIAN REGULATIONS

Till July 1994, the discharge of produced waters in the Italian offshore was regulated by the Presidential Decree No.886 of May 24, 1979 "Integration and Adjustment of the Police Regulations for Mines and Quarries covered by the Presidential Decree No.128 of April 9, 1959, regarding hydrocarbon prospecting, exploration and exploitation activities in territorial seas and the continental shelf". Article No.62 of the above-mentioned Decree allowed the discharge of non-polluted water into the sea after an adequate treatment of the effluents aboard the platforms and provided that the hydrocarbon concentration was lower than 50 ppm. Other types of hydrocarbon-contaminated waters, like for instance bilge water, were not covered by this Decree, but fell under the London Conventions of 1973 and 1978 (ratified by Italy in 1980 with the Law No.662), which fixed in 15 ppm the maximum allowable hydrocarbon concentration.

The permit for discharging produced waters into the sea, initially released by the Ministry of the Navy and subsequently by the Ministry of the Environment after the latter was set up in 1986, had a validity of 5 years when given by the Ministry of the Navy and then of only one year when the Ministry of the Environment took its place. In order to obtain the permit, the oil companies had to present an application and a technical attachment with a lot of information describing the type of disposal, the characteristics of the receiving environment and some chemical data as the hydrocarbon content.

Finally, the legislative aspect was defined with the enactment of the already recalled Ministry of the Environment's Decree of July 28,1994, entitled "Procedures for the issue of the permit to discharge into the sea materials originating from liquid or gaseous hydrocarbon prospecting, exploration and exploitation"; this Decree is referred, in particular, to wastes resulting from drilling (whole fresh water-base muds and cuttings from both fresh water-base and low toxicity oil-base muds) and production (produced and/or process waters) operations.

For what regards produced waters, they are classified, according to the Decree, in the following categories:

- *formation water*: fossil water present in the geological formations and associated with the hydrocarbons;
- *production water*: water which separates from the hydrocarbons during production;
- *process water*: water injected during production in order to maintain reservoir pressure.

The Decree allows the discharge of such waters provided that their maximum mineral oil content does not exceed 40 mg/L (as measured with the infrared spectrophotometric method or another method calibrated to it) and the application requesting the authorization

Table 1. Produced water parameters

- Temperature
- Salinity
- Total material in suspension
- T.O.C. (total organic carbon)
- Total oil
- Total aromatic hydrocarbons
- Paraffins
- Total inorganic nitrogen
- Ammoniacal nitrogen
- B.O.D. 5
- Heavy metals: Cr, Pb, Cd, Hg, As, Cu and Ni.

to discharge is positively accepted. The application has to be accompanied by a technical annex, covering the following four main areas:

1. *General Information*: on the company wich applies for obtaining the discharge permit.
2. *Plant & Equipment Information*: description of water treatment and disposal systems.
3. *Analytical Data*: a) qualitative and quantitative chemical characterization of the chemicals to be used during the production life of the field. Also biological information on the same chemicals is required (acute and chronic toxicity, bioaccumulation, biodegradation); b) qualitative and quantitative chemical-physical characterization (Table 1) of the water to be disposed of, both before and after the treatment to which it is subject aboard the platform
4. *Information on the Receiving Environment*: a) data on thermal, haline, current metric and anemometric regime of the area surrounding the platform; b) sea floor sediment characteristics; c) eventual presence of natural sensitive areas, tourist resort facilities, fish species, zoobenthonic associations, etc. within a 5 nautical miles radius from the platform; d) dispersion behaviour forecasts of the discharged water into the sea.

The application and the technical form have to be sent to the Ministry of the Environment, through the head of the competent maritime districts where the operations take place. The Ministry of the Environment, before giving any official reply, carries out a stringent technical evaluation of the proposal by means of national authorized institutions, such as the Water Research Institute of the National Research Council (IRSA-CNR), the Central Hydrobiological Laboratory, the Central Institute for Scientific and Technological Research Applied to the Sea and the Superior Health Institute. The authorization to discharge, once released, remains valid for four years and may be further renewed upon request. Surveillance and control of the waste characteristics and quantities are carried out by the competent maritime districts representatives or by technical bodies in charge of the matter and the results are sent to the Ministry of the Environment, which assures the compliance of the data initially declared by the oil company with the real situation.

PRODUCED WATERS TREATMENT

Once obtained the authorization to discharge, the production activities can then be performed in the routine way, adopting, among the other things, the best available techniques capable to maintain as low as possible the potential impact of the discharges on the sea environment.

On all platforms, waters undergo treatment before being disposed of in order to reduce, within the acceptable limits, their content in hydrocarbons, solids, organic matter, heavy metals, etc. Generally, the treatment facilities consist of an atmospheric degasser connected to a low pressure vent pipe, where gas fractions eventually present are separated from the formation water coming out of the well. The liquid effluents are, then, sent to a three-phase separator, that may be either rotary (in more recent platforms) or static with dividers (in older platforms). The separation of liquid hydrocarbons and solids (usually small rock fragments produced with oil or gas) from formation water takes place in these separators; the solids, which are allowed to settle in a section of the separator, are removed and brought onshore for disposal, while liquid hydrocarbons are conveyed onshore by injecting them into the gas transportation sealines. The water fraction passes, then, through active carbon filters with the purpose to reduce, by adsorption, dissolved and undissolved hydrocarbons eventually present in the water phase. Some wells can produce water with high volumes of clay and silt (this happens when poorly consolidated formations are producing) and in this case it becomes necessary to further improve the treatment capabilities of the platform by installing quartzite or cartridge filtering equipments upstream the active carbon filtering system in order to safeguard the active carbon's working potential and life. The waters, after having undergone this treatment process which is tailored mainly for controlling and removing hydrocarbons, suspended solids and total organic matter, are, finally, disposed of directly into the sea. Table 2 shows the efficiency of the above-described system, illustrating the average characteristics of the waters upstream and downstream the treatment facilities.

CHEMICAL AND BIOLOGICAL CHARACTERIZATION OF PRODUCED WATERS

The efficiency of the treatment plants and the compliance of both waters and additives with the chemical, physical and biological characteristics declared by the oil company when the request for the authorization to discharge was submitted to the Ministry of the Environment, are monitored and guaranteed by means of a careful evaluation of the equipment performances and a strict analytical control on produced waters.

To cope with this particular problem and all other environment-related activities, AGIP in 1990 has set up its own Environmental Laboratories with the purpose to support operations through a continuous and reliable monitoring of the wastes produced during drilling and production phases. This support is provided by carrying out systematic sampling, chemical and biological analytical campaigns to check for the eventual presence of potential contaminants and their quantities; this service is assured by using sophisticated laboratory

Table 2. Produced water treatment

	Average values upstream treatment, *mg/L	Average values downstream treatment, *mg/L
Hydrocarbons	500	< 40
Suspended solids	1300	< 600
TOC	1100	< 700

* Values are expressed in mg/L.

equipment and by following the most up-dated procedures and standards suggested or imposed by national and international organizations.The results of these campaigns constitute the basis for maintaining a complete and up-dated knowledge on the status of the waste characteristics and for taking the consequent corrective actions.

The monitoring of produced waters is systematically assured by AGIP from many years and has been recently further improved in such a way that nowadays it has been introduced into AGIP routine activities.

In the next paragraphs, the results coming out from the last campaign, carried out in 1994 on 49 Italian platforms, are shown and briefly discussed.

Monitoring of Inorganic and Organic Compounds

Produced waters can contain a large variety of dissolved inorganic and organic compounds, that are easily found in reservoir formation rocks and waters (Tibbets *et al.*, 1992); it follows that in terms of compounds number, and not always necessarily in terms of concentrations, the produced water's chemical composition can be considerably complex and extremely variable, changing from reservoir to reservoir and sometimes, within the same reservoir, from one platform to another platform, depending on the production history of the wells.

The inorganic compounds are usually alkaline and alkaline-earth metal salts and often, though generally in very small quantities, heavy metals salts.

The analyses, carried out by AGIP on its produced waters, are addressed in determining, in particular, the presence of heavy metals, as required by the Decree's Attachment B/2. The heavy metals to search for are: lead, copper, cadmium, chromium, mercury, arsenic and nickel. The results of the analyses, performed with a Perkin Elmer 5100 PC atomic absorption spectrophotometer and following the ASTM D-3919 recommended procedures, point out that in the great majority of cases the concentration of heavy metals in the waters fall very often below the instrument precision (Table 3). But in spite of the usual low concentrations at which they are present, the constant monitoring of heavy metals is essential to prevent possible biological consequences (in terms, mainly, of toxicity and bioaccumulation) on marine organisms. It has been observed that heavy metals can cause environmental harassment, mostly because of their toxicity-related effects. However, taking into account also the normally quite high dispersion velocities of the effluents into the ecosystem, it seems reasonable to expect that this category of chemical compounds do not affect too much the aquatic environment (considering also the fact that the gas fields, which represent the vast majority of AGIP hydrocabon fields in the Adriatic offshore, are producing low volumes of associated water).

In addition to the heavy metals, also other inorganic compounds have to be investigated, that is the potential sources of nitrogen (total, nitric and nitrous inorganic nitrogen;

Table 3. Average concentrations of metals in produced water

Heavy metals	Concentrations (mg/L)
Lead	0.5
Copper	0.5
Cadmium	0.05
Total chromium	0.2
Mercury	0.005
Arsenic	0.5
Nickel	0.5

Table 4. Average concentrations of salts in produced water

Inorganic salts	Concentrations (mg/L)
Nitric nitrogen	25
Nitrous nitrogen	0.5
Ammoniacal nitrogen	80
Sodium chloride	58,000

ammoniacal nitrogen) and sodium chloride. The analyses, at this regard, were carried out by using a DR-2000 direct reading spectrophotometer at wavelengths specific for the type of compound to be analyzed. Table 4 illustrates the average concentration levels of these salts and it appears clearly from it that the average values assumed by them can be considered acceptable; only the sodium chloride concentration is high (but one has to remember that the produced waters are discharged into the sea!).

For what is concerning the organic compounds determination, this point essentially regards total oil, total aromatic hydrocarbons and paraffins. The analyses were conducted by means of the solvent extraction procedure followed by infrared ray spectroscopy, as foreseen by regulations in force. This analytical technique allows the measurement of the total oil content into the water, with particular attention given to the dispersed oil phase (consisting mostly in aliphatic hydrocarbons) and to the dissolved oil fraction characterized by aromatic components.

The average data obtained during the produced water characterization campaign are collected in Table 5, which shows that the total oil is around 30 mg/L, therefore below the maximum allowable limit of 40 mg/L, and is constituted by aromatic hydrocarbons in low concentrations (2.5 mg/L) and mainly by aliphatic hydrocarbons (27.5 mg/L).

As resulted by the analyses, the total oil content can be considered as one of the few parameters, which requires particular attention, because if the efficiency of the treatment installations is, for any reason, compromised, its values can rapidly pass the limit. Consequently particular care has to be devoted in order to maintain the equipment as efficient as possible. AGIP, at this regard, is launching a project, together with other international operators, in order to assure the continuous and on-line monitoring of this parameter (as well as of other parameters which appear to be, in certain circumstances, critical) and a further improvement of the treatment plants with the final target to reach values far below the maximum allowable limit foreseen by the current legislation.

As mentioned above, a critical role can be played, in particular cases, by other parameters such as total suspended solids, total organic carbon content (TOC) and the biochemical oxygen demand (BOD_5). Usually their values, as resulted from waters sampled and analyzed in 1994, range as indicated in Table 6.

Total suspended solids are measured through filtration, total organic carbon with a gas-infrared analysis method and the biochemical oxygen demand by means of the five-day

Table 5. Average concentrations of organic compounds in produced water

Organic compounds	Concentrations (mg/L)
Total oil	30
Aromatic hydrocarbons	2.5
Aliphatic hydrocarbons	27.5

Table 6. Average concentrations of chemical parameters in produced water

Chemical parameters	Concentrations (mg/L)
Suspended solids	100
T.O.C.	400
B.O.D. 5	200

closed bottle test, as recommended by the 301/D O.E.C.D. (Organization for Economic Cooperation and Development) Guidelines.

Biological Characterization

The Attachment B/2 of the July 28, 1994 Decree foresees that the chemicals to be used during the production life of a well be characterized not only from a chemical and physical point of view but also from a biological one. Chemicals are specific products added into the gas or oil stream and fulfill many important functions; they, in fact, could be classified as corrosion inhibitors, biocides, coagulants, flocculants, antifoamers, emulsion breakers and antifouling agents.

In AGIP gas fields the most commonly used products are methanol and diethylene glycol; methanol has the function to inhibit the formation of hydrates, while diethylene glycol serves as the basic gas dehydration agent.

The biological tests, aimed at evaluating and predicting possible effects of these products on marine organisms, are essentially the acute toxicity, the bioaccumulation and the biodegradability tests.

Acute Toxicity Tests. The acute toxicity test, run by AGIP Laboratories, is based on recommended national and international practices (Schiff et al., 1992); for instance, the short-term toxicity test follows the procedures devised by the Laboratory of Biological Research in Aquatic Pollution of Ghent University, Belgium, which contemplate the adoption as test organism of the salt-water crustacean Artemia Salina and the 202 O.E.C.D. method, where the fresh-water crustacean Daphnia Magna is exposed to the effects of potential pollutants.

The tests are usually carried out following two subsequent stages: a preliminary test and a final test. The results are expressed as median lethal concentration (LC50, in ppm) or, in other words, as the concentration of the chemical causing 50% of the organisms' death. After having established the concentration range to investigate, the solutions containing various concentrations of the chemical under evaluation are poured in several beakers; the organisms are, then, introduced into these containers starting with a blank solution (that is without testing substance) and proceeding from the lowest to the highest selected concentration. Based on this preliminary test, the interval of concentrations, within which the approximate LC50 falls, is established; within such an interval (defined by the maximum concentration that causes 0-10% death rate and by the minimum one that causes 90-100% death rate), a number of concentrations is identified, in geometric progression, on which the final test is carried out. After the required exposure time has elapsed, the effects provoked by the substance on the organisms are evaluated by counting the number of dead organisms in each beaker. The death rate should not exceed 10% in the blank solution for the test to be considered valid. The LC50 value is estimated through an EPA (Environmental Protection Agency) probit analysis; values of LC50 after 24 hours and relative confidence limits are calculated by means of a statistical analysis allowing, usually, a straight-line correlation

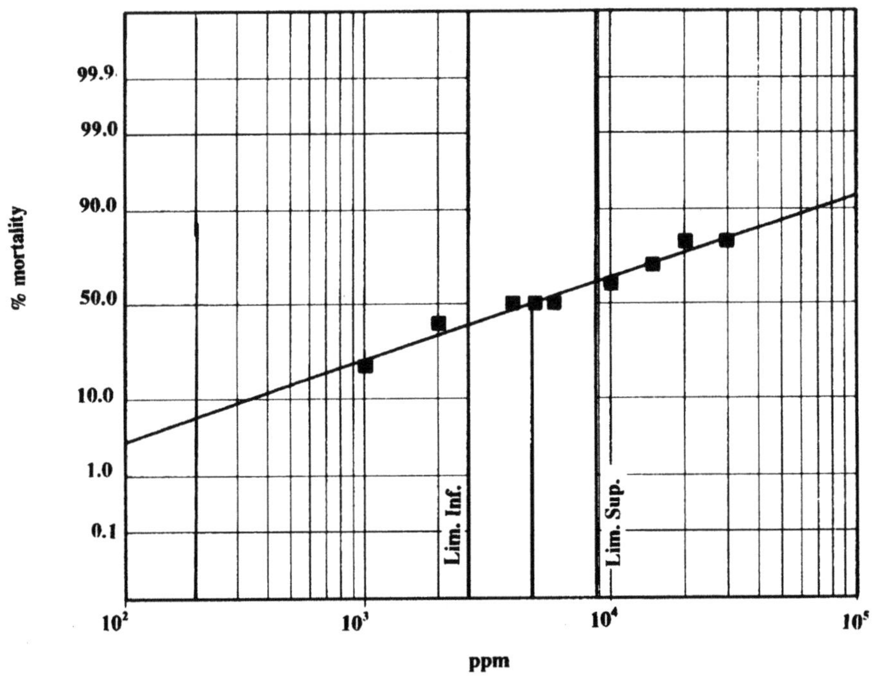

Figure 1. Acute toxicity test on methanol using Artemia Salina.

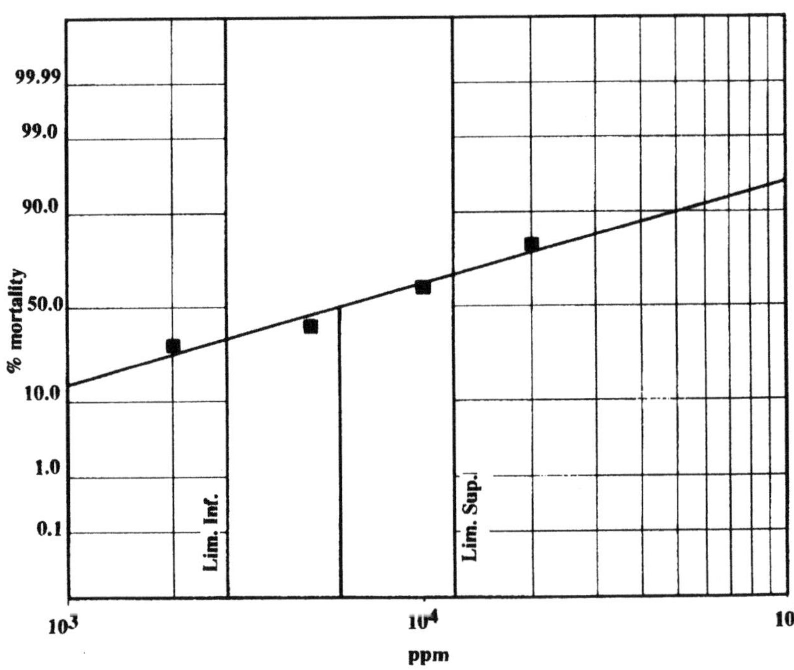

Figure 2. Acute toxicity test on methanol using Daphnia Magna.

between substance concentrations, expressed in ppm, and the organisms'death rate percentages. In this way, the concentration of the product that causes 50% of the organisms'death (LC50) can be determined.

Figures 1 and 2 illustrate the LC50 values obtained for the methanol, using as testing organisms both Artemia Salina (Fig.1) and Daphnia Magna (Fig.2); in the first case the LC50 resulted to be 4760 ppm, while in the second case it was around 5850 ppm; according to the study "The Impact of Water-Based Drilling Mud Discharges on the Environment: Overview" of Industry&Environment Office, United Nations Environment Programme, 1st Edition 1985, this product can be considered as "slightly toxic".

Analogously, Figures 3 and 4 show the results regarding the diethylene glycol, which gave an LC50 around 11850 ppm in presence of Artemia Salina and 11930 ppm when using the crustacean Daphnia Magna; in this case, following the classification of the above-mentioned study, the product can be considered as "practically non-toxic".

Bioaccumulation Test. The bioaccumulation test, adopted by AGIP Laboratories, is the method "No.107 O.E.C.D. Guideline for Testing of Chemicals: Partition Coefficient (n-Octanol-Water)". The partition coefficient (Ko/w) of a product between the aqueous and the lipophilic (n-octanol) phases is used to measure the transfer of a substance from a water-based environment to an organism and, also, the bioaccumulation potential of the same substance.

The Ko/w coefficient may be calculated only if the substances under testing are soluble in both phases, because in order to make possible the calculation of the concentration of the solute in the two solvents, a balance condition must be reached. The Ko/w coefficient is measured by means of a high performance liquid chromatography (HPLC) quantitative analysis.

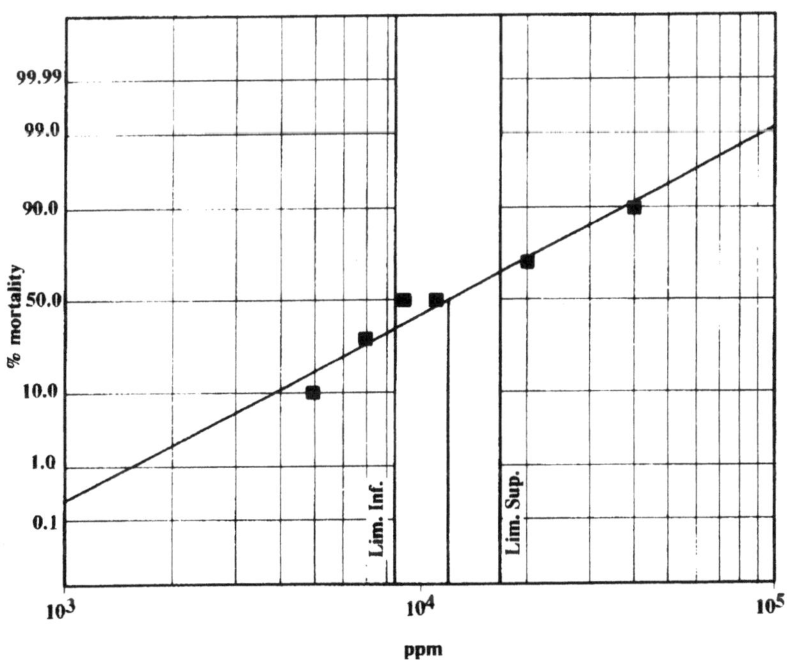

Figure 3. Acute toxicity test on diethylene glycol using Artemia Salina.

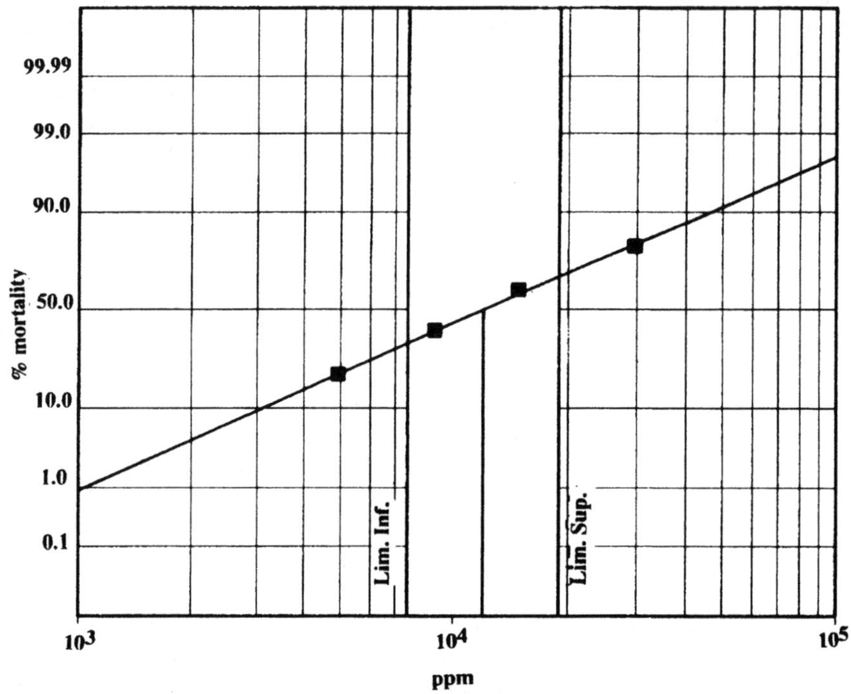

Figure 4. Acute toxicity test on diethlylene glycol using Daphnia Magna.

In the case of the two already cited chemicals the Ko/w coefficients resulted to be 0.855 for methanol and 0.386 for diethylene glycol or, in terms of log P, -0.07 and -0.4, respectively. This leads to conclude that both chemicals exhibit a low tendency to bioaccumulate in the organisms.

Biodegradability. The method, applied for the evaluation of the biodegradability of a product, is the "No.301/D O.E.C.D. Guideline for the Testing of Chemicals' Ready Biodegradability: Closed Bottle Test".

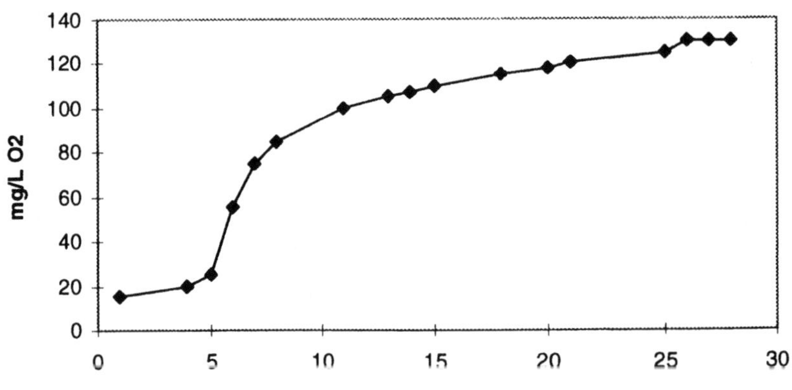

Figure 5. Biodegradation test on sodium acetate.

The test consists in introducing the substance under evaluation in a series of bottles, which in turn are put into a thermostatic bath at a temperature of 20°C for 28 days; every 24 hours the BOD (Biochemical Oxygen Demand) value is recorded, that is the quantity of oxygen required by the bacteria, inoculated into the bottles, to degrade the organic compound. For a correct test, it is recommended to measure, also, the biodegradability of a reference substance with a known biodegradation percentage. Generally, in AGIP tests, sodium acetate, which exhibits a degradation around 62.5% (Fig.5), is used as reference; according to the No.301/D O.E.C.D. method, the tests are considered valid only if the reference substance presents a biodegradation percentage greater than 60% within a reasonable short time span.

The biodegradation percentage at 28 days is easily calculated through the following equation:

$$\text{biodegradation }\% = 100 \times [\text{mg } O_2/\text{mg substance}]/[\text{COD/mg substance}] \qquad (1)$$

The COD (Chemical Oxygen Demand) value is obtained by making the heated substance react with a strong oxidizer (usually, potassium dichromate: $K_2Cr_2O_7$) and then conducting a photometric analysis.

The measurements, run on methanol and diethylene glycol, showed that the BOD28 for methanol was below 60%, indicating, therefore, that this compound can be considered "not readily biodegradable"; furthermore, only 30 mg/L of O_2 were required by the microorganisms in the period of 28 days (Fig. 6). On the contrary, diethylene glycol, showing a biodegradation percentage greater than 60% (Fig. 6), with an O_2 consumption of 320 mg/L, can be considered as a "readily biodegradable compound".

CONCLUSIONS

Several considerations and recommendations for future actions can be drawn as a result of the extensive campaign carried out by AGIP and aimed at characterizing from chemical, physical and biological points of view the waters produced; in particular, it results that:

- heavy metals (lead, copper cadmium, chromium, mercury, arsenic, nickel) are usually present in the produced waters in traces or at very low concentrations;

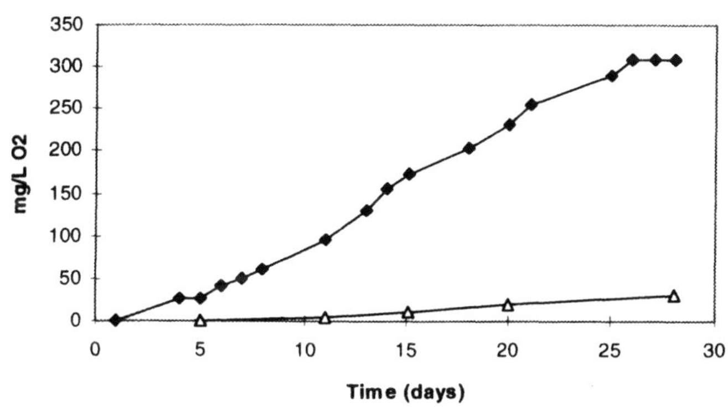

Figure 6. Biodegradation test on Methanol (Δ) and Diethylene Glycol (◆).

- the different forms of nitrogen (nitric, nitrous and ammoniacal) do not represent a real problem;
- the total oil content is in the vast majority of the cases below the 40 ppm limit, though a very careful maintenance of the treatment facilities is fundamental in order to assure the respect of this limit;
- the same consideration applies for the total solid and organic matter content;
- the chemicals, used during oil and gas production, are generally limited to methanol for preventing hydrates formation and diethylene glycol as main gas dehydration agent. Both products are usually added to the gas stream in low concentrations.The biological characterization of these two products pointed out that:
  - methanol shows: a) a limited toxicity (LC50~5000 ppm) and can be classified as "slightly toxic"; b) its tendency to bioaccumulate is low (Ko/w=0.855); c) the biodegradation percentage is not particularly high (<60%), thus rendering the product as "not readily biodegradable";
  - diethylene glycol shows: a) a low toxicity (LC50~11000 ppm) and can be considered as "practically non-toxic"; b) a very low partition coefficient (Ko/w=0.386), indicating a very poor tendency of the compound to bioaccumulate in the organisms; c) the biodegradation percentage is high (>60%) and this means that the product is "readily biodegradable".
- further efforts in optimizing the existing treatment installations on board the platforms has been recognized by AGIP as highly recommended. At this regard a project is under way in order to assure a continuous and on-line monitoring of those parameters (especially total oil, solids and organic matter), which result to play a critical role if not adequately controlled, making in such a way possible a prompt and targeted maintenance of the equipment. At the same time new and more efficiently performing devices (hydrocyclones, centrifuges, oil-absorbent polymers, etc.) are under evaluation in order to further minimize the impact on the environment of this typology of waste.

ACKNOWLEDGMENTS

The authors thank D. Giacca (Agip) and A. Festa (Agip) for their contributions to this work, and Agip SpA for the permission to publish it.

REFERENCES

E & P Forum. (1993). North sea produced water fate and effects in the marine environment. *E & P Report., No. X.XX/XXX*, 68pp.

Johnsen,S., Smith, A.T., Brandehaug, J. (1994). Identification of acute toxicity sources in produced water. In: Proceedings of the Second International Conference on Health, Safety & Environment in oil & gas exploration & production. *SPE 27138*, 383-390.

Marsid, M., Siwito,S. (1994). Produced water treatment. In: Proceedings of the Second International Conference on Health, Safety & Environment in oil & gas exploration & production. *SPE 27177*, 699-705.

OECD. (1981). OECD guidelines for testing of chemicals. Partition coefficient n-octanol/water. OECD Method No. 107.

OECD. (1984). OECD guidelines for testing of chemicals. Daphnia Magna, acute immobilization test and reproduction test. OECD Method 202.

OECD. (1990). OECD guidelines for testing of chemical's biodegradability in seawater. Closed Bottle Test. OECD Method No. 301 D.

Schiff,K.C., Reish,D.J., Anderson, J.W., Bay, S.M. (1992). A comparative evaluation of produced water toxicity. *Enviro., Sci. Research., 46,* 199-207.

Sorstrom, S.E., Aunaas, T., Bjerken, S., Brakstad, O.G., Johansen, O., Kaarstad,I., Mork, G., Stromgren, T., Thendrup, A. (1992). Produced water. Chemical and toxicity study. *Report of IKU., No. 222059.00/01/02*, 75pp.

Stephenson,M.T. (1992). Components of produced water: a compilation of results from several industrystudies. *Petroleum Technology*., 44, 548-603.

Tibbets, P.J.C., Buchanan,I.T., Gawel, L.J., Large,R. (1992). A comprehensive determination of produced water composition. *Enviro., Sci. Research., 46*, 97-112.

Vanhaecke,P., Persoone,G. (1984). The ARC-test: a standardized short-term routine toxicity test with Artemia nauplii. Methodology and evaluation. In Persoone,G., Jaspers,E., Claus,C., eds., *Ecotoxicological testing for the marine environment., 2*, 143-157.

BIOLOGICAL TREATMENT PROCESS FOR REMOVING PETROLEUM HYDROCARBONS FROM OILFIELD PRODUCED WATERS

Gilbert Tellez and Nirmala Khandan

US Environmental Protection Agency
New Mexico State University

INTRODUCTION

"Produced water" is the single largest volume of waste generated by the oil and gas industry. These waters are generated during crude oil and natural gas production, at both onshore and offshore operations. The amount of produced water generated is dependent upon the method of recovery and the nature of formation. In some formations, large volumes of water are pumped to the surface with the oil and gas in the early stages of production; in others, water is not produced until the formation has been significantly depleted; and in some, water is never generated.

During onshore recovery, the crude oil/gas/water mixture is processed by gravity separation at the well site and the water extract is discharged into surface pits or above ground storage facilities for evaporation or subsequent reinjection into the subsurface for permanent disposal/secondary recovery. In offshore operations, these waters are discharged directly into the sea. Since produced waters are laden with a range of petroleum hydrocarbons, salinity, etc., such treatment and disposal practices have resulted in contamination of surface, ground, and coastal waterways.

Problems can occur during final disposal in subsurface reinjection, earthen impoundments used for evaporation, and surface-stream/offshore discharges. For example, surface water discharge practices were determined to be detrimental to a Nueces County, Texas creek during a continuous produced water discharge (approximately 8,500 gal/day) from an adjacent oilfield site (Texas Parks and Wildlife Department, 1981). During discharge, the creek maintained an oil sheen and was devoid of native vegetation and fish, which were abundant in the upstream stretches of the creek. Total petroleum hydrocarbon concentrations in excess of 9.0 mg/L were extracted from the downstream waters.

The effects of produced waters discharged from oil and gas operations on shallow water coastal wetlands have recently emerged to be of significant environmental concern (Rayle and Mulino, 1992). Direct discharge is permitted for produced waters from offshore oil and gas production platforms or shore-side treatment facilities designated for territorial sea discharge. Many of these facilities have been unable to comply with the Best Practicable

Produced Water2: Environmental Issues and Mitigation Technologies
Edited by M. Reed and S. Johnsen, Plenum Press, New York, 1996

Technology (BPT) methods for petroleum hydrocarbon separation from produced waters (Brown et al., 1992). If not removed, hydrocarbon concentrations in produced water are known to impact species diversity in benthic communities close to shallow water discharges (Middleditch, 1984, and Neff, 1987). Neff et al. (1989) found elevated TPH concentrations less than 300 meters from a shallow water discharge, and less than 100 meters from a deep water discharge in Louisiana coastal waters. Neff(1987) concluded that benthic communities within 20 meters of both discharges were influenced by sediment contamination, probably due to produced water discharge.

Produced waters contain both soluble and insoluble petroleum fractions of complex mixtures of organic compounds similar to those found in crude oils and natural gases. The individual constituents cover a broad range of boiling points, carbon numbers, chemical families, and structural isomers (Senn and Johnson, 1985). The hydrocarbon fractional groups commonly found in produced water include: alkanes, alkenes, alkynes, aromatics, polynuclear aromatics, and complex hydrocarbon compounds containing oxygen, nitrogen, and sulfur.

The most commonly detected hydrocarbon groups are C_{10} to C_{30} straight-chain alkanes (Chapelle, 1993). The n-alkanes which are present at the highest concentration are C_{14} through C_{18} alkanes, with concentrations gradually decreasing with increasing chain length to C_{34} (Neff et al., 1992). Only 25% of the n-alkanes of produced water are of the higher molecular weight C_{21} through C_{34} n-alkanes. Produced waters from gas production operations generally contain higher levels of BTEX than waters from oil production (Fillo et al., 1992). Stephenson (1991) obtained a mean benzene concentration range from 5.8 to 12.2 mg/L in gas operations and 1.3 to 8.7 mg/L in oil productions. Polynuclear aromatic hydrocarbons (PAHs) are present in crude oils and produced waters in moderate amounts. PAHs are considered to be hazardous compounds because of their carcinogenic characteristics in animal studies and potential health risks to man.

Literature on economic treatment and/or disposal of oilfield produced water is very limited. The Gas Research Institute (GRI) of Chicago, Illinois, published two reports evaluating the relative cost of produced water treatment (Decker, 1991, and Lee Ryan and Fillo, 1978). The range of cost covers nearly two orders of magnitude, from \$0.03 to \$3.00/barrel. These extremes were derived for treatment applications utilizing the least economical burden (surface discharge) preceded by the most basic treatment (aeration followed by sedimentation), to the most costly (evaporation). The cost range for deep well injection including the cost for transporting the produced water to the disposal sites is approximately \$1.25 to \$1.75/barrel. Deep well injection costs may reflect regional differences as well as technical differences associated with converting producer wells versus new well development. Cost for biological treatment followed by surface discharge, ranged from \$0.03 to \$0.25/barrel. However, this technique is seldom used because of the insufficient technical and operational information available to the oil and gas industry.

While the biodegradability of petroleum hydrocarbons has been well demonstrated, biological treatment of produced waters laden with such chemicals has not been attempted because of the very high TDS levels often found in produced waters. Our previous studies under laboratory conditions using produced water samples with TDS levels up to 100,000 mg/L, demonstrated that biodegradation was possible after adequate acclimation (Tellez, 1991). In this study, those findings are extended in designing and evaluating the performance of a field scale biological treatment system in removing petroleum hydrocarbons from two oilfield produced water sources. The performance of the proposed system is demonstrated in terms of removals of total petroleum hydrocarbon (TPH), total n-alkanes (TNA), and aromatic hydrocarbons (BTEX).

MATERIALS AND METHODS

Produced Water

This field study was conducted on produced waters generated from oil and gas production wells located in the mineral rich Permian Basin situated in portions of Lea, Eddy, Chaves, and Roosevelt Counties of New Mexico. In 1985, these four counties produced over 70.5 million barrels of crude oil and 462.3 billion cubic feet of natural gas from a total of 26,143 wells, while generating 4.4 barrels of produced water per barrel of oil (Tellez, 1991). Two distinct sources of produced waters were evaluated: in one case, the produced water was drawn from an open, above ground storage tank; in the other, produced water was drawn from a closed storage tank. The TDS levels of both these sources were in excess of 50,000 mg/L, and their BTEX fractions were significantly different: 3 mg/L in the open tank vs. 8 mg/L in the closed tank due to volatilization to the atmosphere.

Biological Reactor Configuration

The reactor configuration consisted of a 4 x 10 x 4 ft. 12 gauge steel tank with three compartments in series, followed by two 4 x 4 x 4 ft. slow sand filters in parallel. The first compartment (945-L; 250 gal) served as a completely-mixed (no cell recycle) pre-aeration tank designed to increase the dissolved oxygen content of the produced water to an operating concentration of approximately 3 to 4 mg/L. This would allow for stimulation of any naturally existing aerobic microbial cultures. An air flow rate of 0.345 m^3/min. (12 cfm, equivalent to 305 lbs/day of oxygen) was provided using a Fuji ring compressor to ensure complete mixing and adequate oxygen transfer.

The second compartment (945-L; 250 gal), designated as the aeration tank, received the effluent from the pre-aeration tank. This tank was initially bioaugmented with a commercial culture (Petrobac-S®). Complete mix of the cells was provided by delivering 0.431 m^3/min. (15 cfm) of air (equivalent to 382 lbs/day of oxygen) which maintained the bacterial cells in total suspension while providing an operating dissolved oxygen concentration in the range of 3 to 4 mg/L.

The effluent from the aeration tank then entered the third compartment (1,050-L; 278 gal) which served as a clarifier. A biomass recycle system was installed between the clarifier and the aeration tank to maintain the desired MLSS concentration in the aeration tank. The clarified effluent was discharged into one of the two parallel slow sand filters to remove any carryover of bacterial cells. The two filters were operated on an alternating basis, allowing for continuous flow of the waste stream during clean-up and backwashing of the filter media.

Naturally Occurring and Bioaugmented Cultures

Two types of microbial cultures were evaluated in this study. The pre-aeration tank was operated aerobically on a continuous-flow basis allowing for naturally occurring bacteria to grow. An initial bioaugmented culture product, Petrobac-S®, was utilized in the aeration tank. This product is a hydrocarbon utilizing consortium specially formulated for degrading crude or refined hydrocarbons in moderate saline environments under aerobic conditions.

As presented by previous research, acclimation of this commercial culture to the high TDS levels in the produced water is essential (Tellez, 1991). The acclimation procedure consisted of adding the commercial Petrobac-S® seed to 945-L of fresh water in the aeration tank. Acclimation was achieved by daily additions of 95-L of produced water from the

Hobbs, New Mexico closed tank system. By discarding biomass at 95-L per day, a 10-day solids detention time was maintained at an MLSS concentration of approximately 700 mg/L (dry cell weight). A cell yield of 0.38 mg MLSS/mg TNA was obtained during the acclimation period.

Total N-Alkanes (TNA) Analysis

Gas Chromatography/Mass Spectrophotometry (GC/MS) techniques were used in the quantitation of the n-alkane and aromatic hydrocarbons. Daily GC/MS analysis were initiated by collecting 1-L completely mixed produced water sample volumes across the treatment train. Samples were then vigorously agitated in a separatory funnel for two minutes with a 10:1 water to solvent ratio. Methylene chloride was used as the solvent. After allowing a 10 minute separation time between the water and solvent phases, the bottom organic solvent layer was then removed. The organic solvent extract was concentrated with dry nitrogen to bring the final volume to 10-mL. This small final extract volume ensured continuity of the extract by eliminating the possibility of having a dilute hydrocarbon sample or the amount of sample injected into the GC being to small, both of which could produce false conclusions about the extent of biodegradation.

Upon completion of the extraction procedures, 2-μL of the extract was injected into the gas chromatograph/ion trap mass spectrometer (Saturn Varian Co.). The GC/MS was equipped with an RTx-5 30m x 0.32 mm dia. capillary column, 5% diphenyl, 95% dimethyl polysiloxane; carrier gas was helium at a flowrate of 1 mL/min. and a split flow of 60 mL/min. GC/MS running conditions were 600 second filament/multiplier delay time; 35°C initial temperature with 5 min. isothermal, 8°C/min. ramp to a final temperature of 280°C with 15-min. isothermal, for a total run time of 50.6 minutes per sample.

Positive identification of the n-alkanes ranging from decane to tricotane (unbranched carbons from C_{10} through C_{30} n-alkanes) was achieved, first by mass spectral identification and then by acquisition of standard n-alkanes and comparison with unknown retention times. Quantitation was performed by measuring peak areas of the individual compounds, and by correlating with a standard calibration curve. A correlation coefficient of 0.99 was used as an acceptance criterion. A quality control sample was run every 10 samples. The quality control was ±10%; if the quality control sample did not pass, a new calibration curve was entered. TNA concentrations were determined by summation of the all n-alkane compounds present. TPH analysis, as described below, was performed on all field evaluations to ensure QA/QC for the GC/MS analysis methods and procedures.

Aromatic Hydrocarbons Analysis

Upon extraction of a 1-L field produced water sample, as described in the above n-alkane section, aromatic GC/MS (Hewlett-Packard 5890 series II) analysis was via split injection with a 2-μL sample volume and a 30:1 split ratio. The injector was held at 280°C and the GC/MS interface at 260°C. The oven temperature was initially held at a temperature of 50°C for 1.5 min., then programmed to 290°C at a ramp rate of 7°C/min with a 15 minute hold at 290°C. The mass selective detector was operated in the selected ion mode at m/Z values of 57.1 for alkanes greater than butane, 78.0 for benzene, and 91.1 for toluene, ethylbenzene, and xylene. Aromatic hydrocarbons were quantified by running a series of 100 μg (600 μg total) BTEX concentration standards (mixture of benzene, ethylbenzene, toluene and xylenes), from which a response curve was obtained. Aromatic hydrocarbon content of the produced water was then run against this BTEX standard to obtain appropriate concentrations.

Total Petroleum Hydrocarbons (TPH) Analysis

Environmental Protection Agency Method 418.1 (Spectrophotometric, Infrared) was used to determine total recoverable petroleum hydrocarbons. Unlike oil and grease determination, petroleum hydrocarbons is the measure of only the mineral oils, which include the aromatic hydrocarbons (i.e. benzene, toluene through naphthalene), n-alkanes, and all non-polar hydrocarbons.

Procedures included collection of a 1-L produced water sample followed by acidification to a pH less than 2. Acidified samples were then serially extracted with freon-113 in a separatory funnel. Polar compounds such as esters, alcohols, aldehydes, fatty acids, and sulfur-bound carbons were co-dissolved with the petroleum hydrocarbons in freon-113, and were removed with a silica gel adsorbent. After the silica gel was added, an infrared spectrum of the extract (solvent plus solute) was recorded by placing it in a 100 mm pathlength cell and exposing it to the infrared beam scanning at the absorption band of 2,930 cm^{-1} (Perkin Elmer Model 1600). The spectrum for the solute alone was

Table 1. Results of water quality analysis across the treatment systems

		Produced water source	
		Open tank	Closed tank
Raw water characteristics			
Dissolved oxygen	[mg/L]	< 1.0	< 1.0
Total dissolved solids	[mg/L]	52,164	55,023
Total n-alkanes	[mg/L]	126	115
Total aromatics	[mg/L]	3.1	7.7
Total petroleum hydrocarbons	[mg/L]	135	126
Effluent from pre-aeration tank			
Dissolved oxygen	[mg/L]	3.1	2.6
Total dissolved solids	[mg/L]	54,256	57,329
Total n-alkanes	[mg/L]	51.1	56.5
Total aromatics	[mg/L]	ND	ND
Total petroleum hydrocarbons	[mg/L]	58.3	61.1
Effluent from aeration tank			
Dissolved oxygen	[mg/L]	3.9	3.8
Total dissolved solids	[mg/L]	54,229	57,128
Total n-alkanes	[mg/L]	0.04	0.16
Total aromatics	[mg/L]	ND	ND
Total petroleum hydrocarbons	[mg/L]	0.27	0.44
Effluent from settling tank			
Dissolved oxygen	[mg/L]	2.3	2.9
Total dissolved solids	[mg/L]	51,637	54,647
Total n-alkanes	[mg/L]	0.01	0.02
Total aromatics	[mg/L]	ND	ND
Total petroleum hydrocarbons	[mg/L]	ND	ND
Effluent from sand filters			
Dissolved oxygen	(mg/L)	2.7	2.8
Total dissolved solids	(mg/L)	51,234	54,196
Total n-alkanes	(mg/L)	0.01	0.03
Total aromatics	(mg/L)	ND	ND
Total petroleum hydrocarbons	[mg/L]	ND	ND

recorded and measured by a simultaneous substraction of the solvent spectra. Produced water TPH quantitation was based upon comparisons to standard calibration curves which consisted of 15.0-mL n-hexadecane (n-alkane), 15.0-mL isooctane (branched alkane), and 10.0-mL chlorobenzene (aromatic).

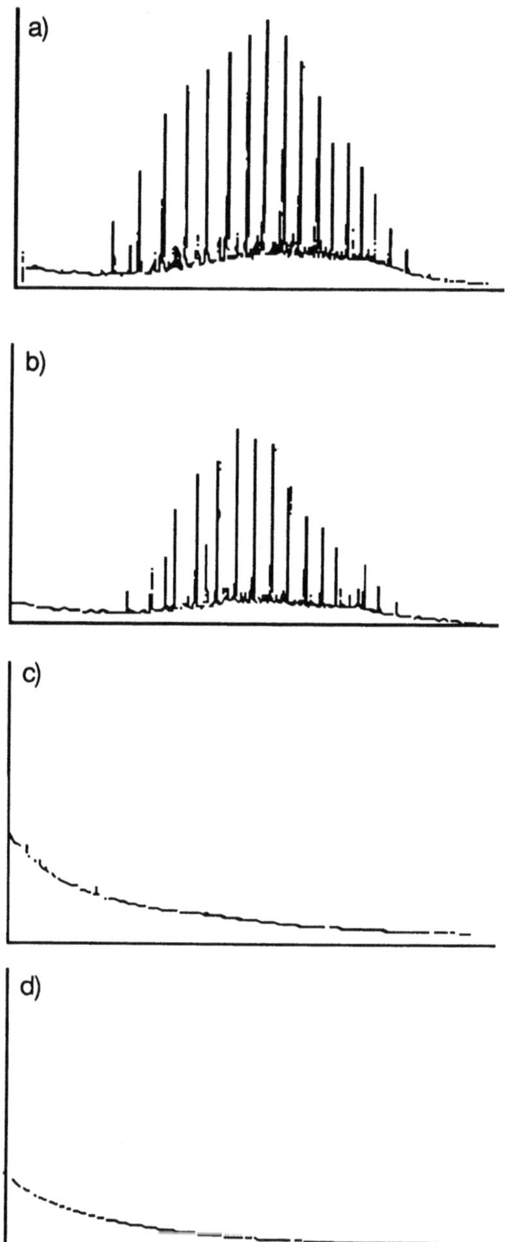

Figure 1. Typical GC/MS Chromatographs: a) Raw produced water; b) effluent from preaeration tank; c) effluent from aeration tank; and d) final effluent.

Continuous-Flow Testing

Upon completion of the 10 day acclimation period as described above, the produced water from the closed tank system was brought online for continuous-flow treatment. Produced water flowrate was set at 1,890 L/day (500 gal/day), to maintain an MLSS concentration of approximately 700 mg/L in the aeration tank. The pre-aeration and aeration tanks reached steady-state conditions in approximately five days. At this time, MLSS, TPH, n-alkane and aromatic removal remained constant on a day-to-day basis. Each day, a sufficient amount of biomass was wasted to maintain fixed MLSS concentrations in the aeration tank. Growth rates in the pre-aeration tank were maintained by its hydraulic detention time.

The following operational and water quality parameters were measured daily for all treatment units from the pre-aeration tank to the sand filter: influent flowrate; influent petroleum hydrocarbons (TPH, n-alkanes, aromatics); effluent petroleum hydrocarbons (TPH, n-alkanes, aromatics); cell return line petroleum hydrocarbons (TPH, n-alkanes, aromatics); and, mixed liquor suspended solids (MLSS).

Table 2. Performance of treatment system under various operating conditions

		Produced water source			
		Open tank		Closed tank	
		Pre-aerat. tank	Aeration tank	Pre-aerat. tank	Aeration tank
RUN 1					
Flow rate	[L/day]	378	378	378	378
MLSS	[mg/L]	61	737	56	737
TNA removal	[%]	92	99	89	99
BTEX removal	[%]	99	99	99	99
TPH removal	[%]	88	99	86	99
RUN 2					
Flow rate	[L/day]	756	756	756	756
MLSS	[mg/L]	52	651	46	632
TNA removal	[%]	72	99	69	99
BTEX removal	[%]	99	99	99	99
TPH removal	[%]	67	99	64	99
RUN 3					
Flow rate	[L/day]	1134	1134	1134	1134
MLSS	[mg/L]	48	550	37	560
TNA removal	[%]	60	99	54	99
BTEX removal	[%]	99	99	99	99
TPH removal	[%]	54	99	49	99
RUN 4					
Flow rate	[L/day]	1512	1512	1512	1512
MLSS	[mg/L]	31	483	23	449
TNA removal	[%]	45	99	35	99
BTEX removal	[%]	99	99	99	99
TPH removal	[%]	33	99	34	99
RUN 5					
Flow rate	[L/day]	1890	1890	1890	1890
MLSS	[mg/L]	22	401	9	408
TNA removal	[%]	28	99	12	99
BTEX removal	[%]	99	99	99	99
TPH removal	[%]	21	99	10	99

TNA- conc. of total n-alkanes; BTEX- conc. of benzene, toluene, ethylbenzene, and xylene; TPH- conc. of total petroleum hydrocarbons.

The dissolved oxygen, pH, and temperature of the treatment units were also checked on a daily basis. After treating the closed tank system for 45-days, the treatment system was again brought to a steady-state initial MLSS concentration of 700 mg/L. At this time, the influent from the closed tank was terminated and the open tank system was brought online at a flow rate of 1,890 L/day. Water quality sampling and analysis were again performed, as described previously for an additional 45-day testing period.

RESULTS

Typical dissolved oxygen (DO), total dissolved solids (TDS), total n-alkanes (TNA), total aromatics, and total petroleum hydrocarbons (TPH) profiles across the various units in the treatment system are presented in Table 1 for the two sources of produced waters tested. While the TDS levels remained the same throughout the system, the overall reductions in TNA, aromatics, and TPH exceeded 99%. The high TDS levels found at this site are common in the Southeastern oil producing regions of New Mexico.

Typical GC/MS chromatographs of the n-alkanes (unbranched hydrocarbons) in the waste stream at various points across the treatment system are shown in Figure 1a to d. These n-alkanes, accounting for more than 90% of the petroleum hydrocarbons in these produced waters, are almost completely removed by the system as illustrated in Figure 1d.

The system was run at five different flow rates and biomass concentrations to evaluate removal efficiencies. Results of these five runs are summarized in Table 2. Again, the system achieved consistently high removals of TNA, BTEX, and TPH. These results were also used to determine the biokinetic parameters that describe the growth of the organisms in the pre-aeration and the aeration tanks. The biokinetic parameters determined form these runs, using standard procedures, are listed in Table 3. These data could be used for process modeling and analysis, reactor sizing, and scale up purposes.

SUMMARY AND CONCLUSIONS

The feasibility of removing petroleum hydrocarbons from oilfield produced waters using biological treatment was evaluated under laboratory and field conditions. Based on previous laboratory studies, a field scale prototype system was designed and operated over a period of four months. Two different sources of produced waters were tested in this field study under various continuous flow rates ranging from 375 L/day to 1,800 L/day. One source of produced water was an open storage pit; the other, a closed storage tank. The TDS concentrations of these sources exceeded 50,000 mg/L; total n-alkanes exceeded 100 mg/L;

Table 3. Biokinetic parameters from field studies

	Produced water source			
	Open tank		Closed tank	
	Pre aerat. tank	Aeration tank	Pre aerat. tank	Aeration tank
Yield, Y [mg/mgTNA]	0.59	0.62	0.64	0.47
Decay coeff., kd [1/day]	0.02	0.01	0.01	0.01
Half velocity const, Ks [mg/L]	47	2	73	4
Max. velocity, μmax [1/day]	1.88	0.27	1.98	0.21

total petroleum hydrocarbons exceeded 125 mg/L; and total BTEX exceeded 3 mg/L. Removals of total n-alkanes, total petroleum hydrocarbons, and BTEX remained consistently high over 99%. During these tests, the energy costs averaged $0.20/barrel at 12 barrels/day.

ACKNOWLEDGMENTS

This work was supported, in part, by the DOE-Waste-management Education Research Consortium (WERC) of New Mexico.

REFERENCES

Brown, J. S., Sour, T. C., Wade, M.J., Neff, J.M., 1992. Chemical and toxicological characterization of produce water freon extracts. *Produced Water*. Plenum Press, New York.

Chapelle, F. H., 1993. *Ground-water microbiology and geochemistry*. John Wiley & Sons, Inc., Somerset, NJ.

Decker, D., 1991.Estimate of water production from Northern San Juan Basin. Natural Gas Supply Project Advisor Group Meeting - Coalbed Methane Project Area (Internal Document - Unpublished), Birmingham, Alabama.

Fillo, J. P., Koraido, S. M., Evans, J.M., 1992. Sources, characteristics, and management of produced water from natural gas production and storage operations. *Produced Water*. Plenum Press, New York.

Lee Ryan, P. B., Fillo, J. P, 1978. Evaluation of management options for coalbed methane produced water. Proceedings of the 1991 Coalbed Methane Symposium, 35, 5-13.

Middleditch, B. S., 1984. Ecological effects of produced water discharges from offshore oil and gas production platforms. Final Report on API Project No. 248, American Petroleum Institute, Washington, D.C., 160.

Neff, J. M., 1987. Biological effects of drilling fluids, drill cuttings and produced waters. *Long-Term Environmental Effects of Offshore Oil and Gas Development*. Elsevier Petroleum Publishers Ltd., London.

Neff, J.M., Sour, T.C., Maciolek, N., 1989. Fate and effects of produced water discharges in nearshore marine waters. American Petroleum Inst. Publication No. 4472, American Petroleum Institute, Washington, D.C., 300.

Neff, J. M., Sour, T. C., Maciolek, N., 1992. Composition, fate and effects of produced water discharges to nearshore marine waters. *Produced Water*. Plenum Press, New York.

Rayle, M. F. and Mulino, M.M., 1992. Produced water impacts in Louisiana coastal waters. *Produced Water*. Plenum Press, New York.

Senn and Johnson, 1985. Interpretation of gas chromatography data as a tool in subsurface hydrocarbon investigations. Petroleum Hydrocarbons and Organic Chemicals in Groundwater - Prevention, Detection, and Restoration.

Stephenson, M. T., 1991. Components of produced water: a compilation of results from several industry studies. Paper No. 23313 presented at the First International Conference on Health, Safety and Environment, Hague, Netherlands, November 10-14.

Tellez, G.T., 1991. Bioreclamation of New Mexico oilfield produced wastewaters: characterization and feasibility study. Masters Thesis, New Mexico State University, NM .

PRETREATMENT OF OIL FIELD AND MINE WASTE WATERS FOR REVERSE OSMOSIS

W. L. Bourcier,[1] H. Brandt,[2] and J. H. Tait[2]

[1] Lawrence Livermore National Laboratory
Livermore, California 94550
[2] Aquatech Services Inc.
P.O. Box 946
Fair Oaks, California 95628

INTRODUCTION

This paper describes a brine treatment process for the Morcinek mine near the city of Katowice in Upper Silesia in Poland. The process was developed jointly by Aquatech Services, Inc. (Aquatech) and the Lawrence Livermore National Laboratory. Support for the research was obtained in part by the U.S. Department of Energy.

The process consists of chemical pretreatment, reverse osmosis treatment and submerged combustion evaporation.[2-5] A schematic diagram of the disposal process is shown in Figure 1. In this figure, brine from the mine is pumped into a storage tank. The brine is produced during the mining of coal. The Aquatech process also could be applied to brines produced during the production of oil or gas.

From the storage tank shown in Figure 1, the brine flows through pretreatment units to remove brine constituents that may harm the membranes of the reverse osmosis unit, or that may adversely affect the quality of the product water. The pretreatment is a critical step in the process. The emphasis of this paper is on the research carried out to develop a viable, robust and cost effective pretreatment process.

In the reverse osmosis process, brine flows under pressure past a porous membrane. The pressure gradient across the membrane causes substantially salt free water to flow through the membrane, while a brine with high salt concentration stays behind.

After the reverse osmosis process, the concentrated brine is delivered to a submerged combustion evaporator. In submerged combustion evaporation, a mixture of gas and air is combusted in a burner that is placed above a brine bath. Combustion takes place above the brine level while the combustion gases are vented through the brine thereby evaporating part of the brine in the evaporator.

The reverse osmosis process reduces the brine volume by approximately 60 per cent, while the combination of the reverse osmosis and submerged combustion processes reduces the volume about 90 per cent. Underground disposal will be used for the remaining

Produced Water2: Environmental Issues and Mitigation Technologies
Edited by M. Reed and S. Johnsen, Plenum Press, New York, 1996

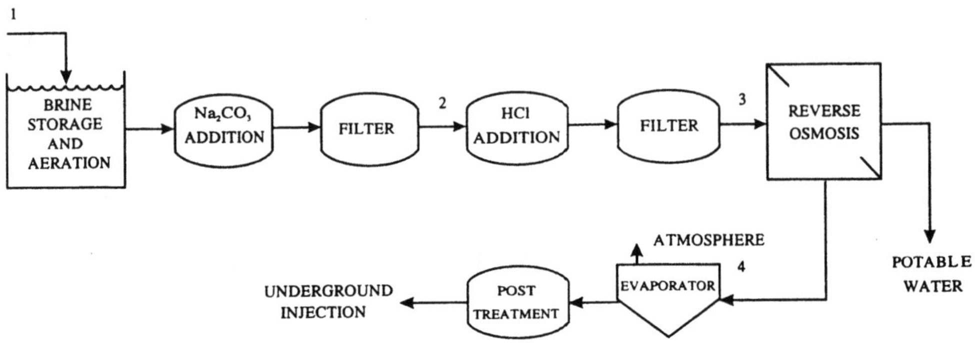

Figure 1. Schematic diagram of brine pretreatment process.

concentrated brine from the process. The potable water from the process can be used for agricultural purposes.

EFFECT OF BRINE CHEMISTRY ON REVERSE OSMOSIS PROCESS

A critical aspect of the use of a reverse osmosis process is the pretreatment of the brine to insure that salts and other constituents will not come out of solution in the reverse osmosis unit.[13,14] Natural fluids such as those produced from mines, oil, or gas fields are unique among many types of waste streams in that they are almost always saturated with many types of solid phases.[11] The reason is that they have remained in contact with minerals underground for extended time periods during which they reach an equilibrium state with these minerals. During production, minor perturbations of these brines such as changes in temperature, pressure, or contact with the atmosphere, may cause precipitation. The important difference between these brines and other waste streams is not that they are more concentrated with dissolved species, but that they are always near, or at, saturation with many solids and readily begin to precipitate with minor thermodynamic perturbations.

Processes associated with brines that may potentially precipitate solids from these brines are: (1) oxidation by air which precipitates iron, manganese, and other metal oxides less soluble under oxidizing conditions; (2) loss or gain of gases such as CO_2 while the brines are in contact with the atmosphere; (3) temperature changes during production; and (4) increases in concentration of dissolved species during the reverse osmosis process.

When relatively pure water is removed from the brine during the reverse osmosis process, the brine becomes more concentrated in dissolved species and will precipitate minerals if they become supersaturated. Previous experience together with computer modeling calculations indicate that carbonates, sulfates, hydroxides and clay minerals commonly precipitate.[8,9]

GEOCHEMICAL MODELING OF PROCESS USING EQ3/6 COMPUTER CODE

To develop the pretreatment strategy, the brine saturation state is analyzed with the EQ3/6 geochemical modeling codes developed at the Lawrence Livermore National Laboratory.[1,17-19]

The geochemical modeling codes contain provisions for calculating aqueous phase speciation, mineral saturation states, kinetically controlled mineral dissolution and precipitation reactions, redox disequilibria, surface charge and complexation, and gaseous equilibria. The thermodynamic database which supports the codes has data for 850 aqueous species, 2,000 solids, and 80 gases, and includes Pitzer data for brines.[12] The computer codes are limited in some applications by the lack of: (1) thermodynamic data at high temperatures and high concentrations for some species; (2) good kinetic data for some mineral precipitation processes; and (3) data for surface complexation reactions.

The use of geochemical modeling codes provides important advantages over less rigorous methods of estimating scaling potential of brines. For example, methods which use the Langlier index [10] or Stiff-Davis index [16] are valid only for a narrow compositional range of "standard" brines. These types of methods do not work well with fluids that contain significant amounts of species which form strong ion pairs with Ca^{2+}, and also do not account for other species, such as carboxylic acids, which often are present and erroneously are included in the measured alkalinity.[7] Both of these effects can give rise to substantial error in the estimated carbonate saturation index.

The simulation provides a cost-effective method for evaluating potential pretreatment techniques such as water softening, the addition of anti-scaling agents such as chelating agents, and other types of scale inhibitors. The geochemical modeling codes also can be used to simulate brine passage through the reverse osmosis membrane to predict whether inorganic scale will form. When combined with pilot-scale laboratory experiments, the simulation provides a method for optimizing the treatment process.

Observations of mineral precipitation/dissolution phenomena in natural systems can be generalized as follows: minerals are unlikely to precipitate from solutions that are undersaturated with respect to those minerals except in rare cases where biological activity facilitates it. An example is the precipitation of calcium carbonate by micro-organisms in sea water.

If a brine passes through a reverse osmosis unit and remains undersaturated with respect to all mineral phases, it is unlikely that inorganic mineral scale will form; minerals commonly do not precipitate from brines that are slightly supersaturated. A nucleation barrier to mineral precipitation exists that commonly prevents minerals from precipitating. The degree of supersaturation that can be maintained without precipitating these phases is a complex function of time, temperature and the availability of favorable surface sites.

To perform the geochemical modeling calculations, an initial equilibrium calculation is performed. This calculation provides both the distribution of aqueous species among the potential aqueous complexes and the saturation states for all solids. A reaction path for the brine then is specified. The reaction path is defined as a set of reactants (solid, liquid, or gas) that reacts with the fluid over time. A temperature range from 0°C to 300°C also can be defined. The reaction may be kinetically controlled using one of several rate laws with the user supplying the rate constant, or may be equilibrium controlled where the system of fluid and solids is assumed to be at equilibrium throughout the reaction path.

In the geochemical modeling calculations of the Morcinek brine, equilibrium at all stages along the reaction path was assumed. The brine composition is first input into the code and its aqueous speciation is computed. Sodium carbonate is then added to the brine. The pH rises and carbonate minerals precipitate. After a sufficient amount of carbonate is added, the precipitate phases are removed from the system, simulating the filtering of the precipitate sludge. HCl is then added to the brine to reduce the pH to 6.0. Finally water is removed from the brine to simulate the passage of the brine through the reverse osmosis unit. During the modeling calculations a small amount of NaCl is allowed to pass through the membrane to simulate the fact that the product water contains a few hundred ppm of salt.

Table 1. Morcinek mine brine composition during pretreatment and after reverse osmosis separation (in mg/kg) at locations in figure 1. Data are for adding the minimum amounts of Na_2CO_3 and HCl per liter brine to avoid Nontronite clay – $NaFe_6AlSi_{10}O_{30}(OH)_6$ – from precipitating in the reverse osmosis process

Element	1 Original Brine	2 Na_2CO_3 Addition	3 Acid Addition	4 Brine After RO
Cl-	18,500	18,500	18,520	44,230
Na	10,300	10,250	10,250	24,470
Ca	1,169	961	961	2,295
Mg	511	468	468	1,118
S	80	80	80	191
Sr	44.1	40.5	40.5	96.7
C	9.7	11.2	11.2	26.8
Si	6.5	1.5	1.9	4.5
Ba	1.4	0.07	0.07	0.17
Mn	1.1	5×10^{-6}	5×10^{-6}	1×10^{-5}
Fe	0.3	1×10^{-8}	1×10^{-8}	2×10^{-8}
Al	0.03	1×10^{-6}	9×10^{-6}	2×10^{-5}
pH	7.5	7.5	6.1	5.9
TDS	31,089	31,100	31,108	74,303

DISCUSSION OF THE EQ3/6 GEOCHEMICAL MODELING CODE RESULTS

The pretreatment process consists of three main steps: 1) addition of sodium carbonate (Na_2CO_3) while in contact with the atmosphere (aeration); 2) filtration to remove precipitates and suspended particles generated during Na_2CO_3 addition; 3) addition of hydrochloric acid (HCl), and filtration. The changes in brine chemistry that accompany these steps, along with the initial brine composition, are shown in Table 1. The initial brine composition was obtained from brines of the Morcinek mine. The pretreatment described in this paper is tailored to this one brine. A different brine could require a different pretreatment. The purpose of this paper is to describe a brine treatment methodology, and not to give a treatment solution for all brines.

Aeration causes the precipitation of barium and strontium carbonates through exposure to atmospheric carbon dioxide. Iron and aluminum hydroxide (along with silica in some cases) precipitate through the exposure of the brine to atmospheric oxygen. Although the geochemical computer simulations predict these changes, in natural systems, mineral precipitation may not take place because nucleation barriers exist and mineral surfaces may be poisoned. A higher amount of supersaturation often is needed to initiate the precipitation process.

The addition of sodium carbonate provides this increase in degree of supersaturation. Adding sodium carbonate increases the pH and increases the total carbonate content of the solution. Both factors decrease carbonate mineral solubilities and cause the precipitation of these phases. The concentrations of Ba, Sr, Mg, and Ca decrease as carbonate minerals precipitate from solution (see Figure 2 and Table 1). Addition of sodium carbonate is sufficient to reduce concentration of potential scale formers to levels where scale will not form in the reverse osmosis unit. The amount of sodium carbonate required is arrived at by an iterative method in which the goal is to add a minimum amount of Na_2CO_3 and still have the brine remain undersaturated with respect to all solids after the reverse osmosis treatment. The geochemical modeling code is run using pre-determined amounts of Na_2CO_3 until a minimum amount of the additive needed is found.

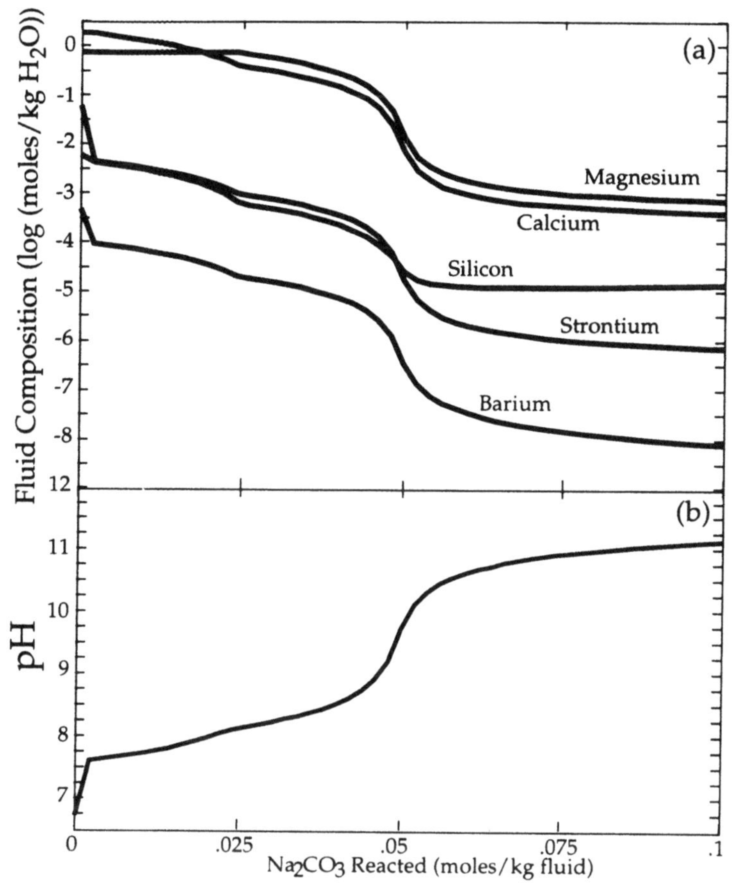

Figure 2. Chemical effects of adding Na_2CO_3 to Morcinek brine.

Filtration with a multimedia filter should be done after addition of Na_2CO_3 to use available suspended solids as nucleation sites for mineral precipitation. Both the existing suspended

particulates and fresh precipitates can then be removed by filtration prior to acidification. This step should remove most of the common solids that foul reverse osmosis membranes (i.e. sulfates, oxides, and carbonates).

After removal of the suspended solids, the solution is acidified to a pH of approximately 6.0 using hydrochloric acid. This value for the pH was chosen as optimal for reverse osmosis treatment both in terms of membrane performance and potential corrosion of the water treatment equipment.

The predicted chemical effects of the reverse osmosis process on the acidified brine are shown in Figure 3. The mineral saturation index (log(Q/K)) in Figure 3a is a measure of the degree to which a mineral is undersaturated (negative value) or supersaturated (positive value). The ratio Q/K refers to the concentration product (Q) of species in solution divided by the concentration product (K) at equilibrium. K also is referred to as the solubility product. If a species is undersaturated $Q<K$ and $\log(Q/K)<0$. On the other hand, if a solid is supersaturated $Q>K$ and $\log(Q/K)>0$. The solubility product therefore can be used as a predictor of the potential for membrane scaling. Figure 3 shows that all the potential

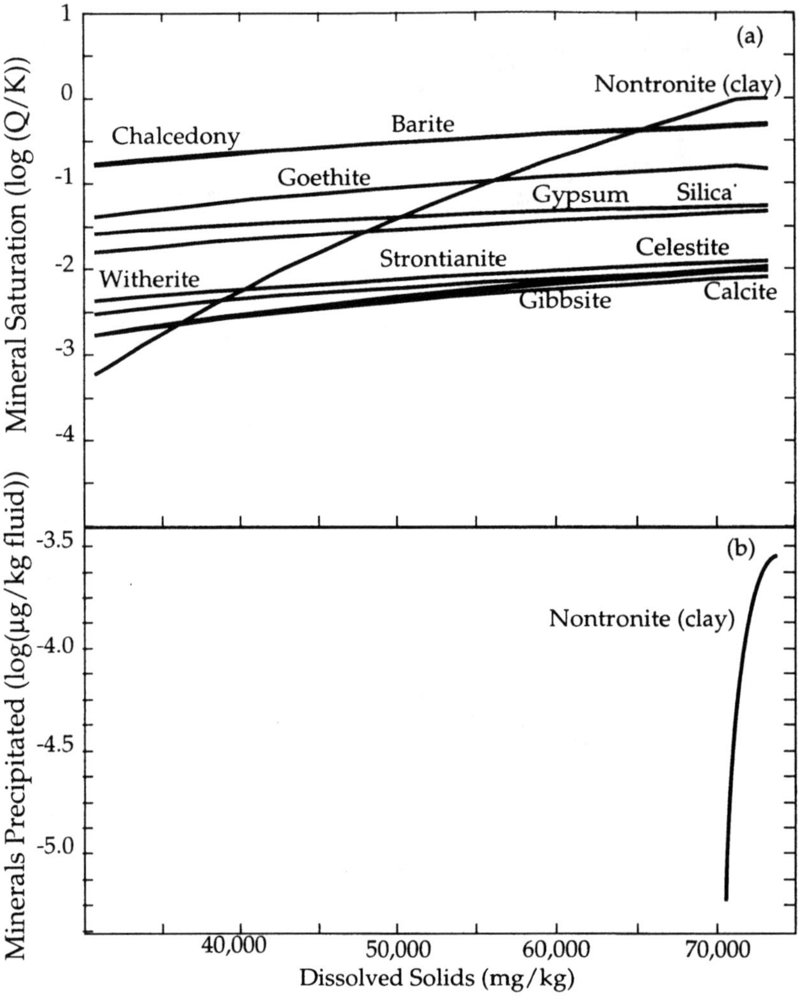

Figure 3. Chemical effects of reverse osmosis on pretreated Morcinek brine. All phases remain undersaturated during the reverse osmosis process. When the total dissolved salts (TDS) increases to more than 74,303 ppm nontronite clay - $NaFe_6AlSi_{10}O_{30}(OH)_6$ - begins to precipitate. The chemical formulae for the various constituents are given in the appendix.

scale-forming solids are significantly undersaturated during the reverse osmosis treatment and therefore will not precipitate.

Simply adding acid to the brine without the step of adding Na_2CO_3 does not prevent mineral precipitation. The solubilities of sulfate minerals such as barium sulfate ($BaSO_4$), calcium sulfate ($CaSO_4.H_2O$) and strontium sulfate ($SrSO_4$) are relatively insensitive to pH. Computer modeling calculations which simulate the addition of HCl to the brine without the intermediate sodium carbonate step show that the brine will continue to precipitate barium and strontium minerals during the reverse osmosis process. These metals, or the sulfate, must be removed from the brine before they enter the reverse osmosis unit in order to prevent scale formation.

HCl is preferred over common acids such as sulfuric acid because the chloride anion forms few insoluble solids. Sulfate forms sparingly soluble solids with numerous cations, such as Ba, Sr,

Ca. Adding sulfuric acid pushes the fluid closer to saturation with respect to these phases than does HCl. However, cost considerations may dictate whether hydrochloric or sulfuric acid (or some mixture) is used. Nitric acid also is not preferred because it has the potential to oxidize the reverse osmosis membranes and nitrate availability may encourage microbial growth.

Ultrafiltration to remove particles as small as 0.1 microns should be used to remove any remaining suspended load before passage into the reverse osmosis unit. Economic factors will determine whether ultrafiltration is performed before or after HCl addition. Ultrafiltration before HCl addition will minimize the amount of HCl that needs to be added to the brine because the suspended load of very fine particles will tend to dissolve in the presence of the acid. More acid will therefore be necessary to obtain the desired pH. On the other hand, ultrafiltration after HCl addition will extend the lifetime of the filters because there will be reduced numbers of fine particles, some of which have reacted with the previously added HCl and dissolved. The tradeoff is between the cost of supplying and cleaning the filters, and the cost of HCl.

Besides inorganic scaling, some types of biological activity are adverse to membrane performance. Biofilms may form on the membrane and interfere with passage of water through the pores. Bacteria also can cause fouling. The proposed pretreatment for the inorganic scalants has a favorable effect on potential biological problems as well. Changes in pH occurring during the addition of Na_2CO_3 and HCl to the brine will kill many of the natural bacteria present because most individual species of bacteria can only thrive within a restricted pH range.

LABORATORY VERIFICATION OF THE EQ3/6 GEOCHEMICAL MODELING CODE RESULTS

Bench-top laboratory experiments were performed to confirm the predictions of the geochemical modeling code results. Approximately 2 liters of simulated mine brine were

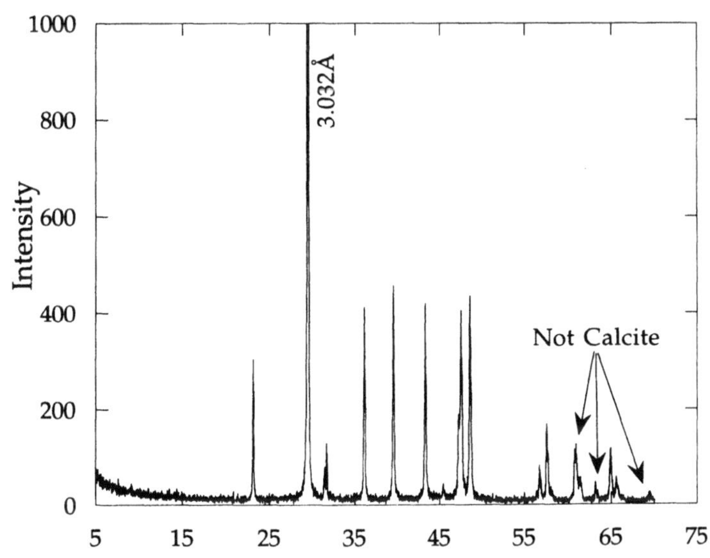

Figure 4. X-ray diffraction spectrum of precipitate generated during pretreatment. All peaks except those indicated correspond to calcite.

generated from soluble salts of all the major components. After addition of the salts, the pH was adjusted to 7.0 using a small amount of HCl to simulate the pH of the brine of the Morcinek mine.

A solution of sodium carbonate was then added to the simulated brine. Immediate precipitation of carbonate took place. The brine turned to a milky white color. Within a few seconds, coagulation of the milky precipitate took place, and the precipitate began to settle to the bottom of the test container. Based an Stoke's settling velocity, most of the precipitating particles had a size in the range from 5 to 15 microns.[13]

X-Ray diffraction of the dried precipitate revealed it to be a well-crystallized solid carbonate. The X-Ray diffraction spectrum is shown in Figure 4. The peaks are sharp and well-defined and closely match the pattern of calcite ($CaCO_3$). Rapid precipitation often gives rise to disordered solids with broad poorly-defined X-Ray peaks. This is not the case for precipitates from the pretreatment process.

Chemical analysis of the digested solid using inductively coupled plasma emission spectroscopy (ICP) confirm that the solid is primarily calcium carbonate with lesser amounts of magnesium and strontium. (See Figure 5a) The measured concentrations of calcium, magnesium, and strontium in the solid compares favorably with predicted values obtained from the EQ3/6 geochemical modeling codes using a simple ideal solid solution model for carbonates with the calcite structure. The ideal model predicts more Mg to be present in the solid than the measured amount. This difference is consistent with the known non-ideal behavior of Ca-Mg carbonates which select against Mg.

Figure 5b shows the predicted and observed decrease in elemental concentration in solution during pretreatment. Predicted and measured values for calcium and strontium show good agreement. However, the geochemical codes predict that more magnesium is removed from solution than the amount actually observed. The reason for this difference is that the codes over-predicts the amount precipitated in the carbonate precipitate. A more accurate non-ideal model for the carbonate solid solution which accounts for the selective partitioning of calcium over magnesium would improve the prediction of magnesium solubility.

Barium is predicted to decrease in concentration by almost 95 per cent, whereas it is observed to decrease only by about 25 per cent. These results are consistent with previous observation of barium in natural waters. Barium minerals tend not to equilibrate fully with the solution during short time periods. The pretreatment process allows precipitation to take place over a few minutes, so that the system does not have time to equilibrate fully with barium minerals. As a results, barium remains supersaturated in the solution.

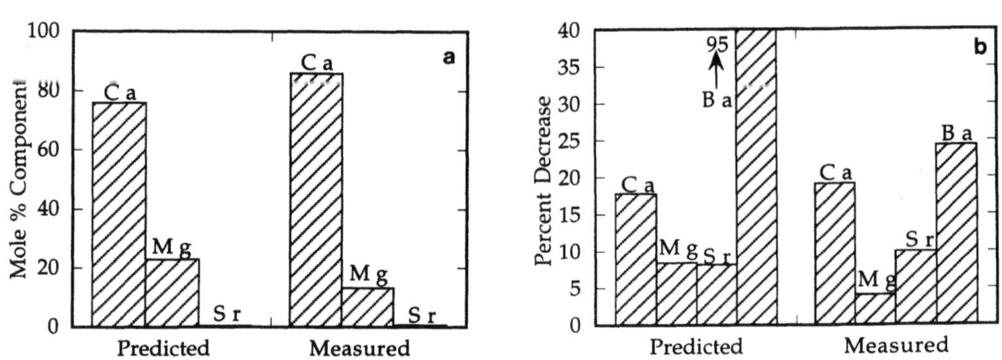

Figure 5. Predicted and measured composition during pretreatment. (a) Carbonate precipitate generated. (b) Decrease in elemental concentrations in solution. Predicted composition refers to EQ3/6 geochemical code calculations.

The EQ3/6 geochemical modeling calculations show fairly good agreement with the results of the laboratory experiments. In spite of the good agreement, it is recommended that the geochemical code predictions never be used alone, but always should be verified in laboratory experiments.

POST TREATMENT OF BRINE FROM REVERSE OSMOSIS PROCESS

The reverse osmosis process can reduce the volume of the Morcinek brine by approximately 60 per cent. The brine volume can be further reduced by use of a submerged combustion evaporator that evaporates water vapor from the brine. In either case, a certain volume of concentrated brine remains. A common disposal method is to post treat this concentrated brine and inject it into a subterranean formation.

Injection of the concentrated brine into a subterranean formation in a manner that does not plug the underground formation requires careful analysis. Plugging of the underground formation would make the injection formation unfit for further use. Furthermore, injected fluids have been known to react with the minerals of the underground formation thereby causing injection wells to blow out, and to result in environmental and operational accidents.[6,14,15]

In the post treatment of the brine, the brine composition, pressure and temperature of the injection formation and the mineral constituents of the formation need to be considered. After the reverse osmosis process, the concentrated brine is near, or slightly above, saturation with respect to one, or more, solids. The brine leaving the reverse osmosis unit is at a relatively low pressure of a few bars and at a temperature of approximately 25°C. On the other hand, an underground injection formation may have a pressure of several hundred bars and a temperature of about 100°C depending on the depth of the formation. These changes in temperature and pressure need to be considered in the analysis of potential scaling during injection.

Usually the most cost-effective post treatment is the dilution of the concentrated brine from the reverse osmosis process with a portion of the original brine feedstream. The EQ3/6 geochemical modeling codes can be used to determine the interaction of the diluted brine mixture with formation minerals to insure that the injected brine does not adversely react with these minerals during the life of the injection formation.

In most cases, filtration, followed by dilution and addition of a small amount of acid would prevent scale formation during injection of the brine into an underground formation. The optimum amounts of dilutant and acid can be determined by use of the EQ3/6 geochemical modeling codes. A minimum amount of dilutant clearly is desirable for economic reasons.

The predictions of the geochemical modeling codes need to be verified by laboratory experiments before injection is started to insure that the code predictions accurately reflect the behavior of the brine mixture during and after the injection process.

SUMMARY CONCLUSIONS

Geochemical modeling is used to simulate the chemical pretreatment of produced waste waters (brines) from oil and gas wells or mines before passage through a reverse osmosis unit. The reverse osmosis process separates waste water into potable water and a more saline brine. Although sea water has been successfully treated using the reverse osmosis process, efforts to treat the brines from oil, gas and mining operations have been less successful due mainly to precipitation of solids and fouling of the reverse osmosis membranes.

The pretreatment method consists of: (1) coarse filtering to remove suspended solids; (2) sodium carbonate addition to precipitate alkaline earth carbonates (the most common scale formers); (3) microfiltration to remove carbonate precipitates; and (4) HCl addition to undersaturate the solution with respect to mineral phases. The fluid then enters the reverse osmosis unit.

Oxides, carbonates, sulfates, clays, and other aluminosilicates can potentially scale the membrane surface. Because these phases have variable sensitivities to pH, redox state, ionic strength, and temperature, the brine flowing to the reverse osmosis unit is modified to avoid supersaturation with respect to all potential precipitates as the solution is concentrated during the reverse osmosis process.

The treatment process has been tested in the laboratory. A pilot scale demonstration is expected to start at the Morcinek mine in Poland in early 1996. A commercial plant would treat 3,000 cubic meters of brine per day.

The advantage of the use of the geochemical codes is that, at each step of the process, the chemical composition of the brine is known. As a result the pretreatment can be optimized to avoid precipitation of salts in the reverse osmosis process.

In the particular case of the treatment of the Morcinek mine brine, the limiting factor is the nontronite clay - $NaFe_6AlSi_{10}O_{30}(OH)_6$ - that would be formed if the brine concentration would exceed 74,303 ppm in the reverse osmosis process. Without the use of the EQ3/6 geochemical code, it is doubtful that one could have foreseen that formation of nontronite clay would be the limiting factor in the reverse osmosis process because the Morcinek brine contains only 0.03 ppm aluminum, 0.3 ppm iron, and 6.5 ppm silica.

The geochemical modeling results were verified in laboratory experiments and good agreement was obtained between the results of the geochemical modeling code and the experiments.

ACKNOWLEDGMENTS

The research presented in this paper was funded, in part, by Small Business Innovation Research Grants No. DE-FG03-91ER 81105 A 002 and A 003 from the Office of Energy Research of the U.S. Department of Energy. The pretreatment modeling was performed under the auspices of the U.S. Department of Energy by the Lawrence Livermore National Laboratory at Livermore, California under contract No. W-7405-Eng. 48 and contract AB05-SFEW-0030 from the Morgantown Energy Technology Center.

The pilot demonstration at the Morcinek mine is being conducted with the Polish co-sponsorship and cooperation of EXBUD-MIMET and the Jastrzebska Coal Company. This program is also co-sponsored by Aquatech Services, Inc., the U.S. Department of Energy and the U.S. Environmental Protection Agency. Aquatech Services, Inc. expresses its appreciation to the U.S. Department of Energy and the Environmental Protection Agency for the support of this effort.

REFERENCES

1. Bethke, C.M., 1992. The geochemist's workbench: A users guide to Rxn, Act2, Tact, React, and Gtplot, *U. of Ill. Press.*

2. Brandt, H., and Tait, J.H., 1992. Process eases coalbed brine disposal. *American Oil and Gas Reporter*. 29 - 33.
3. Brandt, H., and Jackson, K.J., 1993. Process for coalbed brine disposal. U.S. Dept. Energy Fuels Tech. Rev. Mtg.
4. Brandt, H., and Bourcier, W.L., 1994. Treatment process for coal mine waste water. *Second Intern. Symp. on Envir. Contamin. in Centr. and East. Europe*. Budapest, Hungary.
5. Brandt, H., and Bourcier, W.L., 1994. Treatment process for waste water disposal using coalbed methane. *First Silesian Intern. Conf. Coalbed Methane Utiliz. Katowice, Poland; Pol. Found. Energy Effic.*, 116-122.
6. Brower, R.D., Visocky, A.P., Krapac, I.G., Hensel, B.R., Peyton, G.R., Nealon, J.S., and Guthrie, M., 1989. Evaluation of underground injection of industrial waste in Illinois. *Illin. Scient. Surv. Joint Rep.* 2.
7. Butler, J.N., 1982. Carbon dioxide equilibria and their applications, *Add. Wesl. Publ.*, 259.
8. Fountoukidis, E., Maroulis, Z.B., and Marinos-Kouris, D., 1990. Crystallization of calcium sulfate on reverse osmosis membranes. *Desal. 79*, 47-63.
9. Isner, J.D., and Williams. R.C., 1993. Analytical techniques for identifying reverse osmosis foulants, reverse osmosis: membrane technology, water chemistry, and industrial applications, ed. Zahid Amjad. *Van Nostr. Reinh.*, 237-273.
10. Langlier, W.F., 1946. Chemical equilibrium in water", *J. Am. Water Works Assn.*
11. Morel, F.M.M., and Hering, J.G., 1993. Principles and applications of aquatic chemistry, *Wiley & Sons*.
12. Pitzer, K.S., 1973. Thermodynamics of electrolytes - I. Theoretical basis and general equations. *J. Phys. Chem. 77*, 268-277.
13. Pontius, F., 1990 Water quality and treatment: A handbook of community water supplies. *McGraw-Hill.*
14. Roy, W.R., Mravik, S.C., Krapac, I.G., Dickerson, D.R., and Griffith, R.A., 1989. Geochemical interaction of hazardous wastes with geological formation in deep-well systems. *Env. Geol. Notes, 130 Illin. State Geol. Surv.*
15. Roy, W.R., Seyler, B., Steele, J.D., Mravik, S.C., Moore, D.M., Krapac, I.G., Peden, J.M., and Griffith, R.A. 1991. Geochemical transformations and modeling of two deep-well injected hazardous wastes. *Ground Water, 29*, 671-677.
16. Stiff, H.A., and L.E. Davis, 1952. A method for predicting the tendency of oil field waters to deposit calcium carbonate. *Petr. Trans. AIME*, 195.
17. Wolery, T.J.,1992 EQ3/6. A software package for geochemical modeling of aqueous systems: Package overview and installation guide. *UCRL-MA-110662 PT1, Lawr. Liv. Nat. Lab.*, 1-66.
18. Wolery, T.J., 1992. EQ3NR. A computer program for geochemical aqueous speciation-solubility calculations: Theoretical manual, user's guide, and related documentation. *UCRL-MA-110662 PTIII. Lawr. Liv. Nat. Lab., 246.*
19. Wolery, T.J. and Daveler, S.A., 1992. EQ6, A computer code for reaction-path modeling of aqueous geochemical systems; theoretical manual, user's guide and related documentation," *UCRL-MA 110662 PTIV, Lawr. Liv. Nat. Lab.*, 338.

APPENDIX

The chemical formulae for the constituents shown in Figure 3 are given below.

Constituent	Formula	Constituent	Formula
Barite	$BaSO_4$	Goethite	$FeO(OH)$
Calcite	$CaCO_3$	Nontronite (clay)	$NaFe_6AlSi_{10}O_{30}(OH)_6$
Celestite	$SrSO_4$	Strontianite	$SrCO_3$
Chalcedony	SiO_2	Witherite	$BaCO_3$
Gibbsite	$Al(OH)_3$		

47

COMPARISON OF INJECTIVITY PREDICTION MODELS TO ESTIMATE ULA FIELD INJECTOR PERFORMANCE FOR PRODUCED WATER REINJECTION

Claas van der Zwaag and Lars Øyno

Reservoir Laboratories AS
Pirsenteret, N-7005 Trondheim, Norway

ABSTRACT

Waterflood injectivity prediction models suggested by Barkman and Davidson (1972), Eylander (1987) and van Velzen and Leerlooijer (1992) were investigated and evaluated. Ula field data, water quality analyses of field samples and core flood experiments using reservoir samples provided the input datasets to the models. A post-perforation injection field trial on Ula field injector A-03 was evaluated and compared to predicted results. Different field scenarios were used to compare results of injectivity prediction models.

Many assumptions regarding the input datasets were necessary and made a straight-forward application of models difficult. The study showed, however, that the discrepancy between observed field injectivity and predicted injectivity could be related to a lack of field data with respect to the accurate particle concentration in the downhole situation. Also, a lack of accurate input values regarding the specific model parameters describing particle invasion depth may yield misleading, overestimated injector performance for the studied field case.

It has, moreover, been shown that injectivity prediction models can be used to analyze the effect of different downhole injection situations on injectivity. This is demonstrated by three scenarios addressing the transition from test-scale seawater injection to full-scale produced water reinjection and the effect of thermally induced fracturing on produced water reinjection.

The injectivities predicted by the individual models differ considerably.

INTRODUCTION

When injecting low quality water into a reservoir or aquifer for pressure maintainance the success of an operation may be jeopardized by a loss of injectivity due to permeability

Produced Water2: Environmental Issues and Mitigation Technologies
Edited by M. Reed and S. Johnsen, Plenum Press, New York, 1996

impairment. Many investigations focusing on the impact of injection water quality on injector field performance have been conducted. A comprehensive review has been presented by Vetter et al. (1987). Numerical models were suggested and applied to describe mechanisms of surface and deep-bed filtration of particles suspended in the injection water (Abrams (1977), Barkman and Davidson (1972), Ershagi et al. (1986), Eylander (1987), Gruesbeck and Collins (1982), Hofsäss and Kleinitz (1995), Hsi et al. (1994), Kumar and Todd (1988), Pautz et al. (1989), Roque et al. (1995), Sharma and Yortsos (1987), Tang (1982), Todd et al. (1984), van Velzen and Leerlooijer (1992), van Oort et al. (1993)). The application of such models provides predictions of the lifetime performance of water injectors. Predictions, or simulations, can help to create a background for decisions for the design of topside water processing equipment, to estimate reservoir pressure development, or to plan stimulation measures.

The motivation behind this work was to select models that help to predict the water injector performance on Ula-field (Norwegian Continental Shelf).

Existing prediction models and experimental procedures to define water quality and estimate injector lifetime performance were evaluated. Three models were selected as potential candidates. These were the models of :

1. Barkman & Davidson (1972)
2. Eylander (1987)
3. van Velzen & Leerlooijer (1992).

To provide input data and to test these models the post-perforation, pre-fracturing field performance of the Ula water injector A-03 was evaluated. Furthermore, the topside water quality of the injection water prior to injection was analyzed and core flood experiments using Ula reservoir cores and field water samples were carried out.

FIELD CASE

Since the start-up of oil production in autumn 1986 water has been injected into the Ula-field reservoir to supplement the weak aquifer support. Initially seawater was used as the injection fluid. In spring 1994 tests were initiated to evaluate the use of produced formation water as part of the injection fluid. These tests were motivated by environmental concern regarding the discharge of pollutants contained in the formation water. After successful field trials, formation water is presently reinjected into the Ula reservoir on full scale (Vik et al. (1994) . Commonly, a mixture of equal parts of produced water (PW) and seawater (SW) is used. Average rates of 200.000 BPD (31800 m^3) are injected by seven injectors.

The Ula water injector well A 03 was the first injector drilled and brought on-line. An extensive series of field tests was carried out at injection start-up. Tests were evaluated with respect to well injectivity in two phases of the start-up phase. These two phases were:

1. the post-perforation, pre-fracturing performance, and
2. injector performance after thermally induced fracturing (TIF).

Results of investigations focusing on TIF were presented by Svendsen et.al. (1991). Their study shows that TIF is the main mechanism that guarantees the success of Ula water floods.

In this study, however, the results of a post-perforation, pre-fracturing injection trial were evaluated to create a baseline for the validation of injectivity prediction models.

Data describing the seawater injection trial is summarized in Figure 1. For further evaluation this trial was subdivided into three runs. It was assumed that the first run reflects

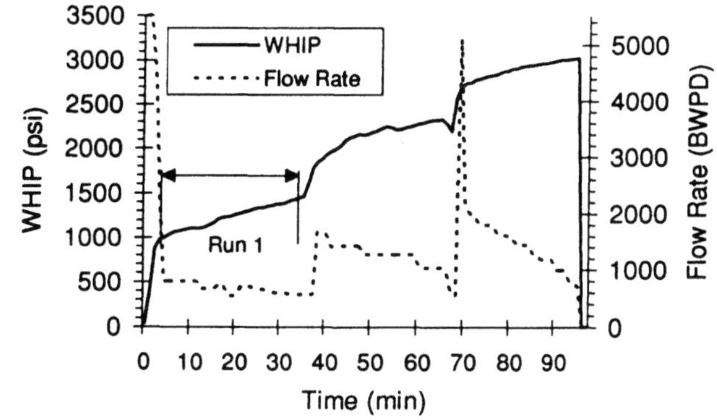

Figure 1. Post perforation injection trial for well A-03 showing wellhead injection pressure (WHIP) and flow rate over time.

initial injection conditions best, since the formation was not affected by previous injections. Figure 2 shows the normalized injectivity index for the first run. A dramatic injectivity decline can be observed. The main result obtained is that injection half-life (i.e. a 50 % reduction of injectivity) occured after only 32 minutes. The cumulative volume injected at this point was 2.46 m^3 (15.5 bbl).

INJECTION WATER ANALYSIS AND CORE FLOOD EXPERIMENTS

The water quality of the injected seawater has been recorded on a regular basis since 1990. Records show that seawater properties are stable with respect to the content of suspended solids. It is assumed that no changes in water quality found place between the time of the field trial (1988) and 1990.

A mixture of seawater (SW) and produced water (PW) is currently injected into the reservoir. Waters are mixed at an average ratio of 1:1. This ratio was also used when running core flood experiments. Table 1 summarizes selected water quality parameters for SW and PW.

Constant rate core flood experiments were conducted using Hassler-type flow cells and representative Ula field cores. The pressure drop over a core plug was measured. Plugs

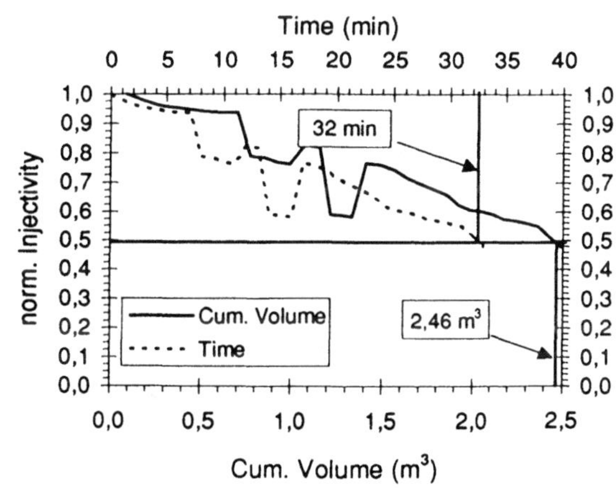

Figure 2. Normalized injectivity index versus cumulative injected volume and time for the post-perforation injection phase, run 1, extracted from Figure 1.

Table 1. Water analysis data

	Seawater	50% SW/50% PW
Total Solids (mg/l)[1] (w/ oil droplets)		0.6-13.0
Suspended Solids (mg/l)[1]	0.1-4.6	2.6-4.6
Mean Particle Diameter (μm)[1]	3.0	N/A
Density (kg/m^3)[2]	1023	1035
Viscosity (cP)[2]	1.011	1.116

[1]Measured onsite in the period from February to May 1994 (after Vik et al. (1994))
[2]Laboratory data of samples obtained in January 1995

(3.8 cm diameter, 8 cm length) were cut, trimmed and cleaned from hydrocarbons by flooding solvent (Isopar L) at ca. 60°C prior to the water flooding experiment.

A core flood consisted of two phases. First, injection water, filtered through 0.45 μm micropore filters, was flooded until the solvent was removed and stable flow conditions were observed (Phase 1). From the pressure drop reading in this test phase the relative water permeability at irreducible oil content was calculated. In the second test phase unfiltered mixture of Ula field PW and SW was injected and the pressure drop was continuously recorded.

A typical result of a core flood is presented in Figure 4. Normalized permeability, which is calculated as the ratio of the relative permeability during the flood in phase 2 and the initial permeability measured in test phase 1, is plotted versus time and the cumulative injected volume.

PREDICTION MODELS

A common approach to predict injectivity decline in a field situation is to analyze the results core flood experiments using representative core material and water samples, and extrapolate them to field scale.

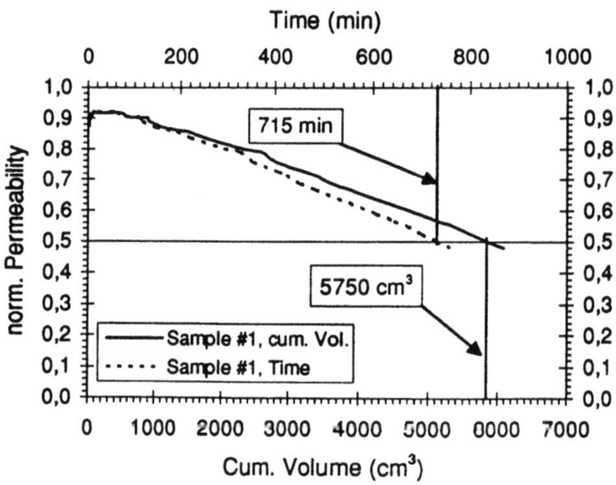

Figure 3. Example demonstrating results of laboratory core flood experiments.

Hofsäss and Kleinitz (1995) distinguish between a phenomenological and a statistical approach to describe filtration processes in a downhole situation. Basically, the phenomenological approach deals with parameters describing the filtration process in terms of filter cake/filter medium properties and a framework dataset consisting of solid concentrations in the injection water, viscosities, flow rates, core/formation dimensions etc.. On the other hand, the statistical approach, as for example network models (cf. Hofsäss and Kleinitz (1995)), attempts to derive probabilities for particle entrainment based on pore and particle size distributions. Several authors (Hsi et al. (1994), Hofsäss and Kleinitz (1995)) state that the phenomenological approach requires less numerical efforts.

Three phenomenological models were selected and compared. These were the models of:

1. Barkman and Davidson (1972);
2. Eylander (1987);
3. van Velzen and Leerlooijer (1992).

These models are formally based on Darcy's Law and equations describing material balances for water and suspended solids. Van Velzen and Leerlooijer (1992) implement also a numerical description of deep-bed filtration dynamics (Iwasaki's equation).

Barkman and Davidson's theory is frequently applied in oilfield waterflood performance prediction, although several shortcomings with respect to the numerical description of filtercake deposition were identified by several authors (Eylander (1987), van Velzen and Leerlooijer (1992) and Hofsäss and Kleinitz (1995)). The model considers linear flow conditions only. Under radial flow, as in the downhole situation, pressure differentials and flow velocities decrease more rapid in the flow direction than in the linear case.

Eylander (1987) modified the Barkman and Davidson model. Flow conditions are still considered linear. Emphasis is put on a description of the specific impairment mechanisms. Impairment mechanisms are identified using constant rate core flood data. An example is given in Figure 5, which shows a plot of the "pressure differential corrected" reciprocal flow rate versus the cumulative volume. The figure shows a filtration curve reflecting internal filtercake deposition. The slope "a" and the intercept "b" (see Figure 5) are measures for the filtercake permeability and the effective brine permeability of the porous medium, respectively.

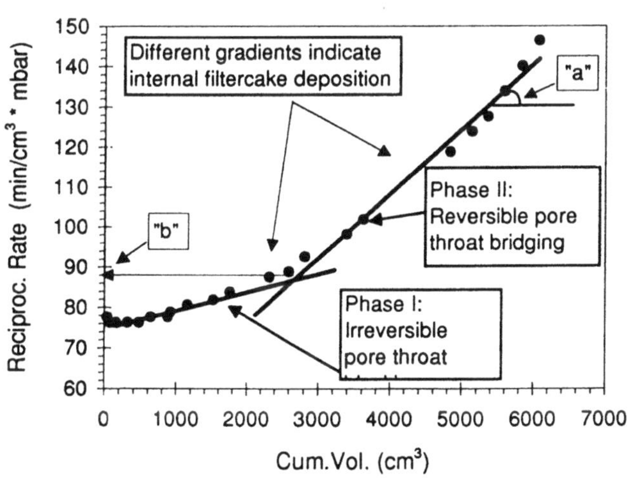

Figure 4. Eylander plot of constant rate core flood experiments indicating internal filtercake deposition.

Table 2. Input field data

Operator:	BP	
Field:	Ula	
Injector:	A-03	
Injection rate (q_0)[1]	622 bbl/day	99 m^3/day
Reservoir height (h_0)	292.7 ft	89.2 m
Hole radius (r_w)	4.25	0.108 m
Drainage factor (r_e/r_w)	1800	
Perforations	4 per ft	13.1232 per m
Perforation diameter	0.5	0.0127 m
Perforation area	229.9 sq ft	0.1483 m^2
Corr. Reservoir height (h)	0.183 ft	0.005 m
Reservoir permeability (k_f)	173.15 mD	1.71E-13 m^2
Solids content injection fluid	4.6 ppm	

[1]Average for field case

Van Velzen and Leerlooijer (1992), and in a later publication van Oort, van Velzen and Leerlooijer (1993) developed a model considering radial flow conditions. They designed radial core flow experiments which showed good agreement between measured and predicted data. The authors suggest a semi-empirical approach to describe filtration characteristics, including a "damage factor", δ_f, and a volume filter coefficient, λ_V. Both factors are obtained from core flow experiments. For the derivation of the volume filter coefficient the authors suggest multi-port core floods.

APPLICATION OF PREDICTION MODELS

For the models of Barkman and Davidson, Eylander and Van Velzen and Leerlooijer the abbreviations BD, EYL and VVLL, respectively, will be used subsequently.

The dataset describing the Ula field injector A-03 (Table 2) was extracted from drilling, completion and production reports. An example for the dataset characterizing the core flood experiment is given in Table 3.

Table 3. Input experimental data

Density water (r_w)	1035 kg/m^3
Density filtercake (r_c)	2000 kg/m^3
Radius filtermedium	0.0189 m
Area filtercake (a_c)	0.00112 m^2
Core length (l)	0.0809 m
Temperature (t)	22°C
Viscosity	1.066 cP
Calc. Viscosity (m)	0.00107 Ns/m^2
Flow rate exp. (q)	480 ml/h
Filter cake porosity (φ_c)	0.200
Filter/core porosity (φ_m)	0.206
Experimental suspended solids content (W_e)	13 ppm

To use the prediction models several assumptions were made:
BD and EYL:

1. The permeability of the filtercake, k_c, was defined by equation 4a in Eylander's model,
2. The porosity of the filtercake, φ_c, was set to 20%
3. The particle bridging depth (BD) or particle invasion depth (EYL) was set to 2 cm, i.e. the particle bridging or invasion radius, r_a, or r_d, was calculated as the wellbore radius plus particle bridging or invasion depth, $r_a = r_w + 0.02\ m$ or $r_d = r_w + 0.02\ m$.

EYL:

4. The permeability of the core material, k_m, was calculated according to equation 2 in Eylander's study, i.e. not using the experimentally determined initial brine permeability.

VVLL:

5. The half-invasion depth, $L_{1/2}$, was set to 0.25 cm.

To predict the effect of variations in particle concentrations in field scenarios it was for all models equally assumed that:

6. The filtercake permeability remains constant for different particle concentrations.
7. Seawater (used in the field test) and the 1:1 mixture of seawater and produced water had the same particle size distribution, so that the filtercake permeability was not affected by the different suspensions.

Finally, water was in the downhole situation injected through perforations. To apply all models equally, the following assumption was made:

8. The injection surface area offered by perforations over the complete reservoir section is equal to the area of an open hole section having the same wellbore radius but a corrected (smaller) section height (see Table 1).

COMPARISON OF FIELD DATA AND PREDICTED DATA

For a comparison of prediction models, three scenarios were evaluated. These were:

1. Comparison of the post-perforation, pre-fracturing field case and predicted injector performance,
2. Full scale produced water reinjection into the unfractured reservoir,
3. Injection into a fractured reservoir using full scale injection rates and assuming a fracture surface area determined by a Hydraulic Impedance Test.

Scenario 1: Comparison of Post-Perforation/Pre-Fracturing and Predicted Injector Performance. Table 4 shows the results of the injectivity prediction. It can be seen that the calculated injector performance is higher than the values observed in the field case. Even when reducing the particle bridging depth (BD), the particle invasion depth (EYL) or the "Half-Invasion" depth (VVLL), predicted injectivity is at least 7 times higher than the observed field values (see Table 4).

Table 4. Post-perforation/pre-fracturing and predicted injector performance

	Injector Halflife, $t_{0.5}$ (Field Case 32 min)			Cum. Injected Volume, $V_{0.5}$ (Field Case 32 2.46 m^3)		
Model	BD (min)	Eyl (min)	VVLL (min)	BD (m^3)	Eyl (m^3)	VVLL (m^3)
Scenario 1	550.1	425.3	964.4	40.2	10.4	70.5
Scen. 1. Min. Invasion Depth	380.0	254.7	224.0	27.8	6.2	16.4

As a first approach, it was assumed that the deviation between predicted and observed values is due to an unexpectedly higher particle content as caused by corrosion products, rock debris or cement which can be abundant in the start-up phase of an injection well.

To match field injectivity with values calculated by the prediction models, an increased solid content in the injection fluid was simulated. Figure 6 presents the concentrations of solids at which calculated half-life performance matches the field data. Values between 50 ppm and approximately 101 ppm were found for the different models.

Scenario 2: Full Scale Produced Water Reinjection into the Unfractured Reservoir. Reinjection of PW/SW into the unfractured reservoir at full-scale injection rates (30000 bbl/d) was simulated. Results are shown in Table 7. It can be seen that the calculated half-life performance is very low. The data indicates that injectivity would be reduced to 50% after 3 to 24 minutes.

Scenario 3: Injection Into a Fractured Reservoir Using Full Scale Injection Rates. The Ula reservoir is thermally fractured. Hydraulic Impedance Tests (see Millar et al. (1989), Paige et al. (1992)) were carried out 45 days after injection start-up. Fractures with 2.2 m full height and 20 m to 34 m half length were measured. This would yield a total fracture surface of approximately 300 m^2.

To simulate this surface gain the wellbore radius in the prediction models was increased and it was assumed that filtration processes along the fracture surface are similar

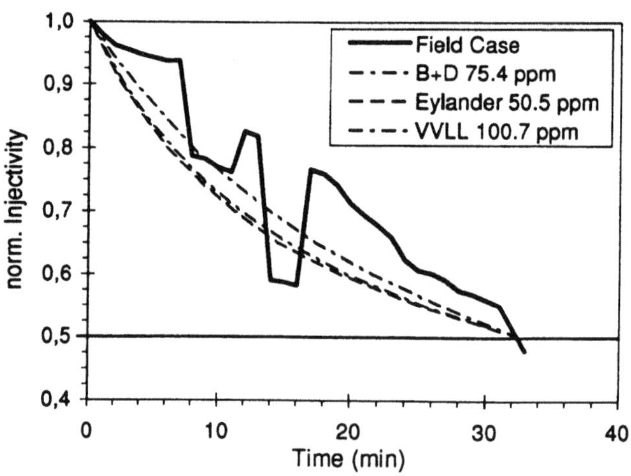

Figure 5. Predicted injectivity after matching with field data.

Table 5. Injector performance simulating full-scale injection into the unfractured reservoir

Scenario 2	Injector Halflife, $t_{0.5}$			Cum. Injected Volume, $V_{0.5}$		
Model	BD (min)	Eyl (min)	VVLL (min)	BD (m^3)	Eyl (m^3)	VVLL (m^3)
Perforated	4.2	2.8	22.9	13.8	3.1	75.9

to those in the wellbore. The simulation (Table 6) yielded injector half-lifes between 4 days (EYL) and 32 days (BD).

DISCUSSION

The application of injectivity prediction models requires many assumptions regarding the datasets describing field data and water quality data, the representativeness of core properties as well as individual model parameters as for example particle invasion depth or filtercake porosity. A detailed discussion of the influence of each parameter is beyond the scope of this paper.

It has, however, been shown, that injectivity prediction models can be used to analyze the impact of different downhole injection situations. The scenarios presented address the validation of injectivity models based on a field case using seawater as the injection fluid, the transition from test-scale seawater injection to full-scale produced water reinjection and the effect of thermally induced fracturing on produced water reinjection.

In the following, the most important error sources for the modeling of injector performance are identified. Furthermore, the results of the injection scenarios will be evaluated relative to each other.

The determination of the particle bridging or invasion depth has been a major concern of discussions in the papers of Barkman and Davidson (1972) and Eylander (1987). Van Velzen and Leerlooijer (1992), on the other hand, suggest the determination of the "Half-invasion depth" using multi-port core floods. The standard setup used in this study does not allow to evaluate this parameter. However, the impact of the particle invasion depth on injector performance prediction can be simulated. Using the Eylander plot (Figure 4) it becomes evident that particles invade the core material and deposit an internal filtercake. A simulated increase in the particle bridging radius (BD) or the invasion radius (EYL) using the wellbore radius plus core length as the upper limit leads to a steady increase in injector half-life resulting in doubled injectivity half-life at the maximum.

The model of Van Velzen and Leerlooijer reacts more complex on variations in the half-invasion depth within the limits given by the dimensions of the core plug. Injector half-life increases towards a maximum at approximately 3 cm. At this location, the half-life time $t_{0.5}$ is 2.5 times the calculated value for an earlier assumed half-invasion depth of 2.5 mm.

Table 6. Injector performance simulating full-scale injection into the fractured reservoir

Scenario 3	Injector Halflife, $t_{0.5}$			Cum. Injected Volume, $V_{0.5}$		
Model	BD (d)	Eyl (d)	VVLL (d)	BD (m^3)	Eyl (m^3)	VVLL (m^3)
Fractured Reservoir	22.1	3.0	12.3	105,378	4,759	58,885

In view to the application of prediction models it has been demonstrated that the discrepancy between the injectivity observed for a field trial run on well A-03 on Ula-field and the predicted injectivity may be related to differences in the particle concentration measured at the topside and the actual solid concentration downhole. This example shows that prediction models may be applied to analyze unexpected field situations.

The simulation of produced water reinjection under full scale injection rates considering an unfractured and a fractured reservoir (Scenario 2 and 3) demonstrates that injection performance increases by a factor of 750 (VVLL) to 7500 (BD). A fracture surface of 300 m^2 was simulated by increasing the wellbore diameter from 8.5 inches to 42 inches using an open hole reservoir section. Furthermore, sceanario 2 demonstrates that injections through perforations into an the unfractured reservoir would not be feasible.

A comparison of the half-life performance predicted by the different models shows that the Eylander model is most conservative in all three cases considered. Van Velzen and Leerlooijers model predicts highest performance in scenario 1 and 2, while longest half-life in scenario 3 is predicted by the Barkman and Davidson model.

CONCLUSIONS

Three injectivity prediction models, 1) Barkman and Davidson (1972), 2) Eylander (1987) and 3) van Velzen and Leerlooijer (1992) were compared, using Ula field data, water quality analysis and core flood experiments to provide the input datasets. A post-perforation injection field trial on Ula field injector A-03 was evaluated and compared to predicted results. Different field scenarios were used to compare injectivity prediction models among each other. The following results were obtained:

1. The application of injectivity prediction models requires many assumptions regarding the input datasets, the representativeness of core properties, as well as individual model parameters as for example particle invasion depth or filtercake porosity.
2. The study showed that observed field injectivity is lower than predicted injectivity. This was related to a lack of data with respect to the accurate particle concentration in the downhole situation. The discrepancy between field data and model values may also partly be induced by too optimistic assumptions regarding the different parameters describing particle invasion depth suggested by the individual models.
3. Injectivity prediction models can be used to analyze the influence of different downhole injection situations, as demonstrated by three scenarios addressing the transition from test-scale seawater injection to full-scale produced water reinjection and the effect of thermally induced fracturing on produced water reinjection.
4. Eylander's model shows most conservative estimates in all scenarios investigated. Compared to the other models Van Velzen and Leerlooijer's model predicted highest injector performance under adverse downhole injection conditions (injection through perforations at high rates), while Barkman and Davidson's model indicated best injectivity for simulated fracturing of the reservoir.

FUTURE WORK

Obviously, future work should be directed to minimize uncertainties in the input dataset. Based on the experience gathered in this study it is suggested that, first of all, downhole injection water sampling and improved multi port core flood tests, as suggested by Van Velzen and Leerlooijer (1992) may contribute to higher quality prediction data.

Ula field waterfloods are as initially mentioned strongly favored by thermally induced fracturing (TIF). Prediction models may be used to analyze the surface area gain induced through TIF by a comparison of actual field injectivity and fitting predicted injectivity to field values. Results obtained by such a study may supplement records obtained by Hydraulic Impedance Tests and may also help to keep track of the continuous surface gain caused by TIF. The application of prediction models needs more field studies for correlation and identification of errors in the models.

ACKNOWLEDGMENT

The financial support from the Norwegian Research Council and BP Norway Ltd. are gratefully appreciated. We would like to thank Gjert Laading and Eirik Normann (Research Council of Norway), Siri Bakke and Eilen A. Vik (Aquateam), Thor A. Hjelmaas and Hans Grüner (BP Norway), and Tor Eirik Stranna (Aker Engineering) for their work and cooperation within the MUST-PWRI research program of the Norwegian Research Council. Special thanks for her contributions to this work are directed to Ragnhild "Rags" Bjørnå.

REFERENCES

Abrams, A., 1977. Mud design to minimize rock impairment due to particle invasion. *Journal of Petroleum Technology*. May. 586-592.

Barkman, J. H., and Davidson, D. H., 1972. Measuring water quality and predicting well impairment. *Journal of Petroleum Technology*. July. 865-873.

Ershaghi, I., Hashemi, R., Caothien, S. C., and Abdassah, D., 1986. Injectivity losses under particle cake buildup and particle invasion. SPE paper 15073 presented at the 56th California Regional Meeting of the SPE, Oakland, Ca. 273-279

Eylander, J. G. R., 1987. Suspended solids specification for water injection from coreflood tests. SPE paper 16256 presented at the Intl. Symposium on Oilfield Chemistry, San Antonio, Tx. 121-133.

Gruesbeck, C., and Collins, R. E., 1982. Entrainment and deposition of fine particles in porous media. *SPE Journal*. December. 847-855.

Hofsäss, T. and Kleinitz, W., 1995. Infiltration von Feststoffen in Wasserinjektionsträger. *Erdöl, Erdgas, Kohle*. **111**, 9, 363-370.

Hsi, C. D., Dudzik, D. S., Lane, R. H., Buettner, J. W., and Neira, R. D., 1994. Formation injectivity damage due to produced water reinjection. SPE paper 27395 presented at the International Symposium on Formation Damage Control, Lafayette. 505-513.

Kumar, T., and Todd, A. C., 1988. A new approach for mathematical modeling of formation damage due to invasion of solid suspension. SPE paper 18203 presented at the 63rd Annual Conference and Exhibition of the SPE, Houston, Tx. 221-232.

Millar, J.W.A., Roberts, J.D. and Mellor, D.W., 1989. Ula Field: Water Injector A-03 Hydraulic Impedance Measurements, Internal Report ID 10 84 50, BP Research, Sunbury Research Center.

Paige, R. W., Roberts, J. D. M., Murray, L. R., and Mellor, D. W., 1992. Fracture measurement using Hydraulic Impedance Testing. SPE paper 24824 presented at the 67th Annual Technical Conference and Exhibition of the SPE, Washington. 605-614.

Pautz, J. F., Crocker, M. E., and Walton, C. G., 1989. Relating water quality and formation permeability to loss of injectivity. SPE paper 18888 presented at the Production Operations Symposium, Oklahoma City. 565-576.

Roque, C., Chauveteau, G., Renard, M., Thibault, G., Bouteca, M., and Rochon, J., 1995. Mechanisms of formation damage by retention of particles suspended in injection water. SPE paper 30110 presented at the European Formation Damage Conference, The Hague. 329-343.

Sharma, M. M., and Yortsos, Y. C., 1987. Fines migration in porous media. *AIChE Journal*. **33**. October. 1654-1662.

Svendsen, A. P., Wright, M. S., Clifford, P. J., and Berry, P. J., 1991. thermally induced fracturing of Ula water injectors. *SPE Production Engineering.* **November. 384-390.**

Tang, J., 1982. Injectivity damage caused by reinjection of oily produced water. CIM paper No. 82.33.55 presented at the 33rd Annual Technical Meeting of the Petroleum Society of CIM, Calgary.

Todd, A. C., Somerville, J. E., and Scott, G., 1984. The application of depth of formation damage measurements in predicting water injectivity decline. SPE paper 12498 presented at the Formation Damage Control Symposium, Bakersfield, Ca. 233-245.

van Velzen, J. F. G., and Leerlooijer, K., 1992. Impairment of a water injection well by suspended solids: Testing and prediction. SPE paper 23822 presented at the Intl. Symposium on Formation Damage Control, Lafayette. 477-490.

Vetter, O. J., Kandarpa, V., Stratton, M., and Veith, E., 1987. Particle invasion into porous medium and related injectivity problems. SPE paper 16255 presented at the Intl. Symposium on Oilfield Chemistry, San Antonio, Tx. 101-120.

Vik, E. A., Bakke, S., Øyno, L., and Elliot, G, 1994. Produced Water Reinjection (PWRI): Single well trial at Ula. Aquateam AS, Oslo. Report # 93-045 (Part 1 and 2). June, 20.

INDEX

MIX
Papier aus verantwortungsvollen Quellen
Paper from responsible sources
FSC® C105338

If you have any concerns about our products,
you can contact us on
ProductSafety@springernature.com

In case Publisher is established outside the EU,
the EU authorized representative is:
Springer Nature Customer Service Center GmbH
Europaplatz 3, 69115 Heidelberg, Germany

Printed by Libri Plureos GmbH
in Hamburg, Germany